Junzo Otera and
Joji Nishikido
Esterification

Related Titles

Mizuno, N. (ed.)

Modern Heterogeneous Oxidation Catalysis

Design, Reactions and Characterization

2009

ISBN: 978-3-527-31859-9

Carreira, E. M., Kvaerno, L.

Classics in Stereoselective Synthesis

2009

ISBN: 978-3-527-32452-1

Ackermann, L. (ed.)

Modern Arylation Methods

2009

ISBN: 978-3-527-31937-4

Bandini, M., Umani-Ronchi, A. (eds.)

Catalytic Asymmetric Friedel-Crafts Alkylations

2009

ISBN: 978-3-527-32380-7

Andersson, P. G., Munslow, I. J. (ed.)

Modern Reduction Methods

2008

ISBN: 978-3-527-31862-9

de Vries, J. G., Elsevier, C. J. (eds.)

The Handbook of Homogeneous Hydrogenation

2007

ISBN: 978-3-527-31161-3

Junzo Otera and Joji Nishikido

Esterification

Methods, Reactions, and Applications

Second, Completely Revised, and Enlarged Edition

WILEY-VCH Verlag GmbH & Co. KGaA

The Authors

Prof. Dr. Junzo Otera
Okayama University of Science
Department of Applied Chemistry
1-1 Ridai-cho
Okayama 700-0005
Japan

Dr. Joji Nishikido
The Noguchi Institute
1-8-1 Kaga Itabashi-ku
Tokyo 173-0003
Japan

Library of Congress Card No.: applied for

British Library Cataloguing-in-Publication Data
A catalogue record for this book is available from the British Library.

Bibliographic information published by the Deutsche Nationalbibliothek
The Deutsche Nationalbibliothek lists this publication in the Deutsche Nationalbibliografie; detailed bibliographic data are available on the Internet at <http://dnb.d-nb.de>.

Composition Toppan Best-set Premedia Limited
Printing Strauss GmbH, Moerlenbach
Bookbinding Litges & Dopf GmbH, Heppenheim
Cover Design Schulz Grafik-Design, Fußgönheim

Printed in the Federal Republic of Germany
Printed on acid-free paper

ISBN: 978-3-527-32289-3

Contents

Esterification. Methods, Reactions, and Applications. 2nd Ed. J. Otera and J. Nishikido

ISBN: 978-3-527-32289-3

Preface

Few would dispute that the synthesis of esters has played a most important role in organic synthesis from its infancy. This importance stemmed from its utility in diverse fields both in the laboratory and in industry. Ester moieties, irrespective of whether acyclic and cyclic, constitute major backbones, as well as functional groups of chemical significance, in numerous natural products and synthetic compounds. The essential feature of esterification that particularly distinguishes it from other reactions lies in its broad utilization in industry. Just a brief chronological look quickly reminds us of aspirin (acetyl salicylic acid), fatty acid esters, polyesters, macrolides, and so on. In addition to being essential molecular components in their own right, ester groups also play versatile temporary roles in organic synthesis for protection of carboxylic acids and hydroxy groups. The synthesis of natural products, especially macrolides, sugars, and peptides, depends heavily on acylation technology.

Being carboxylic acid derivatives, esters are largely produced from the reactions between the corresponding acids and alcohols. Transformation from one ester into another (transesterification) is also useful. On the other hand, since esters are also derivatives of alcohols, ester synthesis is also important from the standpoint of alcohol chemistry, such as acylation. A variety of routes to arrive at esters are therefore feasible, and numerous methods have been reported. Surprisingly, though, no book focused solely on 'esters' has been available up to now, esterification or transesterification usually being included in many books as a sub-class of functional group transformations. Obviously, this is not a fair treatment if the central position of (trans)esterification in organic synthesis is taken into account. Why did such biased circumstances arise? A number of reasons can be identified immediately, but only a few representatives among them are given here. Since (trans)esterification has such a long history and the reaction itself is simple, many people, especially in academia, take it for granted that little room is left for further scientific improvements. In industry, on the other hand, (trans)esterification still has continuing significance, and many new technologies, being classed as know-how, therefore remain undisclosed. Since the utility of (trans)esterification has spread into diverse fields, it is indeed laborious to cover the whole, as even those people who are deeply involved in the (trans)esterification fields, regardless of whether in academia or in industry, have rather limited knowledge about what is

Esterification. Methods, Reactions, and Applications. 2nd Ed. J. Otera and J. Nishikido

ISBN: 978-3-527-32289-3

going on outside the very narrow disciplines close to them. Despite such undesirable circumstances, (trans)esterification has in fact been, and is still undergoing, extensive innovations. It is the aim of this book to inform a broad range of chemists and technicians on the state of the art relating to both fundamental ideas and practical aspects of (trans)esterification.

The book consists of two parts. The first part thoroughly reviews the (trans) esterification reaction, from conventional approaches to the most up-to-date progress in terms of reaction patterns, catalysts, reaction media, etc., so that readers may acquire general, basic knowledge of the reaction. In addition, those wanting to survey suitable methods for a specific target will find great help from consulting this part. A number of 'Experimental Procedures' given may help readers judge which reactions are suitable for their purpose.

Synthetic applications of (trans)esterification are the subject of Part II. These reactions, many of which may have already appeared in Part I, are reorganized according to their respective synthetic purposes. Various aspects of interest to synthetic chemists are summarized, followed by an overview of industrial utilization.

Because of its long history, it has been impracticable to survey the literature of esterification completely from the beginning. A full survey from 1990 has therefore been made by use of commercial databases, while reference works appearing before 1990 have been selected arbitrarily depending on their importance. I believe that this treatment is fully acceptable, namely to cover the literature that is of basic significance and to represent recent progress to meet the requirements for 'modern esterification.' Resulting from this, we encountered more than 5000 references, but for reasons of space not all of them could be accommodated in the text of this book. Only examples selected in terms of fundamentality and generality have been taken, to provide a comprehensive view of the overall aspects as broadly as possible. All collected references have been placed in a database library, a copy of which is provided on disc at the back of this book. Those who wish to obtain more detailed information will be able to consult this library through keyword access.

Last, but not least, I would like to express my sincere appreciation to Miss Masayo Kajitani, who contributed greatly to the literature survey and the illustrations. Without her collaboration and patience, this book might have not been completed.

Okayama, November 2002 *Junzo Otera*

Preface to the Second Edition

More than five years have passed since the first edition was published. Esterification has continued to advance in the meantime, proving its continuing importance in organic synthesis and chemical industry. A literature survey covering the period 2002–2007 performed with the aid of SciFinder revealed about 5300 relevant references, simple enumeration of which is not the purpose of the second edition, of course. Only about 500 of these references, which have particular scientific or practical merits, have been newly included in this volume, which follows the original format of the first edition. It is of interest to note that nearly half of them are associated with green chemistry, reflecting rapidly growing environmental concerns. This trend is clearly seen in the accelerated development of solid acid catalysts and enzymes. Replacement of conventional organic solvents by new reaction media such as water, fluorous solvents, ionic liquids, and supercritical solvents has also become very popular for the same reason. Esterification has acquired further innovation from the engineering side. As a consequence of its remarkable progress, Section 7.4, 'New Technologies,' now appears in this edition.

Chapter 8, 'Industrial Uses,' has also been expanded. First, ethyl acetate and acrylic esters as fundamental raw materials are briefly surveyed (Sections 8.1 and 8.2, respectively). As biodiesel fuel has recently undergone dramatic industrial development, an overview of this topic is presented in Section 8.5. Industrial aspects of macrolide synthesis are also newly covered in a separate section (Section 8.7). Furthermore, a number of additions have been made to other sections in this chapter.

For this new edition, a thorough literature survey and many new illustrations were required. These painstaking tasks were carried out to perfection by Ms Satoko Kira and Ms Hisaji Dozen, to whom the authors would like to express their sincere appreciation.

Okayama, November 2009

On behalf of the authors,
Junzo Otera

Esterification. Methods, Reactions, and Applications. 2nd Ed. J. Otera and J. Nishikido

ISBN: 978-3-527-32289-3

Introduction

The biggest problem frequently encountered in (trans)esterification technology arises from the fact that in the majority of cases a reversible reaction is involved. To bias the equilibrium to the product side, one of the reactants must be used in excess and/or one of the products must be removed constantly during the reaction. Use of a nonequilibrium reaction approach, with the aid of activated reactants such as acid anhydrides and halides or alkoxides, can be effective to circumvent the problem on some occasions, but is not always general. Ester synthesis reactions are usually conducted with the aid of acid or base catalysts, and so the employment of catalysts or promoters that are suitably active but also compatible with other functional groups is of great importance. Progress has been made to overcome these problems in (trans)esterification reactions.

Esterification can be regarded as the transformation of carboxylic acids or their derivatives into esters, a procedure which is carried out in many natural products syntheses, in the protection or kinetic resolution of carboxylic acids, and in the fatty acids industry. However, the converse reaction–transformation of alcohols into esters, as in protective acylation of hydroxy groups, kinetic resolution of alcohols, and so on – is equally important. The normal substrate/reagent relationship cannot therefore be straightforwardly applied to esterification. Moreover, in intramolecular cases (lactonization) and polycondensation, both function as equal partners. In Chapter 1 the alcohol component is regarded as the substrate, because modifications of carboxylic acids are available in greater variety. Of course, such classification is not strict: the carboxylic acid component might as well be taken as a substrate in, for instance, the direct reaction between neat alcohol and carboxylic acid. In any event, the chapter is subdivided into sections according to the means by which the carboxylic acid is modified. Each section is then further sub-classified according to the activation modes.

Tin alkoxides, together with some other metal alkoxides, are useful for selective acylation of polyols and, in particular, play an important role in sugar chemistry. This subject is grouped separately in Chapter 2.

In Chapter 3, the carboxylic acid component is treated as the substrate, reacting with various reagents other than alcohols. Chapter 4 deals with interconversion between different esters.

Esterification. Methods, Reactions, and Applications. 2nd Ed. J. Otera and J. Nishikido

ISBN: 978-3-527-32289-3

The first two chapters in Part II, Chapters 5 and 6, are both associated with chirality. In response to the increasing need for optically active compounds in modern synthetic chemistry, great progress has been achieved in ester technology, serving for the production of enantiomerically enriched or enantiomerically pure alcohols and carboxylic acids through kinetic resolution and desymmetrization.

Chapter 7 covers miscellaneous topics of great significance in terms of synthetic utility and various selectivities. Natural product syntheses in which esterification has played a crucial role are also described.

Finally, in Chapter 8, industrial uses of esterification technology are examined. Since many currently operational (especially state-of-the-art) processes are 'know-how,' which means that they are effectively veiled in darkness, it is not an easy task to give any kind of detailed account of them here. At best, we can outline salient points that are either common knowledge or available from the literature. Despite such limitations, the reader should be able to form an idea of how and where esterification is utilized in practice.

Part One Methodology

Esterification. Methods, Reactions, and Applications. 2nd Ed. J. Otera and J. Nishikido

ISBN: 978-3-527-32289-3

1
Reaction of Alcohols with Carboxylic Acids and their Derivatives

1.1
Reaction with Carboxylic Acids

1.1.1
Without Activator

Although the direct reaction between alcohol and carboxylic acid is conventionally conducted under acid or base catalysis conditions, the catalyst-free reaction is more desirable. This requirement is satisfied when the reaction is carried out at high temperatures. For example, propanol or hexanol can be treated with various aliphatic carboxylic acids (1.35 equiv.) in an autoclave at 150 °C to furnish esters in poor to excellent yields (Scheme 1.1) [1]. The reaction is strongly influenced by the reaction temperature; the yield of propyl acetate is only 18% at 85 °C.

$RCOOH + MeOH \xrightarrow{CBr_4,\ h\nu} RCOOMe$

R= C_9H_{19} (90%)
R= $(CH_2)_7Br$ (98%)
R= CH_2CH_2cHex (96%)
R= cHex (40%)
R= Bn (96%)
R= $CH(CH_3)Ph$ (95%)
R= Ph (14%)
R= $CH{=}CHCH{=}CHCH_3$ (99%)

Scheme 1.1

Interestingly, condensation between sugars and α-hydroxycarboxylic acids can be performed in water (Scheme 1.2) [2]. The reaction takes place at 60 °C, regioselectively on the primary hydroxy groups of mannose, galactose and glucose. The avoidance of any catalysts or additives allows the instant application of the products in food technology and cosmetic formulation.

Esterification. Methods, Reactions, and Applications. 2nd Ed. J. Otera and J. Nishikido

ISBN: 978-3-527-32289-3

Experimental Procedure Scheme 1.2 [2] General procedure: A mixture of hydroxycarboxylic acid, carbohydrate, and water is heated to 60 °C in air for 24 h. To isolate the product as a pure compound for characterization, the reaction mixture is extracted twice with diisopropyl ether, the solvent is removed *in vacuo*, and the residue is chromatographed on silica gel (CH_2Cl_2/MeOH; 85 : 15).

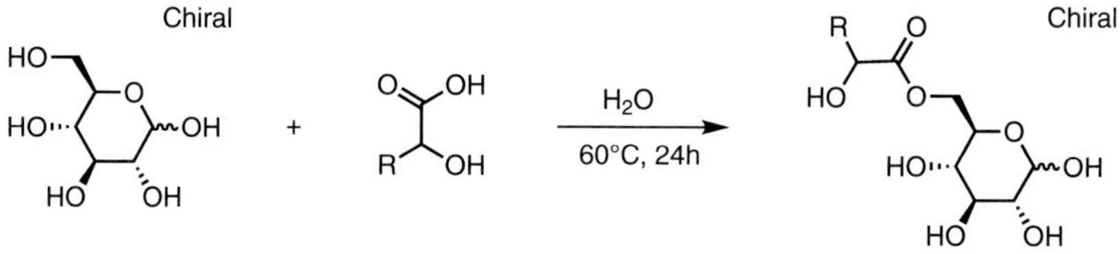

Scheme 1.2

The equilibrium in the reaction between ethanol and acetic acid can be shifted in favor of the ester by application of CO_2 pressure (Scheme 1.3) [3]. The ester yield is increased from 63% in neat solution to 72% in CO_2 at 333 K/58.6 bar. This outcome is far from satisfactory, though the possibility does suggest itself that the equilibrium may be improved by changing the reaction conditions.

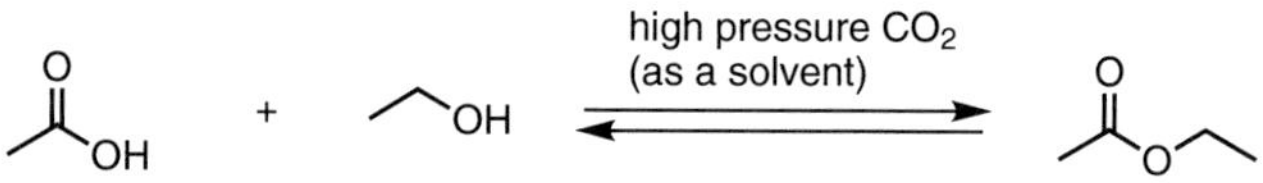

Scheme 1.3

On the whole, the catalyst-free reaction is ideal but difficult to achieve. Some special conditions are necessary, and employable reactants are rather limited. Nonetheless, it is obvious that this line of technology should be advanced more extensively in the context of green chemistry.

1.1.2 Acid Catalysts

1.1.2.1 Brønsted Acids

Since acid catalysis is one of the most popular methods for esterification, numerous papers are available. When the substrates are acid-resistant, the reaction is usually carried out in the presence of a Brønsted acid such as HCl, HBr, H_2SO_4, $NaHSO_4$, $ClSO_3H$, NH_2SO_3H, H_3PO_4, HBF_4, AcOH, camphorsulfonic acid, etc. (Scheme 1.4).

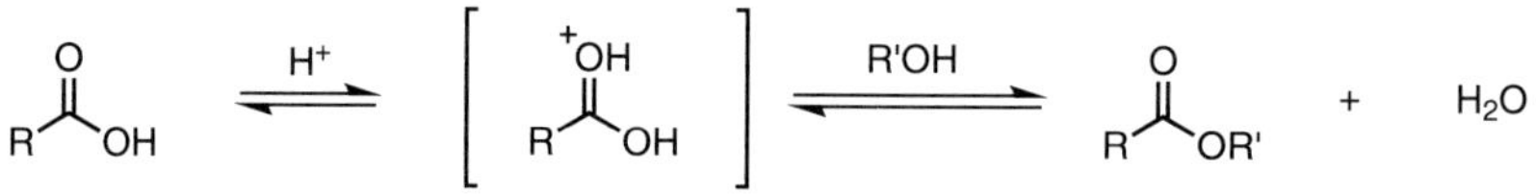

Scheme 1.4

In cases in which the acidity is not high enough to trigger the desired reaction, the acid is combined with an activator. For example, the lactonization shown in Scheme 1.5 proceeds sluggishly with HCl only, but the reaction is effected smoothly in the presence of HCl and 3A molecular sieves [4]. The esterification of phenols with both aliphatic and aromatic carboxylic acids – difficult to achieve under normal conditions – can be catalyzed by a combination of H_3BO_3 and H_2SO_4 (Scheme 1.6) [5].

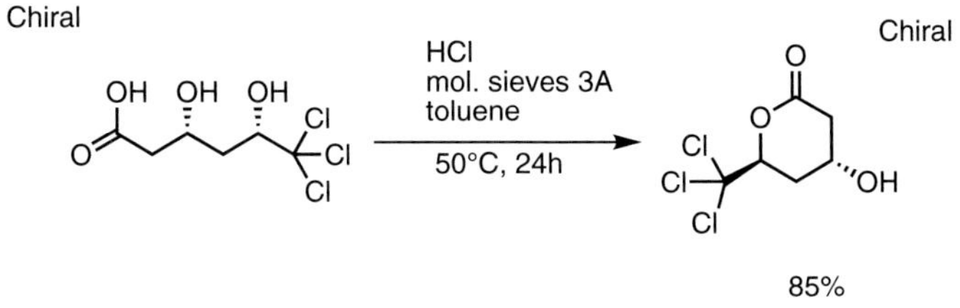

Scheme 1.5

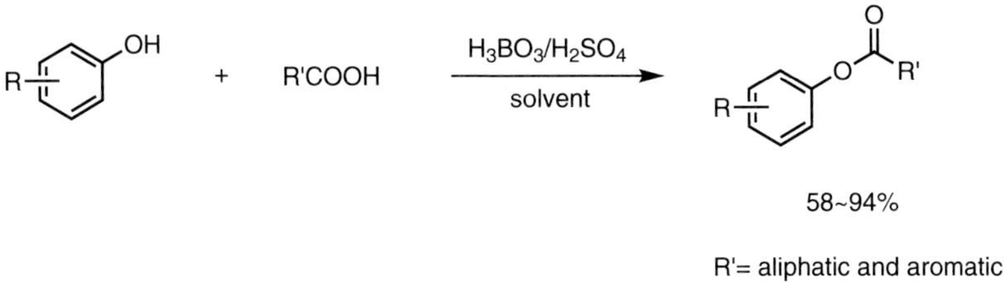

Scheme 1.6

Other ways to activate the acid catalysts are provided by the use of ultrasound and microwave. H_2SO_4-catalyzed esterification, which usually requires a long reaction time under refluxing conditions, is complete at room temperature in several hours on exposure to ultrasonic waves [6]. Microwave irradiation accelerates the *p*-toluenesulfonic acid-catalyzed esterification, the reaction finishing within 10 min [7].

Aqueous HCl is not employable for water-sensitive compounds. In such cases, dry HCl gas must be used, but generation of this is not operationally simple. Alternatively, generation of HCl under anhydrous conditions is conveniently feasible by addition of acetyl chloride to methanol or ethanol. Treatment of alcohol and carboxylic acid in the HCl solution obtained provides the desired ester (Scheme 1.7) [8]. By this method, the concentration of HCl can be readily adjusted by changing the amount of acetyl chloride.

Experimental Procedure Scheme 1.7 [8] A typical experimental procedure involves the addition of a known amount of acetyl chloride, usually from a weighed syringe, to an ice-cold solution of an equivalent or excess amount of methanol (or ethanol) in an inert organic solvent, such as ethyl acetate, containing an equivalent amount of the compound to be treated. The acidic solution may also be prepared in the pure alcohol. Ice-cold solutions are used in order to increase the solubility of the HCl and to prevent its escape, the initial generation of the HCl being exothermic. In cases in which simple esterifications are to be carried out, excess acetyl chloride may be used without detrimental effects, since the workup involves simple evaporation of the solvent(s) and excess HCl. The solutions are allowed to warm to room temperature, and the reactions are complete within 0.5–24 h.

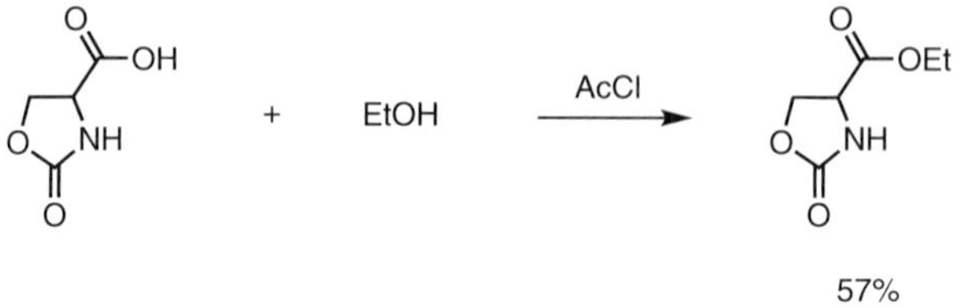

OH
NH
O
O
O
+
EtOH
AcCl
OEt
NH
57%

Scheme 1.7

A similar protocol is available with TMSCl (trimethylsilyl chloride) (Scheme 1.8) [9]. In this case, TMSCl is added to a mixture of alcohol and carboxylic acid. It has been suggested that the reactant alcohol works as a proton donor as well. On the other hand, the initial formation of the silyl ester is another proposed mechanism for a similar reaction (Scheme 1.9) [10]. Steroid-based alcohols also can be esterified by the use of TMSCl or TMSBr [11].

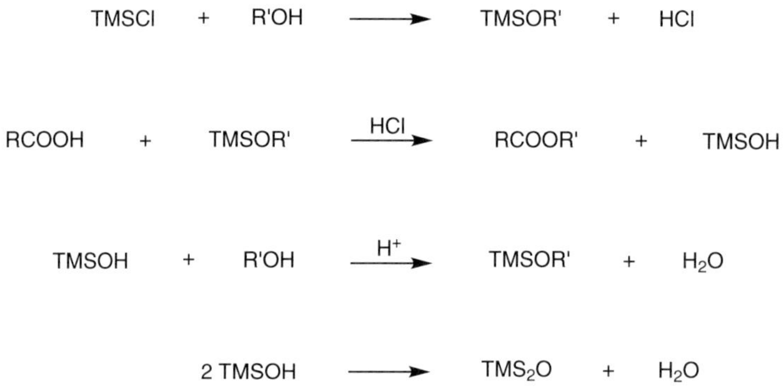

TMSCl + R'OH → TMSOR' + HCl
RCOOH + TMSOR' HCl→ RCOOR' + TMSOH
TMSOH + R'OH H+→ TMSOR' + H2O
2 TMSOH → TMS2O + H2O

Scheme 1.8

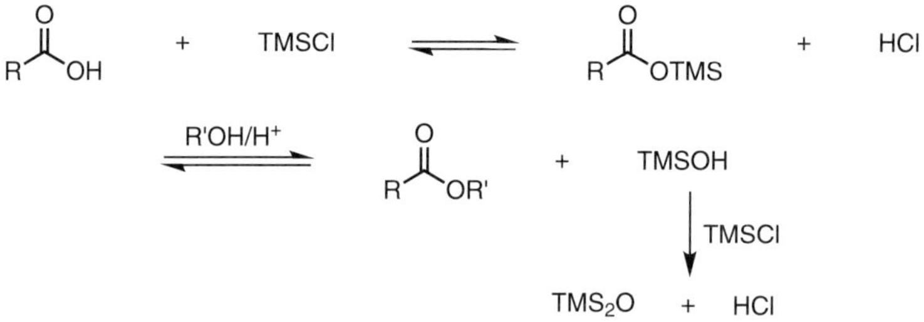

R
OH
+
TMSCl
O
OTMS
+
HCl
R'OH/H+
OR'
+
TMSOH
TMSCl
TMS2O
+
HCl

Scheme 1.9

In situ generation of catalytic HCl is accessible photolytically. Photoirradiation of carboxylic acids in methanol containing CBr_4 (0.05 equiv.) furnishes the corresponding methyl esters (Scheme 1.10) [12]. Interestingly, sp^3 carbon-tethered carboxylic acids undergo esterification smoothly under these conditions, while sp^2 or sp carbon-tethered carboxylic acids are not esterified. Similar photolytic esterification occurs in CCl_4 or $BrCCl_3$ in place of CBr_4 [13]. It has been proposed that HCl generated by abstraction of an α-hydrogen of alcohol by Cl radical is the real catalytic species in this reaction.

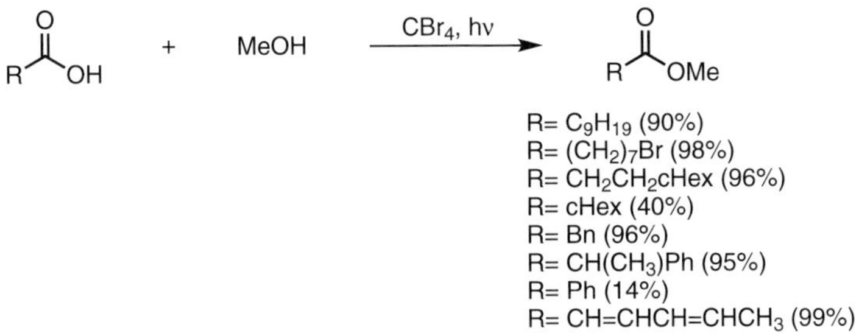

RCOOH + MeOH → (CBr_4, hν) → RCOOMe

R= C_9H_{19} (90%)
R= $(CH_2)_7Br$ (98%)
R= CH_2CH_2cHex (96%)
R= cHex (40%)
R= Bn (96%)
R= $CH(CH_3)Ph$ (95%)
R= Ph (14%)
R= $CH{=}CHCH{=}CHCH_3$ (99%)

Scheme 1.10

Despite rather harsh conditions, the Brønsted acid-catalyzed reaction sometimes enjoys selectivities. Continuous extraction technology enables selective monoacetylation in reasonable yields upon treatment of 1,*n*-diol with acetic acid in the presence of H_2SO_4 (Scheme 1.11) [14]. Stereoselectivity is also attained in TFA-catalyzed esterification (TFA = trifluoroacetic acid), as shown in Scheme 1.12 [15]. In this reaction, the inversion of the stereochemistry at C-4 proceeds effectively via the carbocation, the nucleophilic attack of an acetate anion on the carbocation taking place preferentially from the opposite side of the bulky 3,4-(methylenedioxy) benzoyl group in the Felkin-like model to afford the *anti*-acetate.

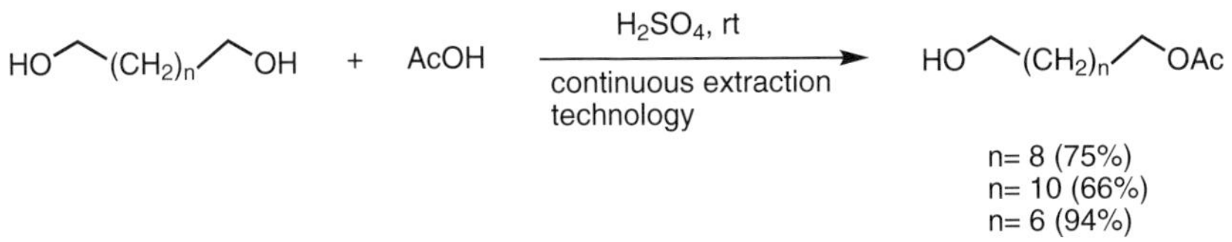

$HO(CH_2)_nOH$ + AcOH → (H_2SO_4, rt; continuous extraction technology) → $HO(CH_2)_nOAc$

n= 8 (75%)
n= 10 (66%)
n= 6 (94%)

Scheme 1.11

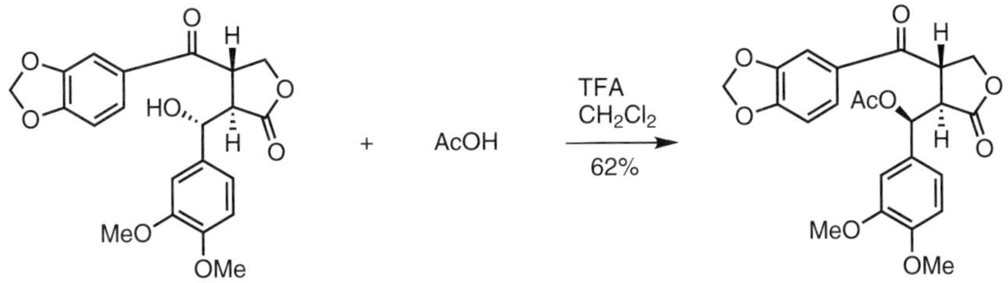

+ AcOH → (TFA, CH_2Cl_2, 62%)

Scheme 1.12

A unique formation of *tert*-butyl esters is notable. When a mixture of carboxylic acid and *tert*-butyl alcohol is exposed to H_2SO_4 absorbed on $MgSO_4$, esterification takes place smoothly (Scheme 1.13) [16]. The reaction is successful for various aromatic, aliphatic, olefinic, heteroaromatic, and protected amino acids. No reaction occurs with the use of anhydrous $MgSO_4$ or H_2SO_4 alone. The reaction is initiated by dehydration of *tert*-butyl alcohol followed by addition of carboxylic acid to the resulting isobutylene.

Experimental Procedure Scheme 1.13 [16] In a typical small-scale experiment, concentrated sulfuric acid (0.55 mL, 10 mmol) is added to a vigorously stirred suspension of anhydrous magnesium sulfate (4.81 g, 40 mmol) in 40 mL of solvent. The mixture is stirred for 15 min, after which the carboxylic acid (10 mmol) is added. *tert*-Butyl alcohol (4.78 mL, 50 mmol) is added last. The mixture is stoppered tightly and stirred for 18 h at 25 °C, or until the reaction is complete by TLC analysis. The reaction mixture is then quenched with 75 mL of saturated sodium bicarbonate solution and stirred until all magnesium sulfate has dissolved. The solvent phase is separated, washed with brine, dried ($MgSO_4$), and concentrated to afford the crude *tert*-butyl ester, which is purified by chromatography, distillation, or recrystallization as appropriate.

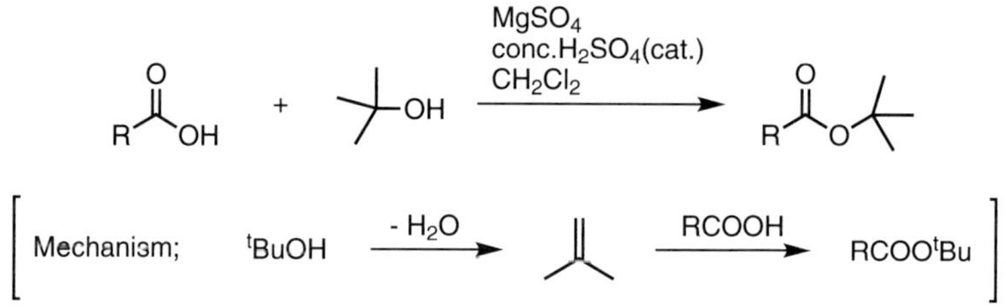

RCOOH + tBuOH —[$MgSO_4$, conc.H_2SO_4(cat.), CH_2Cl_2]→ RCOOtBu

[Mechanism; ^tBuOH —(- H_2O)→ isobutylene —(RCOOH)→ $RCOO^tBu$]

Scheme 1.13

Hydrophobic polystyrene-supported sulfonic acids catalyze reaction between carboxylic acid and alcohol in water [17]. The catalysts are recovered and reused for further reactions.

The acidity of strong acids is moderated by forming the corresponding ammonium salts. Diphenylammonium triflate is an efficient catalyst for mediation of condensation between alcohol and carboxylic acid in a 1 : 1 ratio [18]. The reaction usually affords greater than 90% yields of esters simply on treatment of the reactants with 1 mol% of the catalyst in refluxing toluene. After the reaction is complete, the solvent is evaporated and column chromatography of the residue furnishes the esters. When the reaction is performed in a fluorocarbon solvent such as perfluorohexane, the yield is increased [19]. Pentafluorophenylammonium triflate is also a good esterification catalyst [20]. The pentafluorophenyl group causes formation of a hydrophobic environment around the catalytic center, so that dehydration techniques are required. Use of bulky ammonium groups together with arenesulfonyl anion results in highly efficient esterification [21]. Dimesitylammonium pentafluorobenzenesulfonate is one of the most useful catalysts, effecting condensation between carboxylic acid and alcohol in a 1:1 ratio without use of Dean-Stark apparatus.

Experimental Procedure [18] 3-Phenylpropanoic acid (150 mg, 1.0 mmol), 1-octanol (130 mg, 1.0 mmol), and diphenylammonium triflate (3.2 mg, 0.01 mmol) are heated (80 °C) in toluene for 4 h. Evaporation of toluene (ca. 40 °C) under reduced pressure gives the crude material, which is purified by column chromatography (hexane/ether 10:1) to give the desired carboxylic ester (244 mg, 93%).

Polyaniline salts with HCl, HNO_3, H_2SO_4, H_3PO_4, *p*-tolSO_3H, etc. catalyze reactions between carboxylic acids and alcohols, and can be separated easily from the reaction mixture by filtration [22, 23].

Brønsted acidic ionic liquids function as dual solvents/catalysts for condensation between carboxylic acid and alcohol (see Section 7.3) [24, 25]. Immobilized acidic ionic liquids can be recycled in catalysis for esterification [26]. Acidic ionic liquids can catalyze reactions of carboxylic acids with alcohols in water [27].

1.1.2.2 Lewis Acids

Lewis acids are another important class of acid catalyst. In general, they are milder than Brønsted acids and, more importantly, template effects are to be expected as they are sterically bulkier than a proton; the utilization of Lewis acids is therefore rapidly increasing. They are classified as follows, according to elements:

B	$BF_3 \cdot OEt_2$[28–37]; BCl_3[38]; 3,4,5-$F_3C_6H_2B(OH)_2$[39]
Al	$AlCl_3$ (immobilized) [40]; Al/NaI/CH_3CN [41]; $AlCl_3/ZnCl_2$[42]
Zn	ZnO [43]; $ZnCl_2$/microwave [44]; $Zn(OTf)_2$/microwave [45]; $Zn(ClO_4)_2 \cdot 6H_2O$ [46]
In	$InCl_3$[47]
Sn	$SnCl_2$[48]; Bu_2SnO [49, 50]; $(XR_2SnOSnR_2Y)_2$[51–53]; $(XRf_2SnOSnRf_2X)_2$[54]; Ph_2SnCl_2[55]
Ti	$TiO(acac)_2$[56]
Mn	$Mn(OAc)_3 \cdot 2H_2O$ [57]
Fe	$Fe(ClO_4)_3$[58–60]; $Fe_2(SO_4)_3 \cdot H_2O$ [61]; $Fe_2(SO_4)_3 \cdot nH_2O/H_2SO_4$[62]; $FeCl_3$[63, 64]
Ni	$NiCl_2 \cdot 6H_2O$ [65]
Cu	$CuCl_2$[66]; $Cu(NO)_3 \cdot 3H_2O$ [67]; $Cu(OTf)_3$[68]; $Cu(OMs)_2$[69]
Mo	$MoO(acac)_2$[70]
Sc	$Sc(OTf)_3$[71, 72]
Zr;Hf	$ZrCl_4 \cdot 2THF$ and $HfCl_4 \cdot 2THF$ [73–75]; $ZrOCl_2 \cdot 8H_2O$ and $HfOCl_2 \cdot 8H_2O$ [76, 77]; $Zr(O^iPr)_4$ or $Hf(O^iPr)_4$/ $Fe(O^iPr)_3$[78, 79]
I	I_2[80]

$BF_3 \cdot OEt_2$ is the oldest Lewis acid to have been employed as an esterification catalyst, since the BF_3/CH_3OH complex had been known to be used for conversion of simple carboxylic acids to their methyl esters prior to GLC analysis. Although excess $BF_3 \cdot OEt_2$ (2–3 equiv.) and alcohol (>10 equiv.) for 1 equiv. of carboxylic acid should be used, esterification is feasible for 4-aminobenzoic acid, unsaturated organic acids, biphenyl-4,4′-dicarboxylic acid, 1,4-dihydrobenzoic acid, and heterocyclic carboxylic acids.

Experimental Procedure [31] The reaction mixture comprising the acid (0.1 mol), boron trifluoride etherate (0.1 or 0.2 mol, depending on the number of carboxyl groups in the acid), and the appropriate alcohol (ten times in excess of the boron trifluoride etherate) is heated at reflux for a period of time not exceeding 24 h. The esters are precipitated by dilution with a 5% solution of sodium carbonate, followed by filtration or extraction with ether, and purified by crystallization from appropriate solvents or by distillation under reduced pressure.

BCl_3 is also useful for esterification with primary alcohols, but yields are not so high with secondary and tertiary alcohols. The disadvantage of this method is the cleavage of coexisting methyl ether function.

3,4,5-Trifluorobenzeneboronic acid is claimed to be the most effective catalyst among boronic acids (Scheme 1.14). Esterification takes place smoothly if heavy alcohols such as 1-butanol are employed. The reaction is presumed to proceed via a carboxylate intermediate.

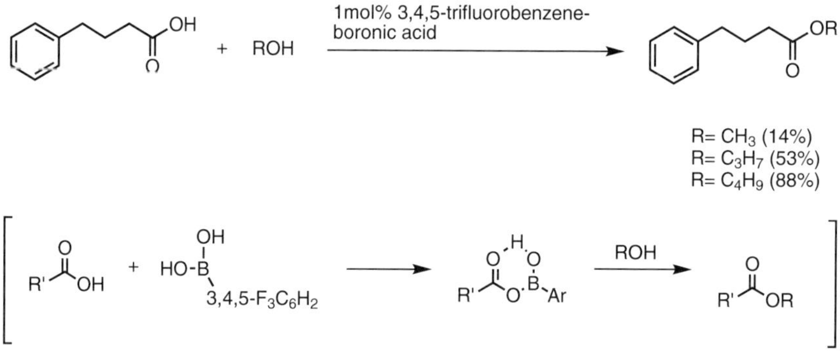

+ ROH → 1 mol% 3,4,5-trifluorobenzeneboronic acid

R= CH_3 (14%)
R= C_3H_7 (53%)
R= C_4H_9 (88%)

[$R'COOH$ + HO-B(OH)-3,4,5-$F_3C_6H_2$ → intermediate (Ar) → ROH → $R'COOR$]

Scheme 1.14

$AlCl_3$ is one of the most popular Lewis acids, but it is not employed in esterification because of its too strong acidity. However, polymer-supported $AlCl_3$ works as a milder catalyst for esterification although the yields are not always as high as those obtained by other methods. The advantage lies in the ease of separation of the catalyst by filtration. The Lewis acidity can be moderated in combination with a soft nucleophile, NaI, in CH_3CN. An equimolar mixture of acid and alcohol is smoothly converted into the desired ester under reflux, but the yield is not high in general (77% at highest). Phenyl esters, which are otherwise rather difficult to prepare, can be obtained by reaction between aromatic or benzylic carboxylic acids with phenol in the presence of a catalytic amount of $AlCl_3$ and one equivalent of $ZnCl_2$. The strong acidity of $AlCl_3$ is responsible for efficient esterification, while $ZnCl_2$ serves for dehydration of the reaction mixture.

Treatment of pentaerythritol with oleic acid in the presence of ZnO as catalyst provides a triester. Production of commercially important *p*-hydroxybenzoic acid ester (paraben) from *p*-hydroxybenzaldehyde and alcohol is catalyzed by $ZnCl_2$ under microwave irradiation conditions. The microwave irradiation is effective for esterification catalyzed by $Zn(OTf)_2$. An equimolar mixture of carboxylic acid and alcohol is esterified, but yields are less than 90%. Smooth esterification is catalyzed by $Zn(ClO_4)_2 \cdot 6H_2O$ in the presence of $MgSO_4$ as a dehydrating agent.

Methyl esterification is feasible by heating a MeOH solution of carboxylic acid in the presence of $InCl_3$ (20 mol%). The reaction proceeds at room temperature under sonication as well.

Another popular Lewis acid, $SnCl_4$, is also not usually employed in esterification, although the milder Lewis acid, $SnCl_2$, can catalyze reaction between carboxylic acids and solvent PrOH. Organotin compounds work quite well, however, because the acidity is moderated by the replacement of chlorine with electron-donating alkyl groups. Bu_2SnO catalyzes lactonization of *seco* acids on continuous dehydration with a Dean-Stark apparatus (Scheme 1.15). This method is effective for large-sized lactones but not for medium-sized ones.

Experimental Procedure Scheme 1.15 [50] Preparation of hexadecanolide. A mixture of 16-hydroxyhexadecanoic acid (817.3 mg, 3.0 mmol) and dibutyltin oxide (74.7 mg, 0.3 mmol) is stirred in refluxing mesitylene (100 mL) for 19 h with use of a Dean-Stark apparatus for the continuous removal of water. Removal of the solvent *in vacuo* (40 °C/0.2 mmHg) yields a yellow oily residue, which is Kugelrohr distilled (60 °C/0.2 mmHg) to give 457.9 mg (60%) of hexadecanolide.

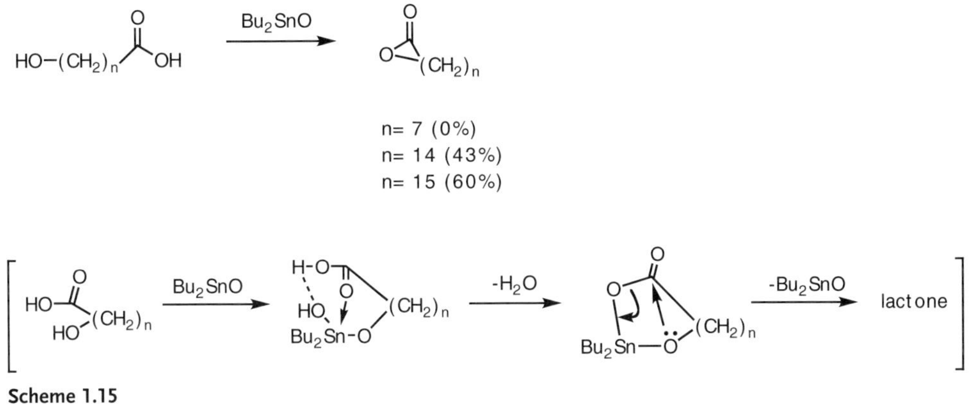

$HO-(CH_2)_n-COOH \xrightarrow{Bu_2SnO}$ lactone $(CH_2)_n$

n= 7 (0%)
n= 14 (43%)
n= 15 (60%)

[$\xrightarrow{Bu_2SnO}$ $\xrightarrow{-H_2O}$ $\xrightarrow{-Bu_2SnO}$ lactone]

Scheme 1.15

Good yields of esters are obtained when carboxylic acids are treated with a catalytic amount of 1,3-disubstituted tetraalkyldistannoxanes, $(XR_2SnOSnR_2X)_2$, in alcohol solvent (Scheme 1.16). The catalysts are very mild, and the reaction is sensitive to the steric bulk of the reactants. The catalysts are also effective for lactonization (Scheme 1.17). The reaction proceeds simply on heating of a decane solution of *seco* acids. The convenience of operation is apparent from the lack of any need for dehydration apparatus and/or dehydration agents. Irreversible esterification apparently takes place because of the hydrophobicity of the surface alkyl groups surrounding the stannoxane core. Interestingly, alkyl side chains on the hydroxyl-carrying carbon of the ω-hydroxy acids exert a profound effect on lactonization. Lactone rings with fewer than 14 members are obtained in poor yields, while incorporation of R groups with more than four carbon atoms dramatically increases the yield.

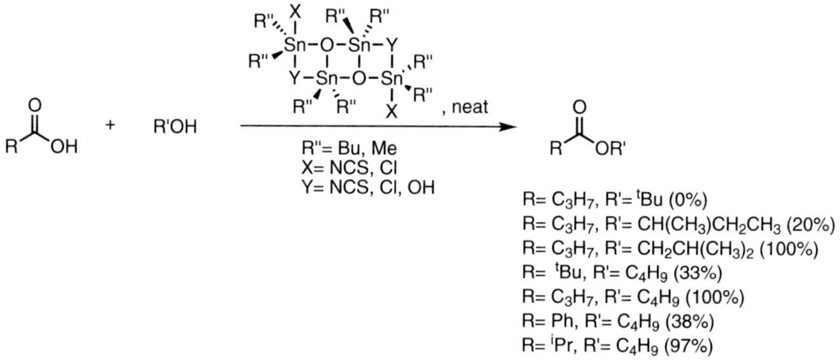

RCOOH + R'OH → (catalyst), neat → RCOOR'

R''= Bu, Me
X= NCS, Cl
Y= NCS, Cl, OH

R= C_3H_7, R'= ^tBu (0%)
R= C_3H_7, R'= $CH(CH_3)CH_2CH_3$ (20%)
R= C_3H_7, R'= $CH_2CH(CH_3)_2$ (100%)
R= ^tBu, R'= C_4H_9 (33%)
R= C_3H_7, R'= C_4H_9 (100%)
R= Ph, R'= C_4H_9 (38%)
R= ^iPr, R'= C_4H_9 (97%)

Scheme 1.16

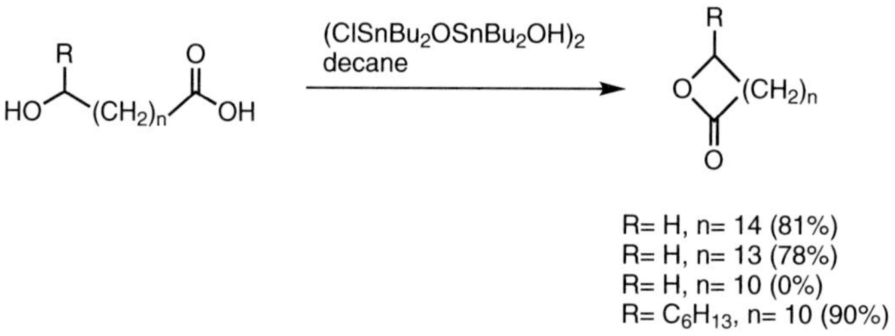

$(ClSnBu_2OSnBu_2OH)_2$
decane

R= H, n= 14 (81%)
R= H, n= 13 (78%)
R= H, n= 10 (0%)
R= C_6H_{13}, n= 10 (90%)

Scheme 1.17

Use of a fluoroalkyldistannoxane catalyst $(XRf_2SnOSnRf_2X)_2$ achieves a highly atom-efficient process in the context of fluorous biphasic technology (see Sections 7.2 and 7.3). Virtually 100% yields are achievable by the use of carboxylic acid and alcohol in a strict 1 : 1 ratio. The catalyst is recovered quantitatively and the catalyst solution in perfluoro-organic solvent is recycled repeatedly.

Experimental Procedure [54] A test tube (50 mL) is charged with 3-phenylpropanonic acid (300 mg, 2.0 mmol), benzyl alcohol (216 mg, 2.0 mmol), $(ClRf_2SnOSnRf_2Cl)_2$ ($Rf = C_6F_{13}C_2H_4$) (172 mg, 0.10 mmol, 5 mol%), and FC-72 (5.0 mL). The test tube is placed in a stainless steel pressure bottle and heated at 150 °C for 10 h. The pressure bottle is cooled, and toluene (5.0 mL) is added to the reaction mixture. The toluene and FC-72 layers are separated, and the latter layer is extracted with toluene (2.0 mL × 2). The combined organic layer is analyzed by GC to provide a quantitative yield of benzyl 3-phenylpropanoate.

Simple dimethyl- and diphenyltin dichlorides catalyze esterification of carboxylic acids in refluxing C_1~C_3 alcohol.

High-yielding ester synthesis from an equimolar mixture of carboxylic acid and alcohol is accessible by the use of $TiO(acac)_2$ (3 mol%). The catalyst is water-tolerant and neutral to leave various functional groups intact. Alcohols are acetylated by heating at reflux with $Mn(OAc)_3 \cdot 2H_2O$ (4 mol%) in acetic acid.

Fe(III) salts are also effective. $Fe(ClO_4)_3 \cdot 9H_2O$ promotes esterification of carboxylic acids in alcohol. The reaction proceeds at room temperature, but a stoichiometric amount of the salt is needed. A catalytic version is available with $Fe_2(SO_4)_3 \cdot nH_2O$ (2 wt% cat. per acid) and $FeCl_3$ (5 mol% per acid). Addition of a small amount of H_2SO_4 greatly increases the catalytic activity of $Fe_2(SO_4)_3 \cdot nH_2O$. Moreover, anhydrous $Fe_2(SO_4)_3$ is highly active for catalyzing acetylation of alcohols in acetic acid. $FeCl_3 \cdot 6H_2O$ (2 mol%) effects esterification of an equimolar mixture of long-chain acids and alcohols in high yields. The reaction requires an excess amount of one reaction component in refluxing benzene or toluene. A similar outcome is obtained with $NiCl_2 \cdot 6H_2O$ catalyst.

Experimental Procedure [61] The Esterification of Adipic Acid with Ethanol in the presence of Ferric Sulfate: A mixture of adipic acid (14.5 g, 0.1 mol), absolute ethyl alcohol (18.5 g, 0.4 mol), dry benzene (35 mL), and commercial ferric sulfate (0.3 g) is placed in a flask equipped with an automatic water separator fitted with an efficient reflux condenser at its upper end. The mixture is heated at reflux on a steam bath for 3 h or until water no longer collects in appreciable amounts in the water separator. The catalyst is filtered off and washed with two 20 mL portions of ether. The combined filtrate is washed with saturated sodium carbonate solution and then with cool water, and is dried with anhydrous magnesium sulfate. Most of the ether and benzene are removed by distillation under normal pressure, and the residue is then evaporated under reduced pressure to give the diethyl adipate at 116–117/9 mm. The yield is 19.4 g (96%).

Cupric salts are another class of species that work as catalysts. $CuCl_2 \cdot nH_2O$ catalyzes conversion of carboxylic acids in methanol solvent at 130 °C, while $Cu(NO_3)_2 \cdot 3H_2O$ effects acetylation of alcohols in refluxing acetic acid. $Cu(OTf)_2$ is used for acetylation of alcohols but to a somewhat limited extent. Cupric methanesulfonate ($Cu(OMs)_2$) is also effective. Treatment of carboxylic acid with alcohol (1.1 equiv.) in the presence of 1 mol% of the catalyst in refluxing cyclohexane affords the desired ester in excellent yield. $MoO(acac)_2$ also catalyzes transformation of propanoic acid into esters in refluxing alcohols.

$Sc(OTf)_3$ can be employed as a catalyst for acylation of high-molecular weight polyethylene glycols. Polycondensation between aliphatic dicarboxylic acids and diols is also achievable with this catalyst to furnish polyesters.

$HfCl_4 \cdot 2THF$ in the presence of 4A molecular sieves enables the use of equimolar amounts of alcohol and carboxylic acid to afford good to excellent yields of the desired esters (see Part Two). This commercially available catalyst is highly active (usually 0.1–0.2 mol% loading) and hydrolytically stable. Polycondensations of ω-hydroxy acids or between dicarboxylic acids and diols to furnish polyesters are also feasible. The selective esterification of primary alcohols in the presence of secondary alcohols or phenol can be achieved with this catalyst (Scheme 1.18). Similar results are obtained with the zirconium analog. Unfortunately, however, these metal chlorides are moisture-sensitive. This drawback is overcome by the use of water-tolerant $ZrOCl_2 \cdot 8H_2O$ and $HfOCl_2 \cdot 8H_2O$. Combination of $Zr(O^iPr)_4$ with $Fe(O^iPr)_3$ exerts a synergistic effect, giving rise to increased catalytic activity as compared to the respective metal alkoxides alone. This combined catalyst is recovered by extraction with ionic liquid, so that recycling of the catalyst is feasible. Another method to recycle the catalyst is immobilization on *N*-(polystyrylbutyl)pyridinium triflylimide. The catalyst can be recycled at least 10 times with this technology.

Experimental Procedure Scheme 1.18 [73]

The typical polycondensation procedure is as follows. A flame-dried, 5 mL, single-necked, round-bottomed flask fitted with a Teflon-coated magnetic stirring bar and a 5 mL pressure-equalized addition funnel [containing a cotton plug and 4 Å molecular sieves (~1.5 g)] surmounted by a reflux condenser is charged with ω-hydroxycarboxylic acid (10 mmol) or α, ω-dicarboxylic acid (10.0 mmol) and α, ω-diol (10.0 mmol) as substrates and $HfCl_4 \cdot 2THF$ (0.200 mmol) as a catalyst in *o*-xylene (2 mL). The mixture is brought to reflux with the removal of water. After 1 day, the resulting mixture is cooled to ambient temperature, dissolved in chloroform, and precipitated with acetone or methanol to furnish pure polyester as a white solid in quantitative yield.

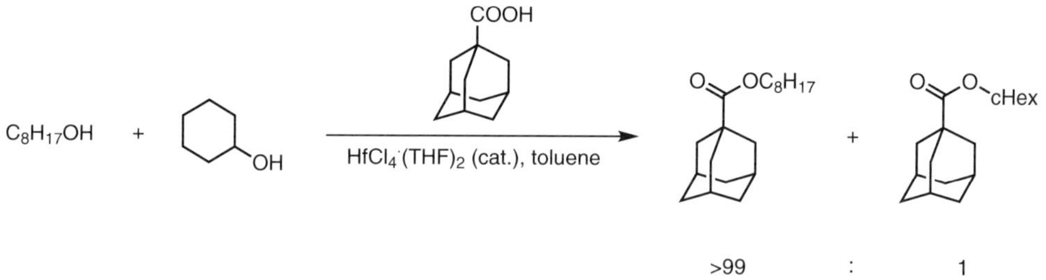

$C_8H_{17}OH$ + cyclohexanol (OH); COOH (adamantane); $HfCl_4(THF)_2$ (cat.), toluene → OC_8H_{17} ester + cHex ester

>99 : 1

Scheme 1.18

When a carboxylic acid is heated in alcohol with a catalytic amount of iodine, esterification takes place [80]. Primary, secondary, and even tertirary alcohols are employable, although the yields are rather low (56%) in the last case. The reaction is tolerant of high amounts of water. It is claimed that the iodine works as a Lewis acid.

Experimental Procedure [80]
Stearic acid (5 g, 17.6 mmol), methanol (10 mL), and iodine (50 mg,) are heated at reflux for the specified time, the progress of the reaction being monitored by TLC. After the reaction, excess alcohol is removed under reduced pressure and the residue is extracted with diethyl ether. The ether extract is washed with a solution of sodium thiosulfate and subsequently with distilled water, dried over anhydrous sodium sulfate, and concentrated *in vacuo* to yield the crude product, which is purified by column chromatography (hexane/ether 9:1) to give the desired carboxylic ester (5.1 g, 98%).

1.1.2.3 Solid Acids

Various solid acids are utilized for esterification, although the substrates that can be employed suffer from considerable limitations due to the strong acidity. Nevertheless, solid acids have a great advantage in that they can be removed from the reaction mixture by filtration and thus applied to large-scale production.

Nafion-H Nafion-H is the oldest solid acid to have been utilized as an esterification catalyst [81]. When a mixture of carboxylic acid and alcohol is allowed to flow over this catalyst at 95–125 °C, high yields of the corresponding esters are obtained with a contact time of ~5 s. A batch reaction is also employable [82, 83].

Experimental Procedure [81] Typical Esterification Procedure: The reactor is charged with activated Nafion-H catalyst (2.0 g). Carrier nitrogen gas is passed through the catalyst at a rate of 30 mL min^{-1}. A mixture of hexanoic (caproic) acid (2.6 g, 0.025 mol) and ethanol (2.9 g, 0.062 mol) is passed through the catalyst at 125 °C at a rate of 0.082 mL min^{-1}, corresponding to contact time of 5–7 s. The two-phase product mixture is diluted with ether (30 mL) and washed with 5% sodium hydrogen carbonate solution (2 × 20 mL), and then with water (2 × 20 mL). The organic layer is dried with magnesium sulfate and the solvent is evaporated. The residue is reasonably pure ethyl hexanoate, which may be distilled for further purification; yield: 3.5 g (98%); b.p. 167 °.

Amberlyst 15 α-Hydroxy esters [84] and α-amino acids [85] are successfully converted into the corresponding esters with this catalyst, while catechol undergoes esterification with acrylic acid to afford 7-hydroxy coumarin (Scheme 1.19) [86]. A detailed kinetic study on esterification of acetic acid with 1-butanol with Amberlyst 15 has appeared more recently [87].

Experimental Procedure [84] A solution of γ-butyrolactone (11.6 mmol) in anhydrous methanol (25 mL) is stored on Amberlyst-15 (25 g) with occasional shaking for 20 h. Methanol is decanted and the Amberlyst is washed with methanol (2 × 20 mL). The combined methanol fractions are evaporated and the residue is distilled to give methyl 4-hydroxybutanoate, b.p. 110–114 °C/ 8–10 mm.

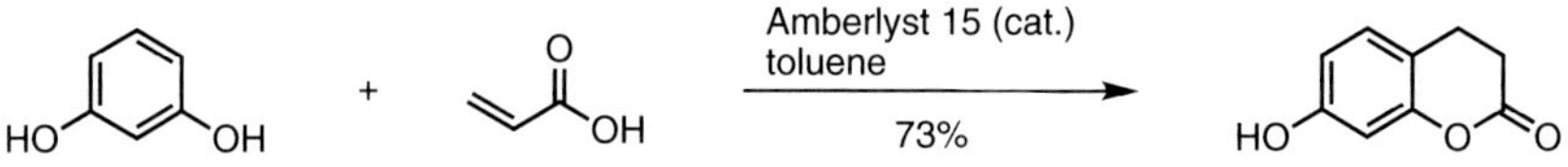

Amberlyst 15 (cat.)
toluene
73%

Scheme 1.19

Amberlite IR120 Various substrates with hydroxy and related functions, such as sugars [88, 89] and shikimic and quinic acids [90], are esterified with this resin.

Experimental Procedure [88] 5-*O*-(α-D-Glucopyranosyl)-D-arabinono-1,4-lactone: A solution of potassium 5-*O*-(α-D-Glucopyranosyl)-D-arabinonate (7.7 g, 20 mmol) in water (20 mL) is passed through an ion-exchange column (Amberlite IR-120 H^+, 200 mL) and eluted with water (500 mL). Concentration of the eluent, followed by drying *in vacuo* (10^{-2} Torr), gives 5-*O*-(α-D-Glucopyranosyl)-D-arabinono-1,4-lactone (3.2 g, 99%) as an amorphous solid, softening around 88–90 °C.

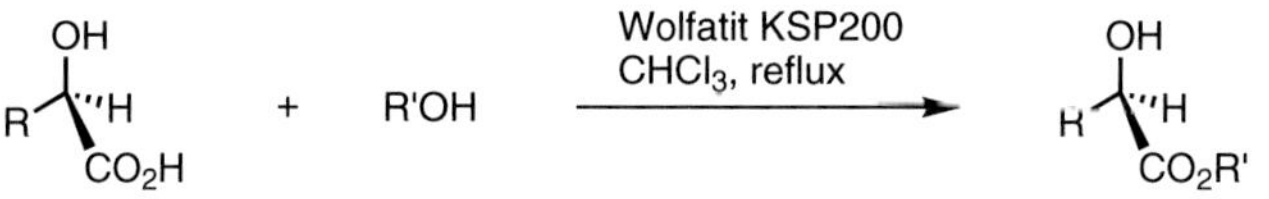

Wolfatit KSP200
$CHCl_3$, reflux

Scheme 1.20

Wolfatit KSP200 Esterification of chiral α-hydroxy carboxylic acids without racemization is feasible by heating in EtOH or MeOH/$CHCl_3$ in the presence of the ion-exchange resin Wolfatit KSP200 (Scheme 1.20) [91]. The products are useful intermediates for synthesis of the corresponding α-hydroxy aldehydes.

Zeolite The rare earth-exchanged RE H-Y zeolite is the best of the various zeolite catalysts [92]. Heating of alcohol solutions of carboxylic acids in the presence of the freshly activated zeolite at 150 °C provides good to excellent yields of esters. The same type of zeolite is also useful for lactonization of *seco* acids [93]. Zeolite catalysts for petroleum cracking are employable for synthesis of α-amino acid esters [94] and phenyl benzoates [95]. Reactions between acetic acid and C_2–C_4 alcohols with Hβ, HY and ZSM-5 zeolites are the subject of extensive studies [96–98].

Experimental Procedure [92] A mixture of phenylacetic acid (5 g, 0.036 mol) and ethanol (50 mL) is placed in a Parr reactor, and freshly activated zeolite (RE H-Y, 5 g) is slowly added. It is then heated at 150 °C under autogeneous pressure for 8 h. The reactor is allowed to cool to room temperature and the catalyst is filtered off and washed with ethanol (2 × 25 mL). The ethanol is removed from the filtrate by distillation. The residue is diluted with dichloromethane (50 mL) and washed with 5% aq. sodium carbonate solution (2 × 25 mL) to remove the unreacted acid, then with water (2 × 25 mL) and finally with brine (20 mL) and dried over anhydrous sodium sulfate. Removal of the solvent provides pure ethyl phenylacetate (5.51 g, 91%).

Mesoporous Silica Mesoporous silica has received extensive attention recently. Al-MCM-41 molecular sieves effect reactions between various acids and alcohols in the vapor phase: acetic acid/amyl alcohol [99]; acetic acid/butyl alcohols [100, 101]; terephthalic acid/methanol [102]; butyric acid/1-pentanol [103]. Microporous titanosilicate ETS-10 molecular sieves are also effective for esterification of long-chain carboxylic acids with alcohols [104]. Sulfonic acid-functionalized mesoporous silicas are utilized for esterification of fatty acids with methanol and glycerol [105–107]. Hydrophobic sulfonic acid-functionalized mesoporous benzene/silica is more active than Nafion-H for esterification of acetic acid with ethanol [108]. A similar hydrophobic SO_3H-modified catalyst is accessible from mesoporous ethylene/silica [109]. Mesoporous MCM-41 and SBA-15 functionalized with perfluoroalkanesulfonic acid are more active for esterification of long-chain fatty acids with alcohols than Nafion/silica composite [110, 111].

Comparison of commercial solid acid catalysts is now available in some reports [112–114].

Modification of Silica and Alumina Treatment of silica or alumina with $ClSO_3H$ results in immobilization of sulfuric acid on the surface of silica or alumina, which catalyzes esterification of aryloxyacetic acid or aromatic carboxylic acid [115, 116]. Silica chloride obtained from silica and thionyl chloride effects esterification of amino acids [117]. Sulfate-, phosphate-, and borate-modified silica, alumina, and zirconia furnish benzyl acetate from acetic acid and benzyl alcohol concomitant with only a small amount of dibenzyl ether [118].

$Nb_2O_5 \cdot nH_2O$ This catalyst is claimed to be more active than cation-exchange resin, $SiO_2 \cdot AlO_3$, and solid super acids [119]. Interestingly, supermicroporous niobium oxide, synthesized using a nonionic block copolymer as a structural directing reagent, is employable for gas-phase esterification of acetic acid with ethanol [120].

$ZrO_2 \cdot nH_2O$ and Mo–ZrO_2 Hydrous ZrO_2, which catalyzes reactions between carboxylic acids and alcohols, exhibits the following advantages: (i) the catalyst is easily prepared and stable in air, and (ii) the reaction does not require water-free conditions [121]. The catalytic activity is further improved by use of Mo–ZrO_2

mixed oxide, because electron-deficient sites are formed by introduction of Mo cations into the lattice of the solid ZrO_2. [122].

Experimental Procedure [121] General Procedures for Vapor-Phase Reactions: The catalytic esterification is carried out in a glass-flow reactor (6.5 mm in diameter) with a fixed-bed catalyst: flow rate of nitrogen gas = 60 mL min^{-1}; catalyst = 2.0 g, 24–60 mesh; reaction temperature = 135–280 °C. A mixture of a carboxylic acid, an alcohol, and a hydrocarbon as an internal standard is fed into the reactor (5 or 10 mL h^{-1}) with the aid of a microfeeder. In some cases benzene is also added, to dilute the reaction mixture. The activity and selectivity of the reaction are determined after a steady state has been reached. The products are then analyzed by gas chromatography (a capillary column, PEG 20 M 30 m). The products are identified by comparison of their retention times with those of authentic samples.

General Procedures for Liquid-Phase Reaction: The catalyst (2.0 g), a carboxylic acid, an alcohol, and a hydrocarbon as an internal standard are placed in a 25 mL round-bottomed flask fitted with a reflux condenser. The contents are then heated under gentle reflux. In some reactions toluene is added to the solution, in order to raise the reaction temperature. In some cases the reaction mixture is placed in an oil bath kept at 77.0 ± 0.1 °C. The reaction mixture is worked up after 5 h, and the products are analyzed by gas chromatography.

Strongly Acidic Carbon Materials Graphite bisulfate ($C_{24}^{+}HSO_4^{-}2H_2SO_4$), which can be prepared by electrolysis of 98% H_2SO_4 with a graphite anode, brings about reaction between alcohol and carboxylic acid in a 1:1 ratio at room temperature [123]. The yields are usually over 90%. Sulfonation of incompletely carbonized D-glucose results in amorphous carbon consisting of small polycyclic carbon sheets with high density of SO_3H groups [124]. This carbon material exhibits remarkable catalytic performance for esterification of higher fatty acids. Poly(vinyl alcohol) membranes crosslinked with sulfosuccinic acid catalyze esterification of acetic acid with isoamyl alcohol [125].

Natural Montmorillonite Another intercalation compound, natural montmorillonite, is useful for selective acylation of various functionalized primary and secondary alcohols [126].

Experimental Procedure [126] In a typical procedure, 1-phenylethanol (5 mmol) and glacial acetic acid (50 mmol, corresponding to a 1:10 molar ratio; 1:3 alcohol/carboxylic acid ratio for other solvent systems) are heated at reflux in the presence of montmorillonite catalyst (100 mg) with stirring for 15 min. After completion of the reaction, monitored by TLC or GC, the reaction mixture is filtered and the filtrate is concentrated to obtain the pure product. The product is analyzed by ^{1}H NMR, while the catalyst is washed with ethyl acetate and dried in an oven at 120 °C for 1 h and then reused.

Metal-Exchanged Montmorillonite and Bentonite Montmorillonites enwrapped with various metal cations such as Na^+, Al^{3+}, Fe^{3+}, Cr^{3+}, Zn^{2+}, Mn^{2+}, Ni^{2+}, Ti^{4+} are active catalysts for esterification [127–131]. Acid-activated bentonite catalyzes reactions between various carboxylic acids and alcohols [132]. A superacid SO_4^{2-}/TiO_2 catalyst on Al^{3+}-modified bentonite brings about esterification of benzoic acid with 1-pentanol [133].

Phosphorus Oxides Phosphorus pentoxide can be used for dehydration between carboxylic acid and alcohol. Heating a mixture of alcohol, carboxylic acid, and P_4O_{10} is the simplest treatment [134]. In addition to intermolecular esterification, lactonization is also achievable [135, 136]. This procedure is modified by initial treatment of P_4O_{10} with alcohol to furnish an equimolar mixture of mono- and dialkylphosphates (Scheme 1.21) [137]. In practice, isolation of these compounds is not necessary, a carboxylic acid being added to the mixture to produce the ester.

Experimental Procedure Scheme 1.21 [137] Esterification of a Liquid Carboxylic Acid: Glacial acetic acid (0.6 mol, 36.0 g) is added to the alkyl phosphate reagent (0.1 mol equivalent), and the reaction mixture is heated at reflux on a water bath for 3 h, with ice-cold water being circulated through the condenser. The reaction mixture is allowed to come to room temperature and extracted with ether (2 × 100 mL), and the organic layer is washed (aq. $NaHCO_3$, 2 × 100 mL) and dried (Na_2SO_4). After removal of the solvent, the residual liquid is distilled through a fractionating column to yield methyl acetate (39 g, 90%), b.p. 54–56°.

Esterification of Solid Acid: Phenylacetic acid (82.2 g, 0.6 mol) is added to the alkyl phosphate reagent. In this case, any required alkanol is added to ensure homogeneous solution. The reaction mixture is heated at reflux for 3 h. It is then diluted with water (100 mL) and extracted with ether (2 × 100 mL), and the organic layer is washed (aq. $NaHCO_3$, 2 × 100 mL) and dried (Na_2SO_4). After removal of solvent, the residual liquid is distilled off to yield ethyl phenylacetate (86%, 85 g), b.p. 224–226°.

$$P_4O_{10} + 6\ ROH \longrightarrow 2\ (RO)P(O)(OH)_2 + 2\ (RO)_2P(O)(OH)$$

Scheme 1.21

A flow system that uses a vertical column is available, although a mixture of $P_4O_{10}/CuSO_4/Na_2SO_4$ is better than simple P_4O_{10} for this purpose [134]. $CuSO_4$ serves both as a water scavenger and, through its color change, as an indicator for the progress of the reaction and the duration of the reactivity of the column, while Na_2SO_4 retains the desired porosity and is useful for sustained reactivity of the column with its water-absorbing property. This packing reagent is also used for a batch reaction [134, 138].

Experimental Procedure [134] Mixtures of various organic acids (0.05 mol), freshly prepared packing reagent (2.5 g), and ethanol (50 mL) are placed in Erlenmeyer flasks and left at room temperature for 20 h with occasional shaking. Removal of the solvent (ca. 30 mL) on a steam bath (15–20 min) leaves residue, which furnishes the ethyl esters on conventional workup.

Ph_3SbO/P_4S_{10} The characteristic feature of this catalyst system is that the reaction temperature (25–85 °C) is lower than those in other procedures [139].

Inorganic Sn- or Ti-Based Solid Acids Amorphous M(IV) tungstates (M = Sn, Ti) are useful for synthesis of dioctyl phthalate [140]. Methyl ester synthesis from octanoic acid is feasible with a ceramic acid obtained by impregnating $SnO_2 \cdot H_2O$ and $(NH_4)_6(H_2W_{12}O_{40})$ followed by calcination [141]. Solid superacid of sulfated tin oxide, SO_4^{2-}/SnO_2, is a highly active catalyst for condensation between acids and alcohols [142, 143]. Similarly, titanium superacid, SO_4^{2-}/TiO_2, is capable of esterifying chemically labile mandelic acid [144].

Heteropolyacids Various bromoacetates are obtained by treatment of bromoacetic acids (1.0 mol) with alcohols (1.1 mol) in the presence of 12-tungstophosphoric acid, $H_3PO_{40}W_{12} \cdot H_2O$, [145]. Its partially substituted Cs and K salts are also useful catalysts for esterification [146]. The corresponding ammonium salt catalyzes selective reactions between aliphatic carboxylic acids and alcohols in the presence of aromatic carboxylic acids [147, 148]. $H_{14}[NaP_5W_{30}O_{110}]$ can be employed for esterification of salicylic acid with aliphatic and benzylic alcohols [149]. Cobalt-containing polyoxometalate, $K_5CoW_{12}O_{40} \cdot 3H_2O$, is suitable for esterification of mandelic acid [150].

Experimental Procedure [145] Bromoacetic acid (1.0 mol), alcohol (1.1 mol), benzene (70 mL), and $H_3PO_{40}W_{12} \cdot nH_2O$ (0.4 g) are placed in a 250 mL round-bottomed flask fitted with a Dean-Stark condenser and the mixture is heated at reflux for 3–4 h until 15–18 mL of water have been collected. The crude solution is separated, washed once with water, twice with a saturated sodium bicarbonate solution, and finally again with water, and is then dried over magnesium sulfate and sodium sulfate (1 : 1), filtered, and distilled under normal pressure or under vacuum.

Acid Catalysts on Inorganic Solid Support Heteropoly acids often leak out of catalyst supports, because these acids are extraordinary soluble in water and several organic solvents. $H_3PW_{12}O_{40}$ can be immobilized by hydrous zirconia, which catalyzes reactions between glacial acetic acid and cyclohexanol [151] and between acetic acid and isoamyl alcohol [152]. $H_4SiW_{12}O_{40}$ on hydrous zirconia brings about esterification of primary and secondary alcohols with C_1—C_3 carboxylic acids [153]. Zirconia is also employable to support WO_3, which catalyzes esterification of palmitic acid with methanol [154]. Porous zirconium phosphate is also employable to support WO_3 [155].

Silica gel is employable for supporting various acid catalysts: P_2O_5 [156]; $H_3PMo_{12}O_{40}$ [157]; $H_3PW_{12}O_{40}$ [158, 159]. All of these supported catalysts are effective for esterification. Grinding $Fe(ClO_4)_3(ROH)_6/SiO_2$ with an equimolar amount of carboxylic acid provides esters [160]. This protocol is operationally simple, but requires a stoichiometric amount of the promoter. Aliphatic carboxylic acids are esterified preferentially over aromatic ones at room temperature with the aid of $NaHSO_4$ supported on silica gel (Scheme 1.22) [161]. $Hf[N(SO_2C_8F_{17})_4]$ supported on fluorous reverse-phase silica gel efficiently catalyzes esterification of methacrylic acid with methanol [162].

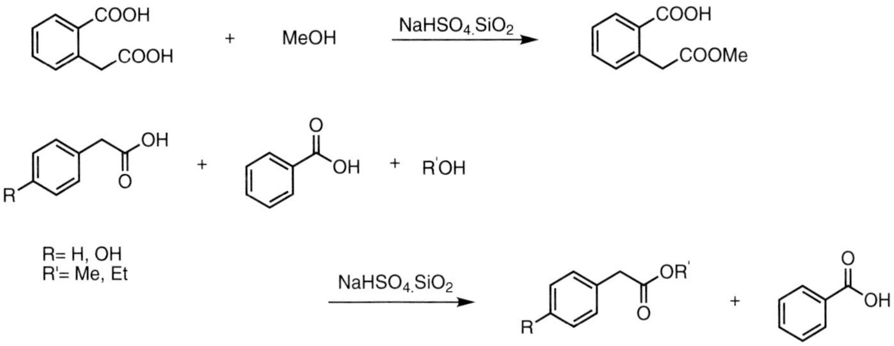

Scheme 1.22

Supporting $ZrOCl_2 \cdot 8H_2O$ on mesoporous silica MCM-41 enhances the catalytic activity for esterification of C_{10}–C_{18} normal acid with alcohols [163]. 12-Phosphotungstic acid and its cesium salts supported on a dealuminated Y zeolite catalyze reaction between 1-butanol and acetic acid in high yield [164].

$H_3PW_{12}O_{40}$ can be supported on neutral alumina, catalyzing esterification of aliphatic carboxylic acids with primary and secondary alcohols [165].

Activated carbon can tightly immobilize or entrap a certain amount of the acids. With $H_4SiW_{12}O_{40}$ entrapped in carbon, vapor-phase esterification of acetic acid with ethanol can be conducted efficiently [166]. Zirconium sulfate supported on activated carbon exhibits higher activity for esterification of oleic acid with 1-butanol [167].

Acid Catalysts on Organic Solid Support Heteropolyacids supported on ion-exchange resin accelerate the rates of reaction between lactic acid and ethanol [168]. Polyaniline-supported acid catalysts are effective for esterification of carboxylic acids with alcohols [169, 170]. Triphenylphosphine ditriflate anchored onto cross-linked polystyrene is useful for ester synthesis from functionally substituted carboxylic acids and alcohols [171]. Esterification of amino acids takes place smoothly with polystyrylsulfonyl-3-nitro-1*H*-1,2,4-triazolide resin [172].

Solid Acid under Microwave Irradiation Esterfication by solid acid catalysts is accelerated by microwave irradiation [173]. Microwave irradiation in a continuous flow system is also feasible [174]. More detailed description of microwave irradiation is given in Section 7.4.

1.1.3 Base Activators

Base-mediated reactions to produce esters are catalytic on some occasions but noncatalytic on others, so both cases are dealt with together in this section. The basic catalysts are not suitable for esterification because esters are hydrolyzed when the reaction mixture is subjected to aqueous workup. Nonetheless, a few nonaqueous methods are available for highly functionalized substrates. ω-Hydroxy acids undergo lactonization upon exposure to KOH/KOMe/glycerin [175].

Another technique is the use of DMAP (4-dimethylaminopyridine). A Kemp's triacid derivative is transformed into a monoester by treatment with Et_3N/DMAP (cat.) (Scheme 1.23) [176]. The synthesis of deoxy derivatives of α-mannosidase inhibitor mannostatin A makes use of the DMAP-methodology as a key step (Scheme 1.24) [177].

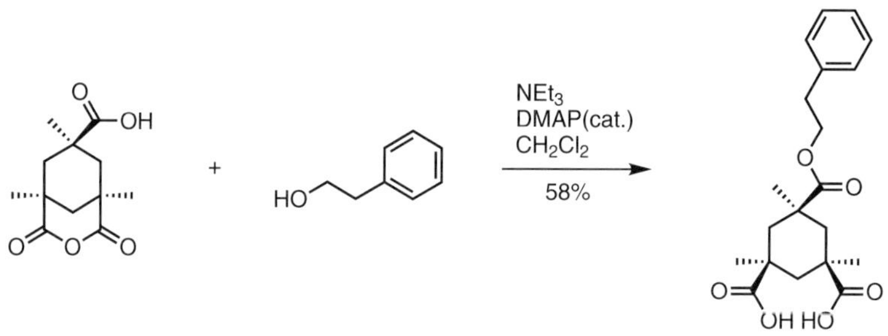

NEt_3
DMAP(cat.)
CH_2Cl_2
58%

Scheme 1.23

Chiral

DMAP
CH_2Cl_2

Chiral
56%

Chiral
5%

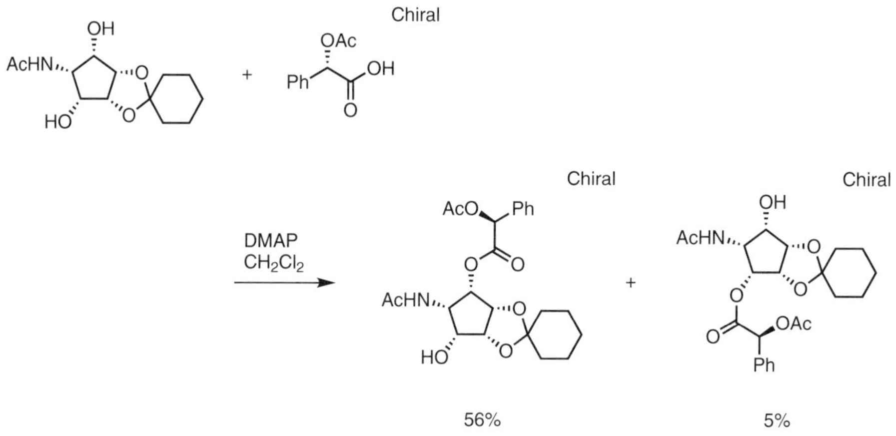

Scheme 1.24

1.1.4
Carbodiimide Activators

The use of DCC (dicyclohexylcarbodiimide) as a promoter represents one of the most versatile esterification methods. Although this reagent is irritant to skin and a stoichiometric dosage or more is necessary, this procedure enjoys various advantages. The reaction usually proceeds at room temperature, and the reaction conditions are so mild that substrates with various functional groups can be employed. The reaction is not sensitive to steric bulk of the reactants, allowing production of esters of tertiary alcohols. As such, a wide range of applications have been achieved in the fields of natural products, peptides, nucleotides, etc.

The application of the DCC method in pure organic synthesis dates back to 1967 [178]. The following mechanism for this reason is suggested (Scheme 1.25).

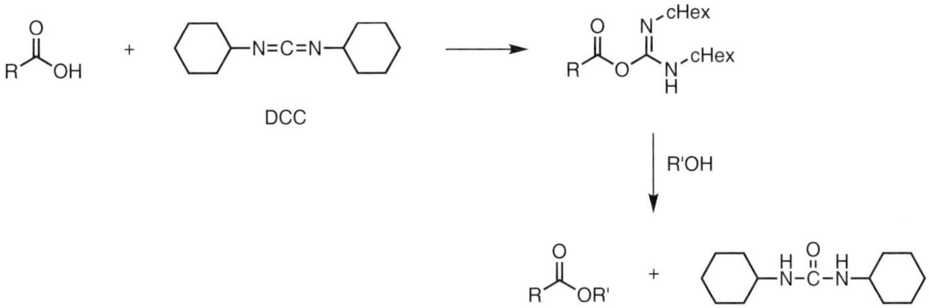

Scheme 1.25

This original procedure, however, unfortunately suffers from some drawbacks: yields are not always high, and undesirable *N*-acylureas are occasionally formed. These drawbacks can be overcome by addition of strong acid such as *p*-toluenesulfonic acid [179, 180]. Alternatively, addition of a catalytic amount of *p*-aminopyridines is more effective [181, 182]. As a result, methyl or *p*-nitrophenyl esters of pivalic and 2,4,6-trimethylbenzoic acids are obtainable. *tert*-Butyl esters of 3,5-dinitrobenzoic acid and glycerol tristearate can also be prepared: these esters are not accessible without the use of *p*-aminopyridines. However, combinations between *tert*-butyl alcohol and more sterically demanding acids such as adamantanecarboxylic acid or 1-phenylcyclohexane-1-carboxylic acid fail to afford the desired esters. The mechanism of the catalyzed reaction is depicted in Scheme 1.26 [182]. The carboxylic acid is first converted by DCC to the anhydride, which then forms an acylpyridinium species with DMAP. Nucleophilic attack on the acyl

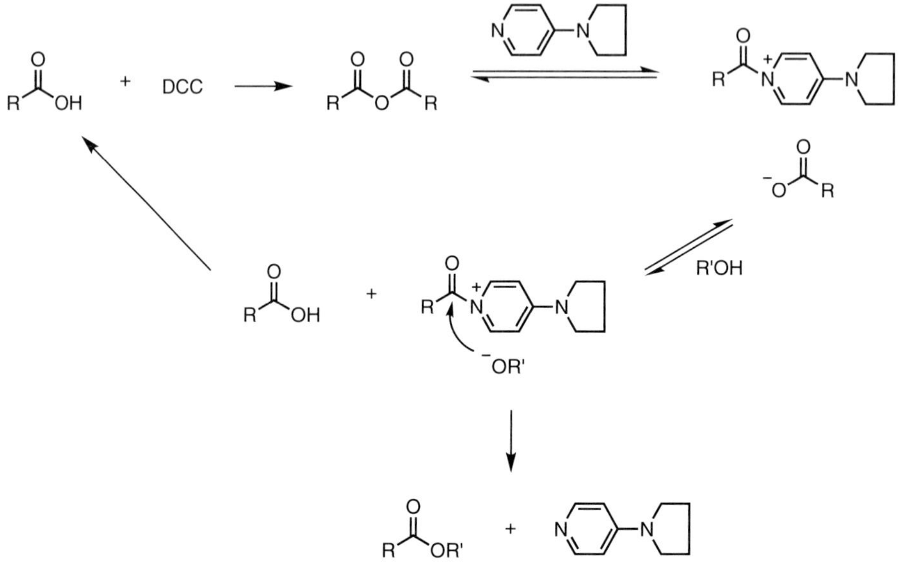

Scheme 1.26

group by $R'O^-$ produces the ester concomitantly with regeneration of DMAP, together with a half quantity of RCOOH, which can be again used for the reaction with DCC. This technique is employable for high-yielding synthesis of anthranilate esters from sterically hindered alcohols [183]. The reaction under high pressure (1.5 GP) effects smooth esterification of monomethylsuccinates with tertiary alcohols [184]. 4-(Pyrrolidin-1-yl)pyridine is also a useful base for deconjugative coupling of 2-cyclohexylideneacetic acid [185].

Experimental Procedure [181] A mixture of DMAP (30–110 mg) and alcohol or thiol (20–40 mmol; 10 mmol with alcohols or thiols that are not easily removable without significant loss; 3.4 mmol with glycerol) is added to a stirred solution of carboxylic acid (10 mmol) in anhydrous CH_2Cl_2 (10 mL, DMF in case of sparingly soluble acids). DCC is added at 0 °C to the reaction mixture, which is then stirred for 5 min at 0 °C and 3 h at 20 °C. Precipitated urea is then filtered off and the filtrate is evaporated *in vacuo*. The residue is taken up in CH_2Cl_2 and, if necessary, filtered free of any further precipitated urea. The CH_2Cl_2 solution is washed twice with 0.5 N HCl and with saturated $NaHCO_3$ solution, and is then dried over $MgSO_4$. The solvent is removed by evaporation, and the ester is isolated by distillation or recrystallization.

When carboxylic acids carry a strongly electron-withdrawing group such as COOR, $P(O)(OEt)_2$, CN, or RSO_2 at the position α to the carboxyl group, sterically hindered alcohols, including tertiary alcohols, undergo smooth esterification

(Scheme 1.27) [186–188]. The reaction can be explained by the intermediacy of the corresponding ketene, which is highly electrophilic but relatively sterically undemanding.

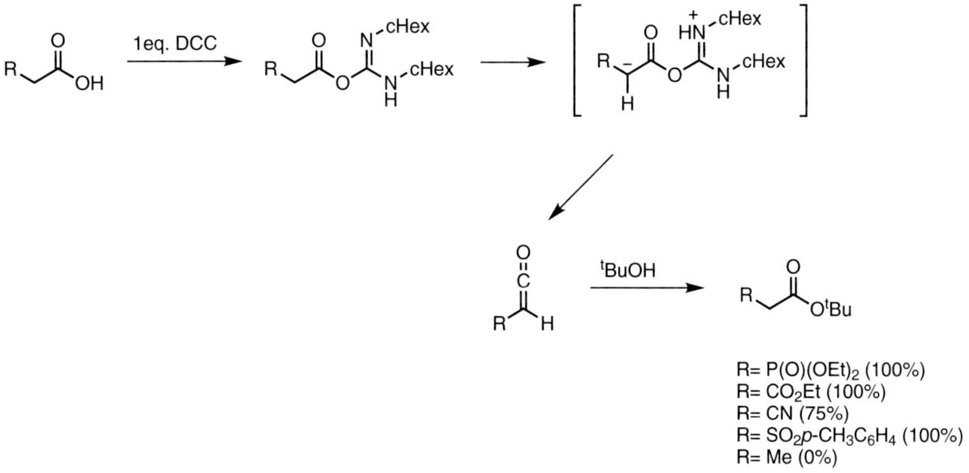

R= $P(O)(OEt)_2$ (100%)
R= CO_2Et (100%)
R= CN (75%)
R= SO_2p-$CH_3C_6H_4$ (100%)
R= Me (0%)

Scheme 1.27

Addition of $Sc(OTf)_3$ to a mixture of DIPC (diisopropylcarbodiimide) or EDC (1-ethyl-3-[3-(dimethylamino)propyl]carbodiimide/DMAP effects reaction between sterically bulkier reactants such as *tert*-butyl alcohol/*t*-Boc-(*S*)-(-)-Glu(OBz) or trimethylbenzoic acid and $Me_2C(OH)COOEt$/4-nitrobenzoic acid [189].

Experimental Procedure [189] Method A (acid in excess): A suspension of *tert*-butyl alcohol (0.094 mL, 1 mmol), scandium triflate (0.30 g, 0.6 mmol), $ClCH_2COOH$ (0.28 g, 3 mmol), and DMAP (0.37 g, 3 mmol) in anhydrous methylene chloride (10 mL) is cooled to −8 °C in an ice-salt bath for 30 min. DIPC (0.49 mL, 3.1 mmol) is added, and the reaction mixture is stirred at −8 °C for 30 min and then allowed to warm to room temperature over 2 h. The reaction mixture is filtered to remove any insoluble material, and the filtrate is washed with HCl (0.1 N, 2 × 20 mL), sodium bicarbonate (0.1 N, 2 × 20 mL), and distilled water (20 mL). The organic phase is dried ($MgSO_4$) and the solvent is removed under reduced pressure. Trace amounts of DIPU are precipitated with ether and removed by filtration. Evaporation of the ether gives the corresponding pure ester (0.14 g, 95%).

Method B (alcohol in excess): A suspension of *tert*-butyl alcohol (1.9 mL, 20 mmol), scandium triflate (0.3 g, 0.6 mmol), *t*-Boc(*s*)-(-)-Glu(OBz) (0.34 g, 1 mmol), and DMAP (0.61 g, 5 mmol) in anhydrous methylene chloride (10 mL) is cooled to −8 °C in an ice-salt bath for 30 min. EDC (0.38 g, 2 mmol) is added, and the reaction mixture is stirred at −8 °C for 30 min and then allowed to warm to room temperature over 2 h. The reaction mixture is filtered to remove any insoluble material, and the filtrate is washed with HCl (0.1 N, 2 × 20 mL), sodium bicarbonate (0.1 N, 2 × 20 mL), and distilled water (20 mL). The organic layer is dried ($MgSO_4$), and the solvent is removed under reduced pressure. The product is further purified by chromatography on a silica gel column with 0–5% methanol in methylene chloride as the eluent to give corresponding pure ester (0.32 g, 80%).

Tributylphosphine is an additive of choice when the reactants are base-sensitive [190], even when the use of DMAP results in poor yield.

The conventional DCC/base esterification of carboxylic acids containing α-halogen atom(s) in a solvent induces complex side reactions due to *N*-acylation. Remarkably, a high yield of the desired ester is obtained when the reaction is conducted in the absence of solvent and base [191].

Since a stoichiometric amount of carbodiimide or more (sometimes 10 equivalents) must be used, immobilization of this reagent is highly convenient for separation from the reaction mixture. EDAC (ethyl dimethylaminopropylcarbodiimide) supported on polystyrene-divinylbenzene resin is effective for synthesis of esters for use in bioconjugation (Scheme 1.28) [192]. A variety of carboxylic acid haptens can be esterified with *N*-hydroxysuccinimide or pentafluorophenol. Of particular significance is the extension of this method to extremely water-soluble active esters that cannot be purified by conventional extraction methods. DCC analogs can be immobilized as well, and as such utilized for macrolactonization of *seco* acids (Scheme 1.29) [193].

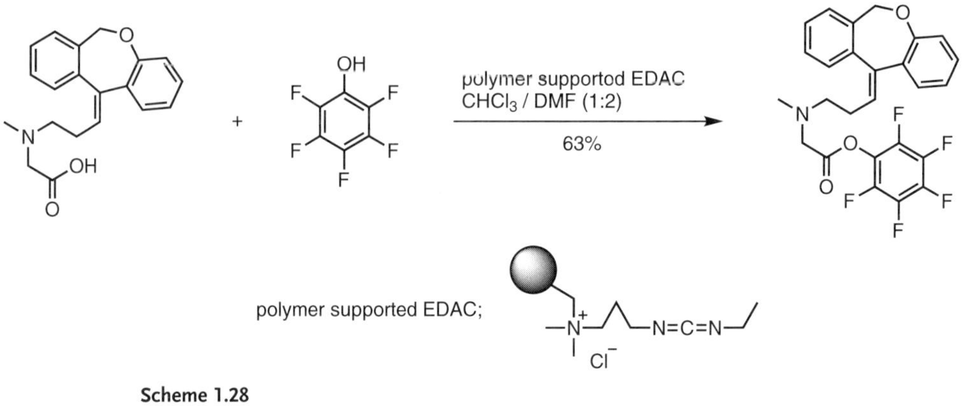

polymer supported EDAC
$CHCl_3$ / DMF (1:2)
63%

polymer supported EDAC;

Scheme 1.28

1.1.5
The Mitsunobu Reaction

The Mitsunobu reaction is another popular technique, although the employment of more than a stoichiometric amount of reagent is necessary [194, 195]. The reaction between alcohols and carboxylic acids proceeds smoothly under neutral

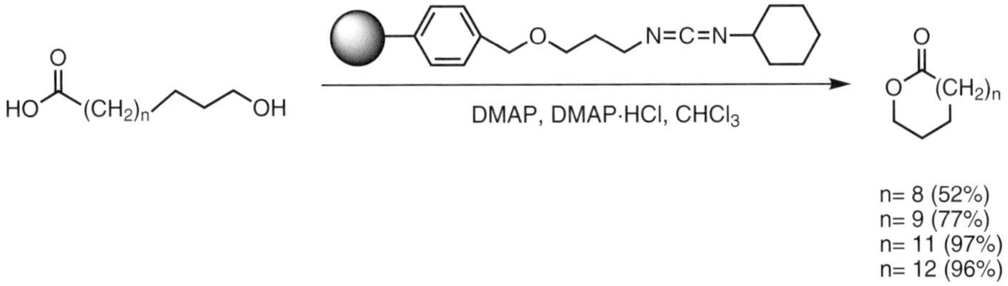

DMAP, DMAP·HCl, $CHCl_3$

n= 8 (52%)
n= 9 (77%)
n= 11 (97%)
n= 12 (96%)

Scheme 1.29

conditions at or below room temperature. Of great significance are its high chemo-, stereo-, and regioselectivities. Typically, a mixture of alcohol and carboxylic acid is treated with DEAD (diethyl azodicarboxylate) and Ph_3P (Scheme 1.30) [196]. The initial step is addition of Ph_3P to DEAD to give a zwitterion, which then reacts with carboxylic acid to afford a phosphonium carboxylate. Reaction of this intermediate with alcohol directly generates a key intermediate, alkoxyphosphonium salt (route A), which undergoes nucleophilic attack by the carboxylate ion to furnish the desired ester. Importantly, the high stereoselectivity arises from the complete inversion at the alkoxy carbon center in the last S_N2 step. The phosphonium carboxylate intermediate is isolable with cyclodiphosphazane [197]. An alternative route is possible (route B) [198]. After addition of a carboxylic acid to DEAD, an acyloxyphosphonium salt is formed, and this is attacked by an alkoxide ion generated by interaction between an alcohol and the hydrazide ion. With the use of primary-secondary diols, almost selective esterification at the primary position can be achieved in preference to the secondary one. The selectivity is explained by the steric hindrance of three bulky phenyl groups attached to the phosphorus atom.

Experimental Procedure [194] General Procedure for Mitsunobu Reaction. A solution of triphenylphosphine (0.01 mol) in ether (10 mL) is added drop by drop into a solution of diethyl azodicarboxylate (0.01 mol) and carboxylic acid (0.01 mol) in ether (10 mL), with vigorous stirring at room temperature. The reaction soon starts, and a white precipitate of triphenylphosphine oxide and diethyl hydrazodicarboxylate appears. After the solution has been kept standing overnight at room temperature, the precipitate is removed by filtration. The ether is removed from the filtrate, and the residue is filtered to remove a small amount of the remaining precipitates. The filtrate is then distilled to give the corresponding esters of the carboxylic acid.

The Mitsunobu reaction has found numerous applications. For instance, benzylic and styryl alcohols are esterified selectively in the presence of (poly)phenolics [199]. Otherwise difficult-to-achieve esterification of benzoic acids with phenols is feasible [200]. Inversion of sterically hindered 17-hydroxy steroids [201] and regioselective esterification of sucrose [202] are feasible.

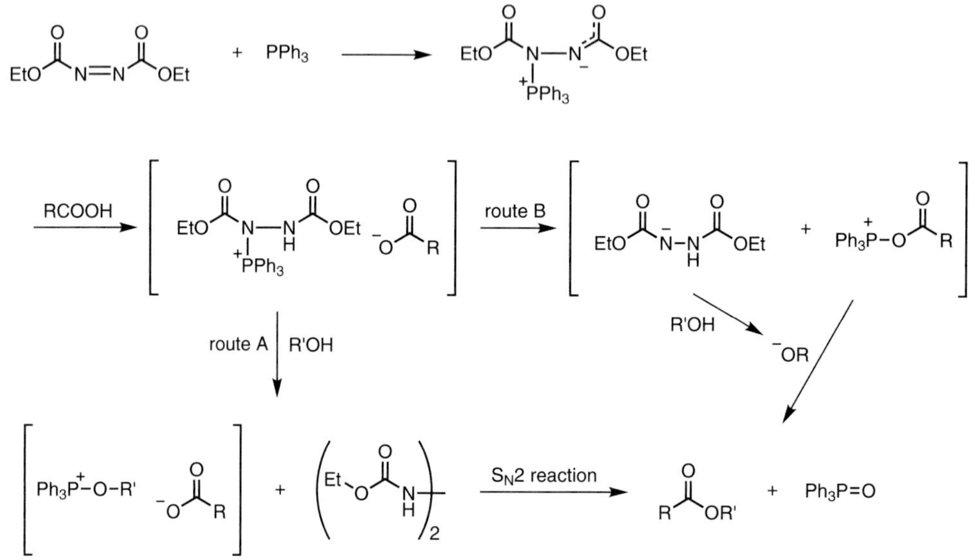

Scheme 1.30

When the alcohol is sterically hindered, the acyloxyphosphonium intermediate in route B plays a key role. The acyloxy-alkoxy interchange is suppressed, and so the acyl carbon of the acyloxyphosphonium species is directly attacked by the alkoxy anion [203, 204]. In such a case the stereochemistry about the alkoxy carbon is retained, as shown in Scheme 1.31.

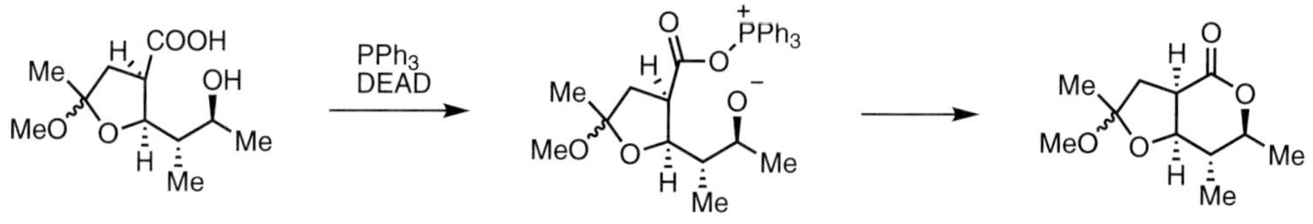

Scheme 1.31

Despite its various advantages, the Mitsunobu process suffers from a serious problem due to the unavoidable use of large amounts of the reagents, the separation of by-products thus occasionally being tedious and problematic. Overcoming these problems is the biggest issue in this procedure. 1,2-Bis(diphenylphosphino) ethane is claimed to be a convenient replacement for Ph_3P because the resulting bis(phosphine oxide) is more readily removed because of its more polar character [205]. Incorporation of an amino function in the phosphine, as in dipheny(2-pyridyl)phosphine [206] and (*p*-dimethylaminophenyl)diphenylphosphine [207] is the next strategy, as acidic workup can be used to remove the phosphine oxides with an amino function. The use of di-*tert*-butyl azodicarboxylate coupled with dipheny(2-pyridyl)phosphine is also more convenient, as the *tert*-butyloxycarbonyl group decomposes to isobutene and CO_2 upon acidic workup [208].

Experimental Procedure [206] Diisopropyl azodicarboxylate (1.97 mL, 10 mmol) is added dropwise with stirring to a cooled (0 °C) solution of diphenyl(2-pyridyl) phosphine (2.63 g, 10 mmol) and ethanol (0.69 g, 15 mmol) in ether (30 mL). Benzoic acid (1.16 g, 9.5 mmol) in the same solvent (10 mL) is then introduced. The reaction mixture is left to stir at room temperature overnight and cooled to −30 °C, and the by-products are removed by filtration. Residual amounts of the oxide remaining in solution are removed by washing with HCl (2 M, 2 × 25 mL). Distillation in a Kugelrohr apparatus affords the required ester (1.15 g, 80%).

Experimental Procedure [208] A mixture of 3-chloro-5-methoxyphenol (79 mg, 0.5 mmol), diphenyl-2-pyridylphosphine (197 mg, 0.75 mmol) and benzyl alcohol (54 mg, 0.5 mmol) is dissolved in anhydrous THF under an atmosphere of nitrogen. Di-*tert*-butyl azodicarboxylate (172 mg, 0.75 mmol) is added to this solution in one portion, and the resulting mixture is stirred at room temperature for one day. GCMS analysis of an aliquot shows complete conversion of starting material into the desired product after 24 h. A solution of hydrogen chloride in dioxane (4 M, 2 mL) is added to the mixture, and after this has been stirred for 1 h the excess solvent is evaporated. The residue is dissolved in ether or dichloromethane and shaken vigorously with magnesium sulfate, and the solvent is evaporated. Flash column chromatography (20% ethyl acetate in hexanes) gives the desired benzyl ester as a pale yellow oil (86 mg, 69%).

Polymer support of alkyl azodicaboxylates is also useful (Scheme 1.32) [209]. Polystyrene-supported methyl azodicarboxylate obtained from 1% cross-linked hydroxymethyl polystyrene resin can be used for various Mitsunobu reactions, with yields comparable to those obtained with soluble dialkyl diazodicarboxylates.

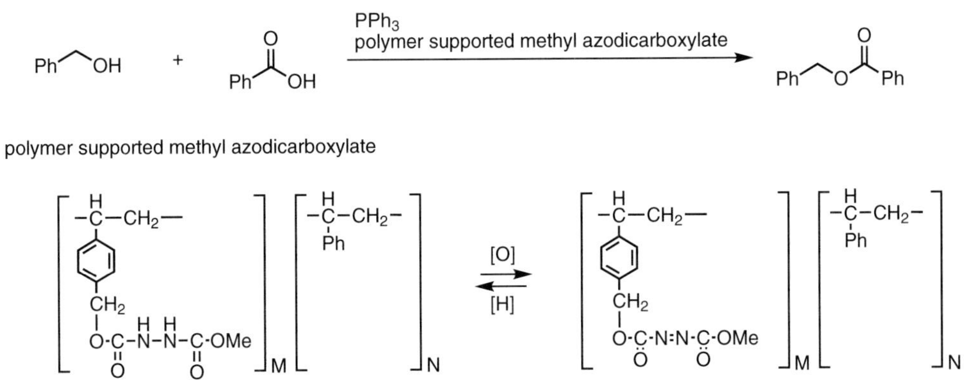

PPh3
polymer supported methyl azodicarboxylate
polymer supported methyl azodicarboxylate
[O]
[H]

Scheme 1.32

Thanks to the lower solubility of the polymer-supported reagent, purification of the resulting esters and recovery of the reagent are very easy. Also noteworthy is that the recovered reduced resin can be re-oxidized to the azodicarboxylate form and used again.

The use of fluorous azodicarboxylate derivatives is another means by which the separation problem may be solved (see Part Two). Bistridecafluorooctyl azodicarboxylate is one such reagent, readily recovered from the reaction mixture simply by extraction with FC-72 (perfluorohexanes) [210]. The combined use of fluorous azidocarboxylate and tertiary phosphine provides further elaboration of separation [211].

The activity of the Mitsunobu reagent may be enhanced. Azodicarboxamides in place of esters can activate less reactive acids. 1,1′- (Azodicarbonyl)dipiperidine/ Ph_3P [212] and *N, N, N′, N′*-terametylazodicarboxamide/Bu_3P [213] effect acylation of various secondary alcohols, including steroids.

Experimental Procedure [212] Under argon atmosphere, alcohol (1 mmol), tributylphosphine (1.5 mmol), and acid (1.5 mmol) are successively dissolved in dry benzene (3 mL) with stirring at 0 °C, and solid 1,1′-(azodicarbonyl)dipiperidine (ADDP, 1.5 mmol) is added to the solution. After 10 min, the reaction mixture is brought to room temperature and the stirring is continued for 24 h. Hexane is added to the reaction mixture, and precipitated dihydro-ADDP is filtered off. The product is purified by SiO_2 column chromatography after evaporation of the solvent *in vacuo*.

Experimental Procedure [213] Under dry Ar atmosphere, solid *N, N, N′, N′*-tetramethylazodicarboxamide (TMAD) (1.5 mmol) is added in one portion, at 0 °C with stirring, to a dry benzene solution (3 mL) of an alcohol (1 mmol), tributylphosphine (1.5 mmol), and a carboxylic acid (1.5 mmol). After 10 min, the reaction mixture is heated at 60 °C and stirred at this temperature for 24 h, during which time dihydro-TMAD crystallizes out. The epimeric mixture of esters obtained is analyzed as follows. The inversion ratios are determined by capillary GLC or ^{1}H-NMR on the crude products obtained by the evaporation of the solvent *in vacuo*. The (combined) yields are obtained after the product isolation by SiO_2 column chromatography.

As shown in Scheme 1.33, treatment of chiral trivalent alkoxyphosphorus compounds with DIAD results in cycloaddition rather than formation of a zwitterion [214]. The newly formed adduct effects the Mitsunobu-like esterification. Unfortunately however, neither yields nor *ee*s of the esters are particularly high: ~50% yields and <39% *ee*s.

Since the alkoxyphosphonium salt is the key intermediate in the standard Mitsunobu reaction, variants to generate this species through other routes have been investigated. Triphenylphosphine-cyclic sulfamide betaine, formed by treating cyclic sulfamide with DEAD/Ph_3P, is stable in the solid state for several months

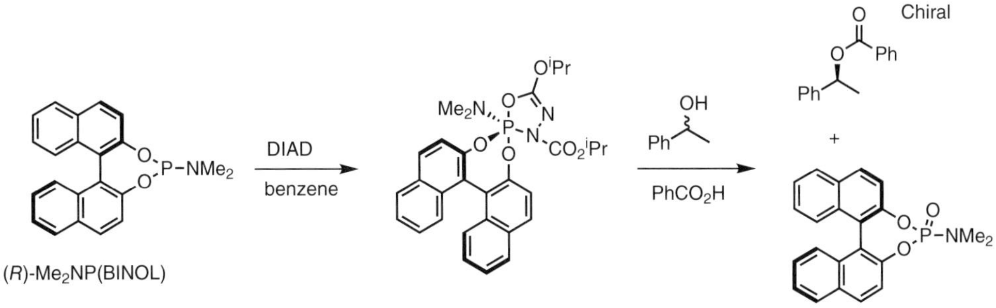

Scheme 1.33

(Scheme 1.34) [215]. This compound can be used for a Mitsunobu-like coupling between alcohol and carboxylic acid.

Experimental Procedure Scheme 1.34 [215] The adduct of triphenylphosphine and 3,3-dimethyl-1,2,5-thiadiazolidine 1,1-dioxide (5.0 mmol) is added portionwise over 10 min to a stirred mixture of the alcohol (3.32 mmol) and carboxylic acid (5.0 mmol) in anhydrous solvent (30 mL), and the resulting clear (milky) solution is stirred at room temperature under nitrogen for several hours. Et_2O (150 mL) is added, and the organic phase is washed with water (40 mL), dilute aqueous K_2CO_3 (40 mL), and brine (40 mL), and is then dried ($MgSO_4$) and concentrated. Products are purified by flash chromatography on silica gel (hexane/CH_2Cl_2 or hexane/EtOAc).

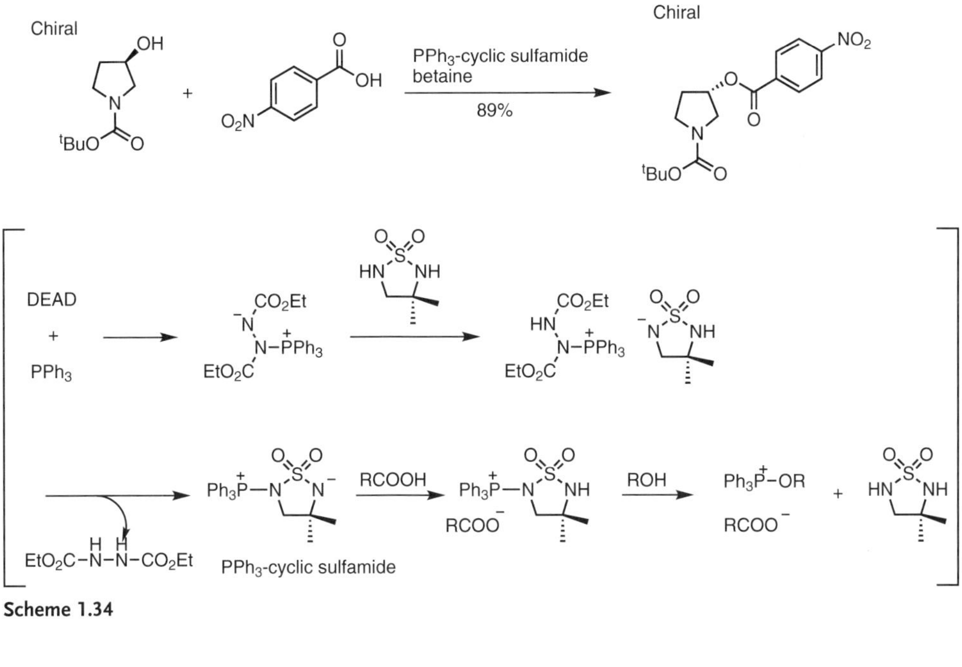

Scheme 1.34

Cyanomethylenetributylphosphorane mediates the direct condensation between alcohol and carboxylic acid through an alkoxy(tributyl)phosphonium salt intermediate (Scheme 1.35) [216]. This reagent is effective even for weak acids ($pK_a > 12$), which are not employable for the authentic Mitsunobu reaction. $Bu_3P{=}C(CO_2Me)_2$ acts similarly [217]. The inversion product prevails in toluene, while the retention is favored in DMF.

Experimental Procedure Scheme 1.35 [216] An alcohol (1 mmol) and a carboxylic acid (1.5 mmol) are successively dissolved in dry benzene (5 mL) with stirring under an argon atmosphere, and cyanomethylenetributylphosphorane (1.5 mmol) is added all at once, by syringe. The reaction mixture is heated with stirring in a sealed tube at 100 °C for 24 h. After evaporation of the solvent *in vacuo*, the product is purified by silica gel column chromatography.

HO O + ROH — $Bu_3P{=}CHCN$, benzene → RO O

$R = C_4H_9$ (99%)
$R = CH_2CH{=}CHCH_2$ (90%)
$R = Bn$ (100%)
$R = CH(CH_3)C_6H_{13}$ (96%)

[$Bu_3\overset{+}{P}\text{-}\overset{-}{C}H\text{-}CN$ — ROH/R'COOH → $R'COO^-$ $Bu_3\overset{+}{P}\text{-}OR$]

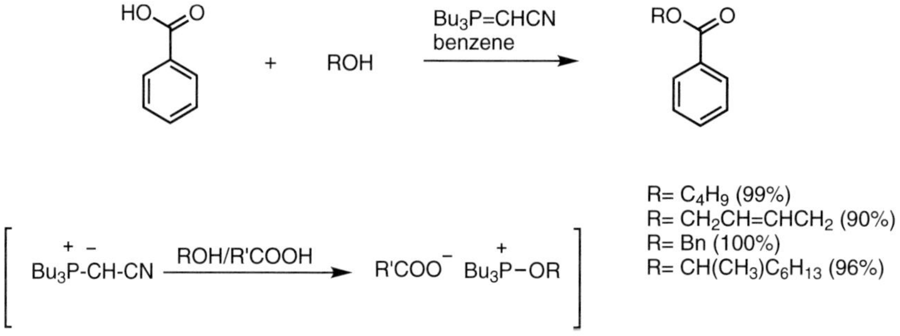

Scheme 1.35

Iminophosphorane functions as an alternative to the alkoxyphosphonium salt [218]. Treatment of benzyl azide with triphenylphosphine (Staudinger reaction) affords an iminophosphorane, the positively charged phosphorus atom of which is an activator of alcohol (Scheme 1.36). Heating of a mixture of alcohol, carboxylic acid, benzyl azide, and triphenylphosphine thus affords the desired esters in good to excellent yields.

Experimental Procedure Scheme 1.36 [218]
A mixture of the alcohol, the carboxylic acid, benzyl azide (1.5 equiv.), and triphenylphosphine (1.5 equiv.) in THF is heated to reflux and concentrated to give a crude product, which is purified by column chromatography.

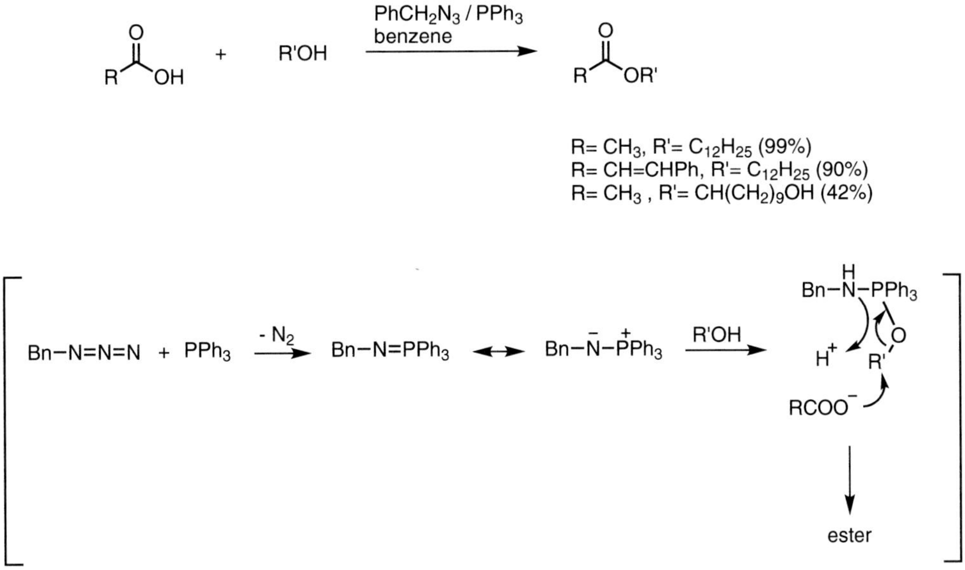

$$RCOOH + R'OH \xrightarrow[\text{benzene}]{PhCH_2N_3 / PPh_3} RCOOR'$$

R= CH_3, R'= $C_{12}H_{25}$ (99%)
R= CH=CHPh, R'= $C_{12}H_{25}$ (90%)
R= CH_3 , R'= $CH(CH_2)_9OH$ (42%)

Bn−N=N=N + PPh_3 —(− N_2)→ Bn−N=PPh_3 ↔ Bn−$\bar{N}$−$\overset{+}{P}Ph_3$ —(R'OH)→ Bn−N(H)−PPh_3 / H^+ / O−R' / $RCOO^-$ → ester

Scheme 1.36

1.1.6
Activation of Carboxylic Acids

2-Halo-1-methylpyridinium salts are another class of templates that induce condensation between alcohols and carboxylic acids (Scheme 1.37) [219, 220]. Thus, treatment of equimolar quantities of alcohol and carboxylic acid with 1.2 moles of the salt in the presence of 2.4 moles of tertiary amine provides good to excellent yields of esters. This procedure is applicable to macrolide synthesis (Scheme 1.38) [221]. Interestingly, the reaction proceeds in refluxing CH_3CN or CH_2Cl_2, at temperatures much lower than required with other techniques. Even medium-sized lactones (except for eight-membered rings) can be obtained in reasonable yields.

Experimental Procedure Scheme 1.37 [219] A mixture of benzyl alcohol (216 mg, 2.0 mmol), phenylacetic acid (272 mg, 2.0 mmol), and tributylamine (888 mg, 4.8 mmol) in CH_2Cl_2 (2 mL) is added under an argon atmosphere to a CH_2Cl_2 (2 mL) suspension of 1-methyl-2-bromopyridinium iodide (720 mg, 2.4 mmol), and the resulting mixture is heated at reflux for 3 h. The dichloromethane-insoluble pyridinium salt is progressively dissolved as the reaction proceeds. After evaporation of the solvent under reduced pressure, the residue is separated by silica gel column or thin-layer chromatography, and benzyl phenylacetate is isolated in 97% yield.

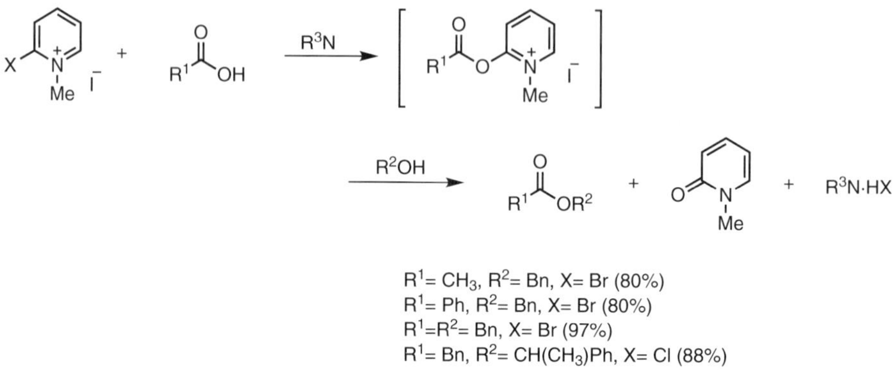
R³N
R²OH
R³N·HX
R¹= CH₃, R²= Bn, X= Br (80%)
R¹= Ph, R²= Bn, X= Br (80%)
R¹=R²= Bn, X= Br (97%)
R¹= Bn, R²= CH(CH₃)Ph, X= Cl (88%)

Scheme 1.37

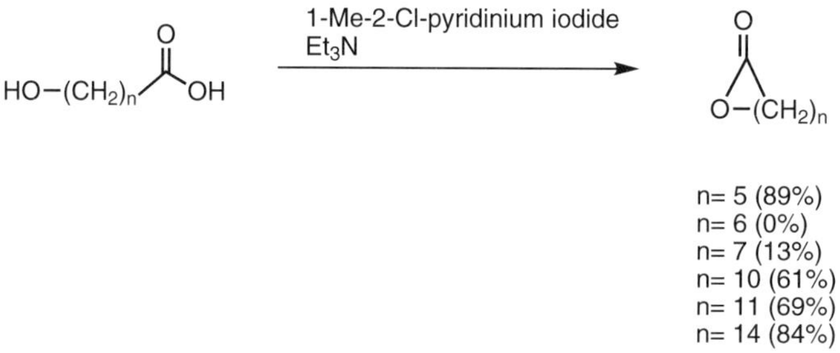
1-Me-2-Cl-pyridinium iodide
Et₃N
HO−(CH₂)n
n= 5 (89%)
n= 6 (0%)
n= 7 (13%)
n= 10 (61%)
n= 11 (69%)
n= 14 (84%)

Scheme 1.38

More efficient and selective acylation is achieved by the use of 2,2′-bipyridyl-6-yl carboxylates (Scheme 1.39) [222]. The reaction is promoted by CsF. Selective acylation on the primary alcohol takes place for primary-secondary diols, as well as for aromatic amino alcohols.

Experimental Procedure Scheme 1.39 [222]
A mixture of 2,2′-bipyridyl-6-yl hexanoate (0.5 mmol), benzyl alcohol (0.6 mmol), and cesium fluoride (2–2.5 mmol, dried well at 140 °C for 3 h *in vacuo* before use) in acetonitrile is stirred for 1 d at room temperature, and benzyl hexanoate is isolated in 90% yield after workup and purification.

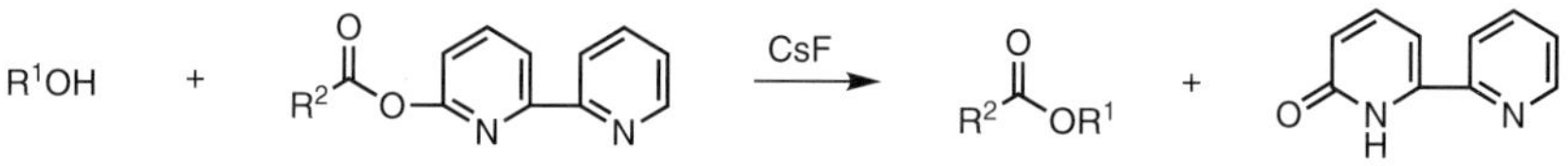
R¹OH
CsF

Scheme 1.39

It is claimed that 4,5-dichloro-1,2,3-dithiazolium chloride works at a lower temperature (–78 °C to room temperature) than the 2-halo-1-methylpyridinium chloride (Scheme 1.40) [223].

Experimental Procedure Scheme 1.40 [223] A solution of the acid (0.87 mmol), the alcohol (0.87 mmol), and 2,6-lutidine (0.233 mL, 2.0 mmol) in dry CH_2Cl_2 (1 mL) is added at –78 °C, under an Ar atmosphere and over a period of 1 min, to a stirred slurry of the dithiazolium salt (0.207 g, 1.0 mmol) in dry CH_2Cl_2 (3 mL). The mixture is stirred at –78 °C for 2 h and allowed to warm to room temperature overnight (12 h). The reaction mixture is quenched with ice (5 g) and poured into CH_2Cl_2 (5 mL). The organic layer is washed with brine (2 × 10 mL), dried over $MgSO_4$, filtered through a plug of silica gel (CH_2Cl_2) and concentrated *in vacuo*. The residue is purified by silica gel chromatography.

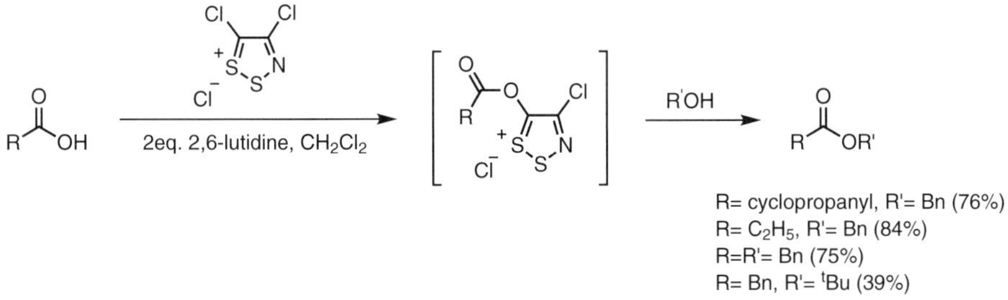

2eq. 2,6-lutidine, CH_2Cl_2

R'OH

R= cyclopropanyl, R'= Bn (76%)
R= C_2H_5, R'= Bn (84%)
R=R'= Bn (75%)
R= Bn, R'= ^{t}Bu (39%)

Scheme 1.40

Various carbonates such as 1,1′-[carbonyldioxy]dibenzotriazole (Scheme 1.41) [224] and di-2-pyridyl carbonate [225] are also useful.

Experimental Procedure Scheme 1.41 [224] A mixture of 1-hydroxybenzotriazole (41 g, 0.3 mol) and trichloromethyl carbonochloridate (18 mL, 0.15 mol) in benzene (200 mL) is heated at reflux with stirring for 2 h. The precipitate is filtered off, washed with benzene, and dried to give 1,1′-(carbonyldioxy)dibenzotriazole; yield: 31 g (70%); m.p. 150 °C (dec.; from benzene). A solution of 1,1′-(carbonyldioxy)dibenzotriazole (0.741 g, 2.5 mmol), benzoic acid (2.5 mmol), and pyridine (2.5 mmol) in *N*-methyl-2-pyrrolidone (4 mL) is stirred at room temperature for 1 h. The alcohol (2.5 mmol) and triethylamine (2.5 mmol) are then added. Stirring is continued for two or three days and the reaction mixture is subjected to conventional workup.

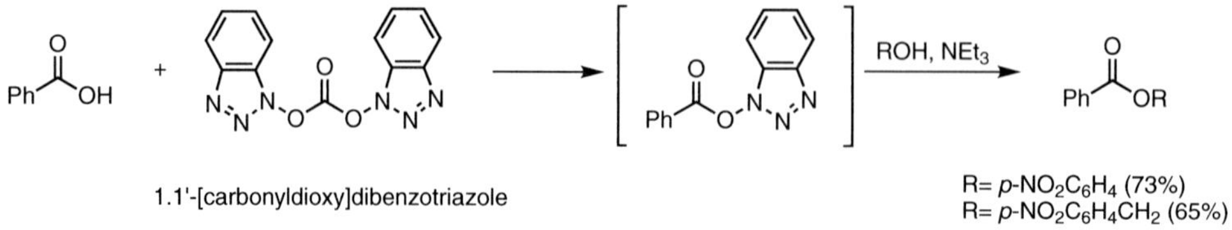

O
Ph OH
+
N N N O O O N N N
ROH, NEt3
Ph O N N N
Ph OR
1.1'-[carbonyldioxy]dibenzotriazole
R= p-NO2C6H4 (73%)
R= p-NO2C6H4CH2 (65%)

Scheme 1.41

Treatment of carboxylic acids with 2-dithienyl carbonate in the presence of DMAP followed by addition of alcohols with $Hf(OTf)_4$ furnishes the corresponding esters (Scheme 1.42) [226, 227].

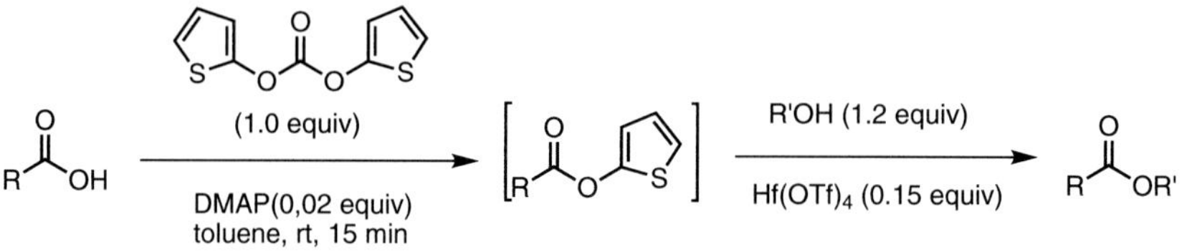

R OH
(1.0 equiv)
DMAP(0,02 equiv)
toluene, rt, 15 min
R'OH (1.2 equiv)
Hf(OTf)4 (0.15 equiv)
R OR'

Scheme 1.42

A succinimidyl group is a good activator for carboxylic moieties, and so phenyl esters of diazoacetate can be obtained by treatment with succinimidyl diazoacetate, prepared from *N*-hydroxysuccinimide and glyoxylic acid tosylhydrazone (Scheme 1.43) [228].

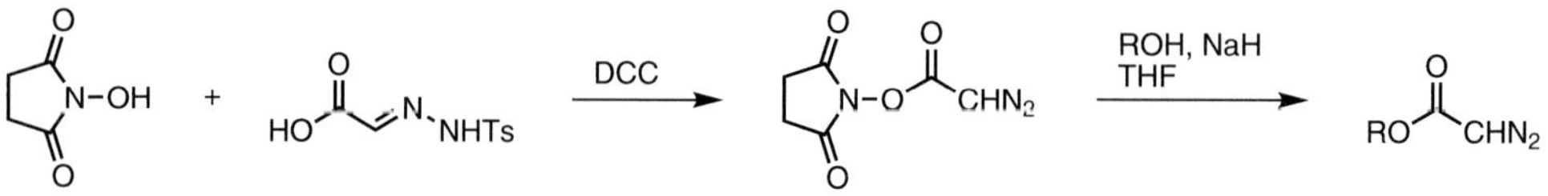

N-OH
+
HO N NHTs
DCC
N-O CHN2
ROH, NaH
THF
RO CHN2

Scheme 1.43

Treatment of carboxylic acids with 4-(4,6-dimethoxy-1,3,5-triazin-2-yl)-4-methylmorpholinium chloride in methanol, ethanol, or isopropyl alcohol in the presence of *N*-methylmorpholine affords the corresponding esters (Scheme 1.44) [229]. The amount of alcohol can be reduced to be nearly equimolar with the acid in THF as solvent.

Experimental Procedure Scheme 1.44 [229] *N*-methylmorpholine (0.24 mmol) is added at room temperature under nitrogen to a solution of carboxylic acid (0.20 mmol) and 4-(4,6-dimethoxy-1,3,5-triazin-2-yl)-4-methylmorpholinium chloride (0.40 mmol) in methanol (1 mL, dried over molecular sieves overnight). After the mixture has been stirred for 1.5 h, the solvent is evaporated, and the residue is extracted with ether. Purification by preparative TLC (hexane/AcOEt 2 : 1) affords methyl ester (0.186 mmol) in 93% yield.

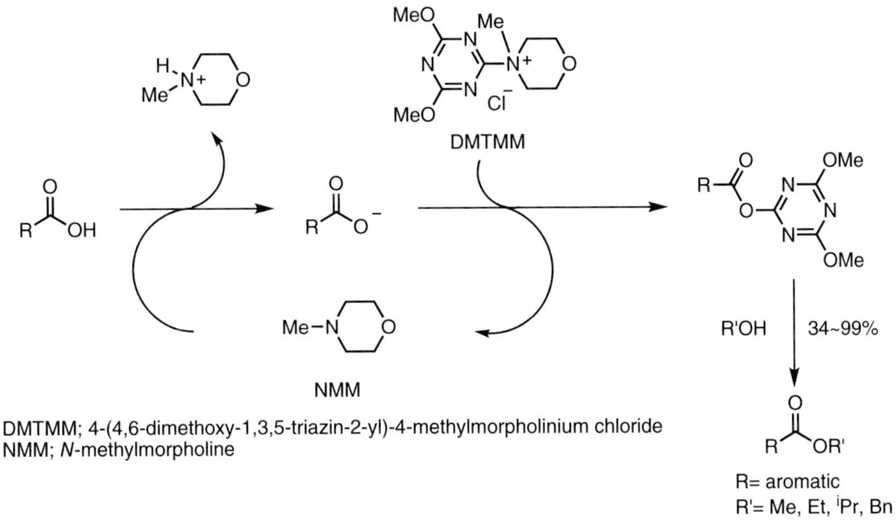

DMTMM; 4-(4,6-dimethoxy-1,3,5-triazin-2-yl)-4-methylmorpholinium chloride
NMM; *N*-methylmorpholine

R= aromatic
R'= Me, Et, ^{i}Pr, Bn

Scheme 1.44

The use of sulfonyl halides for activation of carboxylic acids is a classical procedure. When carboxylic acid and alcohol are mixed in the presence of *p*-toluenesulfonyl chloride (TsCl) in pyridine, a variety of esters are produced [230]. This technique is also employable for the synthesis of β-lactones from β-hydroxy acids [231]. Two mechanisms are plausible, depending on the key intermediate actually reacting with alcohol: (i) an acyl tosyl mixed anhydride or (ii) a symmetric acid anhydride resulting from further reaction of this mixed anhydride with carboxylic acid (Scheme 1.45). The use of TsCl (1.2 equiv.) in the presence of *N*-methylimidazole is effective for reaction between carboxylic acids and alcohols in a 1:1 ratio [232]. 1-Tosylimidazole (1.2 equiv.) activates reaction between alcohol (1.0 equiv.) and RCOONa (2 equiv.) in the presence of Et_3N (1.5 equiv.) and Bu_4NI (cat) to give the structurally diverse esters [233]. Methanesulfonyl chloride/Et_3N (or pyridine) is also usable [234, 235], but yields are not always high because of a side reaction generating sulfene. This drawback can be overcome by use of Me_2N-SO_2Cl, which has no α-hydrogen [236, 237]. Thus, treatment of an equimolar mixture of carboxylic acid and alcohol with the sulfamoyl chloride (2 equiv.) and DMAP (0.2 equiv.) provides esters in good yields. Sulfonyl chloride fluoride [238] and triflic anhydride [239–241] also serve for the direct condensation. *N,N*-Dimethylchlorosulfitemethaniminium chloride [242] and 4,5-dichloro-2-[(4-nitrophenyl) sulfonyl]pyridazin-3(2*H*)-one [243] are other modified activators for esterification (Scheme 1.46).

Experimental Procedure [230] The acid is dissolved in pyridine (20–50 parts, in some cases a salt separates), and benzenesulfonyl or toluenesulfonyl chloride (2 molecular equivalents) is added. The solution is chilled in ice, and the alcohol or phenol (1 molecular equivalent) is added. The solution is kept cold for about 1 h and then poured into three or four volumes of an ice/water mixture. Solid esters are collected by filtration.

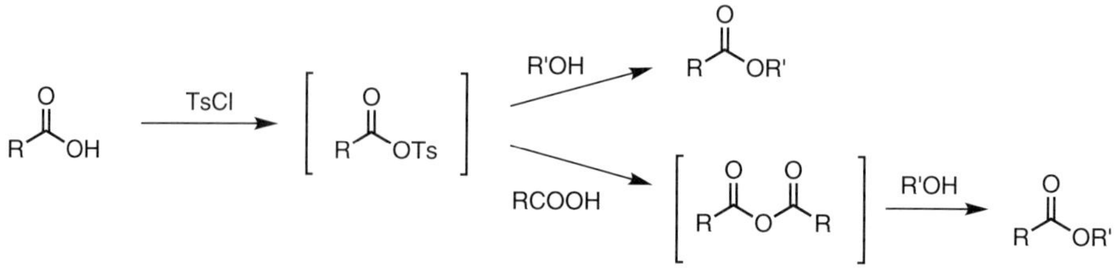

Scheme 1.45

N,N-dimethylchlorosulfitemethanimium chloride

4,5-dichloro-2-[(4-nitrophenyl)sulfonyl]pyridazin-3(2*H*)-one

Scheme 1.46

Commercially available polystyrylsulfonyl chloride resin acts as a solid-supported ester condensation reagent (Scheme 1.47) [244]. The purity of the esters is very good in the reaction between Fmoc-glycinol and carboxylic acids, but no reaction occurs with sterically hindered acids and electron-rich acids. On the other hand, Fmoc-glycine reacts more smoothly except with *tert*-butyl alcohol.

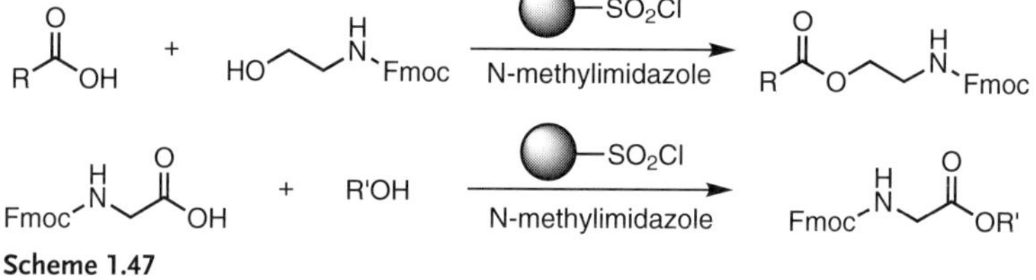

Scheme 1.47

The generation of acyloxy phosphorus intermediates, especially cationic species, is of great use for ester synthesis. Treatment of phenol and carboxylic acid with Ph_3P, CCl_4 and Et_3N provides phenyl carboxylates in reasonable yields, except in cases of reactants with a strongly electron-withdrawing groups (Scheme 1.48) [245]. $Ph_3P/CCl_3CN/DMAP$ also acts similarly [246]. Similar acyloxy phosphonium intermediates are generated by reaction of triphenylphosphine with *N*-bromo or iodo succinimide, treatment of which with alcohol furnishes esters [247].

Experimental Procedure Scheme 1.48 [245] A mixture of benzoic acid (24 mmol), phenol (20 mmol), carbon tetrachloride (24 mmol), triethylamine (24 mmol), and triphenylphosphine (24 mmol) in acetonitrile (30 mL) is stirred at room temperature for 4 h. After the acetonitrile has been evaporated, hexane is added to the residue. The hexane solution is filtered, removing the precipitated triphenylphosphine oxide and triethylamine hydrochloride, washed with an aqueous sodium hydroxide solution, and dried over anhydrous sodium sulfate, and then the hexane is removed. Subsequent distillation or recrystallization of the residual solid gives phenyl benzoate (3.6 g, 91% yield) as white crystals: mp 70–71 °C.

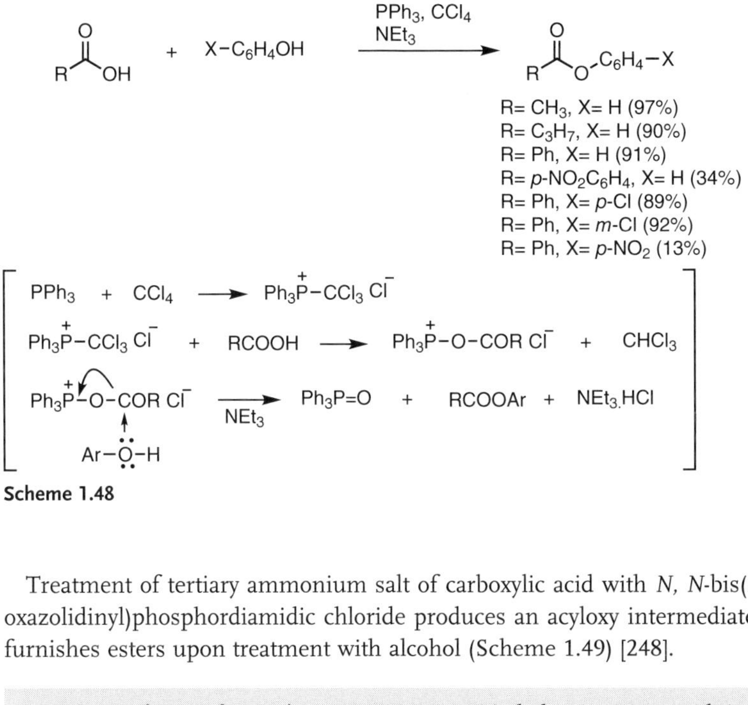

$$RCOOH + X\text{-}C_6H_4OH \xrightarrow[NEt_3]{PPh_3,\ CCl_4} RCOO\text{-}C_6H_4\text{-}X$$

R= CH_3, X= H (97%)
R= C_3H_7, X= H (90%)
R= Ph, X= H (91%)
R= p-$NO_2C_6H_4$, X= H (34%)
R= Ph, X= p-Cl (89%)
R= Ph, X= m-Cl (92%)
R= Ph, X= p-NO_2 (13%)

$$PPh_3 + CCl_4 \longrightarrow Ph_3\overset{+}{P}\text{-}CCl_3\ \overset{-}{Cl}$$

$$Ph_3\overset{+}{P}\text{-}CCl_3\ \overset{-}{Cl} + RCOOH \longrightarrow Ph_3\overset{+}{P}\text{-}O\text{-}COR\ \overset{-}{Cl} + CHCl_3$$

$$Ph_3\overset{+}{P}\text{-}O\text{-}COR\ \overset{-}{Cl}\ (+\ Ar\text{-}\ddot{O}\text{-}H) \xrightarrow[NEt_3]{} Ph_3P{=}O + RCOOAr + NEt_3.HCl$$

Scheme 1.48

Treatment of tertiary ammonium salt of carboxylic acid with *N*, *N*-bis(2-oxo-3-oxazolidinyl)phosphordiamidic chloride produces an acyloxy intermediate which furnishes esters upon treatment with alcohol (Scheme 1.49) [248].

Experimental Procedure Scheme 1.49 [248] Triethylamine (11 mmol) is added to a solution of the acid (5.5 mmol, or 10 mmol, via the anhydride) in 10 mL of solvent (acetonitrile, dichloromethane, dimethylacetamide). Complete dissolution of the mixture takes place easily, except in cases in which 5 mL of solvent are used. Subsequent addition of the alcohol (5 mmol, or 20 mmol) and *N*, *N*-bis(2-oxo-3-oxazolidinyl)phosphordiamidic chloride (5.5 mmol) yields a white precipitate of triethylammonium hydrochloride, which is insoluble under these conditions. All reaction mixtures made in acetonitrile are heated at reflux (1 h~1.5 h) and the triethylammonium salt is dissolved. Reaction mixtures in dichloromethane or dimethylacetamide are kept for 1 h at room temperature. When the reaction time is over, sodium bicarbonate solution (10%, 10–20 mL) is added. Solids are filtered, washed with water until neutral, dried, and identified as the corresponding ester. Some esters are recovered from the organic layer (addition of 10 mL of dichloromethane needed), which is dried with sodium sulfate, filtered and evaporated to dryness.

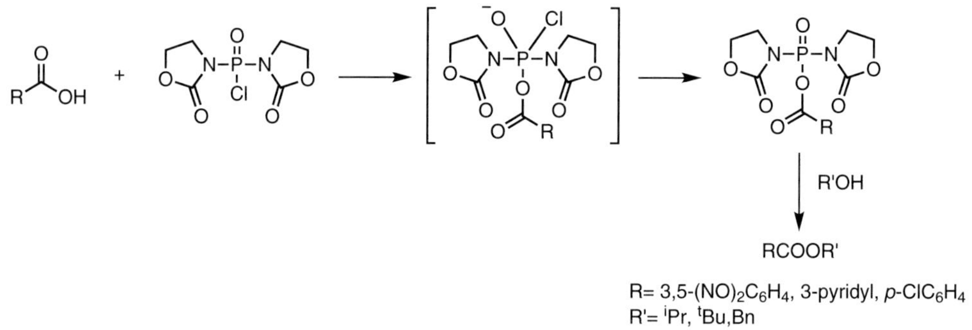

R'OH

RCOOR'

R= 3,5-$(NO)_2C_6H_4$, 3-pyridyl, *p*-ClC_6H_4
R'= ^{i}Pr, ^{t}Bu,Bn

Scheme 1.49

Benzotriazol-1-yloxytris(dimethylamino)phosphonium hexafluorophosphate is a powerful reagent with which to generate an acyloxyphosphonium species (Scheme 1.50) [249]. Amino acids can thus be esterified with a nearly equal amount of alcohol (~1.1 equiv.) except in the case of *tert*-butyl alcohol. The acid- and base-sensitive protecting groups commonly used in peptide chemistry are tolerated in this protocol.

Experimental Procedure Scheme 1.50 [249] An alcohol (1.1 mmol) is added at −20 °C (CCl_4/dry ice bath) under argon to a solution of an *N*-α-protected amino acid (1.0 mmol) and $^i Pr_2NEt$ (0.26 mL, 1.5 mmol) in CH_2Cl_2 (2 mL). After the mixture has been stirred at −20 °C for 15 min, benzotriazol-1-yloxytris(dimethy laminophosphonium hexafluorophosphate) (0.44 g, 1 mmol) is added, and the reaction mixture is left stirring overnight with gradual warming to room temperature. The next day (total reaction time = 10 to 14 h), the reaction mixture is suspended in CH_2Cl_2 and washed sequentially with buffer (pH 4, Aldrich), satd. aq. NaCl, satd. aq. $NaHCO_3$, and satd. aq. NaCl. Drying (Na_2SO_4) and concentration *in vacuo* affords the crude product, which is purified by silica gel flash chromatography.

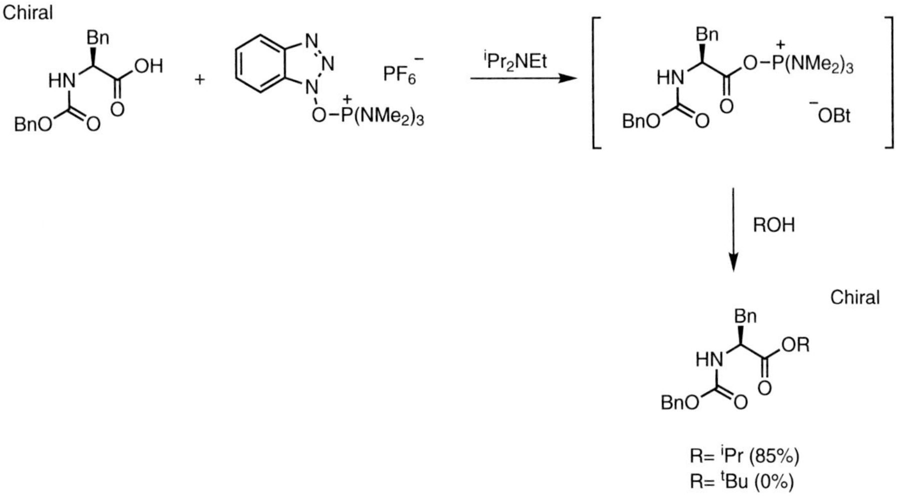

Scheme 1.50

Diphosphonium fluorosulfonate effects dehydration between carboxylic acids and alcohols (Scheme 1.51) [250]. This reagent has been prepared by mixing $Ph_3P{=}O$ and $(FSO_2)O$ and used *in situ* for esterification. It is assumed that the reaction proceeds through an acyloxy- or alkoxyphosphonium intermediate.

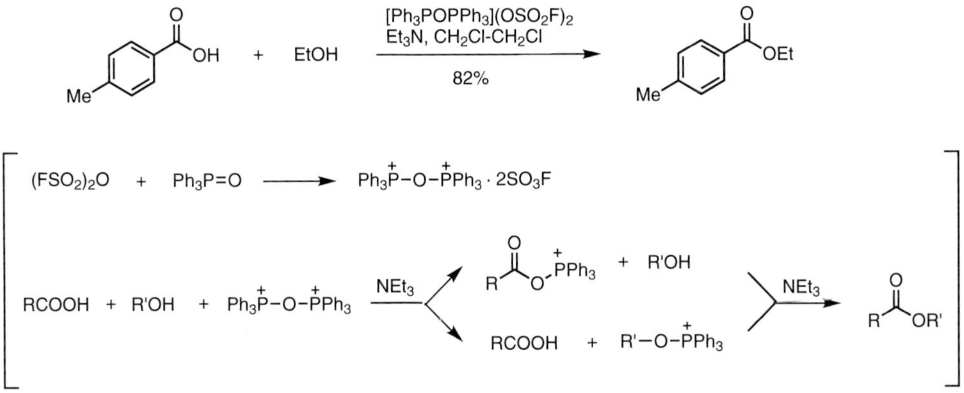

Scheme 1.51

Phase transfer technology has been used to activate carboxylic acids with phosphoric acid diester chlorides generated *in situ* (Scheme 1.52) [251]. The reaction works even for hindered substrates such as pivalic acid, but is not applicable to phenylacetic acid or diphenylacetic acid.

Experimental Procedure Scheme 1.52 [251] Phosphite (10–13 mmol) in toluene (15 mL) is added with stirring to a mixture of carboxylic acid (10 mmol), CCl_4 (100 mmol, 10 mL), K_2CO_3 (5.52 g, 40 mmol), and TEBAC (0.23 g, 1 mmol) in toluene (30 mL). The alcohol (10 mmol) is then added, and stirring at reflux temperature is continued for 10 min. The esters obtained are distilled under reduced pressure or purified by column chromatography on silica gel with benzene/acetone (9 : 1) as eluent.

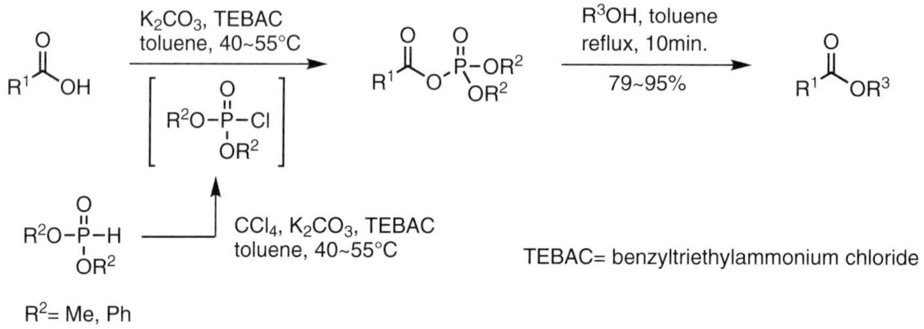

K_2CO_3, TEBAC
toluene, 40~55°C

$R^2O-P(=O)(Cl)-OR^2$

R^3OH, toluene
reflux, 10min.
79~95%

$R^2O-P(=O)(H)-OR^2$

CCl_4, K_2CO_3, TEBAC
toluene, 40~55°C

TEBAC= benzyltriethylammonium chloride

R^2= Me, Ph

Scheme 1.52

Constant-current electrolysis, in an undivided cell, of Ph_3P in the presence of a carboxylic acid in CH_2Cl_2 containing 2,6-lutidinium perchlorate as a supporting electrode produces an acyloxyphosphonium ion, which is converted into the corresponding esters upon treatment with alcohol or phenol (Scheme 1.53) [252].

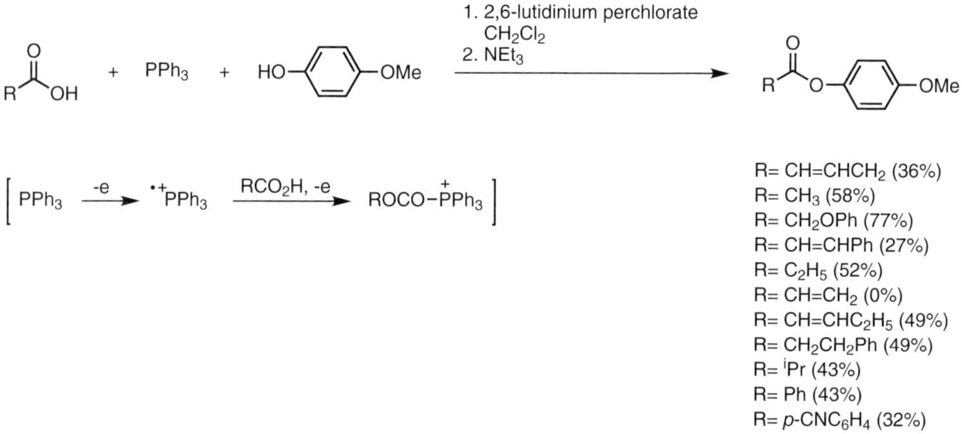

PPh_3 + HO–C$_6$H$_4$–OMe

1. 2,6-lutidinium perchlorate
CH_2Cl_2
2. NEt_3

[PPh_3 –(-e)→ $^{\bullet+}PPh_3$ –(RCO_2H, -e)→ $ROCO-\overset{+}{P}Ph_3$]

R= $CH=CHCH_2$ (36%)
R= CH_3 (58%)
R= CH_2OPh (77%)
R= CH=CHPh (27%)
R= C_2H_5 (52%)
R= $CH=CH_2$ (0%)
R= $CH=CHC_2H_5$ (49%)
R= CH_2CH_2Ph (49%)
R= ^iPr (43%)
R= Ph (43%)
R= p-CNC_6H_4 (32%)

Scheme 1.53

Activation of carboxylic function as a silyl ester is also useful. Direct esterification of equimolar amounts of carboxylic acid and alcohol is achieved with catalysis by $TiCl(OTf)_3$ (0.1 mol %) in the presence of $(Me_2SiO)_4$ (2.0 equiv.) [253]. A silyl carboxylate species is assumed to act as a key intermediate. Medium-sized lactones are accessible through a novel ring-contraction strategy (Scheme 1.54) [254]. ω-Hydroxy acids are trapped by 1,2-bis(dimethylsilyl)benzene with $RhCl(PPh_3)_3$ catalysis, and the contraction of the resulting large-sized rings is achieved by treatment with $Me_2Si(OTf)_2$ (Scheme 1.55). Medium-sized lactones are obtained in high yields, except the seven-membered rings.

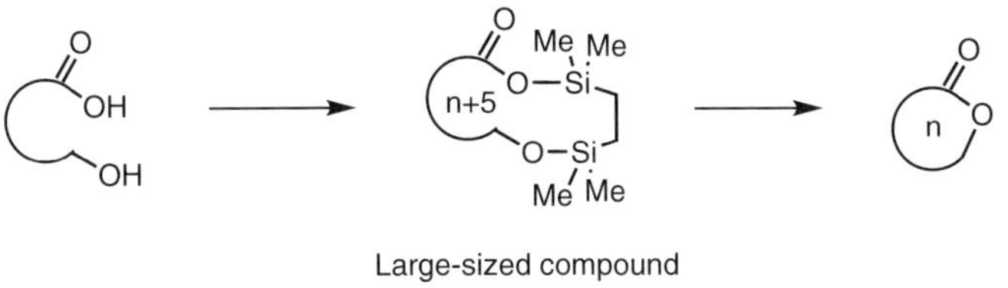

Large-sized compound

Scheme 1.54

20 mM $RhCl(PPh_3)_3$, rt, 6 h

10 mM $RhCl(PPh_3)_3$, 80 °C, 3 h

10 mM $Me_2Si(OTf)_2$, 80 °C, 9 h

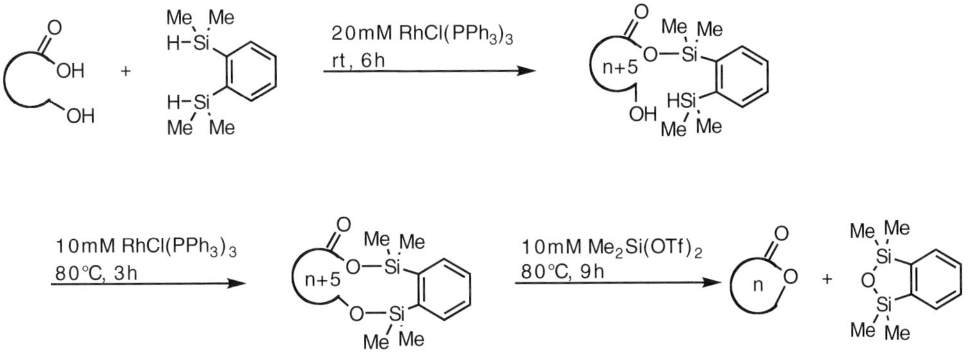

Scheme 1.55

Acyloxyketene acetals, obtainable from carboxylic acids and (trimethylsilyl) ethoxyacetylene by treatment with mercuric ion, afford esters upon treatment with alcohols (Scheme 1.56) [255]. Most conveniently, ketene acetals prepared *in situ* are straightforwardly converted into esters, lactones, and peptides.

Experimental Procedure Scheme 1.56 [255] A solution of alcohol (1.0 mmol) and the ketene acetal (1.2 mmol) in an anhydrous solvent ($ClCH_2CH_2Cl$ or CH_2Cl_2, 2 mL) is stirred under a nitrogen atmosphere for 25 min ~ 2 days. The mixture is concentrated by rotary evaporator and then by use of a high-vacuum pump at 40 °C/0.4 mbar for 1 h to give the pure product.

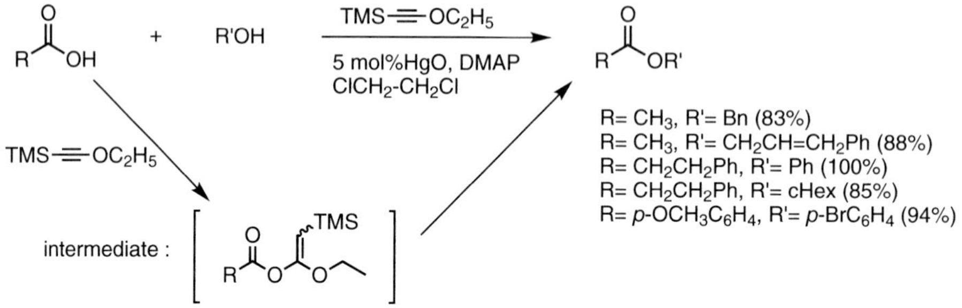

Scheme 1.56

Treatment of carboxylic acid with 1,1′-dimethylstannocene gives tin (II) carboxylates, which react with equimolar amounts of alcohols in refluxing xylene to afford the esters (Scheme 1.57) [256].

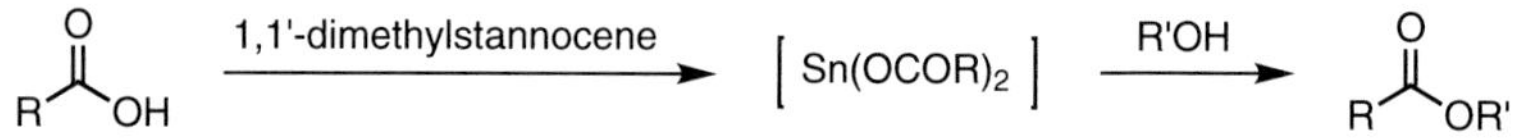

Scheme 1.57

On heating with alcohols in the presence of tris(2-methoxyphenyl)- or tris(2,6-dimethoxyphenyl)bismuthines, carboxylic acids bearing α-hydrogens are readily converted into the corresponding esters in good yields (Scheme 1.58) [257]. It is assumed that the reaction proceeds through ketene formation.

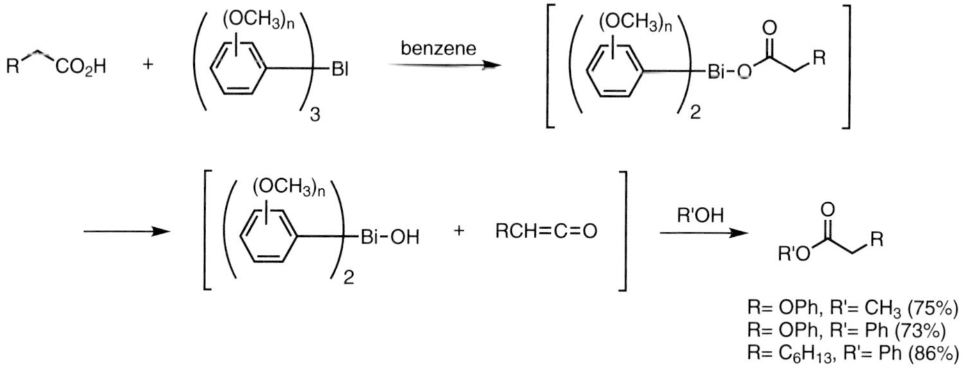

Scheme 1.58

Carboxylic acids are activated by treating with BCl_3 [38], benzeneboronic acids [258], or borane/THF (or Me_2S) [259]. The resulting acyloxyboranes are transformed into esters upon treatment with alcohols. Bismuth (III) carboxylates obtained from Ph_3Bi and carboxylic acids react similarly to give esters [260]. Cerium ammonium nitrate effects reaction between carboxylic acids and alcohols, although the mechanism is not clear [261–263].

1.1.7 Enzymes

Enzymes play an important role in esterification technology. In particular, lipases have been widely used for the resolution of racemic alcohols and carboxylic acids through asymmetric hydrolysis of the corresponding esters. On the other hand, this technology is not straightforwardly applied to esterification because the esters are readily hydrolyzed in the presence of water. A new technology for use of enzymes in anhydrous organic solvent overcomes this difficulty. Yeast lipase (*Candida cylindracea*), for example, almost quantitatively converts a carboxylic acid and an alcohol into the corresponding ester in organic solvent, in a highly stereoselective manner when a chiral acid is employed [264]. The advantages of this technique are that the stabilities of enzymes in organic solvents are much greater than in water, and that some substrates or products are unstable (e.g., toward racemization or other degradation reactions) in aqueous solution but stable in organic solvents. This technology has been extended to the intramolecular reaction of ω-hydroxy acids [265]. Furthermore, the reaction between dicarboxylic acids and diols to arrive at macrolides is feasible through the use of lipase (K-10) (Scheme 1.59) [266].

Experimental Procedure [264] Yeast Lipase-Catalyzed Production of Optically Active Esters: A solution of a racemic acid and an alcohol in a given solvent is treated with powdered lipase from *Candida cylindracea*. The suspension is placed in an Erlenmeyer flask and shaken on an orbit shaker at 250 rpm and at 30 °C to reach a certain degree of conversion. The enzyme is then removed by filtration, and the liquid phase is washed with three portions (each 80 mL) of aqueous $NaHCO_3$ (0.5 M). The obtained organic phase is dried with $MgSO_4$, and the solvent is evaporated in a rotary evaporator. To recover the ester, the remainder is distilled or chromatographed. The aqueous phase is acidified to pH 1 with HCl (6 N), and then the acids are extracted with three portions of CH_2Cl_2 (each 80 mL). The combined methylene chloride fractions are dried with $MgSO_4$, followed by evaporation of the solvent. The acids are isolated from the residue either by distillation or by liquid column chromatography.

Porcine Pancreatic Lipase-Catalyzed Production of Optically Active Esters: A solution of a racemic alcohol and 2,2,2-trichloroethyl butyrate (or 2,2,2-trichloroethyl heptanoate) in ether or heptane is dehydrated and then treated with powdered lipase from porcine pancreas. The suspension is placed in a round-bottomed flask and either shaken on an orbit shaker at 250 rpm or mechanically stirred at 300 rpm. When the degree of conversion reaches 45–50%, and the reaction virtually stops, the enzyme is removed by filtration. The liquid phase is dried with $MgSO_4$, followed by evaporation of the solvent in a rotary evaporator. The residue is subjected to liquid column chromatography; the esters are then separated from 2,2,2-trichloroethyl butyrate by distillation and the alcohols from 2,2,2-trichloroethanol by aqueous extraction of the letter. In another case,

2,2,2-trichloroethyl butyrate and trichloroethanol are first removed by distillation, and the alcohols are then separated from their butyric esters by liquid chromatography, or the remainder is separated by distillation.

lipase (K-10)

a: m= 12, n= 5 (18%)
b: m= 12, n= 16 (30%)
c: m= 10, n= 16 (56%)

a: 17%
b: 7%
c: 15%

Scheme 1.59

Supercritical carbon dioxide functions as an alternative to the organic solvent. Isoamyl acetate is obtained from isoamyl alcohol and ammonium acetate in the presence of lipase in supercritical carbon dioxide [267]. The ammonium salt is crucial because no ester is produced with acetic acid, indicative of delicate reaction conditions. However, isoamyl acetate is obtained in higher yields from the reaction in hexane, so it may be consequently concluded that the supercritical carbon dioxide methodology is not necessarily superior to that in organic solvent. More detailed description of the use of supercritical carbon dioxide is given in Section 7.3.

The enzymatic esterification is applicable to a variety of synthetic uses, for example, selective acylation of a cyclohexane-3,5-diol derivative in *trans* (+)sobrerol synthesis [268], esterification of sterols with fatty acids [269], esterification of phenylacetic and 2-phenylpropanoic acids [270], and esterification of glycosides with lactic acid [271]. PPL-catalyzed esterification (PPL = porcine pancreatic lipase) allows selective formation of monoester of α, ω-dicarboxylic acids (Scheme 1.60) [272].

PPL, CH_2Cl_2

n= 4 (68%)
n= 6 (61%)
n= 7 (93%)
n= 8 (94%)
n= 14 (89%)

Scheme 1.60

The synthesis of amino acids and peptide esters has been achieved. The serine protease substilisin Carlsberg efficiently catalyzes the specific formation of C^{α}-

carboxyl 3-hydroxypropyl or 4-hydroxybutyl esters of certain Boc-amino acids and peptides in high-content 1,3-propanediol or 1,4-butanediol solution, with substrate specificity parallel to that of the normal hydrolytic reaction (Scheme 1.61). This approach can be coupled with kinetic-controlled reverse proteolysis in a two-step enzymatic peptide ligation scheme [273].

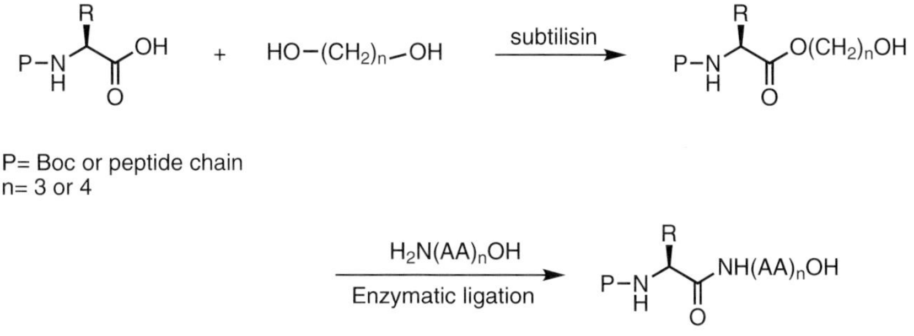

P= Boc or peptide chain
n= 3 or 4

Scheme 1.61

Enzymatic esterification is accelerated by continuous azeotropic removal of water [274, 275]. Removal of water is also achievable by adsorptive control with molecular sieves [276]. On the other hand, when a small amount of water is added to a polar solvent, esterification by *Candida rugosa* exhibits better enantioselectivity and/or yield in esterification of *R*- and *S*-ketoprofen [277]. Modification of enzymes by surfactants improves the esterification process [278–281].

Esterification without solvent is another choice [282–287].

In addition to an advantage of facile separation of the enzyme, higher enantioselectivity and/or higher catalytic activity are also attainable by immobilization of enzymes. Chitosan immobilizes microbial lipases from *Candida rugosa, Pseudomonas fluorescens*, and *Candida antarctica* B to allow repeated uses in esterification of carboxylic acids with various alcohols [288, 289]. Poly(*N*-vinyl-2-pyrrolidone-*co*-styrene) can also immobilize *Candida rugosa*, resulting in enantioselective esterification of (*R, S*)-2-(4-chlorophenoxy)propanoic acid with *n*-tetradecanol [290]. Polypropylene is another immobilizer employable for synthesis of ethyl oleate [291–293].

Various inorganic supports are available for immobilizing enzymes. Silica gels are some of the most popular ones by which to increase the activity and to allow the repeated use of enzymes [294–296]. The stability of *Candida rugosa* can be enhanced when entrapped in hybrid organic-inorganic sol-gel powder prepared by acid-catalyzed polymerization of tetramethoxysilane and alkyltrimethoxysilanes [297]. Gas-phase esterification between acetic acid and ethanol is feasible with lipases immobilized in MCM molecular sieves [298]. Zeolite NaA can be a supporter of Novozyme, catalyzing esterification of geraniol with acetic acid [299]. The layered double hydroxides of Zn/Al—NO_3 hydrotalcite, Zn/Al-dioctyl sodium sulfosuccinate, Ni/Al-sodium dodecyl sulfonate, and Mg/Al-sodium dodecyl sulfate immobilize *Candida rugosa* effectively [300–302]. All immobilized enzymes exhibit higher activities and storage stability. $CaCO_3$ is a good adsorbent for *Rhizopus oryzae* and *Staphylococcus*, increasing the stability of the enzymes for esterification of oleic acid [303, 304]. Continuous esterification is feasible with a packed-bed bioreactor [305, 306].

As seen above, immobilization of enzymes has become popular, and many immobilized enzymes are commercially available. Use of such enzymes is quite common now [307–311].

Kinetic resolution of racemic alcohols is feasible. α-Substituted cyclohexanols [312] and various aliphatic alcohols [313] are resolved through their reactions with alkanoic acids. Notably, the enantioselectivity of the lipase-catalyzed esterification of 2-(4-substituted phenoxy)propanoic acids is dramatically enhanced by addition of aqueous sodium dodecyl sulfate [314]. The effect of this additive is attributed to the increased conformational flexibility of lipase, which has a rigid conformation in organic solvent, preventing its active site from accepting a non-natural substrate with a structure significantly different from that of a natural product. The newly attained flexible conformation allows easier access of the one of the enantiomers through induced fitting.

Experimental Procedure [313] The racemic alcohol (2.5 mmol), octanoic acid (8.0 mL, 50 mmol), and lipozyme (3.0 g) are swirled in either pentane or hexane (3.0 g) at 30 °C for a period of 3–5 weeks. The progress of the resolution is monitored by derivatization of samples of the mixture directly with (*S*)-α-methylbenzyl isocyanate, followed by GLC analysis. The mixture is worked up by suction filtration; the resin is washed thoroughly with solvent and stored at 0–5 °C for future use. The combined organic phase is washed with NaOH (1.25 N) and then with H_2O. After drying ($MgSO_4$), the solvent is removed by careful distillation, and the concentrate is fractionally distilled to obtain the *S* alcohol and *R* octanoate ester. The ester is saponified by heating with KOH (6 N, 25 mL) and methanol (20 mL) under reflux for 16 h. The resulting *R* alcohol is recovered from the saponification step in the usual manner and then distilled.

The synthetic use of biocatalytic kinetic resolution is hampered by the reversible nature of the direct esterification of acids with alcohols. Addition of an orthoester biases the equilibrium in favor of the ester side because of the consumption of the water formed through hydrolysis of the orthoester (Scheme 1.62) [315].

Experimental Procedure [315] Scheme 1.62 Racemic fluorbiprofen (41 mmol, 10 g) is added to a solution of CH_3CN (1 l) containing tripropyl orthoformate (123 mmol, 26.5 mL), 1-propanol (20 mL) and Novozym 435® (100 g). The mixture is incubated by shaking at 45 °C (300 rpm); the degree of conversion and the *ee* of unchanged fluorbiprofen are followed by chiral HPLC analysis. After 6 days, conversion has reached 60% and the reaction is stopped by filtering off the enzyme. Removal of the solvent *in vacuo* leaves a residue that is partitioned between hexane and aq. $NaHCO_3$ (3 g in 200 mL of water). The organic phase is washed with water and dried over Na_2SO_4, and the solvent is removed to afford (*R*)-fluorbiprofen propyl ester (6.8 g, yield 58%, *ee* 64%).

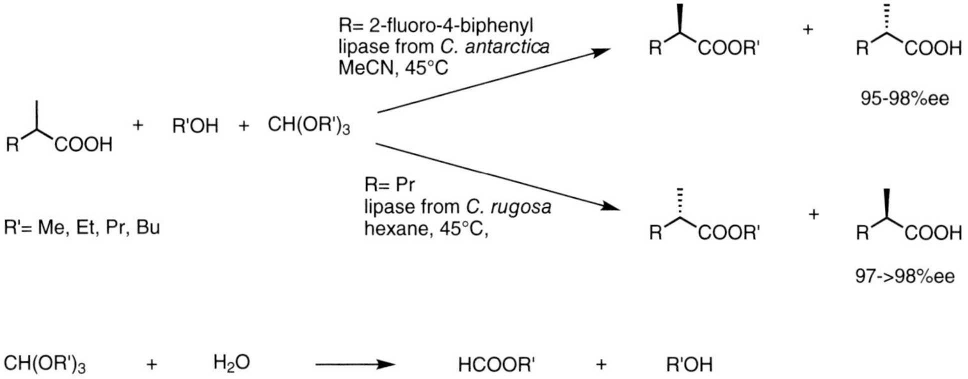

R= 2-fluoro-4-biphenyl
lipase from *C. antarctica*
MeCN, 45°C

R COOH + R'OH + CH(OR')$_3$ → R COOR' + R COOH (95-98%ee)

R'= Me, Et, Pr, Bu

R= Pr
lipase from *C. rugosa*
hexane, 45°C,

→ R COOR' + R COOH (97->98%ee)

$CH(OR')_3$ + H_2O → HCOOR' + R'OH

Scheme 1.62

1.1.8
π-Acids

Treatment of allylic and tertiary benzylic alcohols with a catalytic amount of 2,3-dichloro-5,6-dicyanobenzoquinone in acetic acid affords the corresponding acetates (Scheme 1.63) [316]. It is believed that the reaction proceeds by initial oxidation of allylic alcohol to give a radical cation, which further undergoes carbon-oxygen bond cleavage to form a stable allylic cation species.

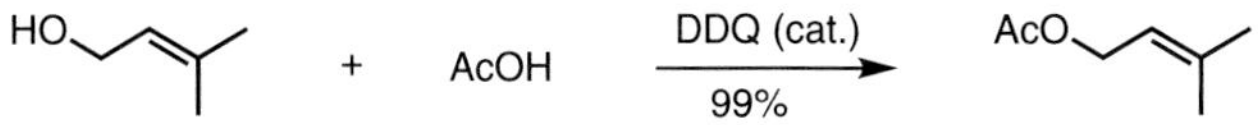

HO + AcOH → (DDQ (cat.), 99%) AcO

Scheme 1.63

Tetracyanoethylene and dicyanoketene dimethyl acetal are more versatile π-acids (Scheme 1.64) [317]. Various aliphatic and aromatic carboxylic acids can be esterified.

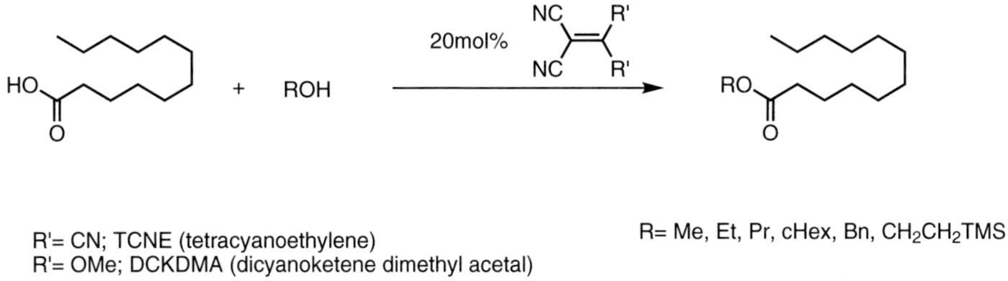

HO + ROH → (20mol% NC, R', NC, R') RO

R'= CN; TCNE (tetracyanoethylene)
R'= OMe; DCKDMA (dicyanoketene dimethyl acetal)

R= Me, Et, Pr, cHex, Bn, CH_2CH_2TMS

Scheme 1.64

1.2
Reaction with Esters: Transesterification

1.2.1
Without Activator

Ester-to-ester transformation through interchange of the alcohol components has been well known for a long time, as 'transesterification.' This reaction occupies a central position in organic synthesis, as important as direct esterification between carboxylic acid and alcohol. Both reactions are equilibrium processes, so what they have most deeply in common is the need to shift the reaction system to the product side to the greatest degree. In this respect, it may be generally said that esterification is more advantageous, because the water co-product is readily separated from the reaction medium because of its incompatibility with organic solvents. This, on the other hand, is also an advantage for transesterification, in that water-sensitive materials are employable. Accordingly, once effective methods for removal of the produced alcohol from the reaction system are established, then transesterification is a better choice of reaction for ester production.

Although acid or base catalysts are usually necessary, transesterification without any promoters is ideal. β-Ketoesters are one such class of compounds, known for a long time to undergo thermal transesterification. Heating of alcohols such as alkanols, menthol, steroid alcohols, etc. in the presence of excess quantities of various β-ketoesters on a steam bath affords the new esters in yields better than 90% [318]. The reaction can be explained as proceeding via an acylketene intermediate (Scheme 1.65) [319]. Accordingly, a more facile reaction takes place when *tert*-butyl acetoacetate is employed as a starting material, because of its high susceptibility to thermolysis [320].

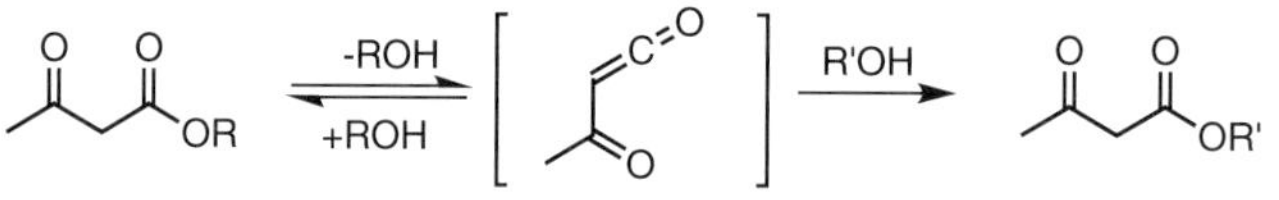

Scheme 1.65

This technique is utilized elegantly in the synthesis of baccatin III derivatives (Scheme 1.66) [321]. Heating of protected baccatin III derivatives in ethyl

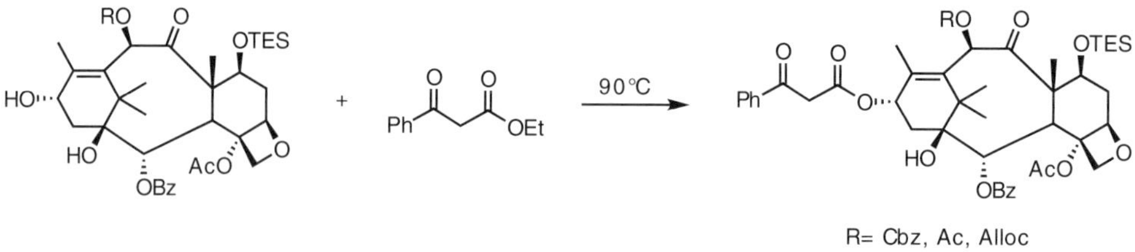

Scheme 1.66

benzoylacetate (20 equiv.) allows smooth transesterification at the sterically hindered α-oriented 13-OH. The reaction is accelerated by continuous removal of the ethanol produced *in vacuo*.

The use of 4A molecular sieves accelerates reaction between ethyl acetoacetate and various alcohols, because selective adsorption of the produced ethanol by the molecular sieves facilitates the equilibrium shift into the product side [322].

Transformation of rapeseed oil, which consists of fatty acid esters of glycerin, into the methyl esters is of practical importance since these lighter esters are useful as biodiesel fuel. Subjection of rapeseed oil to supercritical methanol at 350 °C successfully provides the methyl esters (Scheme 1.67) [323, 324]. The supercritical state of methanol is believed to change the two-phase nature of oil/methanol mixture to a single phase by means of a decrease in the dielectric constant of methanol.

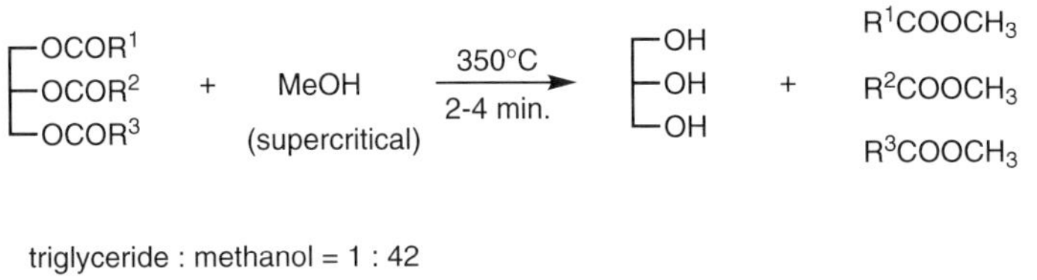

triglyceride : methanol = 1 : 42

Scheme 1.67

Ultrasound irradiation is effective for transesterification of an azalactone. A biologically active amino acid ester is thus accessible in methanol at room temperature (Scheme 1.68) [325].

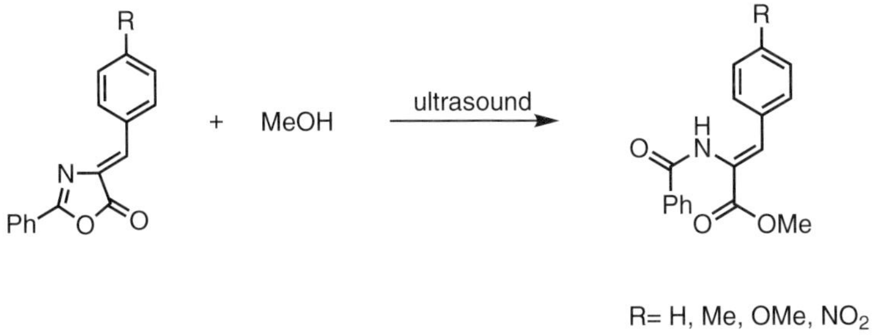

R= H, Me, OMe, NO_2

Scheme 1.68

An α-cyclodextrin clathrate with *m*-nitrophenyl acetate undergoes ester transfer from the guest to the host in the solid state when heated at 117 ° or 140 °C (Scheme 1.69) [326]. No such transesterification occurs with *p*-nitrophenyl acetate complex.

When hydroxy ester intermediates are formed with appropriate conformations, intramolecular esterification takes place spontaneously. Two such examples taking advantage of a radical reaction may be mentioned. As shown in Scheme 1.70, treatment of thionocarbonate diol with Bu_3SnH/AIBN affords an all-*syn* five-membered ring lactone as the major product in 40% yield [327]. The product formed is best explained by a chair conformation in the radical intermediate.

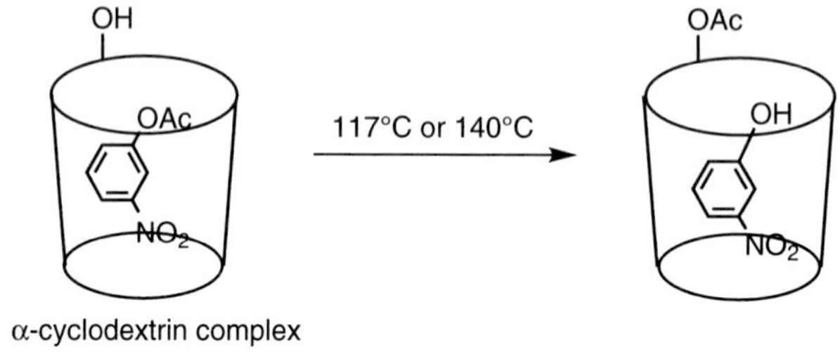
OH
OAc
NO2
117°C or 140°C
OAc
OH
NO2
α-cyclodextrin complex

Scheme 1.69

Hydrogen bonding of the type shown probably contributes in favoring this 1,3-diaxial arrangement of the two OH groups.

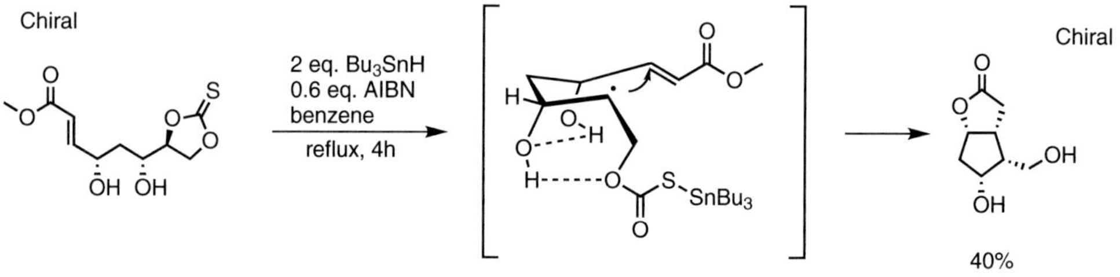
Chiral
2 eq. Bu3SnH
0.6 eq. AIBN
benzene
reflux, 4h
SnBu3
Chiral
40%

Scheme 1.70

When maleate esters with hydroxyl-based substituents on their β-carbons are isomerized to the corresponding fumarate esters by exposure to a thiyl radical, furanones are obtained in one-pot fashion (Scheme 1.71) [328].

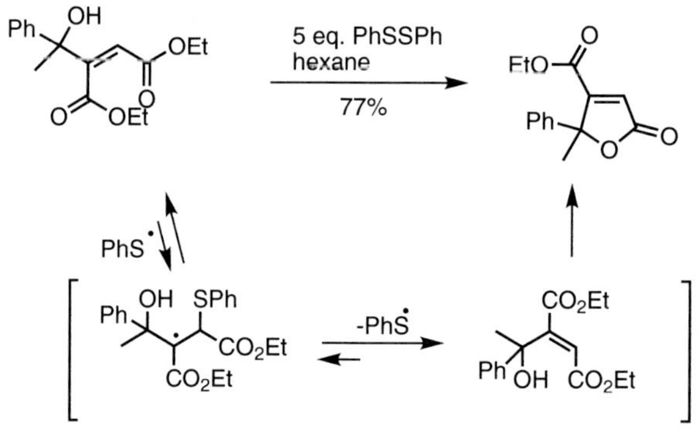
5 eq. PhSSPh
hexane
77%
PhS•
-PhS•

Scheme 1.71

1.2.2 Acid Catalysts

1.2.2.1 Brønsted Acids

Transesterification is most conveniently carried out with the aid of acid catalysts. Brønsted acids are classical but are still employed quite often; among the most popular are HCl, H_2SO_4, and *p*-toluenesulfonic acid. Other acids utilized are HBr, HI, HF, AcOH, trifluoromethanesulfonic acid, camphorsulfonic acid, HBF_4, and $HClO_4$.

Transesterification must be carried out under anhydrous acidic conditions, because the ester is hydrolyzed otherwise. Accordingly, generation of HCl in a quantitative manner through reaction between acetyl chloride and an alcohol as performed for esterification is feasible (see Section 1.1.2.1, Scheme 1.7). For example, a *p*-nitrophenyl ester has been transformed into the methyl ester without destruction of an *N*-Cbz protecting group (Scheme 1.72) [8].

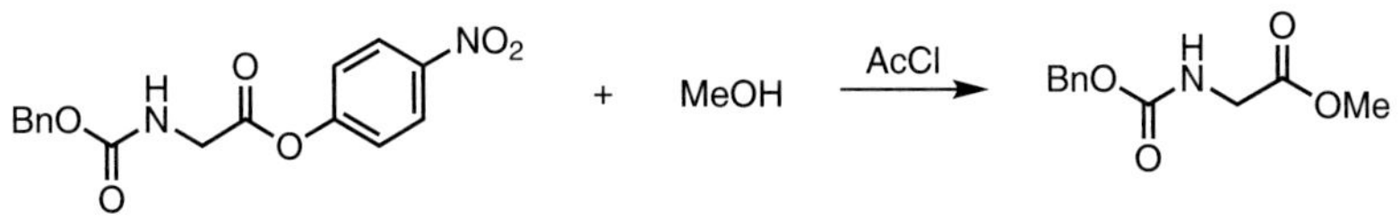

BnO–C(=O)–NH–CH₂–C(=O)–O–C₆H₄–NO₂ + MeOH —AcCl→ BnO–C(=O)–NH–CH₂–C(=O)–OMe

Scheme 1.72

The TMSCl/methanol technique is also applicable to transesterification. A variety of methyl esters are converted into propyl esters upon treatment with TMSCl in propanol (Scheme 1.73) [329].

Experimental Procedure Scheme 1.73 [329] $(CH_3)_3SiCl$ (1 mL) is placed in a round-bottomed glass flask loaded with a mixture of methyl ester (1.5 mmol) and 1-propanol (5 mL). The flask is closed with a glass stopper or fitted with a reflux condenser and a drying tube, and the reaction mixture is stirred either at 25 °C for 6 h or under gentle reflux for 2 h. The reaction mixture is then diluted with diethyl ether (5 mL), transferred to a separatory funnel, and washed with saturated sodium bicarbonate solution (10 mL), followed by brine (10 mL). The organic solution is then dried with anhydrous magnesium sulfate and evaporated to dryness. The residue is weighed. After a standard workup, the residue is redissolved in hexane and the solution is analyzed by both GLC and GLC/MS to identify the product and to determine the degree of conversion.

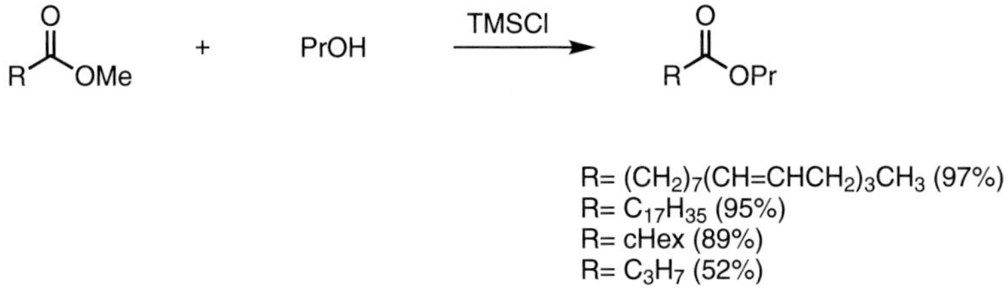

RC(=O)OMe + PrOH —TMSCl→ RC(=O)OPr

R= $(CH_2)_7(CH{=}CHCH_2)_3CH_3$ (97%)
R= $C_{17}H_{35}$ (95%)
R= cHex (89%)
R= C_3H_7 (52%)

Scheme 1.73

In situ generation of HBr from the reaction between Ph_3P and CBr_4 in ester as a solvent has been claimed [330]. Thus, stirring of a solution of alcohol, Ph_3P, and

CBr_4 in 1:0.5:0.5 ratio in ethyl acetate solution at room temperature affords the corresponding acetates (Scheme 1.74). The use of methyl or ethyl formate provides the formates as well. Interestingly, TBS (*tert*-butyldimethylsilyl) and THP groups are directly transformed into the acetates or formates without the need for a deprotection procedure. A proposed mechanism is depicted in the Scheme. The key step is the hydrolysis of a phosphonium intermediate to generate HBr, which acts as an acid catalyst, but it is not clear where the water has come from. Although it is said that addition of 3 equiv. of H_2O is effective in promoting the reaction in some instances, no water is added in other cases.

Experimental Procedure Scheme 1.74 [330] A solution of alcohol (1 mmol), triphenylphosphine (0.5 mmol) and carbon tetrabromide (0.5 mmol) in ethyl acetate (5 mL) is stirred at room temperature under nitrogen atmosphere, and the reaction is monitored by TLC. Evaporation of ethyl acetate followed by flash column chromatography provides the acetate.

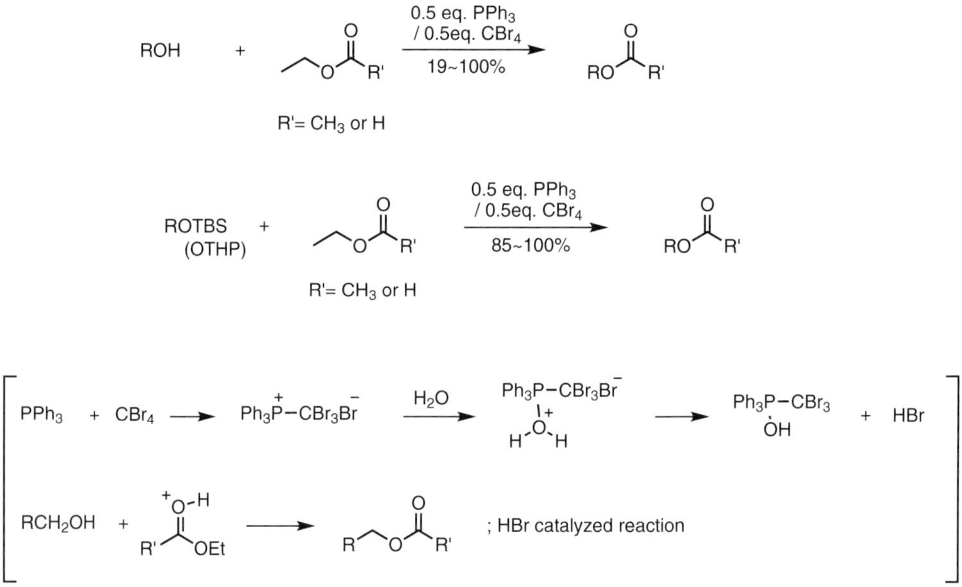

ROH + EtOC(O)R' → ROC(O)R' (0.5 eq. PPh_3 / 0.5eq. CBr_4; 19~100%)

R'= CH_3 or H

ROTBS (OTHP) + EtOC(O)R' → ROC(O)R' (0.5 eq. PPh_3 / 0.5eq. CBr_4; 85~100%)

R'= CH_3 or H

PPh_3 + CBr_4 → $Ph_3P^+-CBr_3Br^-$ →(H_2O) $Ph_3P-CBr_3Br^-$ / $H-O^+-H$ → Ph_3P-CBr_3 / OH + HBr

RCH_2OH + R'C(=O^+-H)OEt → RCH₂OC(O)R' ; HBr catalyzed reaction

Scheme 1.74

A two-step procedure involving silyl ester formation followed by alcoholysis offers a one-pot esterification (Scheme 1.75) [331]. Treatment of ester with TMSI (1.0 equiv.)/I_2 in refluxing $CHCl_3$ furnished the silyl ester. Addition of alcohol (2.5 equiv.) to the reaction mixture, once more followed by heating, again furnishes the new ester.

Experimental Procedure Scheme 1.75 [331] A solution of methyl benzoate (1.4 g, 10 mmol), iodotrimethylsilane (2.0 g, 10 mmol) and iodine (0.25 g, 1 mmol) in chloroform (20 mL) is heated at reflux under nitrogen for 3 h. The mixture is allowed to cool to ambient temperature and stirred with a drop of mercury until the color of iodine disappears. A solution of ethanol (1.15 g, 25 mmol) in chloroform (2 mL) is then added, and the mixture is again heated under reflux overnight. The mixture is then allowed to cool to room temperature, quenched with aqueous sodium hydrogen carbonate solution (5%, 50 mL) and extracted with ether (2 × 50 mL). The ether extract is washed with water and dried with anhydrous sodium sulfate. Evaporation of the solvent and distillation of the residue gives ethyl benzoate.

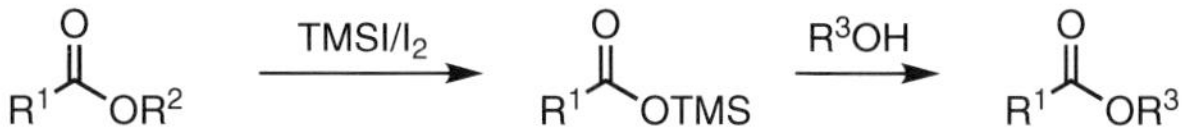

$R^1C(O)OR^2 \xrightarrow{TMSI/I_2} R^1C(O)OTMS \xrightarrow{R^3OH} R^1C(O)OR^3$

Scheme 1.75

Treatment of lactones with TMSI (1.5 equiv.) in the presence of an alcohol (2.5 equiv.) provides a short and convenient route to iodoalkyl esters (Scheme 1.76) [332]. The reaction is effective for β-, γ-, δ-, and ε-lactones. Primary, secondary, and benzyl alcohols, as well as 2-trimethylsilylethanol are employable. HI resulting from reaction between TMSI and alcohol is believed to act as a catalyst in this reaction.

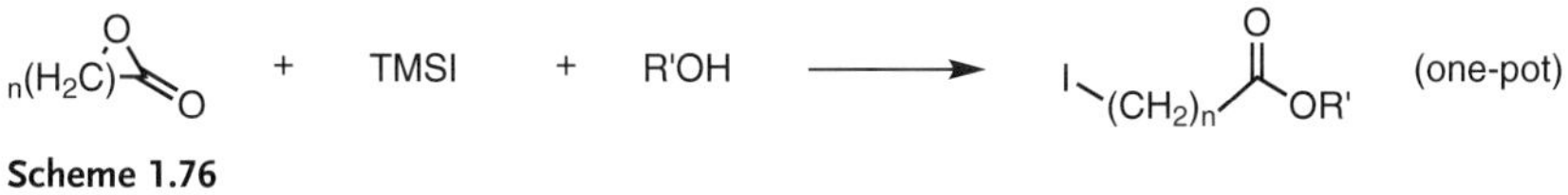

$_n(H_2C)$-lactone + TMSI + R'OH ⟶ $I(CH_2)_nC(O)OR'$ (one-pot)

Scheme 1.76

The lactonization shown in Scheme 1.77 can be successfully run only when an HF/pyridine complex is employed [333]. (+)-Compactin and (+)-3,5-di-*epi*-compactin is accessible by this procedure.

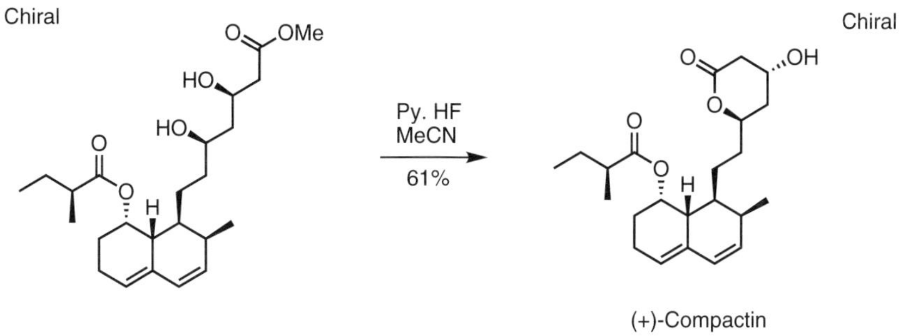

Chiral

OMe, HO, HO

Py. HF
MeCN
61%

Chiral

OH

(+)-Compactin

Scheme 1.77

TsOH (*p*-toluenesulfonic acid) is a useful catalyst for transesterification. As shown in Scheme 1.78, an indole lactone is formed when the depicted hydroxyl ester is heated in benzene in the presence of a catalytic amount of TsOH, N–O acyl transfer occurring under other conditions [334].

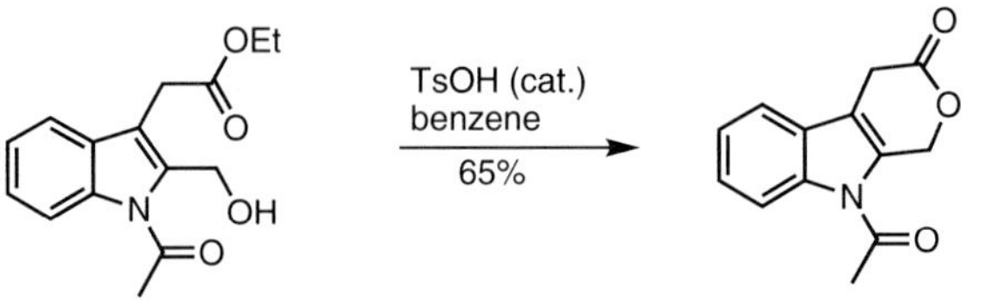

TsOH (cat.)
benzene
65%

Scheme 1.78

A tricyclic lactone undergoes alcoholysis under TsOH catalysis conditions to furnish unique dicyclopropyl β-ketoesters (Scheme 1.79) [335]. Even sterically hindered tertiary alcohols and phenols are employable, though the reaction is slower.

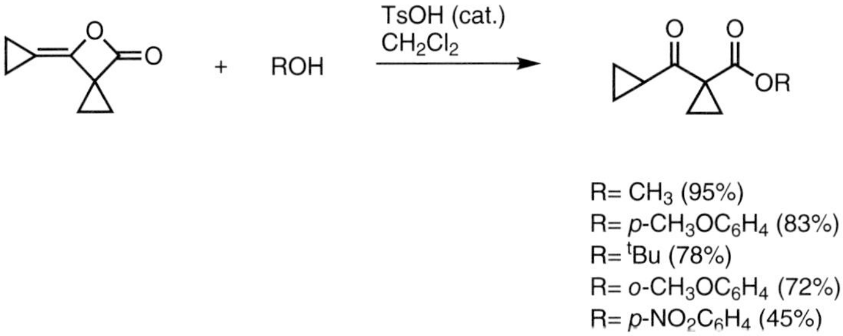

+ ROH

TsOH (cat.)
CH_2Cl_2

R= CH_3 (95%)
R= *p*-$CH_3OC_6H_4$ (83%)
R= ^{t}Bu (78%)
R= *o*-$CH_3OC_6H_4$ (72%)
R= *p*-$NO_2C_6H_4$ (45%)

Scheme 1.79

Upon treatment with alcohols and phenols in the presence of catalytic amounts of TsOH or H_2SO_4, novel unsymmetrical diesters of dicarboxylic acids with an isopropenyl ester moiety on one terminal selectively undergo transesterification at this terminal to provide the monoesters (Scheme 1.80) [336]. This procedure is utilized for synthesis of oxaunomycin derivatives.

1-Ethoxyvinyl esters derived from carboxylic acids and ethoxyacetylene function as efficient acylation reagents (Scheme 1.81) [337]. The reaction is catalyzed by TsOH or H_2SO_4. Not only bulky alcohols, but also phenols, are acylated smoothly. Olefins and nitriles are tolerated under these conditions. This procedure is applicable to macrolactonization [338]. Ethoxyvinylation of ω-hydroxy acids, followed by intramolecular transesterification with camphorsulfonic acid catalyst, affords reasonable yields of 15 to 22-membered lactones.

The reaction between phenols and β-ketoesters furnishes coumarin lactones (Pechmann reaction). This reaction is effected by use of trifluoromethanesulfonic acid and utilized for the synthesis of (±)-calanolide A and C (Scheme 1.82) [339].

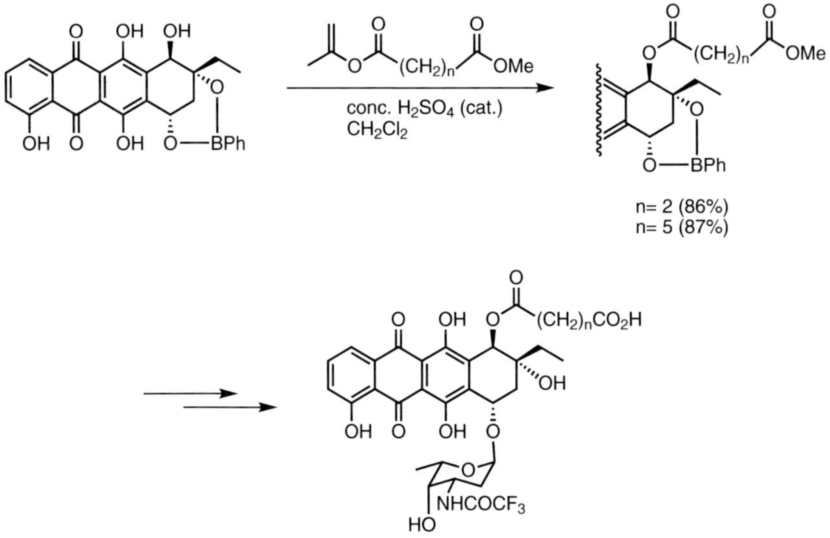

conc. H_2SO_4 (cat.)
CH_2Cl_2

n= 2 (86%)
n= 5 (87%)

Scheme 1.80

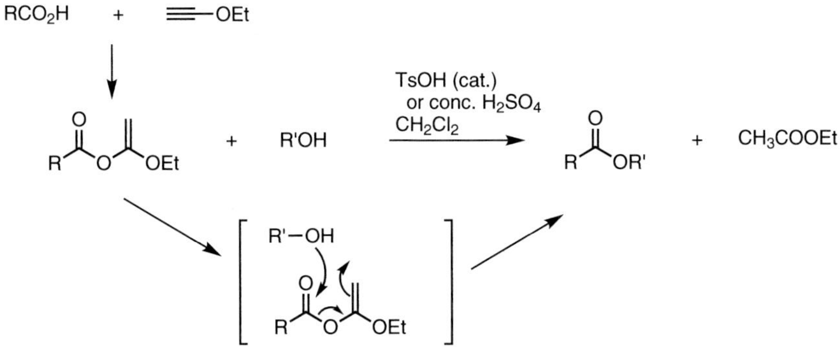

RCO_2H + ≡—OEt

TsOH (cat.)
or conc. H_2SO_4
CH_2Cl_2

+ R'OH → + CH_3COOEt

Scheme 1.81

TFA
neat
99%

(±)-Calanolide A or (±)-Calanolide C

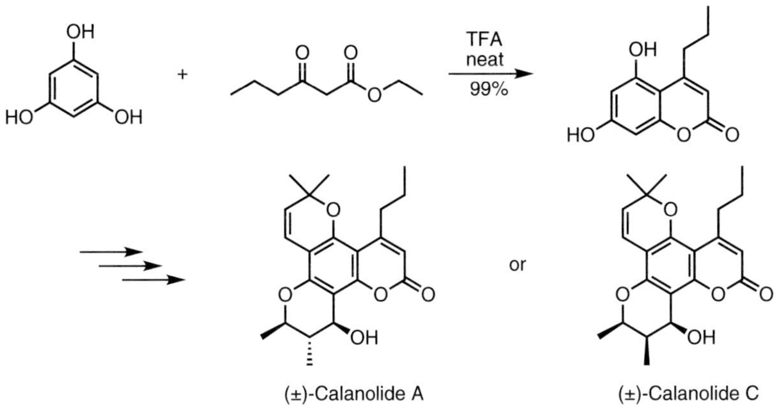

Scheme 1.82

1.2.2.2 Lewis Acids

The most synthetically versatile methodology for transesterification is provided by Lewis acids, due to their mildness, simplicity in operation, catalytic capabilities, and so on. In particular, titanium and tin compounds have long been used in both laboratories and industry. It may be said that these compounds, irrespective of whether they are organic or inorganic, possess more or less catalytic activity for transesterification. Among them, titanium tetraalkoxides are specifically useful in terms of their availability and readiness of handling. Heating an ester in alcohol solvent in the presence of $Ti(OR)_4$ smoothly effects transesterification, although the titanate catalyst usually has to be employed in a rather large quantity (normally 0.2–0.6 mol per ester) (Scheme 1.83) [340]. Since the alcohol is used in excess, the alkoxy group in the titanate need not necessarily be identical with the alcohol component. A number of functional groups are tolerated, and the use of dried solvent is not necessary. It should be noted that methyl esters cannot be prepared, because of the low solubility of tetramethyl titanate. However, this drawback can be overcome by use of an ethylene glycol derivative that is soluble in methanol without formation of $Ti(OMe)_4$ (Scheme 1.84) [341]. Alternatively, ester exchange with methyl propanoate in the presence of $Ti(OEt)_4$ is also effective (Scheme 1.84). This procedure is utilized for synthesis of esters with sterically hindered alcohols, such as menthol, borneol, 2-adamantylmethanol, and so on [342].

Experimental Procedure Scheme 1.83 [340] Ethyl 3-(*tert*-butyldimethylsiloxy) butanoate (2 g, 8.1 mmol) is dissolved in isopropanol (30 mL), tetraisopropyl titanate (1 g, 3.5 mmol) is added, and the mixture is heated at reflux temperature for 6 h. It is then cooled to ~ 45 °C, quenched with hydrochloric acid (1 N, 50 mL), leading to temporary turbidity, and extracted with pentane (2 × 150 mL). The organic extract is washed with saturated aqueous sodium hydrogen carbonate (30 mL), dried with magnesium sulfate, and evaporated under reduced pressure at 60 °C to remove solvent and residual isopropanol to give the isopropyl ester (2 g).

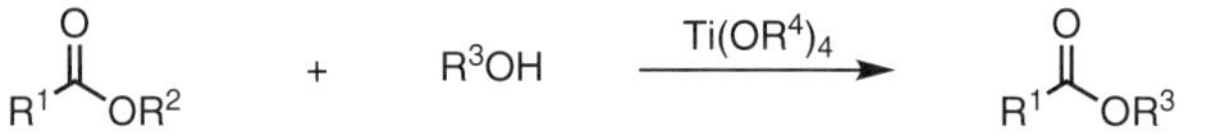

R^1COOR^2 + R^3OH —[$Ti(OR^4)_4$]→ R^1COOR^3

Scheme 1.83

$Ti(OEt)_4$ + $HO(CH_2)_2OH$ → $(EtO)_3TiO(CH_2)_2OTi(OEt)_3$ *Gly-Ti*

RCOOR' + MeOH —[*Gly-Ti*]→ RCOOMe

RCOOEt + EtCOOMe —[$Ti(OEt)_4$]→ RCOOMe

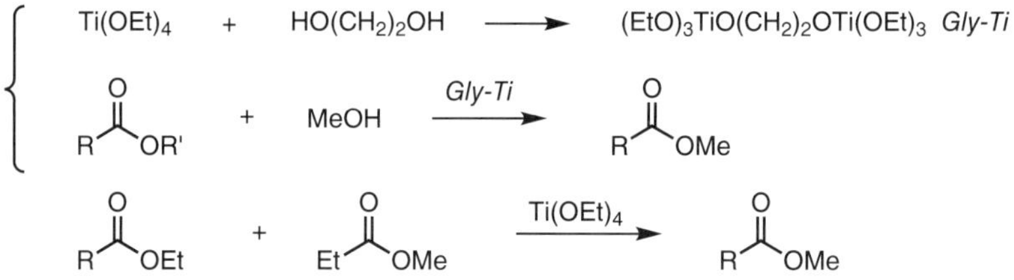

Scheme 1.84

$TiCl_4$ is not usually employed, because of its too strong acidity and hydrolytic instability. It serves, nevertheless, when the strong acidity is required. For example, the Pechman condensation is feasible: treatment of phenol (1.0 equiv.) with β-keto ester (1.5 equiv.) in the presence of $TiCl_4$ (0.5 equiv.) at room temperature provides coumarins in good yields (Scheme 1.85) [343].

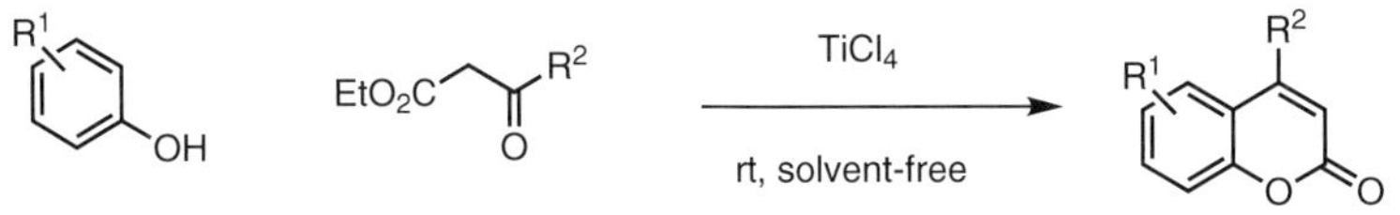

Scheme 1.85

Tandem reactions of transesterification followed by intramolecular 1,3-dipolar cycloaddition of α-methoxycarbonylnitrones with allyl alcohols are promoted by titanium tetraisopropoxide (Scheme 1.86) [344]. Treatment of α-methoxycarbonylnitrone with allylic alcohol (5 equiv.) in the presence of $Ti(O^iPr)_4$ provides stereocontrolled polycyclic compounds in one-pot fashion.

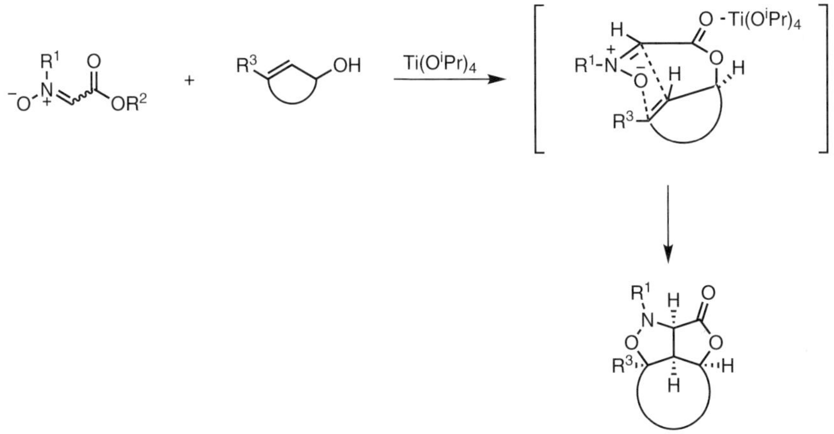

Scheme 1.86

Organotin compounds have long been known as transesterification catalysts [345–347]. However, 1,3-disubstituted tetralkyldistannoxanes are among the most useful [52, 348] (Scheme 1.87). The reaction is conducted by heating an ester in alcohol solvent with a catalytic amount of distannoxane. The catalyst activity is extremely high, so the reaction goes to completion even with a catalyst concentration of only 0.05 mol%. Thanks to the neutral conditions, various functional groups are tolerated and, remarkably, otherwise difficult-to-achieve transesterification of non-enolizable α,α-disubstituted β-keto esters smoothly takes place. An alkoxy distannoxane, on which the substrate ester coordinates, is believed to be a key intermediate. The entropy gain through cooperative *endo*- and *exo*-tins facilitates the alkoxy interchange on the template.

Experimental Procedure Scheme 1.87 [52] Distannoxane-catalyzed Transesterification: A toluene solution (25 mL) of ethyl 2,2-dimethylacetoacetate (791 mg, 5.00 mmol), BnOH (5.40 g, 50.0 mmol), and 1,3-dichlorotetrabutyldistannoxane (278 mg, 0.500 mmol) is heated at reflux for 24 h. The toluene and excess BnOH are evaporated *in vacuo*. The residue is purified by column chromatography on silica gel (hexane/EtOAc 95:5) to give benzyl 2,2-dimethylacetoacetate (968 mg, 88%).

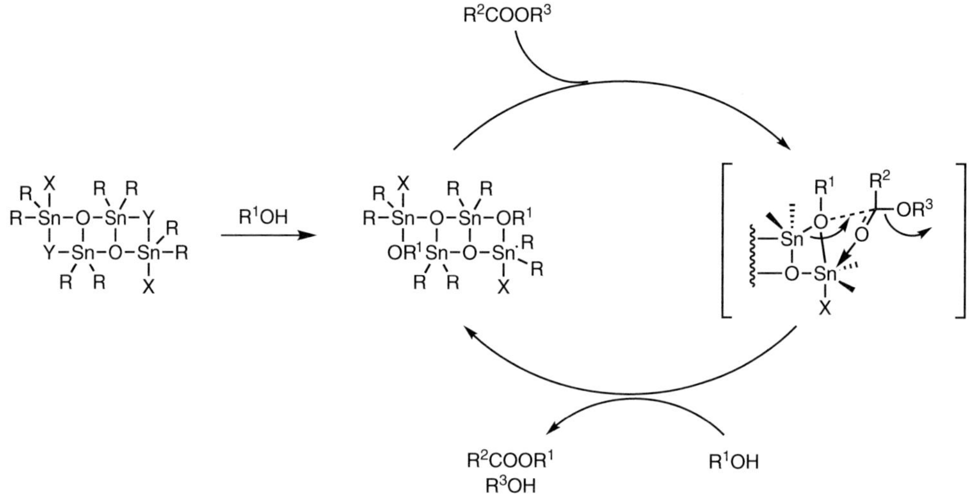

Scheme 1.87

The employment of vinyl or isopropenyl esters as ester components results in highly selective acylation of alcohols (Scheme 1.88) [349, 350] (see Section 7.1). This technique allows selective acylation of primary alcohol in the presence of

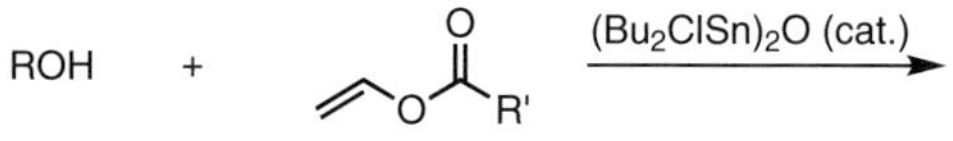

R= Ph, R'= CH_3 (0%)
R= CH_2Bn, R'= CH_3 (98%)
R= $CH(CH_3)Ph$, R'= CH_3 (99%)
R= C_8H_{17}, R'= CH_3 (99%)
R= $(CH_2)_4OCOCF_3$, R'= CH_3 (97%)
R= $(CH_2)S\text{-}S(CH_2)_2OH$, R'= CH_3 (98%)
R= $(CH_2)_4OTBS$, R'= CH_3 (98%)
R= *p*-$OHC_6H_4CH_2CH_2$, R'= CH_3 (99%)
R= $CH(CH_3)C_6H_{13}$, R'= CH_3 (97%)
R= C_6H_{11}, R'= CH_3 (0%)

R= $(CH_2)_4CH(OH)CH_3$, R'= Ph (96%)
R= C_8H_{17}, R'= Ph (98%)
R= C_8H_{17}, R'= ^tBu (97%)
R= C_8H_{17}, R'= $CH{=}CH_2$ (98%)
R= $(CH_2)_4CH(OH)CH_3$, R'= $CH{=}CH_2$ (96%)
R= C_8H_{17}, R'= CH_2Cl (96%)

Scheme 1.88

secondary and tertiary alcohols, leaving various functional groups intact. Transesterification using vinyl acetate is also catalyzed by Et_2Zn/*N*-phenyldiethanolamine [351].

Experimental Procedure [350] Acetylation of Alcohol with Ethyl Acetate: An ethyl acetate solution (10 mL) of 2-phenylethanol (611 mg, 5.0 mmol) and distannoxane (138.2 mg, 0.25 mmol) is heated at reflux for 12 h. After evaporation, the residue is subjected to column chromatography on silica gel to give 2-phenylethyl acetate (772 mg, 93%).

Acylation of Alcohol with Vinyl Acetate: A vinyl acetate solution (5 mL) of 2-phenylethanol (611 mg, 5.0 mmol) and distannoxane (138.2 mg, 0.25 mmol) is stirred at 30 °C for 5 h. After evaporation, the residue is subjected to column chromatography on silica gel to give 2-phenylethyl acetate (788 mg, 96%).

Replacement of the alkyl groups on the distannoxane with fluoroalkyl groups gives fluorous distannoxanes, which are highly soluble in fluorocarbon solvents [352, 353]. Fluorous biphasic transesterification with these catalysts gives 100% yields with use of a 1:1 reactant ratio (see Sections 7.2 and 7.3).

Experimental Procedure [352] Fluorous Biphasic Transesterification: A mixture of ethyl 3-phenylpropanoate (356 mg, 2.0 mmol), 1-octanol (260 mg, 2.0 mmol), and the distannoxane catalyst (171 mg, 0.1 mmol) in FC-72 (4 mL) is placed in a test tube. The test tube is placed in a stainless-steel pressure bottle and heated at 150 °C for 16 h. The reaction mixture is then allowed to cool to room temperature, and toluene (5 mL) is added. The FC-72 layer is washed with toluene (2 × 1 mL), and the combined toluene solution is evaporated to afford pure octyl 3-phenylpropanoate (525 mg, 100%). The FC-72 solution can be reused in the next reaction.

A neutral μ-hydroxy organotin dimer [*tert*-$Bu_2Sn(OH)Cl]_2$ is an efficient catalyst for deacetylation (Scheme 1.89) [354, 355]. Various acetates undergo transesterification to afford the parent alcohols. High selectivities are attained for the combination of primary > secondary, primary > tertiary, secondary > tertiary. Notably, phenyl acetate is deacetylated more readily than the primary acetate. The same primary/secondary selectivity also holds for sugars and nucleosides. Since the reaction is so clean, the product may be used for further reaction without purification. For example, tetraacetylcytidine is deacetylated, and treatment of the crude product with pivaloyl chloride affords cytidine with primary and *N*-pivaloyl groups together with the secondary acetyl groups. Moreover, glycosidation with the crude deacetylation product is also feasible. Treatment of the crude glucose tetracetate obtained by this procedure with tetra-*O*-benzoate-α-D-glucopyranosyl bromide in the presence of AgOTf furnishes the corresponding disaccharide.

Experimental Procedure Scheme 1.89 [355] Deacetylation of 2-Phenylethyl Acetate: 2-Phenylethyl acetate (82.1 mg, 0.5 mmol), [*tert*-$Bu_2Sn(OH)Cl]_2$ (28.6 mg, 0.05 mmol), methanol (5 mL), and THF (5 mL) are stirred at 30 °C for 3.5 h. After addition of ethyl acetate, the reaction mixture is filtered through a thin pad of silica gel, and the yield is determined to be 97% by GC analysis.

ROAc → ROH (R'OH, $[{}^tBu_2Sn(OH)Cl]_2$)

Tetraacetylcytidine + MeOH → (Sn-cat., THF, 30°C, 24h) → 2',3'-di-O-acetylcytidine → (${}^tBuCOCl$, Py, rt, 24h) → 2',3'-di-O-acetyl-5'-O,N-dipivaloylcytidine, 76%

Glucose tetraacetate + tetrabenzoyl glucosyl bromide → (AgOTf, tetramethylurea, MS 3A, CH_2Cl_2, rt, 18h) → disaccharide, 89% based on the bromide

Scheme 1.89

Deacetylation of Cytidine Tetraacetate and Subsequent Pivalation: Tetraacetylcytidine (411 mg, 1.0 mmol), [*tert*-$Bu_2Sn(OH)Cl]_2$ (28.5 mg, 0.05 mmol), methanol (5 mL), and THF (5 mL) are stirred at 30 °C for 24 h. After addition of CH_2Cl_2, the reaction mixture is filtered through a thin pad of silica gel, and the filtrate is concentrated to give a crude 2,3-di-*O*-acetylcitidine. Pivaloyl chloride (4.0 mmol) and pyridine (5 mL) are added to this product, and the mixture is stirred at room temperature for 24 h. After conventional aqueous workup (ethyl acetate/water), the organic layer is dried and evaporated to give a crude product, which is subjected to column chromatography on silica gel to give 2′,3′-di-*O*-acetyl-5′-*O*,*N*-dipivaloylcytidine in 76% yield.

Deacetylation of α-D-Glucose Pentaacetate and Subsequent Glucosylation: The pentaacetate (507 mg, 1.3 mmol), [*tert*-$Bu_2Sn(OH)Cl]_2$ (37.1 mg, 0.065 mmol), methanol (7 mL), and THF (7 mL) are stirred at 30 °C for 4 h. After addition of ethyl acetate, the reaction mixture is filtered through a thin pad of silica gel, and the filtrate is concentrated to give a crude product of the tetraacetate. A mixture of this crude product, tetra-*O*-benzoyl-α-D-glucopyranosyl bromide (660 mg, 1.0 mmol), tetramethylurea (0.25 mL), and molecular sieves (3 Å) in CH_2Cl_2 (5 mL) is stirred at room temperature for 15 min. AgOTf (452 mg, 1.3 mmol) is added. The mixture is stirred in the dark at room temperature for 18 h. After addition of CH_2Cl_2, the reaction mixture is filtered. Conventional aqueous workup followed by column chromatography on silica gel provides the desired disaccharide in 89% yield.

Butylstannoic acid [BuSn(O)OH] mediates transesterification of various substrates, although rather large amounts of the promoter (0.2–1.1 equiv. per substrate) are required [356]. Dibutyltin oxide is useful as well [357]. With this commercially available catalyst, a variety of substrates can be smoothly transesterified.

Experimental Procedure [357] The carboxylic ester (0.6 mmol) is dissolved in the desired alcohol (5 mL). After the addition of dibutyltin oxide (10 mol%; 15 mg, 0.06 mmol) the mixture is heated at reflux for 48–72 h. After completion of the reaction, the mixture is poured into a saturated sodium bicarbonate solution (20 mL) and extracted three times with ethyl acetate. The combined organic layers, which contain the dibutyltin oxide as a fine, white precipitate, are filtered through Celite and dried over sodium sulfate. After removal of the solvent *in vacuo*, the product is purified by flash chromatography.

Tributyltin acetate works for transesterification of a γ-hydroxy α, β-unsaturated ester generated *in situ* (Scheme 1.90) [358]. Palladium coupling between terminal acetylene and γ-hydroxyalkynoate affords butenolide, although furan formation competes. Selective lactonization is achievable by addition of tributyltin acetate (10–40 mol%) to the reaction system to furnish the butenolide exclusively.

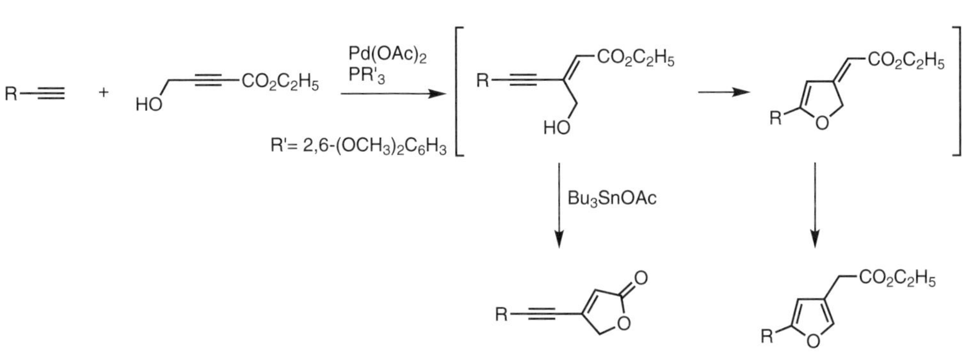

R—≡ + HO—≡—$CO_2C_2H_5$ —[$Pd(OAc)_2$, PR'_3; R'= 2,6-$(OCH_3)_2C_6H_3$]→ [R—≡—C(=CH—$CO_2C_2H_5$)—CH$_2$OH → 2-R-furan-3-ylidene—$CO_2C_2H_5$]

↓ Bu_3SnOAc → R—≡—butenolide; ↓ → R-furan-3-yl—$CH_2CO_2C_2H_5$

Scheme 1.90

Some of the Lewis acids used for esterification are effective for transesterification as well. $Fe(ClO_4)_3$ [359] and iodine [80] are capable of catalyzing transesterification. Treatment of alcohols in refluxing ethyl acetate or formate in the presence of $Cu(NO_3)_2 \cdot 3H_2O$ provides the corresponding acetates or formates [67]. $CuBr_2$ (0.2–1.0 equiv.) catalyzes reaction between 2-pyridyl esters and alcohols (Scheme 1.91) [360]. Despite the disadvantage of requiring a relatively large amount of the copper salt, this procedure is superior to other methods in terms of rapidity of the reaction and its applicability to sterically hindered esters. It is notable that the corresponding thiopyridyl esters are more reactive.

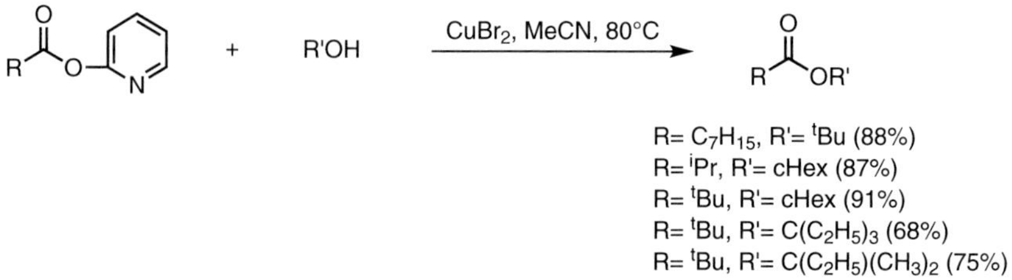

R–C(=O)–O–(2-pyridyl) + R'OH —($CuBr_2$, MeCN, 80°C)→ R–C(=O)–OR'

R= C_7H_{15}, R'= ^{t}Bu (88%)
R= ^{i}Pr, R'= cHex (87%)
R= ^{t}Bu, R'= cHex (91%)
R= ^{t}Bu, R'= $C(C_2H_5)_3$ (68%)
R= ^{t}Bu, R'= $C(C_2H_5)(CH_3)_2$ (75%)

Scheme 1.91

Use of particular esters to activate transesterification is also possible with methoxyacetates (Scheme 1.92) [361]. Treatment of these acetates with $Yb(OTf)_3$ in methanol delivers alcohols, and this approach thus offers a facile deprotection method to recover alcohols.

Experimental Procedure Scheme 1.92 [361] A solution of $Yb(OTf)_3$ (9.5 mg, 30 mol%) in dry methanol (1 mL) is added to a solution of 1,2:3,4-di-*O*-isopropylidene-6-*O*-(methoxyacetyl)-α-D-galactopyranose (19.9 mg, 0.051 mmol) in dry methanol (0.5 mL), and the mixture is stirred for 12 min at 25 °C. A drop of water is added to the reaction mixture, and the solvent is then evaporated under vacuum. Purification of the crude product by preparative TLC on silica gel affords the desired alcohol (13.0 mg, 98%) as a colorless oil.

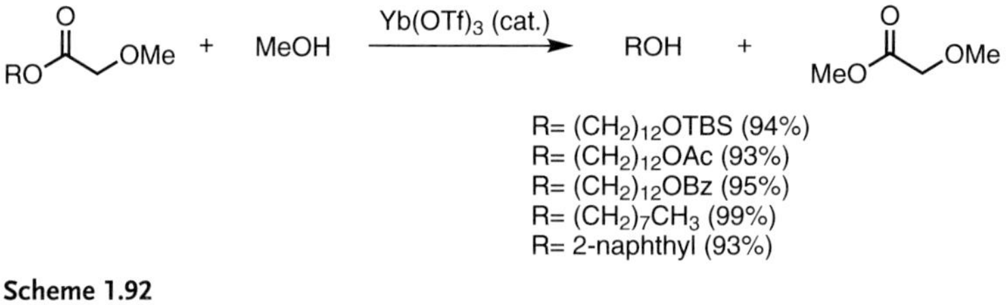

RO–C(=O)–CH$_2$–OMe + MeOH —($Yb(OTf)_3$ (cat.))→ ROH + MeO–C(=O)–CH$_2$–OMe

R= $(CH_2)_{12}OTBS$ (94%)
R= $(CH_2)_{12}OAc$ (93%)
R= $(CH_2)_{12}OBz$ (95%)
R= $(CH_2)_7CH_3$ (99%)
R= 2-naphthyl (93%)

Scheme 1.92

Other activated esters, *N*-hydroxysuccinimidyl esters, are useful for obtaining highly sterically hindered esters (Scheme 1.93) [189]. The reaction is promoted by the $Sc(OTf)_3$/DMAP reagent combination used for activation of the DCC procedure (see Section 1.1.4). DMAP is crucial for this reaction, and a scandium species coordinated by acylium ions is assumed to act as a powerful promoter (Scheme 1.94).

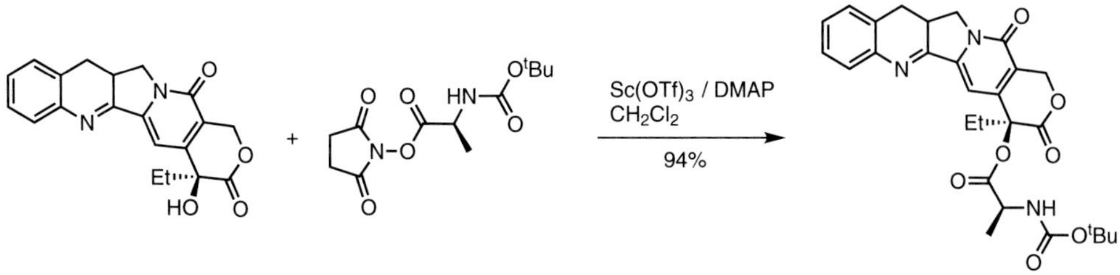

Et''''' HO O N N O O + O O N-O O HN O^{t}Bu O

$Sc(OTf)_3$ / DMAP
CH_2Cl_2
94%

N N O O O Et''''' O O O NH O^{t}Bu O

Scheme 1.93

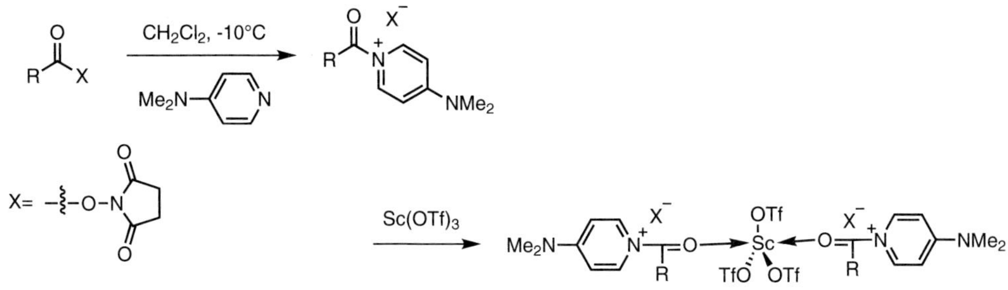

R X O → CH_2Cl_2, -10°C, Me_2N–pyridine → R O N+ X⁻ NMe$_2$

X= –O–N (succinimide)

$Sc(OTf)_3$ → Me_2N N+ X⁻ R O → Sc(OTf) TfO OTf ← O R X⁻ N+ NMe$_2$

Scheme 1.94

Mild $MgBr_2$ is employable for transesterification of activated esters. Transesterification of *N*-alkoxycarbonyl groups of chiral dimethylaminopyridine derivatives is effected by the use of $MgBr_2$ or $ZnCl_2$ [362]. The 4,6-dimethoxy-1,3,5-triazin-2-yl group serves as an active alkoxy group easily replaced by an alcohol in the presence of $MgBr_2$ [363].

Lanthanoid triisopropoxides catalyze transesterification efficiently [364]. The reaction goes to completion on heating esters in refluxing alcohol with 2 mol% of the catalyst. The catalytic activity is much higher than that of $Al(O^iPr)_3$ and $Ti(O^iPr)_4$, decreasing in the order La > Nd > Gd > Yb. $La(O^tBu)_3$ is an efficient catalyst for ring-opening of β-lactones with benzyl alcohol to afford β-hydroxy esters (Scheme 1.95) [365].

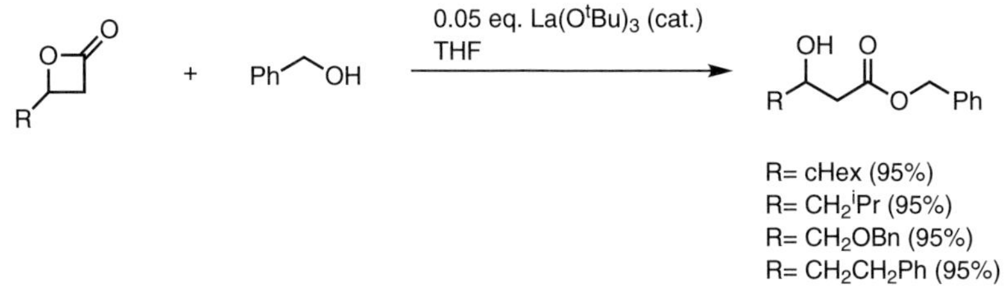

R (β-lactone) + Ph OH → 0.05 eq. $La(O^tBu)_3$ (cat.), THF → R OH O O Ph

R= cHex (95%)
R= $CH_2{}^iPr$ (95%)
R= CH_2OBn (95%)
R= CH_2CH_2Ph (95%)

Scheme 1.95

Enol ester acylation is feasible with an yttrium aggregate [366]. Primary and secondary alcohols react with vinyl or isopropenyl acetate at room temperature in the presence of catalytic amounts (0.05–1 mol%) of $Y_5(O^iPr)_{13}O$ to give the corresponding esters. A chiral version of the yttrium procedure is available (Scheme 1.96) [367]. Yttrium-salen complexes effect transesterification between enol esters and chiral secondary alcohols, resulting in varying degrees of kinetic resolution.

Experimental Procedure Scheme 1.96 [367] A mixture of alcohol (1 mmol) and the yttrium-catalyst in toluene (1.5 mL) under nitrogen is cooled to −3 ~ −25 °C, and the enol acetate (1.27 mmol) is added. When the reaction is complete, the cold solution is poured into water, and the products are extracted with ether. The ether solution is washed with saturated NaCl and dried, and the products are isolated by column chromatography. The *ees* of unreacted alcohols are determined by chiral HPLC.

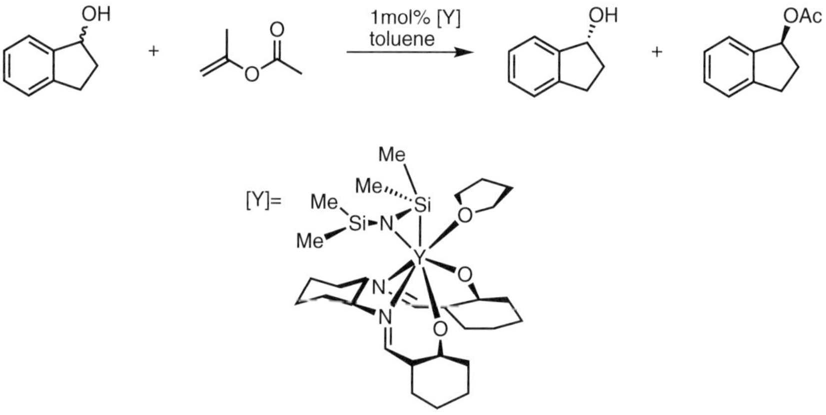

OH + O O → 1mol% [Y] toluene → OH + OAc

[Y]= Me Me Si N Me Me Si Y O O N N O

Scheme 1.96

$Cp^*{}_2Sm(THF)_2$ (Cp^* = pentamethylcyclopentadienyl) is an efficient catalyst for acylation of alcohols with vinyl esters [368]. The acylation of primary alcohols is achievable in excellent yields, while secondary alcohols are acetylated with more difficulty. The deficient reactivity, however, is improved by the presence of cyclohexanone oxime acetate (Scheme 1.97) [369]. Treatment of alcohol with isopropenyl acetate (2 equiv.) and the oxime acetate (0.2–2.0 equiv.) in the presence of $Cp^*{}_2Sm(THF)_2$ (10 mol%) allows transesterification to take place even for tertiary alcohols. A plausible reaction path involves reaction between alcohol and cyclohexanone oxime acetate, followed by acetylation of the oxime generated by isopropenyl acetate.

Experimental Procedure Scheme 1.97 [369] Alcohol (1 mmol) is slowly added under argon to a toluene solution (1 mL) of $Cp^*_2Sm(thf)_2$ (0.1 mmol) in a Schlenk tube, followed by cyclohexanone oxime acetate (0.2–2.0 mmol) and isopropenyl acetate (2 mmol). When the reaction is complete, wet diisopropyl ether (10 mL) is added to the solution, and the catalyst is removed by filtration. Removal of the solvent under reduced pressure affords a yellow liquid, which is purified by column chromatography on silica gel with hexane/ethyl acetate (20/1 v/v) as eluent to give the corresponding acetates.

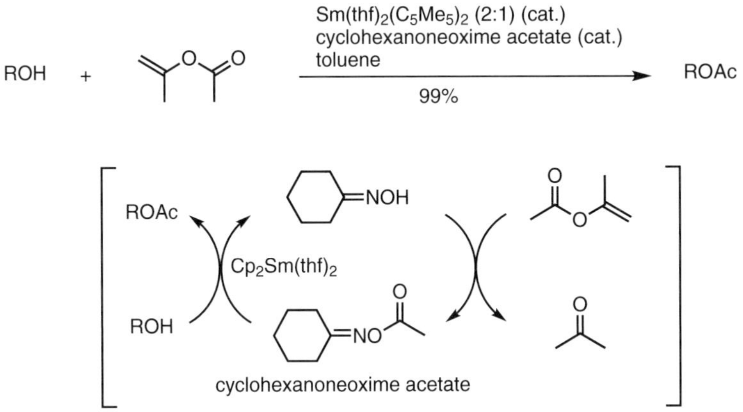

Scheme 1.97

Commercially available methyl or ethyl methacrylates are converted into higher homologs on treatment with alcohols with the aid of a nickel-containing complex reducing agent prepared from nickel acetate, NaH, and ^{t}BuONa [370]. $Sc(OTf)_3$ also acts as a catalyst for transesterification [371].

$Zn(OAc)_2$ immobilized on silica gel exhibits catalytic activity for reaction between dimethylterephthalate and ethylene glycol, and the catalyst can be easily separated from the reaction product and reused [372]. Various transesterification reactions are catalyzed by fluoroapatite or hydroxyapatite-supported $ZnCl_2$ [373].

1.2.2.3 **Solid Acids**

Solid acid catalysts are more important than any other acid catalysts in practical use, and a number of techniques are available. Neutral alumina used for column chromatography can effect transesterification [374]: stirring an alcohol (100 mg) in ethyl acetate (10 mL) over commercially available alumina (10 g) at room temperature affords the corresponding acetate. This reaction is selective for primary alcohols, and primary-secondary diols thus undergo esterification on the primary alcohol almost exclusively (Scheme 1.98) [375]. Under the same reaction conditions phenol is also inert, allowing selective acetylation of the primary alcohol of *p*-(ω-hydroxyalkyl)phenol. Such selectivity holds with sugar diols as well (Scheme

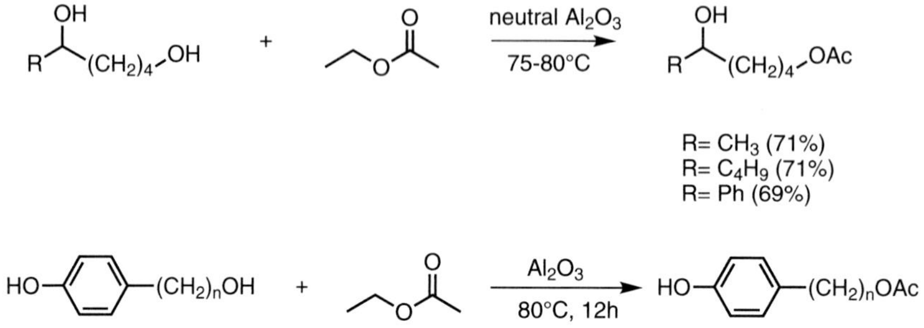

R= CH_3 (71%)
R= C_4H_9 (71%)
R= Ph (69%)

Scheme 1.98

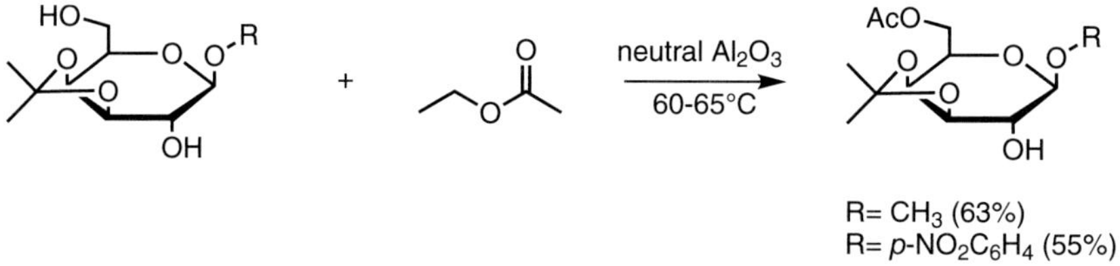

R= CH_3 (63%)
R= p-$NO_2C_6H_4$ (55%)

Scheme 1.99

1.99) [376]. Typically, a solution of the diol (1 mmol) in anhydrous ethyl acetate (100 mL) is heated at 60–65 °C in the presence of neutral alumina (10 g). Both mono- and disaccharides are smoothly acetylated on the primary alcohol.

Experimental Procedure [374] Inside a nitrogen-filled glove bag, Woelm-200-N alumina (~100 g) is transferred from its commercial metal container into a 500 mL, three-necked, round-bottomed flask, which is then fitted with an overhead mechanical stirrer and a stirring rod. An ethyl acetate solution (~150 mL) containing cyclohexylmethanol (1.021 g) is poured into the flask. Rapid stirring is performed for 1 h at 25–30 °C. Ethyl acetate (~100 mL) is then added, and the reaction mixture is poured into a sintered glass funnel containing a pad (~2 cm) of Celite. Slow gravity filtration and rinsing with additional ethyl acetate is followed by rotary evaporation to produce spectroscopically pure cyclohexylmethyl acetate.

Silica gel-supported $NaHSO_4$ mediates monoacylation of symmetrical diols upon treatment with various esters as solvent (Scheme 1.100) [377, 378]. The selective acetylation of primary-secondary diols and also of primary aliphatic-aromatic diols, is also feasible with this catalyst [379]. Smaller amounts of the catalyst (typically 100 mg/L mmol diol) are required than in the alumina procedure.

Experimental Procedure Scheme 1.100 [378] A stirred mixture of 1,4-butanediol (0.44 mL, 5 mmol), $NaHSO_4$—SiO_2 (189 mg, 0.42 mmol), and EtOAc/hexane (2 : 3 v/v, 16 mL) is warmed at 50 °C. The reaction is monitored by TLC (EtOAc/hexane 1 : 2) and GLC. After 5 h, 1,4-diacetoxybutane begins to appear, and the reaction mixture is then filtered. The separated catalyst is washed with CCl_4 (10 mL), and the filtrate and washing are combined and concentrated. Column chromatography of the residue on silica gel (EtOAc/hexane 1 : 4) gives 4-acetoxy-1-butanol (0.54 g, 82%).

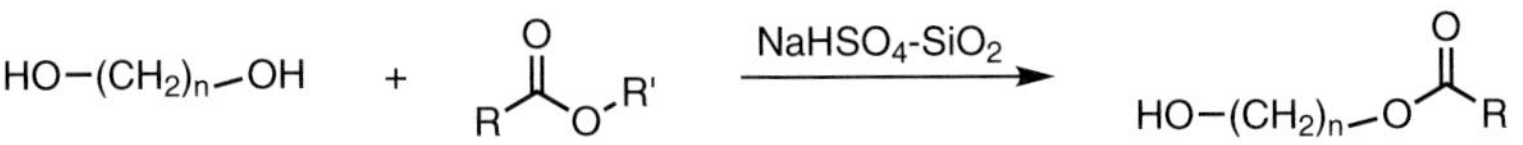

$$HO-(CH_2)_n-OH + RCOOR' \xrightarrow{NaHSO_4\text{-}SiO_2} HO-(CH_2)_n-OCOR$$

Scheme 1.100

Hydrous tin oxide, obtainable from tin metal and conc. HNO_3, can be used for vapor-phase transesterification at 170–210 °C [380]. The catalyst can be recycled for at least 50 runs. Tin oxide incorporated in mesoporous silica molecular sieves SBA-15 catalyzes transesterification of diethyl malonate with various alcohols [381]. PCl_3 immobilized on silica is employable for transesterification, and can be separated easily from the reaction mixture [382]. Hydrous zirconium oxide mediates transesterification in the vapor phase by use of a glass flow-reactor with a fixed-bed catalyst, in the liquid phase, and in the autoclave [121]. Remarkably, this catalyst is effective for lactonization to furnish medium-sized macrolides (Scheme 1.101) [383]. In the vapor phase reaction at 275 °C, ω-hydroxy esters are converted into the corresponding lactones. Otherwise difficult-to-obtain heptanolide and octanolide are accessible in reasonable yields. The catalytic activity is enhanced by modification with trimethylsilyl chloride [384], the yields of heptanolide and octanolide increasing from 49% to 77% and from 35% to 67%, respectively.

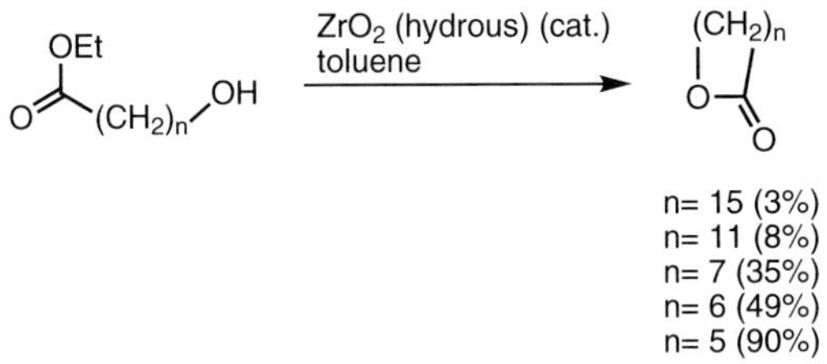

$$EtOOC-(CH_2)_n-OH \xrightarrow[\text{toluene}]{ZrO_2\text{ (hydrous) (cat.)}} \text{lactone } [(CH_2)_n-O-C(=O)]$$

n= 15 (3%)
n= 11 (8%)
n= 7 (35%)
n= 6 (49%)
n= 5 (90%)

Scheme 1.101

β-Keto esters are transesterified by solid acids such as zeolite [385], sulfated SnO_2 [386], natural kolinitic clay [387], Mo—ZrO_2 [388], and yttria-zirconia [389]. The reaction presumably proceeds via an acylketene intermediate. HY-Zeolite is an efficient and reusable catalyst for selective monodeacetylation of symmetrical diol diacetates to give monoacetates [390].

$Fe(ClO_4)_3(EtOAc)_6/SiO_2$ is prepared by adsorption of $Fe(ClO_4)_3(H_2O)_6$ onto chromatographic grade silica gel in the presence of ethyl acetate. Grinding equimolar amounts of this complex with alcohols under solventless conditions produces the corresponding acetates (Scheme 1.102) [391]. The supported reagent also reacts with carboxylic acids to afford the corresponding ethyl esters.

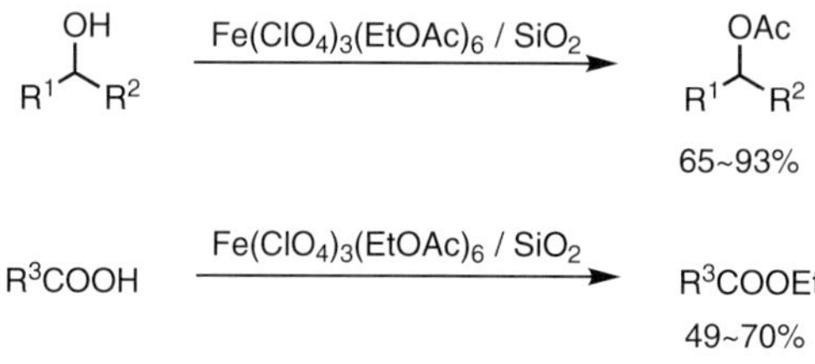

$$R^1CH(OH)R^2 \xrightarrow{Fe(ClO_4)_3(EtOAc)_6\ /\ SiO_2} R^1CH(OAc)R^2 \quad 65\text{–}93\%$$

$$R^3COOH \xrightarrow{Fe(ClO_4)_3(EtOAc)_6\ /\ SiO_2} R^3COOEt \quad 49\text{–}70\%$$

Scheme 1.102

Heteropolyacids such as $H_3PW_{12}O_{40}$, $H_3PMo_{12}O_{40}$, $H_{14}P_5W_{30}O4_{110}$, and $K_5CoW_{12}O_{40}$ are highly active catalysts for transesterification [392, 393]. The catalysts are easily removable and reusable.

1.2.3 Base Activators

1.2.3.1 Metal Salts

As well as acid catalysts, basic reagents also serve for transesterification, and some inorganic metal bases are successfully used. A thorough investigation into the effectiveness of lithium alkoxide is available (Scheme 1.103) [394]. The lithium alkoxide is generated *in situ* by addition of an equivalent amount of BuLi to a THF solution of an alcohol, and treatment of the resulting alkoxide with an ester (mostly methyl esters) results in transesterification. The reaction proceeds at or below ambient temperature, electron-deficient esters generally reacting within minutes. For secondary and tertiary alcohols, equimolar amounts of the ester and alcohol suffice for good yields. Although roughly one molar equivalent of BuLi per mole of alcohol is generally employed, use of smaller amounts suffices, but lengthens reaction time. Alcohols that are insoluble in THF frequently dissolve on addition of BuLi. With primary alcohols, an excess of the alcohol (~3 equiv.) is necessary for good yields. Ethyl esters undergo alcoholysis but with poorer equilibration shifts in the desired direction and lower reaction rates. Phenyl esters appear to offer no advantage, being slower and less effective in the transesterification. Commercial alcohols can be directly employable without drying. Aromatic esters react without problems; α,β-unsaturated esters behave normally, without isomerization, and asymmetric alcohols react without loss of chiral integrity. A major limitation of the method concerns alcohols that complex strongly with the lithium intramolecularly, such as sugar derivatives. Polymerizable compounds such as methyl acrylate are best used in excess. Also, if the alcohol is precious or difficult to separate from the product ester, this problem can be solved by use of an excess of the methyl esters. The reaction is presumably initiated by the strong complexation of the lithium alkoxide to the ester carbonyl group. The lower pK_a of methanol than that of higher homologous alcohols encourages efficient equilibration to produce esters.

Experimental Procedure Scheme 1.103 [394] Butyllithium in hexane (1.6 M; ~0.01 mol) and methyl *p*-methoxybenzoate (0.01 mol) are added under nitrogen to (–)-menthol (0.01 mol) in dry stirred THF (20 mL) in an ice bath. The reaction is monitored by TLC (20% diethyl ether/light petroleum; ~50% in 1 h; 70% in 6 h). After 18 h, water and ether are added and the organic layer is dried, evaporated, and flash chromatographed (20% diethyl ether/light petroleum). The product is further distilled in a Kugelrohr apparatus (215 °C at ~0.5 mmHg) to give menthyl *p*-methoxybenzoate (92%) as a colorless liquid.

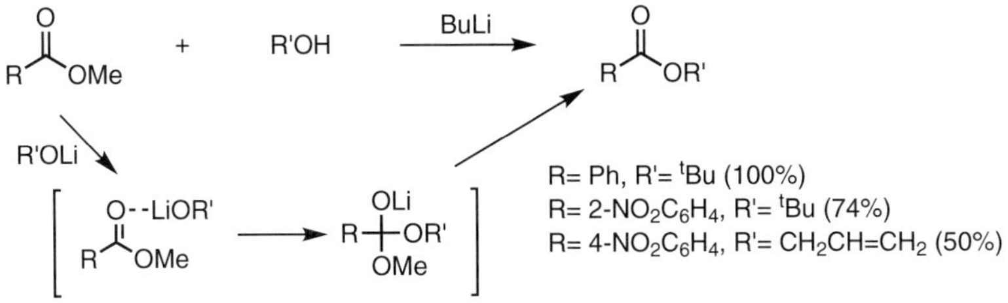
BuLi
R'OH
OMe
R'OLi
O--LiOR'
OLi
OR'
OMe
R= Ph, R'= tBu (100%)
R= 2-NO2C6H4, R'= tBu (74%)
R= 4-NO2C6H4, R'= CH2CH=CH2 (50%)

Scheme 1.103

NaH mediates the reaction between 5-ethoxypentan-1-ol and alkyl 5-halovalerates to furnish transesterification products, no Williamson-type reaction taking place (Scheme 1.104) [395]. Yields of up to 66% are thus obtained from bromovalerates, but considerable amounts of elimination products are formed from the iodo derivatives.

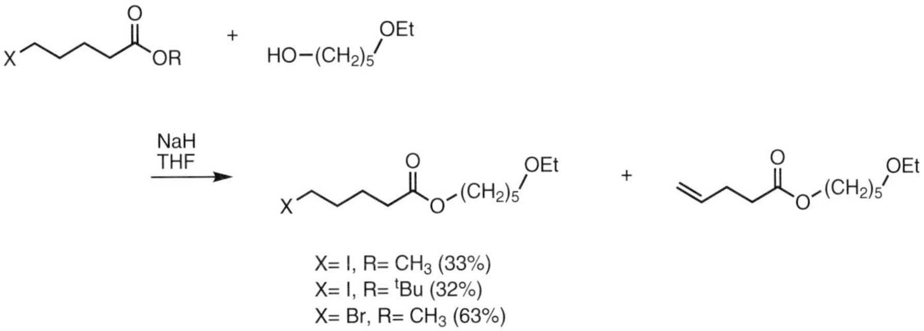
OR
HO–(CH2)5
OEt
NaH
THF
O–(CH2)5
X= I, R= CH3 (33%)
X= I, R= tBu (32%)
X= Br, R= CH3 (63%)

Scheme 1.104

Alcoholysis of esters, particularly acetates, offers a convenient deprotection method. It is possible to cleave primary alcohol acetate selectively in the presence of secondary or tertiary alcohol acetates, and also to cleave secondary alcohol acetates in the presence of tertiary alcohol acetates, with the aid of $Mg(OMe)_2$ (Scheme 1.105) [396]. High selectivities between primary acetates can also be obtained if the β-positions of the acetates offer a different degree of steric bulkiness. This mild reagent is applicable to the selective deprotection of many naturally occurring molecules, including hydroxycitronnellol diacetate, *trans*-sobrerol diacetate, betulin diacetate, and baccatin III. Both aromatic and 4-acetoxy-coumarin acetates undergo deacetylation with zinc-copper couple or activated zinc (Scheme 1.106) [397]. Commercial zinc is completely ineffective, and the reaction is hindered by the addition of water.

Experimental Procedure Scheme 1.105 [396] Magnesium methoxide solution in methanol (10.3%, 0.973 N, 4.1 mL, 4.0 mmol) is added dropwise under nitrogen at room temperature to a stirred solution of 4-methoxyphenylethyl acetate (194.0 mg, 1.0 mmol) and 2-phenyl-2-propyl acetate (187.0 g, 1.0 mmol) in anhydrous methanol (20 mL). The progress of the reaction is monitored by TLC. After the mixture has been stirred at room temperature for 6 h, the 4-methoxyphenylethyl acetate is all consumed. HCl solution (0.2 N) is added until the pH of the mixture is 4 to 5. Some of the methanol is removed under reduced pressure, and the product is extracted with dichloromethane (30 mL × 5). The combined organic layer is dried (Na_2SO_4), filtered, and concentrated. Flash chromatography of the crude residue with hexanes and ethyl acetate (7 : 3) gives 4-methoxyphenylethyl alcohol (144.0 mg, 0.94 mmol) in 94% yield, along with recovered starting material (176.0 mg, 0.94 mmol) in 94% yield. No 2-phenyl-2-propanol is detected.

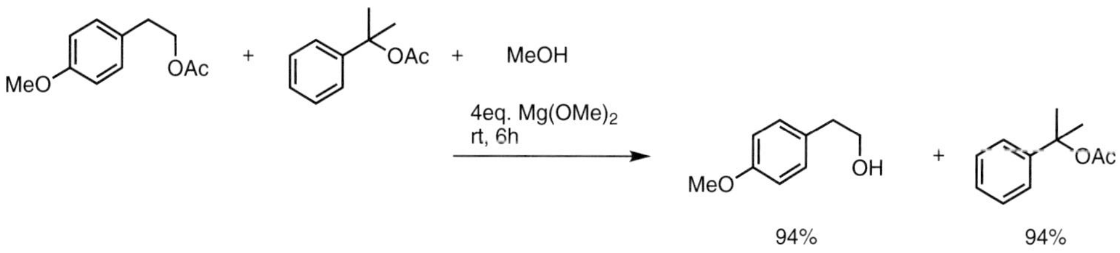

MeO OAc + OAc + MeOH
4eq. $Mg(OMe)_2$
rt, 6h
MeO OH 94% + OAc 94%

Scheme 1.105

OAc O O + MeOH Zn 100% OH O O

AcO AcO OH O O + MeOH Zn 91% HO AcO OH O O

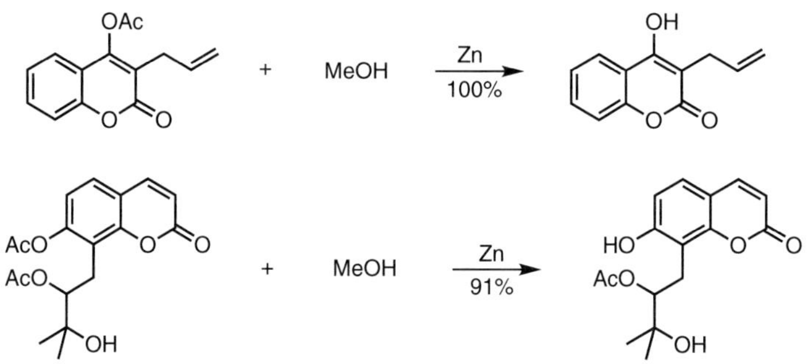

Scheme 1.106

Solid potassium carbonate coated with a phase-transfer catalyst (Carbowqax 6000 or 18-crown-6) can be utilized as a catalytic bed for continuous transesterification under gas-liquid phase-transfer conditions [398].

The effect of alkali metal salts on lactonization of ω-hydroxy ester is the subject of extensive investigation (Scheme 1.107) [399]. The use of K_2CO_3 in THF gives one of the highest yields of the desired macrolide. The rate-determining step is the addition of the alcohol or alkoxide fragment to the carbonyl carbon, followed by the fast breakdown of the resulting tetrahedral intermediate. The conversion of the hydroxy ester to macrolide is an M^+-templated cyclization in which the slow step is the collapse of the tetrahedral intermediate. With the proximity of the hydroxy and ester moieties enforced and the associated entropic costs met by the template effect, the formation of the tetrahedral intermediates is facilitated. The rate of the productive collapse of this intermediate is dependent on the departing alkoxide nucleofuge, hence the superiority of trichloroethyl esters.

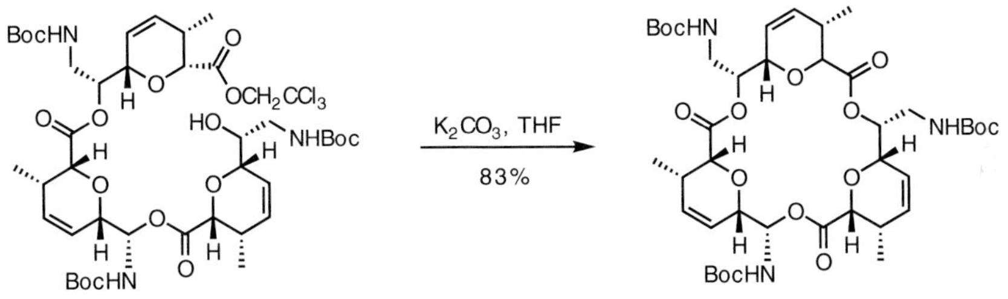

BocHN, O, H, OCH_2CCl_3, HO, NHBoc, H, O, H, O, O, O, O, H, H, BocHN, O

K_2CO_3, THF

83%

BocHN, O, H, O, O, NHBoc, H, H, O, O, O, H, H, BocHN, O

Scheme 1.107

Treatment of β-hydroxy ketones with α-chloro ester in the presence of *tert*-BuOK affords α,β-epoxy-γ-lactones through transesterification followed by Darzens reaction (Scheme 1.108) [400]. Under acidic conditions, the epoxy butenolide is transformed into 3-hydroxy-4,5-dimethyl-2(5H)-furanone, an extremely powerful and very important flavor ingredient.

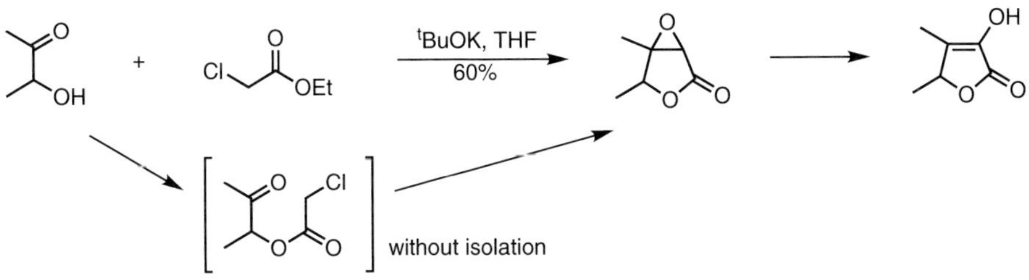

O, OH, +, Cl, O, OEt

^{t}BuOK, THF

60%

O, O, O

OH, O, O

O, Cl, O, O

without isolation

Scheme 1.108

Another phytotoxic butenolide, seiridin – a fungal cypress pathogen produced by three Seiridium species – is accessible through successive Wittig-type olefination and lactonization (Scheme 1.109) [401]. Both reactions are promoted by LiOH with molecular sieves.

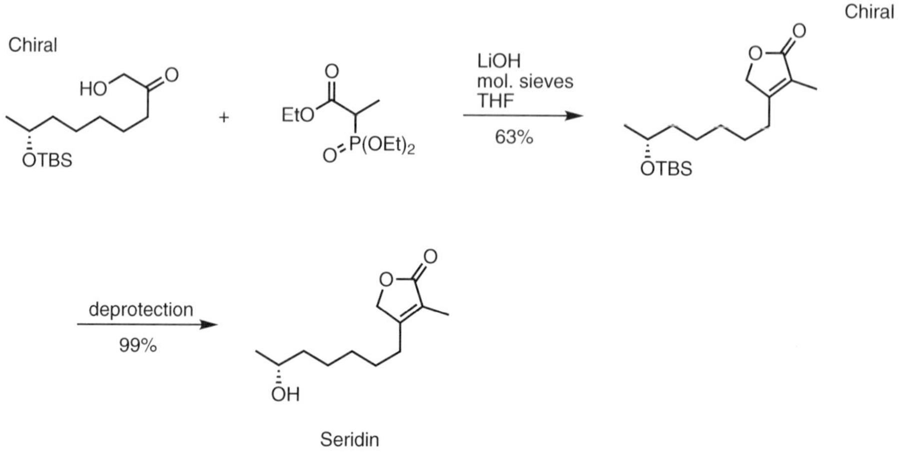

Chiral
HO
O
OTBS
+
O
EtO
P(OEt)2
O
LiOH
mol. sieves
THF
63%
Chiral
O
O
OTBS
deprotection
99%
O
O
OH
Seridin

Scheme 1.109

Interesting ring-expansion and -contraction of macrolides can take place, depending on the base (Scheme 1.110) [402]. Treatmentof the 16-membered lactone, 7,21-di-*O*-trimethylsilyl bafilomycin A_2 with an organocopper reagent prepared from equimolar quantities of methyllithium and cuprous iodide results in ring expansion. A further surprising event occurs when the resulting 18-membered lactone is treated with highly concentrated Bu_4NF, which regenerates bafilomycin A_2, although the use of smaller amounts of the fluoride results in simple desilylation. A plausible explanation is given in Scheme 1.111. The initial

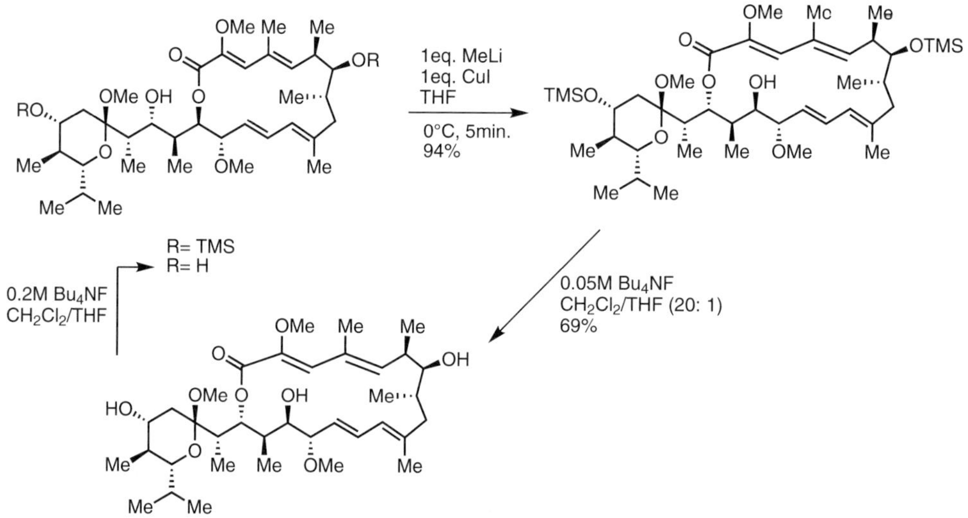

OMe Me Me
O
OR
OMe OH O
Me
RO
Me
O
Me Me OMe
Me
Me Me
1eq. MeLi
1eq. CuI
THF
0°C, 5min.
94%
OMe Me Me
O
OTMS
OMe O OH
Me
TMSO
Me
O
Me Me OMe
Me
Me Me
R= TMS
R= H
0.2M Bu4NF
CH2Cl2/THF
0.05M Bu4NF
CH2Cl2/THF (20: 1)
69%
OMe Me Me
O
OH
OMe O OH
Me
HO
Me
O
Me Me OMe
Me
Me Me

Scheme 1.110

step is the formation of alkoxycopper species suitably situated to undergo a potentially reversible rearrangement involving a copper (I) orthoacetate intermediate. Evidently a kinetically controlled and presumably stereoelectronically allowed process produces the ring-expansion product as the preponderant if not exclusive product. The behavior of this product toward fluoride ion can be explained as follows. In dilute dichloromethane, the effective concentration of Bu_4NF is not high enough to allow reversion to the 16-membered lactone, providing just a desilylation product. In a more concentrated solution of Bu_4NF, the ring contraction presumably takes place via the same orthoacetate, with the tetrabutylammonium species as the cation.

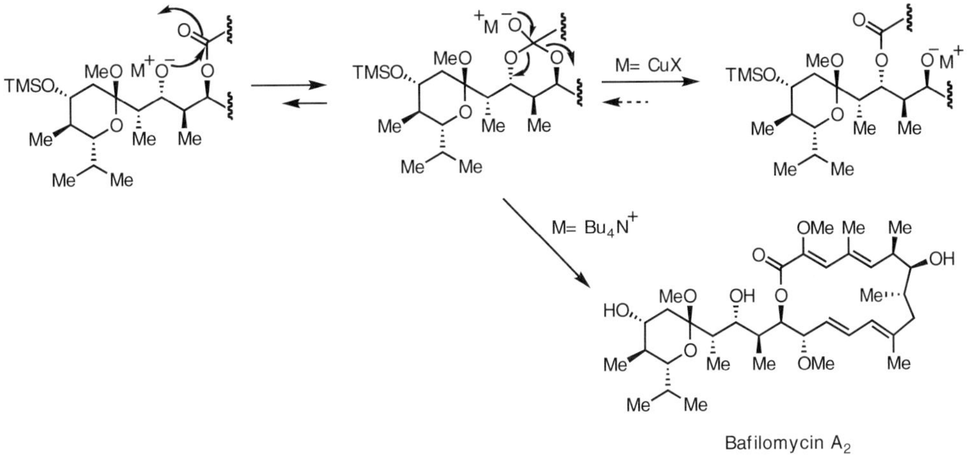

Scheme 1.111

Consecutive Michael addition and lactonization provides useful intermediates for carbapenem (Scheme 1.112) [403]. Treatment of dimethyl malonate with alkyl 2-(1-hydroxyalkyl)propenoates in the presence of NaH furnishes lactones as pairs of diasteromers.

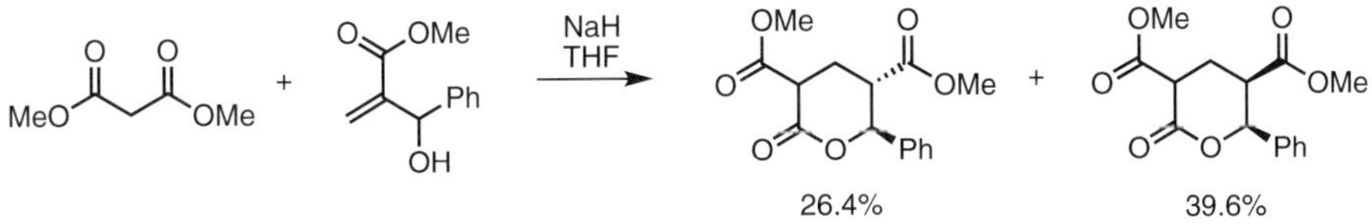

Scheme 1.112

A thermodynamically controlled macrolactonization procedure is feasible with cholic acid derivatives (Scheme 1.113) [404]. The distribution of products, from dimer to pentamer, is basically the same for all the substrates employed, but dependent to some extent on the substituents on C7 and C12. The distinct difference is that a dimer is produced only for the substrate without substituent at C7. The thermodynamic and reversible nature of the obtained distribution is apparent

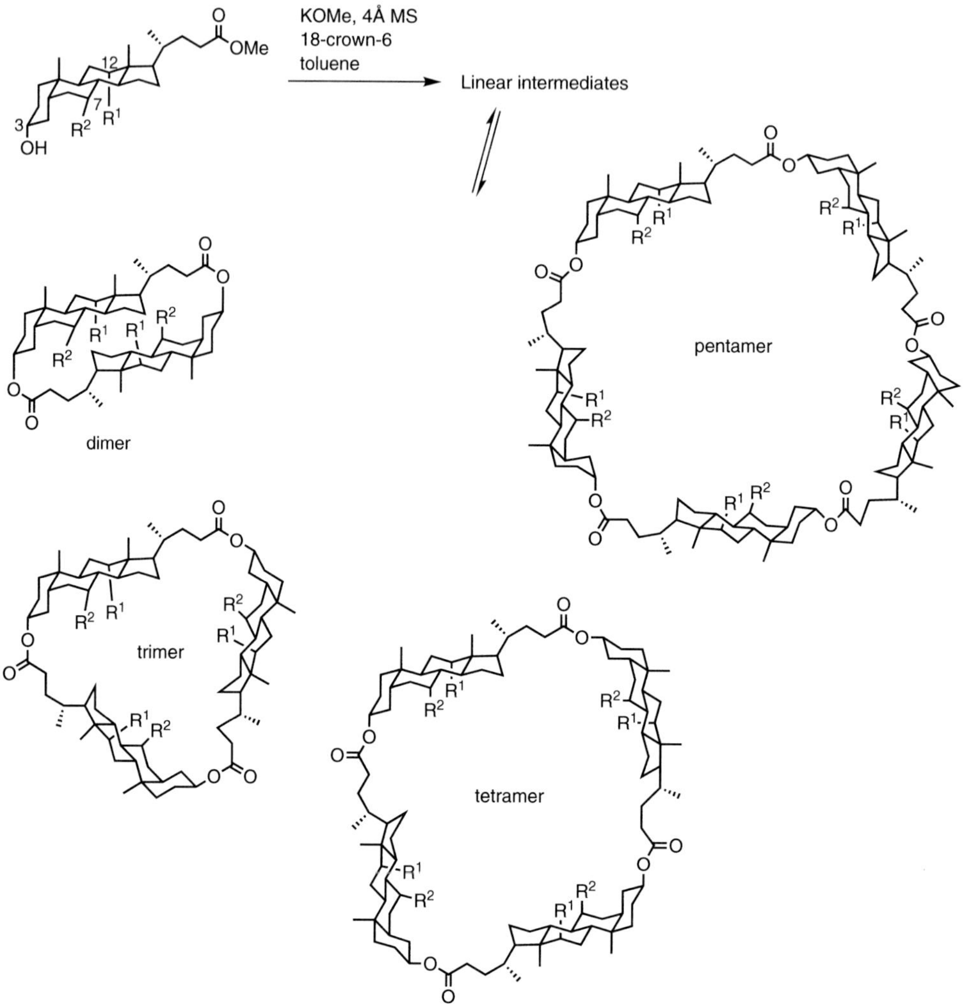

KOMe, 4Å MS
18-crown-6
toluene
Linear intermediates
dimer
trimer
tetramer
pentamer

Scheme 1.113

from the fact that exposure of the pure cyclic oligomers to the reaction conditions gives the same ratio of products irrespective of the starting materials.

Trimerization of hydroxy esters derived from cinchonidine is catalyzed by KOMe/[18]crown-6 (Scheme 1.114) [405, 406].

Transesterificaiton is useful for ring-opening of lactones. Treatment of an α-methylene-β-lactone with sodium methoxide or ethoxide provides β-hydroxy α,β-unsaturated esters (Scheme 1.115) [407]. Alcoholysis of uronolactone with benzyl alcohol in the presence of various bases such as K_2CO_3 and CsF produces a mixture of the desired benzyl ester and its epimer (Scheme 1.116) [408]. The highest ratio (12 : 1) is achieved when 30 equiv. of CsF are employed.

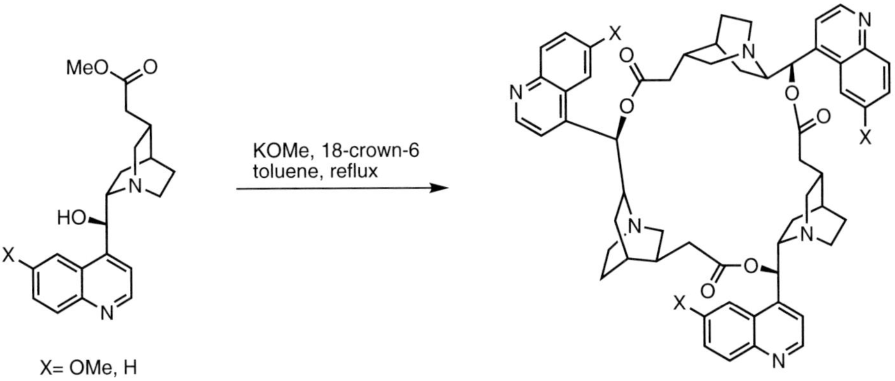

Scheme 1.114

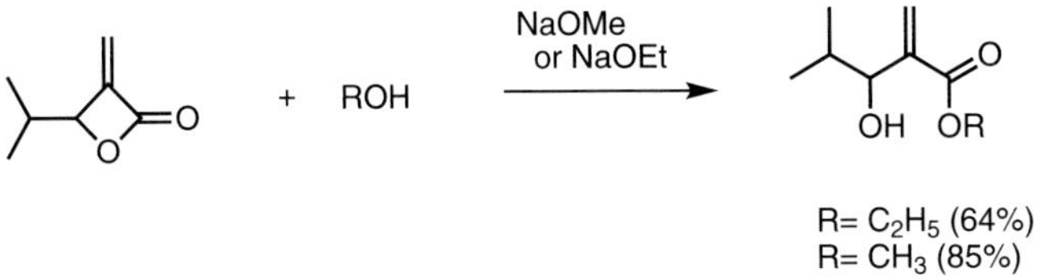

Scheme 1.115

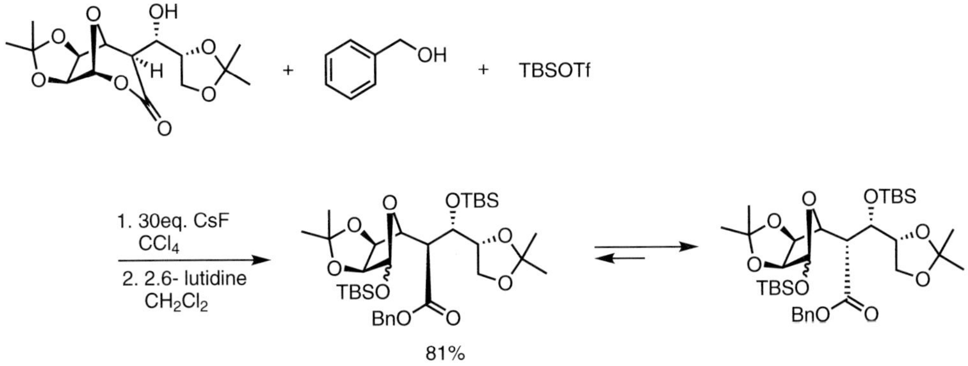

Scheme 1.116

1.2.3.2 **Amines**

Amines are not normally strong enough bases to effect transesterification, but they are employable if the substrates or reagents are reactive. The most common amine, Et_3N (1.5 equiv.), induces ring-opening of 4-arylmethylene-2-phenyl-2-oxazolin-5-ones in refluxing ethanol (Scheme 1.117) [409]

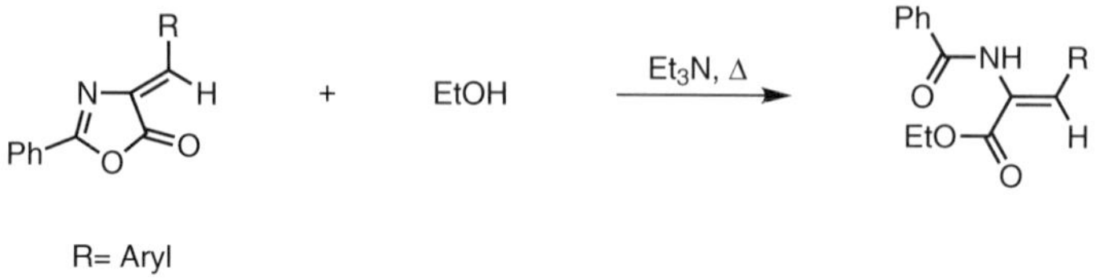

R= Aryl

Scheme 1.117

Ring-opening of diketene is effected by Et_3N. Treatment of diketene and 2-azidothanol (1 equiv.) or ethylene glycol (4 equiv.) in the presence of Et_3N furnishes the corresponding 2-azido- or 2-hydroxyethyl acetoacetate, respectively (Scheme 1.118) [410, 411]. The β-aminoethyl acetoacete obtained by this procedure can be further transformed into diazoacetate (Scheme 1.119) [412].

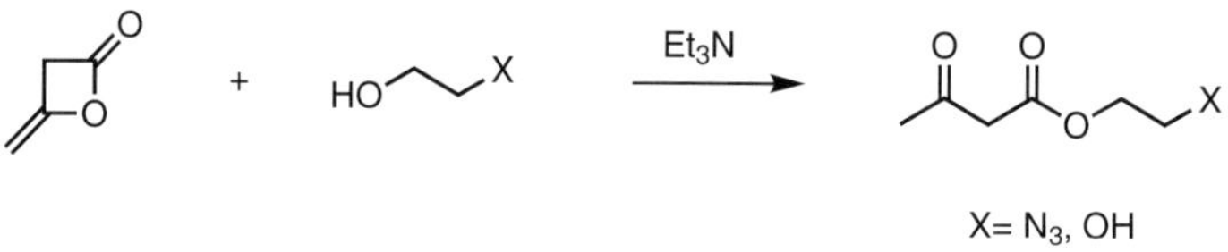

X= N_3, OH

Scheme 1.118

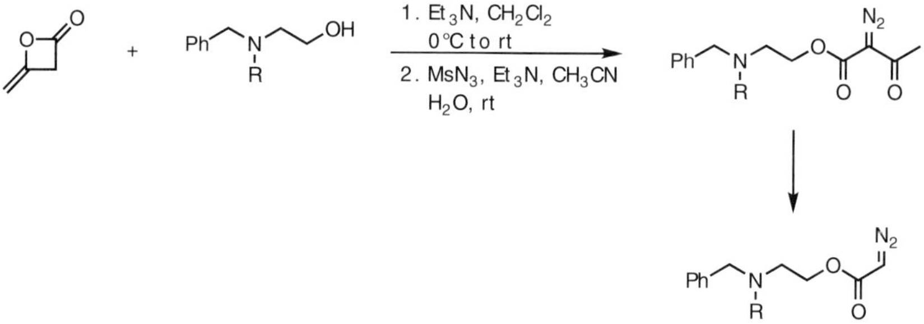

Scheme 1.119

Experimental Procedure [410] Diketene (4.2 g, 50 mmoles) is added dropwise with stirring to preheated (60 °C) 2-azidoethanol (4.35 g, 50 mmoles) containing a catalytic amount of triethylamine (4–5 drops). Diketene is added at such a rate that the temperature of the reaction mixture does not exceed 80 °, and the reaction is allowed to proceed at 80 ° for one additional hour. Removal of the solvent *in vacuo* gives a residue that is purified by silica gel column chromatography with ethyl acetate/hexane (1 : 1, v/v) as eluent to afford 2-azidoethyl acetoacetate as an oil (5.2 g, 60%).

δ-Lactones also undergo ring-opening by alcohol. Treatment of the lactone shown in Scheme 1.120 with methanol in the presence of Et_3N at 23 °C affords the methyl ester [413]. Simlarly, methanolysis of a δ-lactone in the presence of

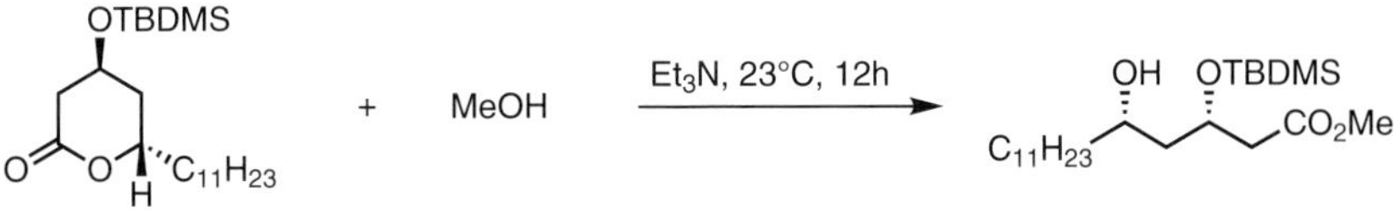

Scheme 1.120

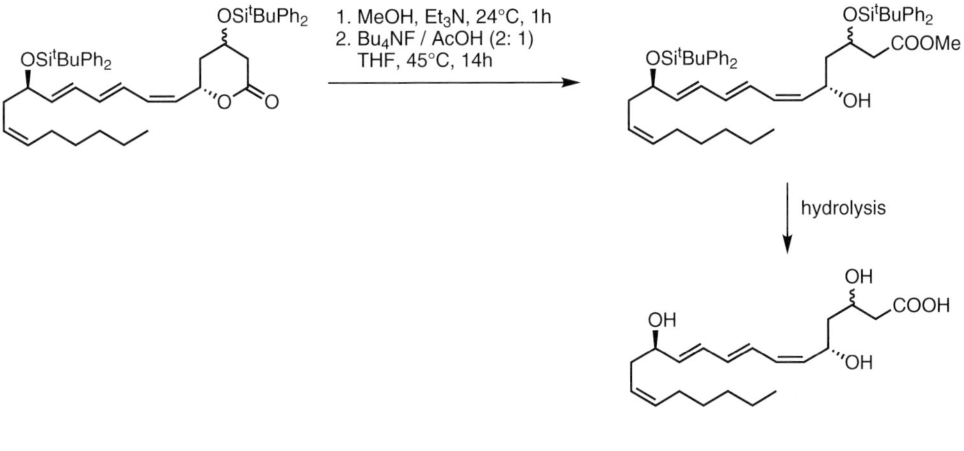

3-Hydroxyleukotriene B_4

Scheme 1.121

Et_3N constitutes the final key step in the synthesis of methyl ester of 3-hydroxyleucotriene B_4 (Scheme 1.121) [414].

Larger rings are quantitatively opened as well (Scheme 1.122) [415].

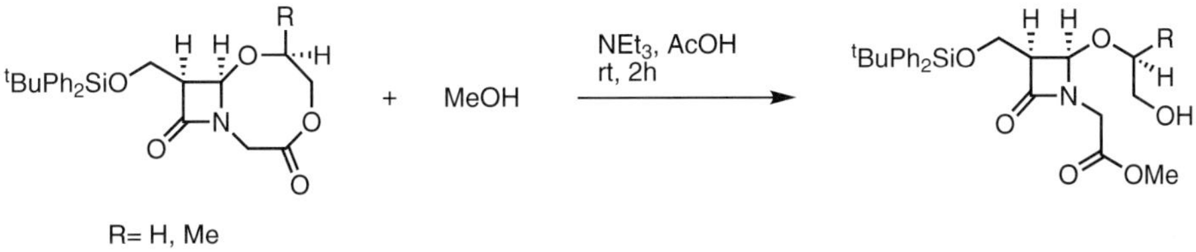

R= H, Me

Scheme 1.122

1-(Benzoyloxy)benzotriazole is a reactive acylation reagent. Selective benzoylation of a primary alcohol over a secondary one in glycols is feasible in the presence of Et_3N in CH_2Cl_2 at room temperature, while the 2-hydroxy groups in glucopyranoside and altropyranoside are selectively benzoylated under the same conditions (Scheme 1.123) [416]. The benzoates obtained are applicable to the synthesis of methyl mono- and di-*O*-α-L-rhamnopyranosyl-α-D-glucopyranosiduronic acids [417].

Experimental Procedure Scheme 1.123 [416] Triethylamine (305 μl, 2.2 mmol) is added at room temperature to a stirred solution of 1-phenyl-1,2-ethanediol (280 mg, 2.0 mmol) and 1-(benzoyloxy)benzotriazole (503 mg, 2.1 mmol) in methylene chloride (8 mL). The reaction mixture is stirred at room temperature for 24 h, diluted with methylene chloride (30 mL), washed with saturated $NaHCO_3$ solution (20 mL) and brine (20 mL), dried over anhydrous $MgSO_4$, and evaporated to dryness. The crude product is subjected to silica gel column chromatography with a hexane/ethyl acetate (6 : 1) mixture as an eluent to afford the dibenzoate (34 mg, 5%), the primary monobenzoate (390 mg, 83%), and the secondary monobenzoate (42 mg, 9%).

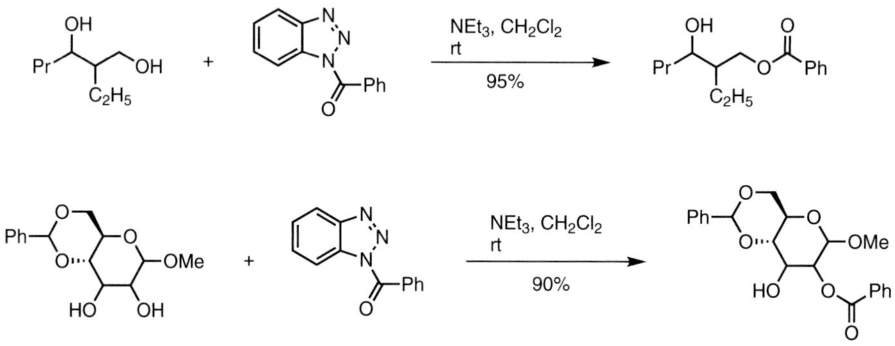

Scheme 1.123

Diisopropylamine is useful for ring contraction from δ-lactones to γ-lactones (Scheme 1.124) [418]. The reaction is conducted in the presence of 2 equiv. of the amine at room temperature to afford γ-lactones, which are converted into 5-methylated 2-deoxysugars.

Experimental Procedure Scheme 1.124 [418] Dry $ZnCl_2$ (4 equiv.) is added at room temperature and under nitrogen atmosphere to a stirred solution of epoxy ester in anhydrous CH_2Cl_2 (15 mL $mmol^{-1}$). When the reaction is complete (20 min), quenching is performed by addition of a saturated solution of $NaHCO_3$, and the mixture is extracted with CH_2Cl_2, dried over $MgSO_4$, and concentrated to give the crude product.

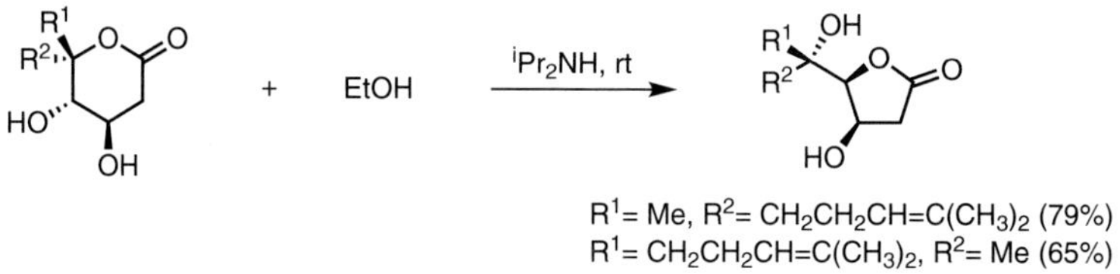

R^1= Me, R^2= $CH_2CH_2CH{=}C(CH_3)_2$ (79%)
R^1= $CH_2CH_2CH{=}C(CH_3)_2$, R^2= Me (65%)

Scheme 1.124

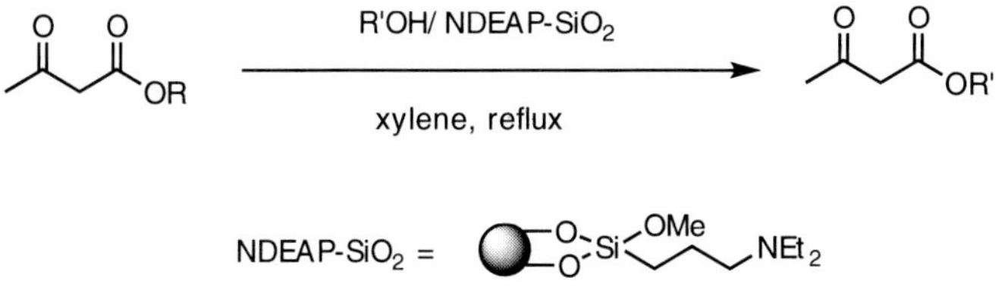

R'OH/ NDEAP-SiO2

xylene, reflux

NDEAP-SiO2 =

Scheme 1.125

Tertiary amine can be immobilized on silica gel (Scheme 1.125) [419]. For example, *N, N*-diethylaminopropylated silica gel (NDEAP-SiO_2) catalyzes transesterification of β-keto esters.

Strongly basic amines find broader applications. DMAP (4-dimethylaminopyridine) is one such. Methyl β-ketoesters are successfully transformed into the corresponding esters with higher alcohol components [420]. This reaction is versatile because it is workable for a variety of alcohols. Tertiary alcohols and non-enolizable β-keto esters are not employable, however, and a large excess (10 equiv.) of β-keto ester usually has to be used. This drawback can be offset by conducting the reaction in the presence of molecular sieves [421]. This new technique provides good yields of the desired esters from the use of β-keto ester, allylic alcohol, and DMAP in 1 : 1 : 1 ratio. It should be noted, though, that non-enolizable α,α-disubstituted keto esters are not employable (see Section 1.2.2.2). The DMAP procedure is applicable to synthesis of phosphonoacetates (Scheme 1.126) [422]. The ester exchange reaction takes place selectively at a carboxylic acid ester moiety rather than at a phosphonic acid ester moiety, and only enolizable phosphonoacetates react effectively with primary and secondary alcohols, but not with tertiary alcohols. This reaction is utilized for synthesis of (-)-pyrenophorin (Scheme 1.127) [423]. Subjection of a lactol to transesterification with phosphonoacetate followed by Horner-Emmons-Wadsworth olefination affords the desired macrolide.

Experimental Procedure [420] A flame-dried, two-necked, 50 mL, round-bottomed flask fitted with a reflux condenser and a nitrogen purge is flushed with N_2 and charged with L-menthol (300 mg, 1.92 mmol), DMAP (70 mg, 0.577 mmol), and toluene (6 mL). The mixture is magnetically stirred until the L-menthol and DMAP are dissolved, and methyl acetoacetate (0.62 mL, 5.77 mmol, 3.0 equiv.) is then added. The mixture is warmed at reflux for 42 h. The reaction mixture is cooled in an ice/water bath and quenched with saturated ammonium chloride solution (20 mL). Extracting solvent (20 mL) is added, and the two layers are separated. The aqueous layer is extracted three times with extracting solvent (25-mL portions). The combined organic layers are dried over anhydrous $MgSO_4$ and concentrated *in vacuo*. The residue is bulb-to-bulb distilled to remove excess methyl acetoacetate (bp_{50} 70 °C). The pot residue is then chromatographed on silica gel (10 g) with EtOAc/petroleum ether (3.0%). The first 30 mL is discarded. The next 60 mL is concentrated *in vacuo* to give L-menthyl acetoacetate (385 mg, 1.6 mmol, 83%) as a clear oil.

Experimental Procedure [421] A mixture of β-keto ester (5 mmol), allylic alcohol (5 mmol), and DMAP (5 mmol) is dissolved in sufficient toluene (ca. 200 mL) to ensure wetting of 25 g of oven-dried molecular sieves (4 Å). The mixture is then heated at reflux until no starting material is detectable by ^{1}H NMR spectroscopy; this typically requires 12–36 h. After being allowed to cool to room temperature, the solution is washed with saturated ammonium chloride (2 × 25 mL) and dried ($MgSO_4$). The toluene is removed by rotary evaporation, and the products are isolated by HPLC (1.5% EtOAc-98.5% Skelly B).

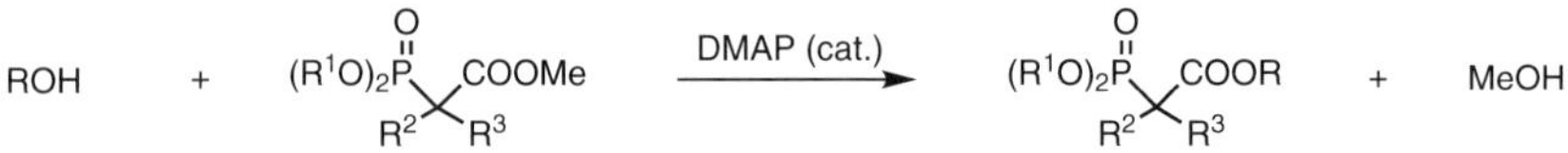

Scheme 1.126

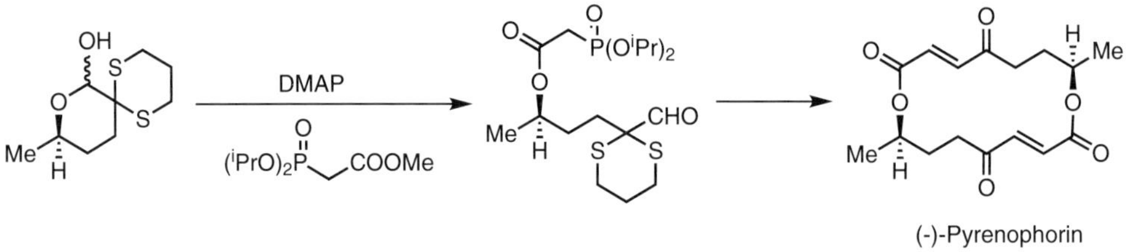

Scheme 1.127

Selective acetylation of a primary alcohol over a secondary alcohol or a phenol is achievable by the use of vinyl acetate in the presence of DMAP, although the yields are modest [424].

DBU (diazabicyclo[5.4.0]undec-7-ene) is another amine that serves for these purposes. Stirring 4-hydroxycyclopentanyl acetate in the presence of a catalytic amount of DBU and molecular sieves 4A in refluxing benzene furnishes a bicyclic lactone (Scheme 1.128) [425].

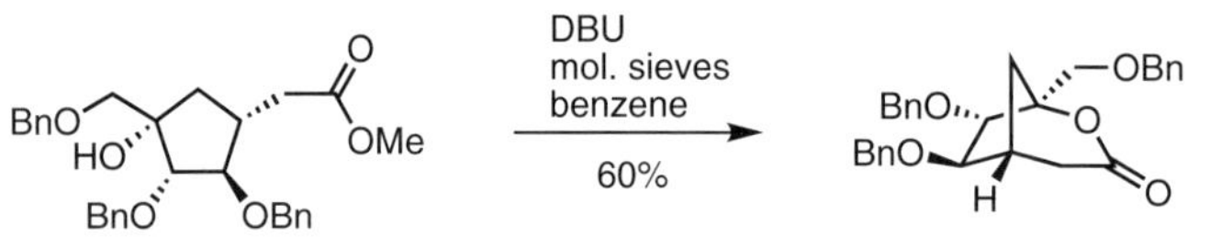

Scheme 1.128

Lactonization of 1-β-hydroxyalkylaziridine-2-carboxylic esters is effected by DBU to furnish 4-oxa-1-azabicyclo[4.1.0]heptane-5-ones (Scheme 1.129) [426].

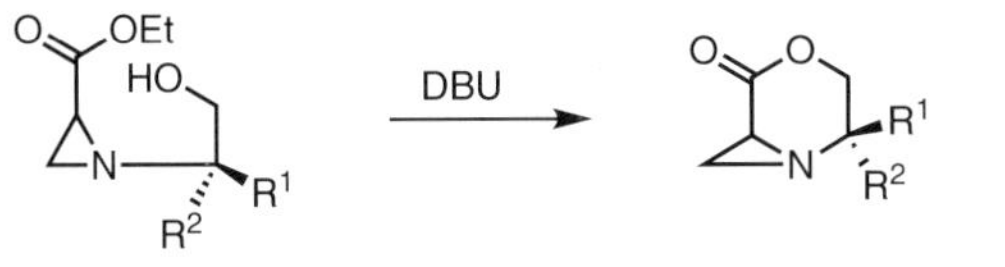

R¹= Et, R²= H (74%)
R¹= H, R²= ⁱPr (81%)
R¹= R²= H (64%)

Scheme 1.129

When stereoisomers of the spirooxazolone shown in Scheme 1.130 are treated with Et_3N in MeOH, only the endo cycloadduct undergoes methanolysis of the oxazolone ring and elimination, the exo isomer remaining unaltered. The use of DBU instead results in clean reaction for both isomers [427].

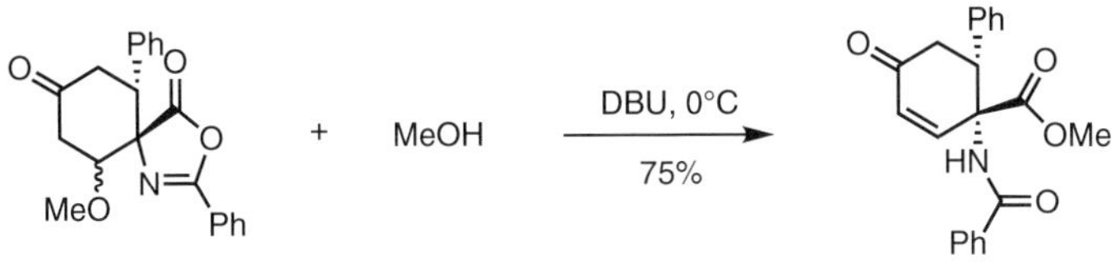

Scheme 1.130

Treatment of triol lactone with DBU results in ring contraction giving (+)-goniofurfurone (Scheme 1.131) [428].

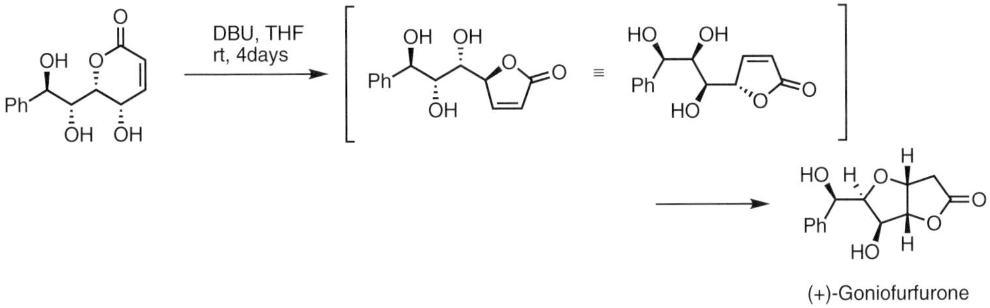

(+)-Goniofurfurone

Scheme 1.131

Concomitant lactone alcoholysis and dehydrochlorination are effected by treatment of 3,5-dichloro-2H-1,4-oxazin-2-ones with alcohol in the presence of DBU, affording 6-chloro-2-pyridinecarboxylic acid esters (Scheme 1.132) [429, 430].

Treatment of oxazolones with methanol/DBU results in ring opening (Scheme 1.133) [431]. Lactonization is induced by treatment of hydroxy ester derivatives of 4,5-*trans*-isoxazoline with DBU (Scheme 1.134) [432]. The 4,5-*trans*-isoxaziline substrate initially undergoes facile isomerization to the cis isomer, thus offering a convenient way to arrive at cis isoxazoline compounds.

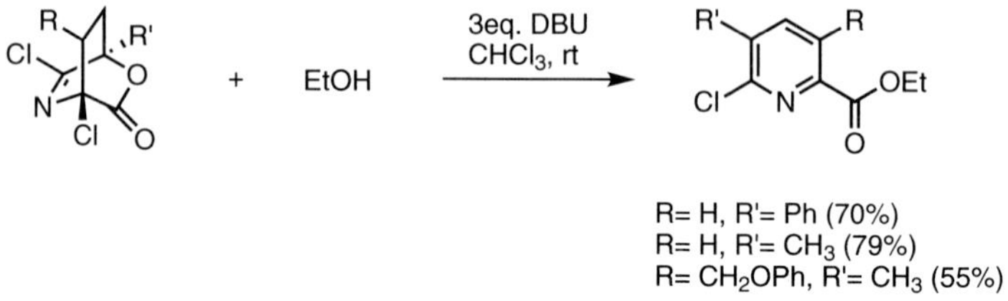

Scheme 1.132

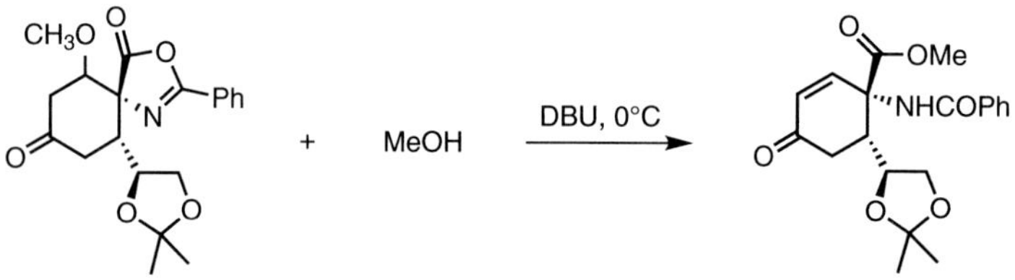

Scheme 1.133

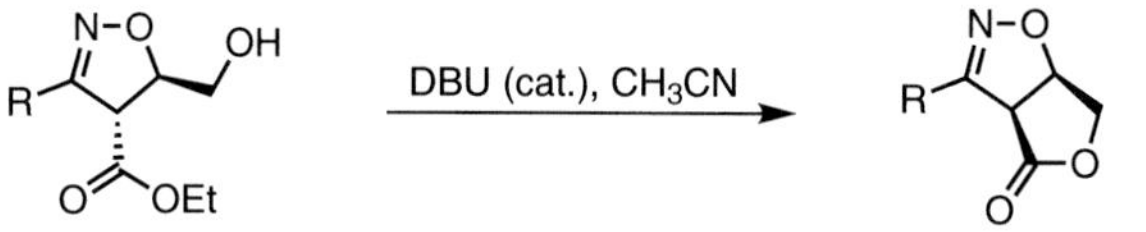

Scheme 1.134

As seen above, DBU is not always active enough to be used as a promoter for transesterification in a general manner. The basicity, and hence activity, are improved by combining it with LiBr [433]. Typically, the use of 0.5 equiv. of DBU together with 2–5 equiv. of LiBr in neat alcohol mediates transesterification of enolizable esters in high yields, except in the case of *tert*-butyl alcohol. Peptide esters are also employable, although elaboration to make the reaction conditions as mild as possible is needed; otherwise epimerization cannot be avoided. This technology is successfully applicable to detach peptides from PAM and Wang resins (Scheme 1.135).

Experimental Procedure Scheme 1.135 [433] LiBr (5 equiv.) and the starting carboxylate ester (1 equiv.) are dissolved under dry Ar in the appropriate quantity of the desired absolute alcohol, so that a 0.2–0.3 M concentration of substrate is obtained. Freshly distilled DBU (0.5 equiv.) is then added to this solution, and stirring is continued at room temperature, the course of the reaction being monitored by TLC or GC. When transesterification is complete, the mixture is evaporated and hydrolyzed with a sat. aq. NH_4Cl solution, or HCl (1 N). The product is extracted twice with Et_2O, the combined organic fractions are washed to neutrality with brine, dried (Na_2SO_4), and evaporated, and the crude product is purified by distillation or flash chromatography.

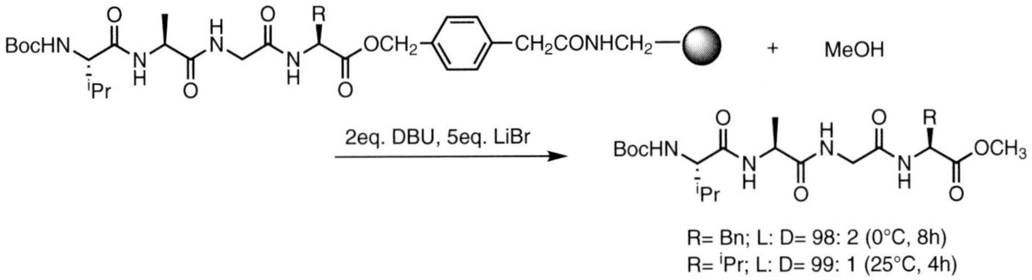

Scheme 1.135

This technique is also applicable to transesterification of the pentafluorophenyl dienoate shown in Scheme 1.136 [434].

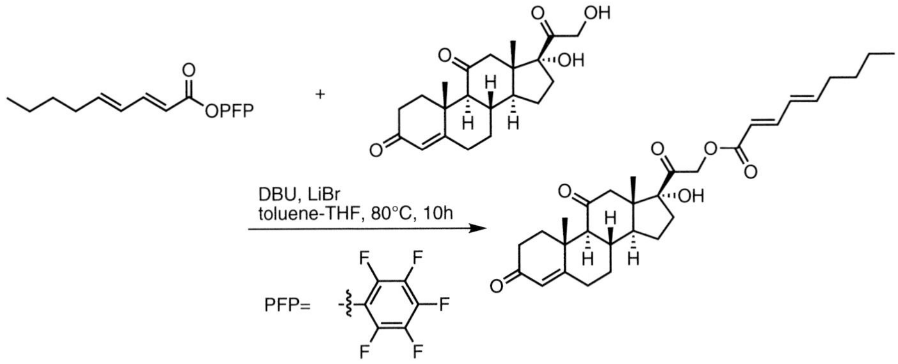

Scheme 1.136

Guanidine-mediated transesterification between acetylated carbohydrates and phenols with alcohol effects deprotection of these substrates [435]. This technique is applied to deacetylation of 3-acetoxycyclopentenone (Scheme 1.137) [436]. However, straightforward application of this procedure to 2-deoxy-2-aminosugars with 2,2,2-trichloroethoxycarbonyl groups results in decomposition of this group to the corresponding carbamate. The use of guanidine/guanidine nitrate in methanol, on the other hand, brings about the desired deacetylation (Scheme 1.138) [437].

Experimental Procedure Scheme 1.137 [436] A stock solution of 0.5 M guanidine in CH_3OH is prepared by addition, under argon atmosphere, of hexane-washed (3×) sodium spheres (1.78 g, 77.4 mmol) to ice-cooled CH_3OH (154 mL). When all the sodium has reacted, guanidine carbonate (14.2 g, 79.0 mmol) is added. This solution is stirred at room temperature for 25 min, and the mixture is allowed to stand, to settle out precipitated salts. Methyl 7-[3-(*R*)-(acetyloxy)-5-oxo-1-cyclopenten-1-yl]-4(*Z*)-heptanoate (93 : 7 *R*/*S* ratio, 12.8 g, 45.6 mmol) in absolute CH_3OH (50 mL) is placed under argon in a separate flask. This is cooled to 0 °C in an ice bath, and the guanidine in CH_3OH prepared above (0.5 M, 100 mL) is added to it by syringe over ~5 min. This mixture is stirred at ~10 °C for 5 min. TLC (80% ethyl acetate in hexane on silica gel) shows complete consumption of acetate. Glacial acetic acid (2.86 mL, 3.0 g, 50.0 mmol) is then added to the reaction mixture to neutralize the guanidine. After the mixture has been stirred for 5 min, solvent is removed at reduced pressure to give a thick slurry. The residue is partitioned between water (100 mL) and toluene/ethyl acetate (1 : 1 v/v, 100 mL). The aqueous layer is further extracted with ethyl acetate (2 × 50 mL). The combined organic layers are washed with water (2 × 50 mL) and brine (50 mL) and dried over sodium sulfate. Removal of solvent at reduced pressure gives a deep amber oil, which is purified by flash chromatography on silica gel with 50% ethyl acetate in hexane to give methyl 7-(3(*R*)-hydroxy-5-oxo-1-cyclopenten-1-yl)-4(*Z*)-heptenoate (8.06 g, 77%) after exhaustive removal of solvent.

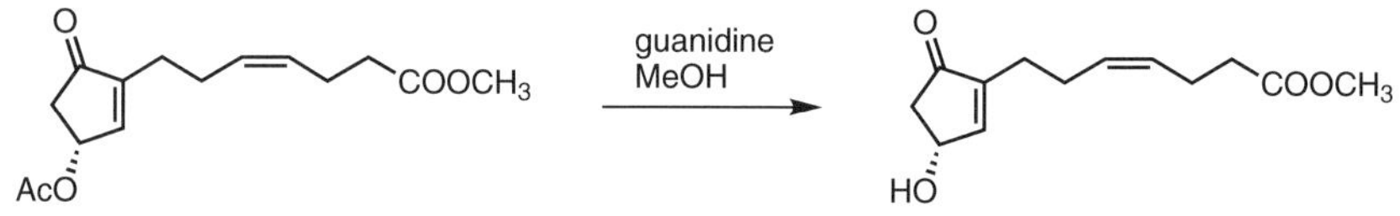

Scheme 1.137

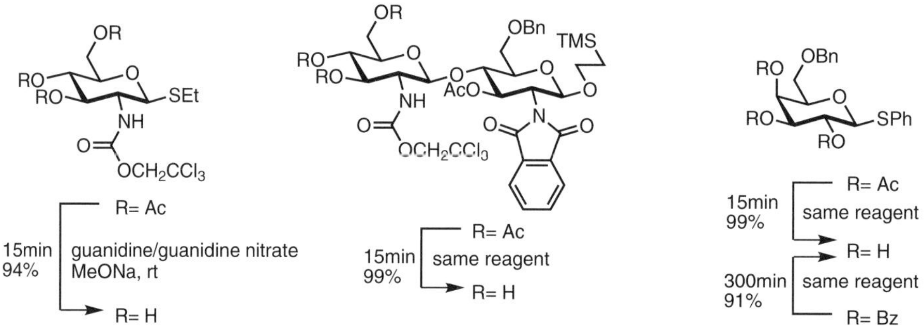

Scheme 1.138

Imidazole is employable as a promoter for lactonization, as shown in Scheme 1.139 [438]. The reaction in CH_3OH at 100 °C allows complete differentiation between the two hydroxy groups. Diisopropenyl oxalate, obtained by addition of

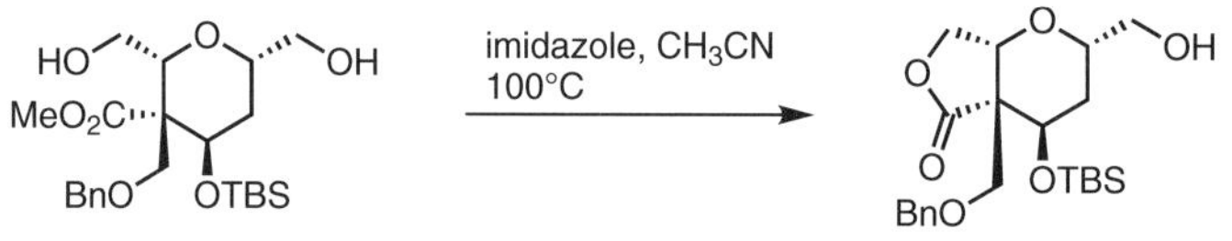

Scheme 1.139

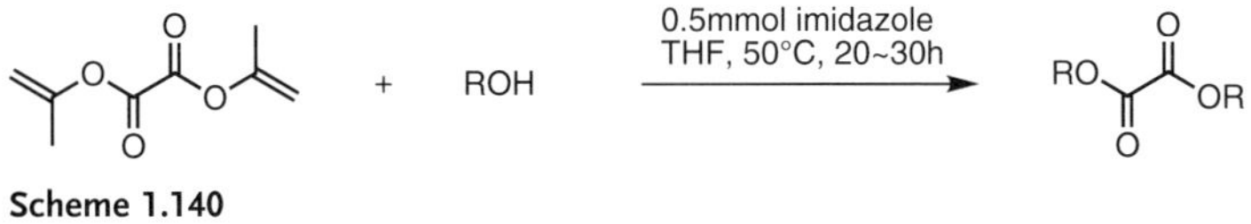

Scheme 1.140

oxalic acid to propyne, is transformed into various oxalate esters upon treatment with alcohol in the presence of imidazole (Scheme 1.140) [439].

1.2.3.3 **Others**

Treatment of a chiral β-lactone with alcohols in the presence of LDA provides chiral half-esters (Scheme 1.141) [440]. Octanol, cyclohexanol, 3-pentanol, farnesol, and gymnoprenol are successfully incorporated by use of 2 equiv. of LDA.

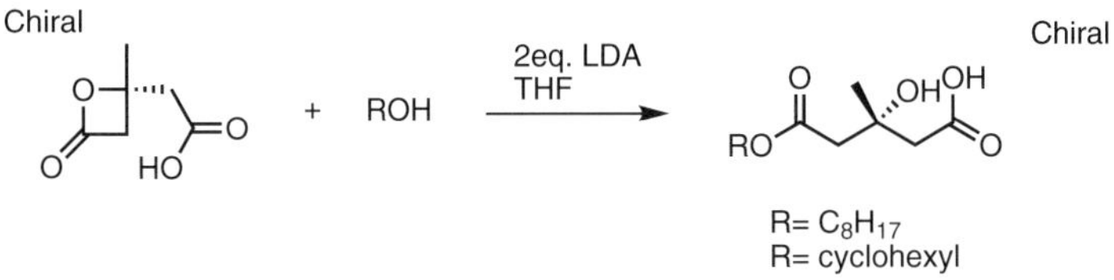

Scheme 1.141

Strongly basic phosphorus compounds are also useful. For example, nonionic superbases of the type $P(RNCH_2CH_2)_3N$ catalyze transesterification of various carboxylic acids with primary, secondary, and allylic alcohols [441]. Acylation of alcohols with enol esters is also feasible. A broad range of substituents – such as epoxide, carbamate, acetal, oxazoline, nitro, and alkynyl groups – are tolerated. *N*-Protected peptides undergo clean transesterification without significant racemization. The iminophosphorane bases $PhCH_2N{=}P(MeNCH_2CH_2)_3N$ and $PhCH_2N{=}P(NMe_2)_3$ catalyze acylation of primary alcohols with enol esters as well [442]. Since these catalysts are not effective for secondary alcohols, these alcohols can be discriminated from primary ones. Immobilization of these catalysts does not gives rise to any decrease in catalytic activity (Scheme 1.142). The new catalysts can be used three times without significant loss of product yield, although the catalyst beads become a fine powder that is more difficult to isolate.

Experimental Procedure [441] Ester (1 mmol) is added at room temperature to a stirred solution of $P(RNCH_2CH_2)_3N$ (R = Me or *i*-Pr, 10–15 mol%) in the alcohol (5 mL). The mixture is stirred for 4–24 h and the alcohol is then evaporated *in vacuo*. The residue is dissolved in diethyl ether and passed through a pad of silica. Solvent evaporation under vacuum gives the product, which is found to be pure by ^{1}H NMR analysis.

Experimental Procedure Scheme 1.142 [442] Enol ester (5 mmol) is added at room temperature to a stirred solution of iminophosphorane base (10 mol %) and alcohol (1 mmol) in THF (0.5 mL). The mixture is stirred at room temperature for 14–38 h, and the solvent and excess enol ester are evaporated *in vacuo*. The residue is purified by column chromatography on a small pad of silica gel, with 0–20% ethyl acetate in hexane as eluent. When polymer-supported catalyst is employed, the polymer is filtered off after the completion of the reaction and washed with ether. The solvent is then evaporated *in vacuo*, and the residue is purified by column chromatography on a small pad of silica gel with 0–20% ethyl acetate in hexane as eluent.

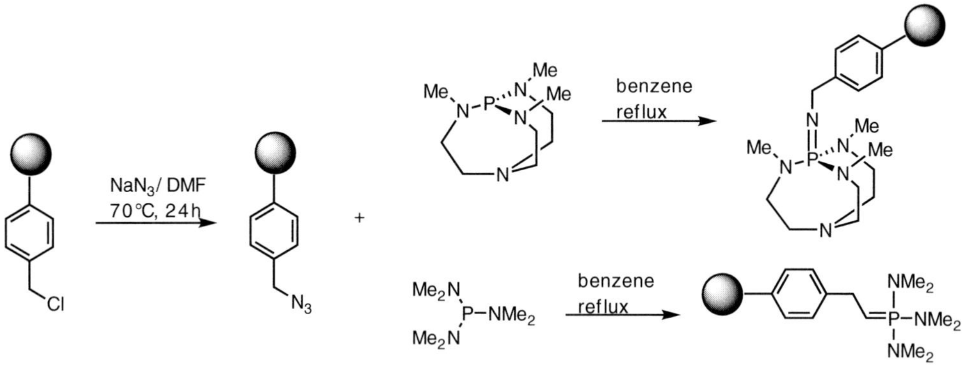

Scheme 1.142

Tris(2,4,6-trimethoxyphenyl)phosphine is utilized for deacetylation [443]. Peracetylated sugars and glucals are deacetylated completely upon treatment with the phosphine in methanol. 2,2,2-Trihaloethylesters are transesterified by the action of Bu_3P/DMAP (Scheme 1.143) [444]; no reaction takes place in the absence of phosphine. The reaction proceeds via an acyloxyphosphonium intermediate, so transesterification does not involve Mitsunobu-type inversion.

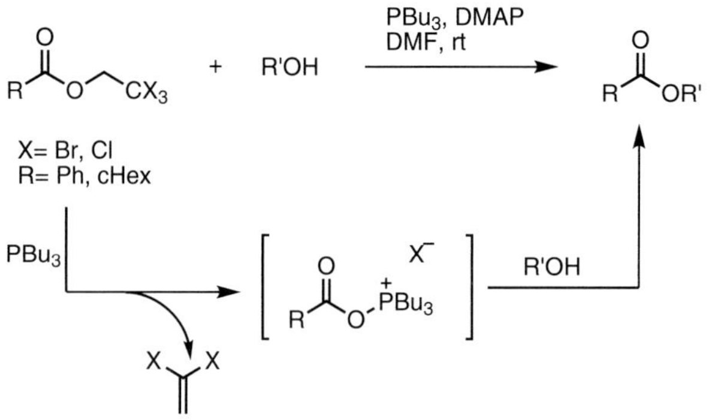

Scheme 1.143

A strong nonionic base, P1 phosphazene, catalyzes selective transesterification between glycerol derivatives and fatty acid methyl esters (Scheme 1.144) [445]. Use of polyols in excess (2–5 equiv.) provides monoesters, while diesters are produced with 0.5 equiv. of polyols.

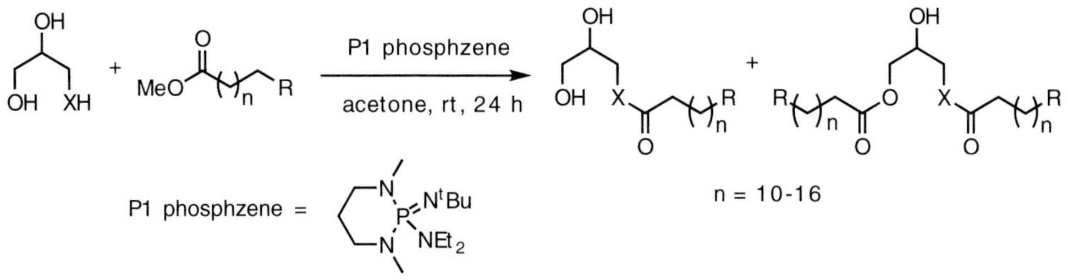

Scheme 1.144

Amberlite IRA 400 (OH) mediates methanolysis of a lactone to give a quantitative yield of *syn*-1,3-diol (Scheme 1.145) [446].

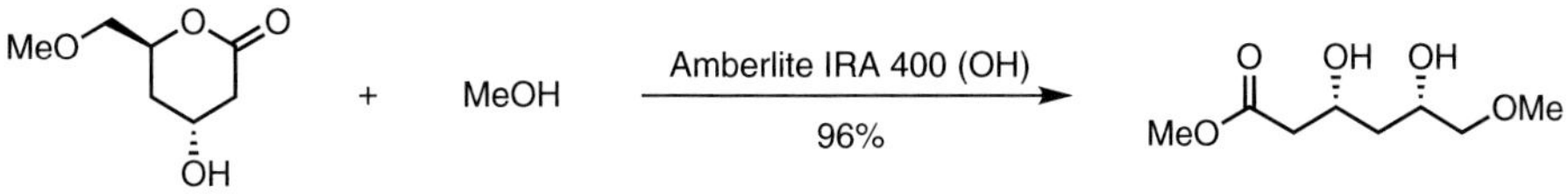

Scheme 1.145

A benzyl to methyl ester transformation of *cis*-octahydroindolone is performed efficiently with the aid of DOWEX 1 × 8, a strongly basic anion exchange resin (Scheme 1.146) [447].

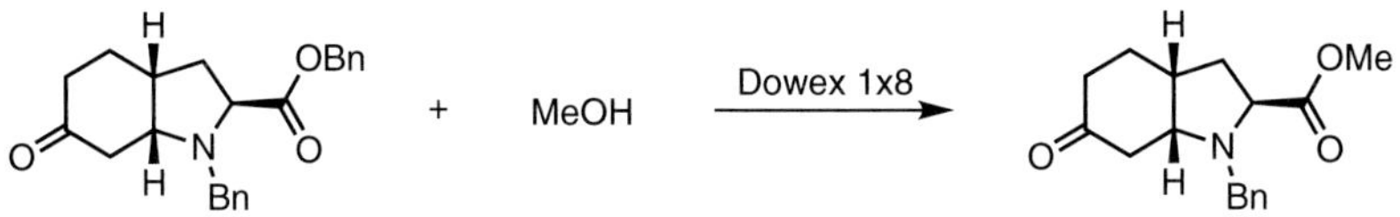

Scheme 1.146

Diphenylammonium triflate, an excellent catalyst for esterification between carboxylic acids and alcohols (see Section 1.1.2.1), also catalyzes transesterification, 1.5 equiv. of alcohol being necessary [18]. Addition of 0.1 equiv. of TMSCl improves catalytic activity.

1.2.4
Other Activators

Transesterification takes place when mixtures of β-keto esters and alcohols are heated in toluene in the presence of NBS (*N*-bromosuccinimide) [448]. NBS is only effective for β-keto esters, no reaction occurring with other esters. N_2O_5 [449] and $PdCl_2/CuCl_2$ are also good catalysts for transesterification of β-keto esters [450].

Samarium/iodine is frequently used in reduction of various functionalities. During these reactions, coexisting esters occasionally undergo transesterification. When conjugate reduction of α, β-unsaturated esters is performed with these reagents in methanol, the ethyl esters are transformed into the corresponding methyl esters (Scheme 1.147) [451]. Reduction of azides [452] and aromatic nitro compounds [453] to the corresponding amines is accompanied by transesterification of coexisting ester functions.

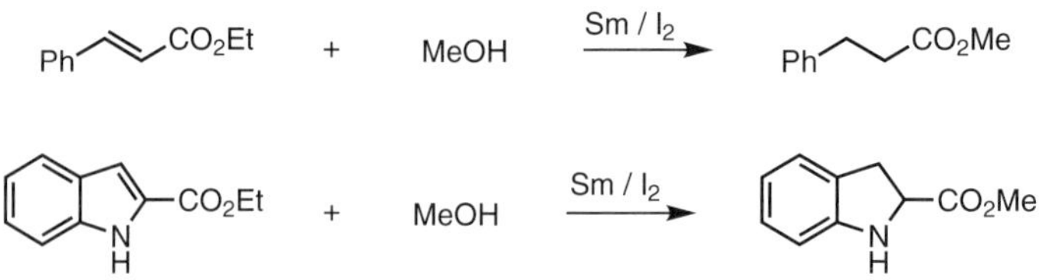

Ph–CH=CH–CO_2Et + MeOH —(Sm / I_2)→ Ph–CH_2CH_2–CO_2Me

indole-2-CO_2Et + MeOH —(Sm / I_2)→ indoline-2-CO_2Me

Scheme 1.147

N-Heterocyclic carbenes are organocatalysts effective for transesterification (Scheme 1.148) [454]. Use of *in situ*-generated carbenes can overcome difficulties associated with handling of the labile carbenes, although the yields of the desired products are lower. The reaction in ionic liquid solvent improves the yield.

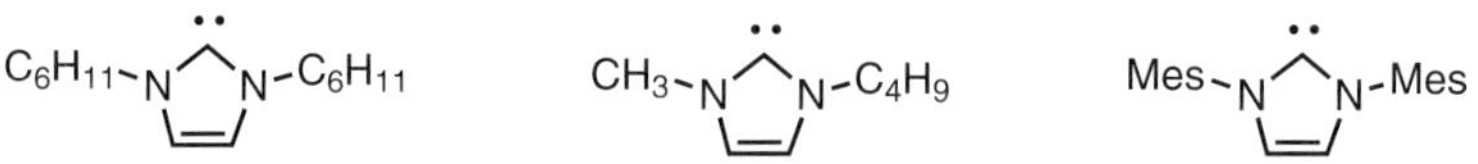

C_6H_{11}–N(imidazol-2-ylidene)N–C_6H_{11} CH_3–N(imidazol-2-ylidene)N–C_4H_9 Mes–N(imidazol-2-ylidene)N–Mes

Scheme 1.148

1.2.5 Enzymes

Transesterification is one of reactions for which enzymes are utilized most effectively. This technology has experienced explosive expansion in the last decade. As in the case of esterification, their utilization was boosted by the development of enzyme-catalyzed reactions in organic solvents. The enzymatic reaction is highly specific to individual reactions and enzymes. In contrast to yeast lipase (see Section 1.1.7), PPL (porcine pancreatic lipase) is not effective for esterification. However, kinetic resolution of secondary alcohols with this lipase by transesterification with 2,2,2-trichloroethyl esters is feasible (Scheme 1.149) [264]. PPL is quite flexible with respect to experimental conditions. The enzymatic process can be carried out in a variety of solvents such as paraffins, toluene, carbon tetrachloride, ether, acetone, acetonitrile, dioxane, and others. In all instances the presence of even a small amount of water severely suppresses the enzymatic transesterification. The lipase can be used as a catalyst over a wide temperature range: the enzymatic transesterification accelerates by more than sevenfold upon a temperature increase from 25 to 60 °C. The enzyme loses virtually no catalytic activity after each transesterification performed at 25 °C and less than a quarter of its initial activity if the reaction is carried out at 60 °C. The PPL/2,2,2-trichloroethyl ester technique is applicable for the preparation of monoacylated sugars [455]. Only the primary hydroxy group can be acylated in pyridine at 45 °C (Scheme 1.150).

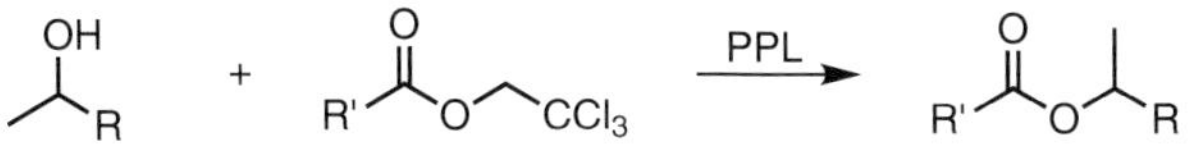

Scheme 1.149

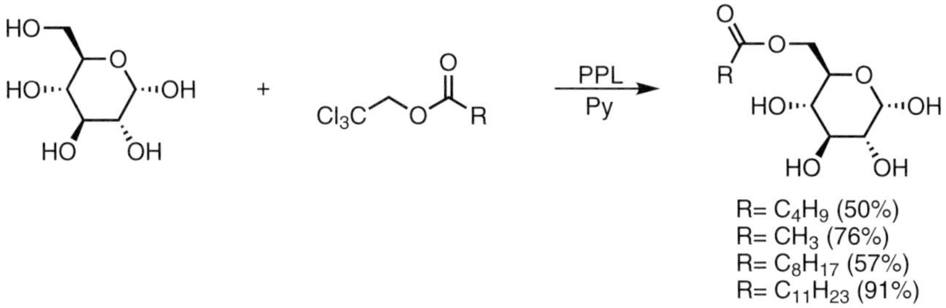

R= C_4H_9 (50%)
R= CH_3 (76%)
R= C_8H_{17} (57%)
R= $C_{11}H_{23}$ (91%)

Scheme 1.150

As well as PPL, lipase P from *Pseudomonas sp.* can also be used for lactonization of methyl esters of ω-hydroxycarboxylic acids (Scheme 1.151) [265]. Subjection of chiral methyl 10-hydroxyundecanoate to the same reaction conditions produces no monolide but a small amount of diolide. However, γ-methyl- and γ-phenylbutyrolactones are obtainable in optically pure form with PPL (Scheme 1.152) [456]. The efficiency of lactone formation from methyl ω-hydroxy carboxylates with lipase Amano P is enhanced by addition of molecular sieves 4A, which serve for removal of the methanol formed [457].

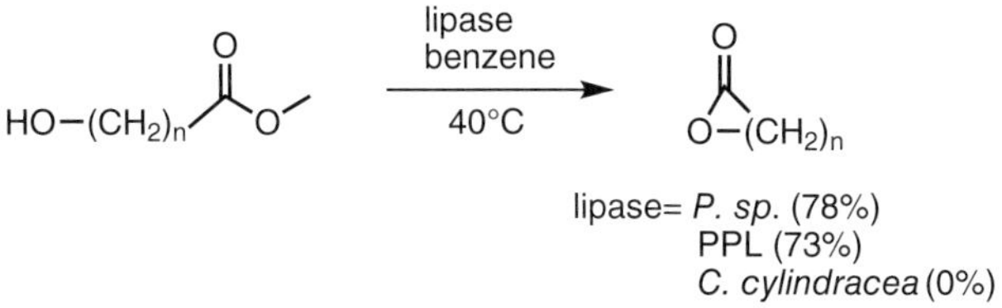

lipase= *P. sp.* (78%)
PPL (73%)
C. cylindracea (0%)

Scheme 1.151

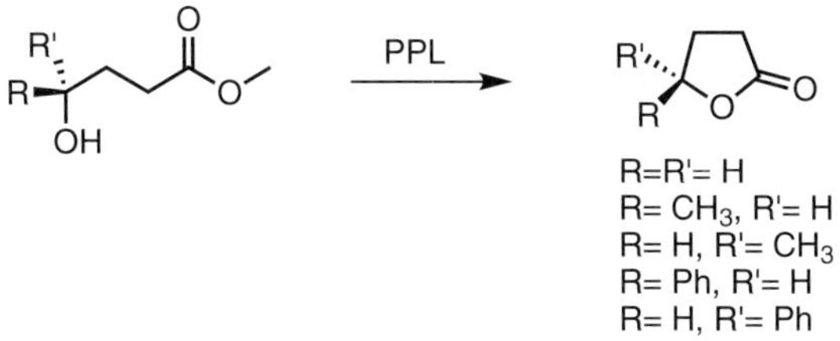

R=R'= H
R= CH_3, R'= H
R= H, R'= CH_3
R= Ph, R'= H
R= H, R'= Ph

Scheme 1.152

There is a unique regioselectivity between primary alcohols [458]. The 4-hydroxy group of the diol depicted in Scheme 1.153 undergoes exclusive esterification by isopropenyl acetate with the aid of *Candidia cylindracea* lipase, while α,α′-alkenediols are acetylated to give *E* isomers selectively with *Pseudomonas cepacia* [459].

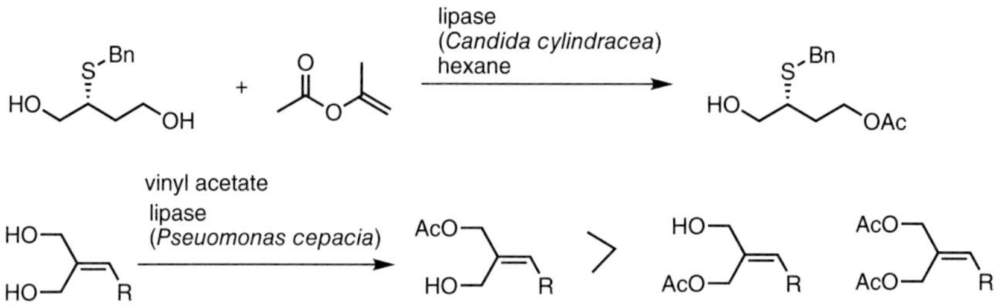

lipase (*Candida cylindracea*) hexane

vinyl acetate lipase (*Pseuomonas cepacia*)

Scheme 1.153

Immobilization of lipase is possible. Thus, CALB (*Candida antarctica* lipase B) attached to a macroporous resin effects transesterification of β-keto esters (Scheme 1.154) [460]. Primary, secondary, allylic and propargylic alcohols are employable. The kinetic resolution of secondary alcohols gives the esters with >90% *ees*. Various kinds of other supports are available: polyurethane foams for *Candida rugosa* [461], polypropylene for *Thermomyces lanuginosus* [462], phyllosilicate sol-gel for *Pseudomonas cepacia* and *Thermomyces lanuginosus* [463], and ceramic particles for *Pseudomonas cepacia* [464]. Commercially available immobilized lipases are also employable [465].

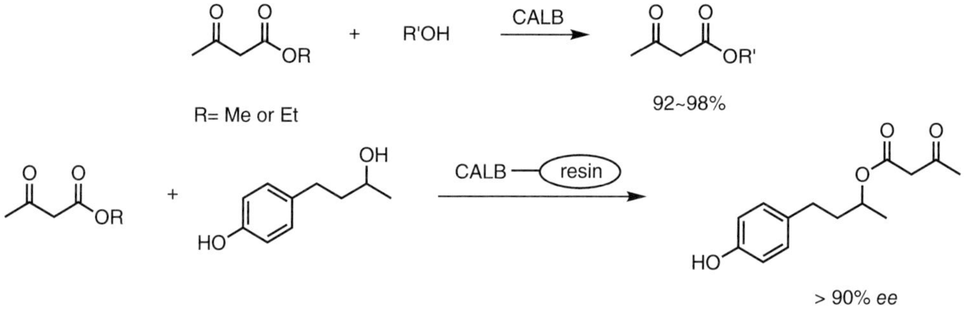

R= Me or Et

92~98%

> 90% *ee*

Scheme 1.154

Immobilization improves the heat resistance of lipases, but reduces the enzymatic activity. This conflict can be counterbalanced with powdered enzymes [466]. The use of cyclodextrins also enhances reaction rates and enantioselectivity of lipase-catalyzed transesterification in organic solvents [467].

The enantioselectivity in kinetic resolution of racemic sulcatol by PPL in ether is increased tenfold by dehydration of the enzyme and use of 2,2,2-trifluoroethyl laurate as the ester (Scheme 1.155) [468].

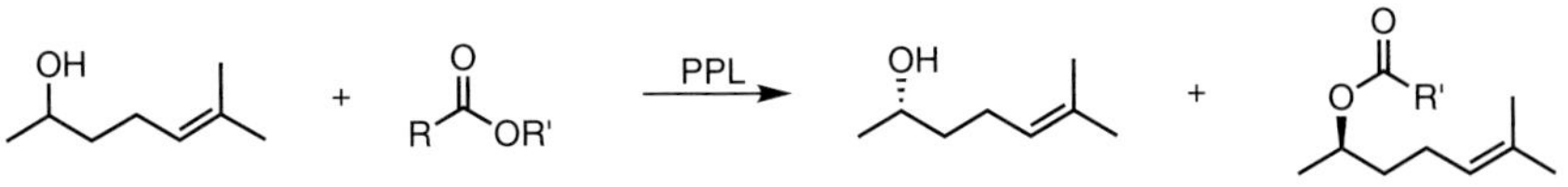

R	R'	E
C_3H_7	$-CHClCHCl_2$	24
	$-CHFCHF_2$	22
C_7H_{15}	$-CHClCHCl_2$	24
	$-CHFCHF_2$	19
$C_{11}H_{23}$	$-CHClCHCl_2$	27
	$-CHFCHF_2$	100

Scheme 1.155

2-Amino alcohols are transesterified successfully, *N*-alkoxycarbonyl derivatives of 2-amino alcohols undergoing kinetic resolution by mammalian lipases in ethyl acetate (Scheme 1.156) [469]. 2-Halo-1-arylethanols are also the substrates that can be successfully kinetically resolved by lipase Amano P from *Pseudomonas fluorescens* [470].

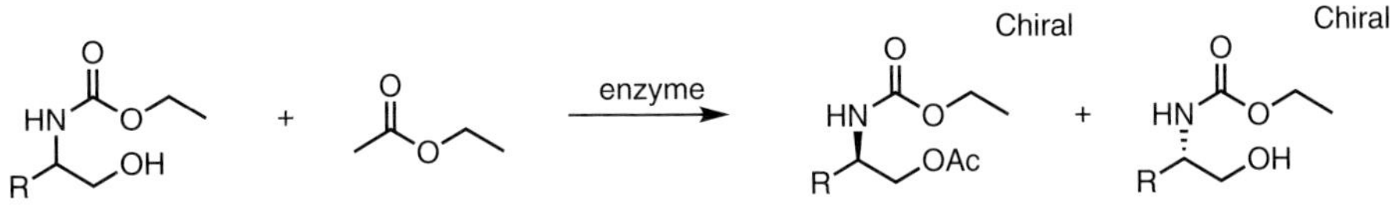

R= Et
R= Me
enzyme= steapsin/Celite or steapsin or pancreatin etc.

Scheme 1.156

The use of amino alcohols in enzyme-catalyzed transesterification is occasionally convenient when the presence of other alcohols results in difficulties in workup operations. Kinetic resolution of racemic methyl *trans*-3-(4-methoxyphenyl)glycidate (a key intermediate for the synthesis of the well-known drug diltiazem hydrochloride) is accomplished with suitable amino alcohols and catalysis by CALB in organic solvents (Scheme 1.157) [471]. The use of normal alcohols in this procedure results in difficulty in separation because of the insolubility of the products.

Protein engineering is useful for improving the activity of lipase. Although 1-halo-2-alkanols are poorly resolved by CALB, mutants display high efficiency of kinetic resolution with vinyl ester [472].

Transesterification reactions between polyesters can be used to prepare copolymers [473]. Transacylation reactions between polyesters are believed to involve intrachain cleavage by the lipase to form an enzyme-activated chain segment,

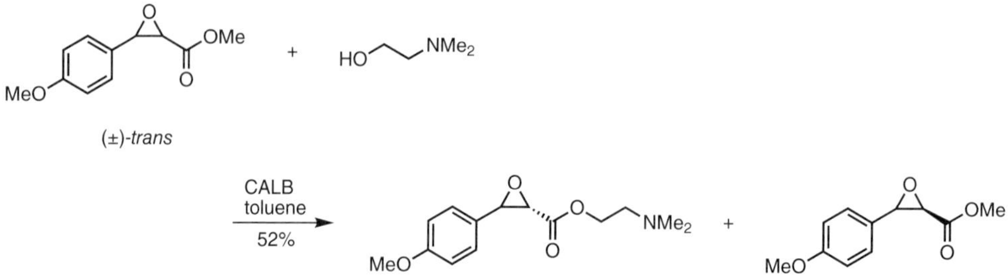

(±)-*trans*

CALB toluene 52%

Scheme 1.157

followed by reaction of this activated segment with the terminal hydroxyl unit of another chain (Scheme 1.158).

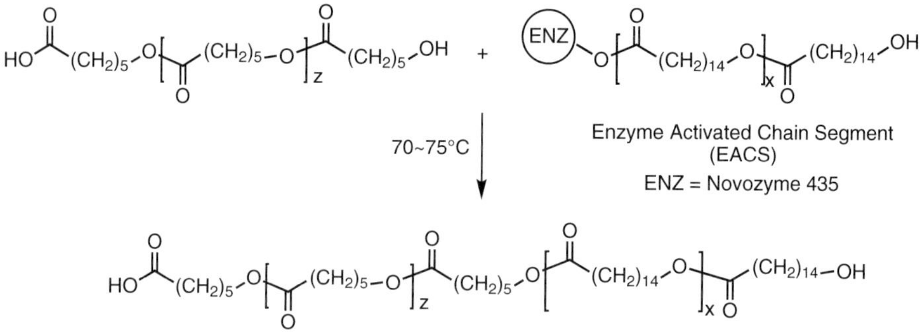

Enzyme Activated Chain Segment (EACS)

ENZ = Novozyme 435

70~75°C

Scheme 1.158

A sequential application of a polymer-supported catalyst for kinetic resolution and a scavenging reagent provides a convenient method for the isolation of chiral secondary alcohols (Scheme 1.159) [474]. After the standard kinetic resolution of racemic secondary alcohols with polymer-supported CALB, the lipase is filtered off. The filtrate containing the ester and unreacted alcohol is then treated with polymer-supported benzoyl choride as a scavenger. Upon filtration, the ester is obtained from the filtrate while hydrolysis of the resin releases the alcohol.

Lipase-catalyzed transesterification in ionic liquids proceeds with markedly enhanced enantioselectivity: the lipases such as CALB and *Pseudomonas cepacia* lipase are up to 25 times more enantioslective in ionic liquids than in conventional organic solvents [475]. Extraction of the CALB-promoted reaction mixture with supercritical carbon dioxide enables facile separation of the products from the lipase/ionic liquid mixture [476]. More detailed description of the use of ionic liquid is given in Section 7.3.

A combination of enzyme and phosphine catalyst allows parallel kinetic resolution [477]. A three-phase system consists of ChiroCLEC-PC (a commercial cross-linked lipase acylation catalyst), polymer-bound mesitoyl anhydride as an insoluble acyl donor, and a soluble chiral phosphine catalyst, together with vinyl pivalate as

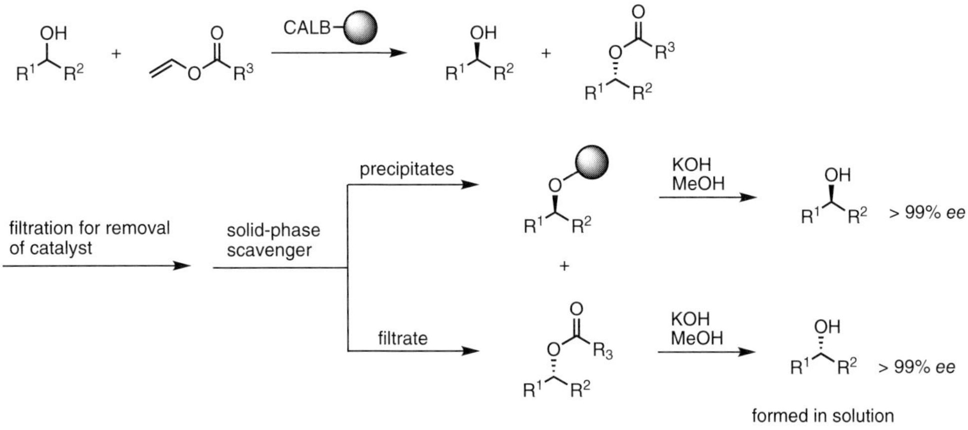

Scheme 1.159

a soluble acyl donor (Scheme 1.160). Such phase separation provides one way to control which reagent is activated by a given catalyst.

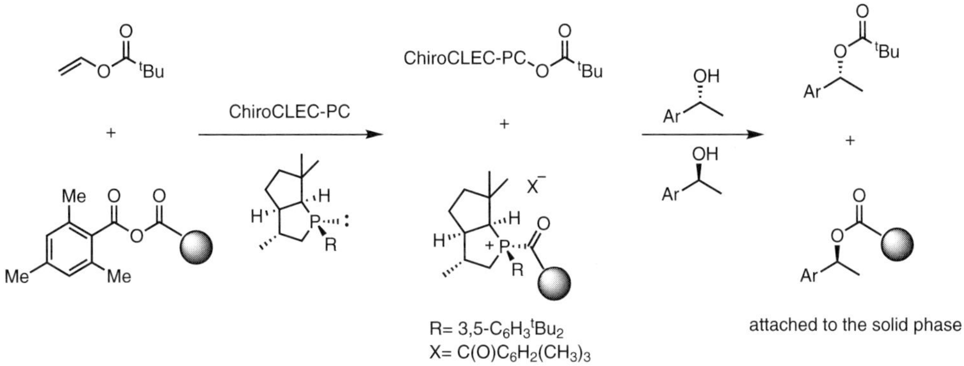

Scheme 1.160

The combination of lipase and artificial catalyst can also provide dynamic kinetic resolution [478]. Treatment of racemic β-hydroxy nitriles with CALB (Novozym-435) and a ruthenium catalyst affords the corresponding (*S*)-acetates in 85% yield with >94% *ee* (Scheme 1.161). The isomerization of the alcohol is catalyzed by the ruthenium catalyst without deterioration in the lipase activity.

Experimental Procedure The ruthenium catalyst (10.8 mg, 4 mol%) and Novozym-435 (20 mg) are placed under argon in a Schlenk flask. A solution of 3-hydroxy-3-phenylpropanonitrile (112 mg, 0.6 mmol) in dry toluene (2 mL) is added under argon. The reaction mixture is stirred at 100 °C for 36 h. The enzyme is filtered off and washed with toluene (3 × 5 mL). The organic layer is evaporated to leave the desired product.

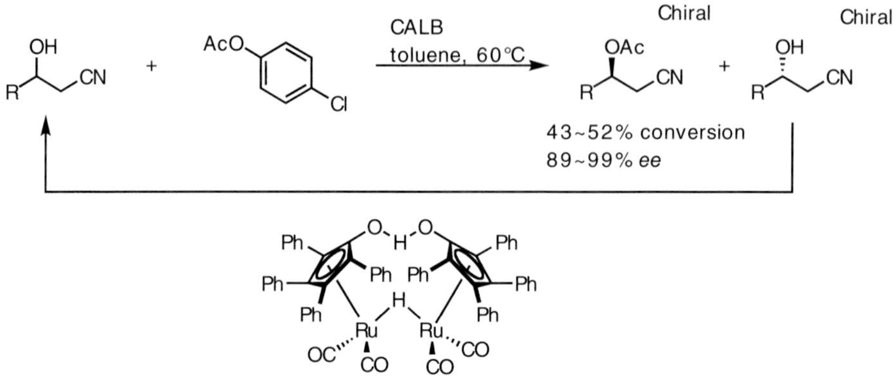

Scheme 1.161

Esterase catalyzes transesterifications carried out in biphasic aqueous/organic mixtures [479]. The aqueous/organic biphase procedure is effective for acylation of glucopyranosides with vinyl acetate catalyzed by *Trichoderma reesei* acetyl esterase [480]. It should be noted, however, that esterification of sucrose derivatives can be carried out in organic solvents as well [481, 482]. Water-insoluble substrates occupy the organic phase, while the enzyme is situated in the aqueous phase. By use of porous supports (Sepharose or Chromosorb) filled with aqueous solutions of hog liver carboxyl esterase as a stereoselective catalyst and methyl propanoate as a matrix ester, the optically active alcohols and their propanoic esters can be produced on a preparative scale. Kinetic resolution of 2-arylpropanoic acid ester is effected by rabbit liver esterase [483].

The proteolytic enzyme subtilisin is catalytically active in anhydrous dimethylformamide. A number of carbohydrates and other sugar-containing compounds can be regioselectively acylated by enzymatic transesterification through taking advantage of the unique dissolving potency and broad substrate specificity of subtilisin (Scheme 1.162) [484]. Monobutyryl esters of the disaccharides maltose, cellobiose, lactose, and sucrose are readily prepared on a gram scale. The presence of a bulky aglycon moiety does not substantially reduce the catalytic efficiency of subtilisin in dimethylformamide, thus permitting preparative-scale enzyme esterification of natural compounds such as riboflavin, salicin, and the nucleosides adenosine and uridine. In addition to the butyryl group, various *N*-acetylamino acid residues can also be introduced onto sugars.

A subtilisin mutant derived from subtilisin BPN′ is 100 times more stable than the wild-type enzyme in aqueous solution at room temperature and 50 times more stable than the wild types in anhydrous dimethylformamide [485]. The mutant enzyme can be applied to regioselective acylation of nucleosides in anhydrous dimethylformamide with 65–100% regioselectivity.

The use of pyridine in place of dimethylformamide allows regioselective esterification of the primary hydroxy groups of ribonucleosides [486]. By this procedure,

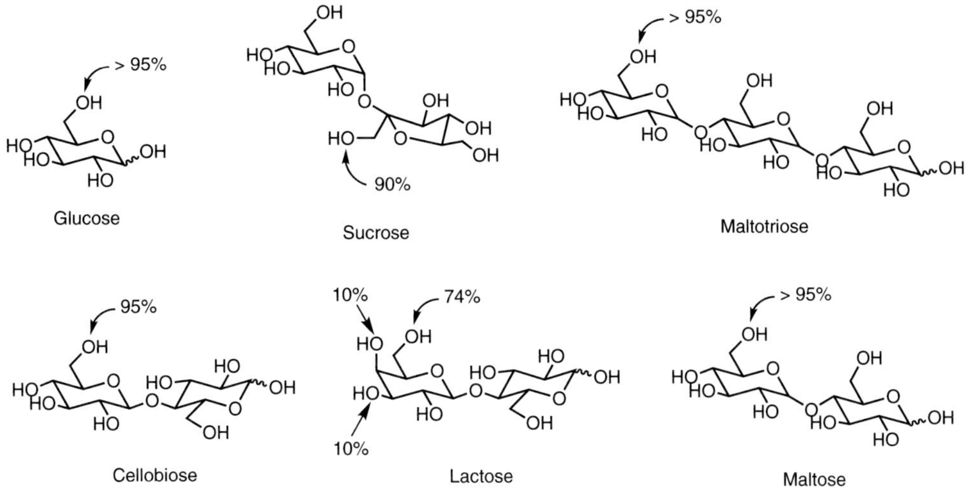

Glucose Sucrose Maltotriose

Cellobiose Lactose Maltose

Scheme 1.162

5′-*O*-acylribonucleosides are obtained in high yields with crude subtilisin and even with cheaper crude protease Proleather.

Cross-linked enzyme crystals of subtilisin, prepared by cross-linking of the crystallized enzyme with glutaraldehyde, are more stable than the original in both aqueous and mixed aqueous/organic solutions [487]. This enzyme is used for transesterification.

The acylase from *Aspergillus* species catalyzes acylation of alcohols with vinyl acetate [488]. This technique can be applied to highly enantioselective kinetic resolution of secondary alcohols [489].

Consecutive enzymatic reactions in one pot are feasible for chemically labile cyanohydrin derivatives (Scheme 1.163) [490]. Racemic *O*-acetylcyanohydrins are resolved by treatment with propanol in the presence of CALB, but isolation of deprotected *S* cyanohydrins as their THP or TBS ethers results in decreases of *ee*. This problem can be solved by esterification *in situ*. After the kinetic resolution, addition of vinylbutyrate to the reaction mixture affords butyrate of the *S* cyanohydrin without any decrease in *ee*.

Experimental Procedure Immobilized Lipase B from *Candida antarctica* CAL-B (368 mg) is added under nitrogen at 25 °C to a solution of *O*-acetylcyanohydrin (3.68 mmol) in dry toluene (36 mL). The reaction mixture is heated to 60 °C and 1-propanol (0.54 mL, 7.36 mmol) is then added. After the mixture has been stirred for 3 h, vinyl butyrate (2.8 mL, 18.4 mmol) is added. The mixture is stirred overnight and filtered. The organic layer is washed with sat. $NaHCO_3$ (25 mL) and dried. Evaporation followed by column chromatography affords the corresponding acetate and butyrate.

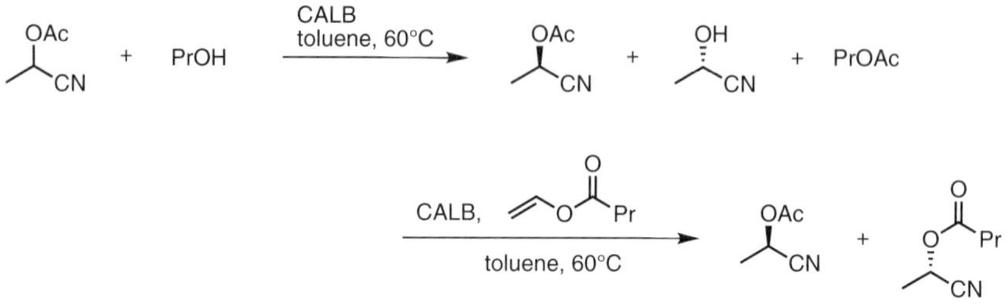

Scheme 1.163

1.3 Reaction with Acid Anhydrides

1.3.1 Without Activator

The acylation of alcohols is most frequently conducted by treatment with acid anhydrides. The use of bases such as amines is a common way to promote the reaction, but acid-based procedures are also available. The reaction without recourse to any activators is more desirable if possible. Many examples have demonstrated that the reaction does indeed proceed, in particular for acetylation. The reaction is best performed by heating an alcohol in acetic anhydride, even secondary and tertiary alcohols successfully undergoing acetylation (Scheme 1.164) [491]. It should be noted that this technique is also useful in terms of conversion of acid anhydrides into half-esters.

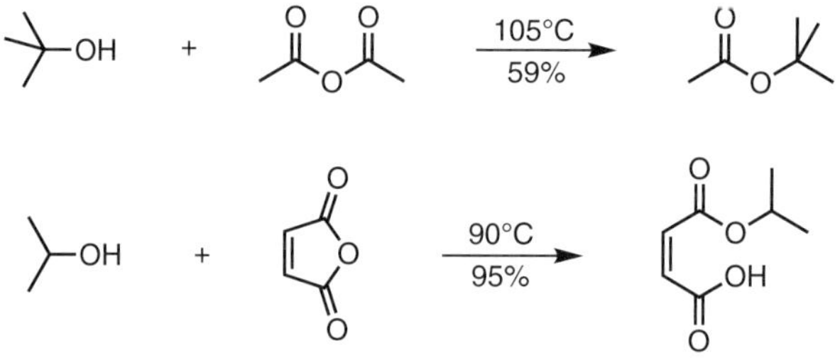

Scheme 1.164

In the synthesis of des-*N*-methyl-*N*-acetylerythromycin derivatives, acetylation at the 2′-position can be best performed simply by treating with Ac_2O in EtOAc at room temperature (Scheme 1.165) [492], addition of amines such as Et_3N, pyridine, or DMAP resulting in formation of complex mixtures. A steroid derivative is also acetylated smoothly by Ac_2O in acetic acid without activator (Scheme 1.166) [493].

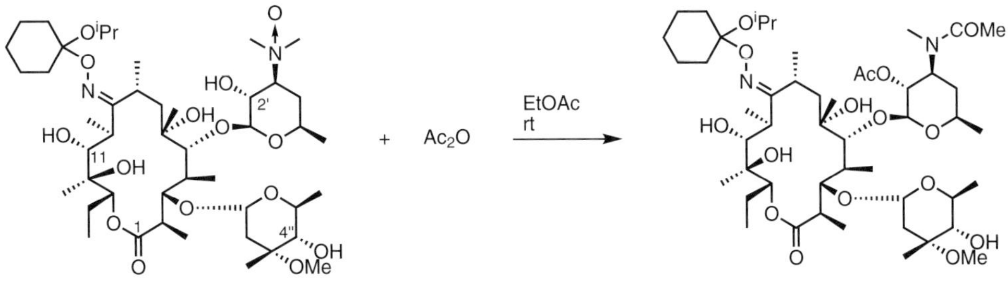

Scheme 1.165

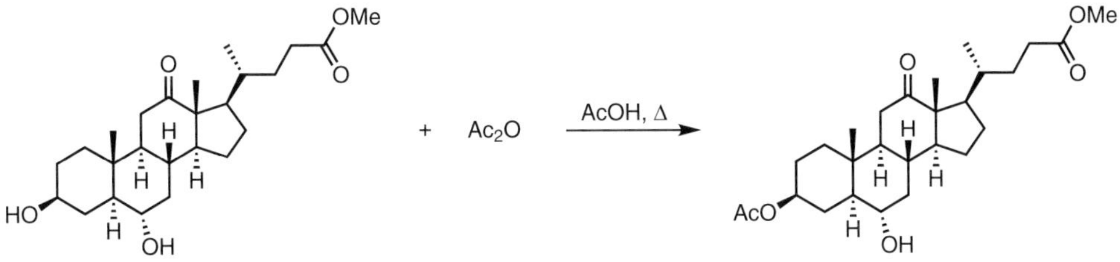

Scheme 1.166

The preparation of valinate esters is successfully achieved by stirring the alcohol with *N*-CBZ-L-Val-*N*-carboxy anhydride at room temperature (Scheme 1.167) [494]. This is a clean reaction because CO_2 is the sole by-product.

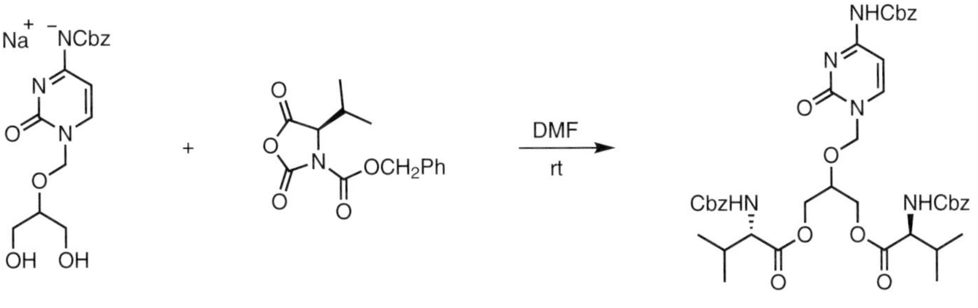

Scheme 1.167

1.3.2 Acid Catalysts

1.3.2.1 Brønsted Acids

Brønsted acids commonly employed for reaction between alcohols and acid anhydrides are sulfonic acids, sulfuric acid, and perchloric acid. There is nothing special to note on the use of these acids, although a number of examples are available. For instance, ethanolysis of *cis*-Δ^4-tetrahydrophthalic anhydride is carried out by the action of catalytic *p*-toluenesulfonic acid [495]. Sulfuric acid effects acetylation of sugars [496] and quinols with acetic anhydride [497]. Sulfamic acid H_2NSO_3H is also employable [498]. Perchloric acid is useful for acetylation of lactones with acetic anhydride [499]. When *N, N*-bis(2-hydroxyethyl)trifluoroacetaminosuccinodiamide is treated with a mixture of trifluoroacetic anhydride and nitric acid, both *O*-acylation and *N*-nitration take place (Scheme 1.168) [500]. When the amine is tertiary, no nitration occurs.

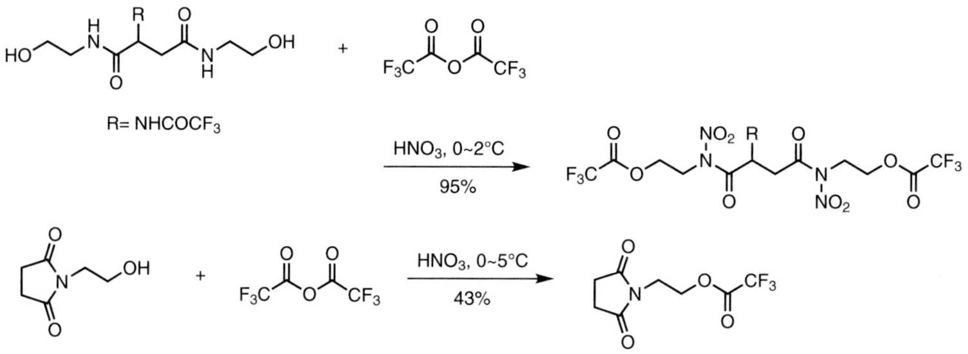

R= NHCOCF$_3$

HNO$_3$, 0~2°C

95%

HNO$_3$, 0~5°C

43%

Scheme 1.168

Pentafluorophenylbis(triflyl)methane is a new, strong, organic Brønsted acid that catalyzes benzoylation of menthol with benzoic anhydride (Scheme 1.169) [501]. Polystyrene-bound tetrafluorophenylbis(triflyl)methane also works similarly as an organic-solvent-swellable catalyst.

Experimental Procedure Scheme 1.169 [501] Synthesis of Polystyrene-Bound 2,3,5,6-Tetrafluorophenylbis(triflyl)methane: BuLi (1.6 M solution in hexanes, 1.9 mL, 3.0 mmol) is added at room temperature to a mixture of poly(4-bromostyrene) [0.37 g, 1 mmol, 2.71 mmol Br/g resin, 2% divinylbenzene (DVB) cross-linked, 200–400 mesh] and benzene (5 mL). The reaction mixture is stirred at 60 °C for 3 h. After the mixture has cooled to ambient temperature, the solvents – together with excess BuLi – are removed by decantation. Benzene (1 mL), THF (1 mL), and lithium pentafluorophenylbis(triflyl)methide (1.36 g, 3 mmol) are added at 0 °C to the residual lithiated resin. The reaction mixture is stirred at room temperature for 0.5 h and further at 70 °C for 6 h. After the mixture has cooled to ambient temperature, water (0.5 mL) and HCl (4 M, 0.5 mL) are added in that order at 0 °C. The polystyrene-bound 2,3,5,6-tetrafluorophenylbis(triflyl)methane resin is filtered off, washed with water (5 mL), a 50% aqueous solution of THF (5 mL), THF (5 mL), and Et$_2$O (5 mL) in that order, and dried at 80 °C under vacuum (ca. 1 Torr) for 12 h to afford polystyrene-bound 2,3,5,6-tetrafluorophenylbis(triflyl)methane (0.46 g). The loading of pentafluorophenylbis(triflyl)methane on the resin polystyrene-bound 2,3,5,6-tetrafluorophenylbis(triflyl)methane is estimated to be 1.01 mmol Tf$_2$CHC$_6$F$_4$-unit/g resin, based on fluorine content as determined by elemental analysis.

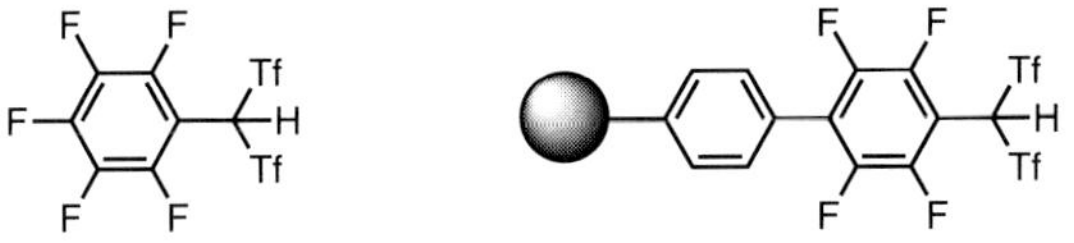

Scheme 1.169

General Procedure for the Acylation of L-Menthol with Carboxylic Anhydride: A mixture of L-menthol (0.47 g, 3 mmol) and acetic anhydride (0.42 mL, 4.5 mmol) or benzoic anhydride (1.02 g, 4.5 mmol) in acetonitrile (7 mL) is stirred at 27 °C in the presence of Brønsted acid catalyst. After the mixture has been stirred at 27 °C for 1–17 h, polystyrene-bound 2,3,5,6-tetrafluorophenylbis(triflyl)methane resin is filtered off, washed with HCl (4 M, 5 mL), water (5 mL), aqueous THF (5 mL), THF (50%, 5 mL), and Et_2O (5 mL) in that order, and dried at 80 °C under vacuum (ca. 1 Torr) for 12 h to be reused. The filtrate is extracted twice with hexane, and the organic layers are dried over magnesium sulfate, filtered, and concentrated under reduced pressure. The crude oil is purified by column chromatography on silica gel (hexane/EtOAc) to afford the corresponding ester.

1.3.2.2 **Lewis Acids**

The last decade has witnessed dramatic progress in the utilization of Lewis acids in the acid anhydride procedure, particularly thanks to the development of metal triflates. Commercially available $Sc(OTf)_3$ is a versatile catalyst for this purpose [502, 503]. The high catalytic activity is apparent from catalyst loading as low as 0.1 mol%. Besides acetylation, pivaloylation and benzoylation are also feasible. Not only primary alcohols but also sterically hindered secondary or tertiary alcohols are acylated. In contrast to the well-known selectivity under basic conditions, aromatic alcohols are less reactive than their aliphatic counterparts in competition reactions with this catalyst (Scheme 1.170). If the reaction rate is slow, mixed anhydrides generated *in situ* serve for improvements. This methodology is elegantly applied to lactonization of ω-hydroxy carboxylic acids (Scheme 1.171). Even medium-sized lactones are obtainable in reasonable yields with contamination with only small amounts of the diolides. Chelation by the acid anhydride moiety concurrent with intramolecular coordination by the terminal hydoxyl is believed to be responsible for the efficient lactonization.

Experimental Procedure Scheme 1.170 [503] *p*-Nitrobenzoic anhydride (253 mg, 0.8 mmol) is dissolved in dry acetonitrile (169 mL), and a cloudy solution of scandium triflate (0. 1 M, 0.8 mL, 0.08 mmol) in acetonitrile is added to the solution at room temperature under argon. A solution of ω-hydroxy carboxylic acid (0.04 M, 10 mL, 0.4 mmol) in THF is added slowly (from a mechanically driven syringe over 15 h) to the mixed solution at reflux under argon, and the reaction mixture is further stirred at reflux for 5 h. After being cooled to room temperature, the solution is quenched with aqueous saturated sodium hydrogen carbonate (4 mL). The resulting mixture is concentrated under reduced pressure and extracted twice with ether. The organic layers are dried over magnesium sulfate, filtered, and concentrated *in vacuo*. Purification is done by column chromatography on silica gel (eluent: hexane/ethyl acetate system) to give the desired lactone in good yield. In some cases, diolide is also afforded as minor product.

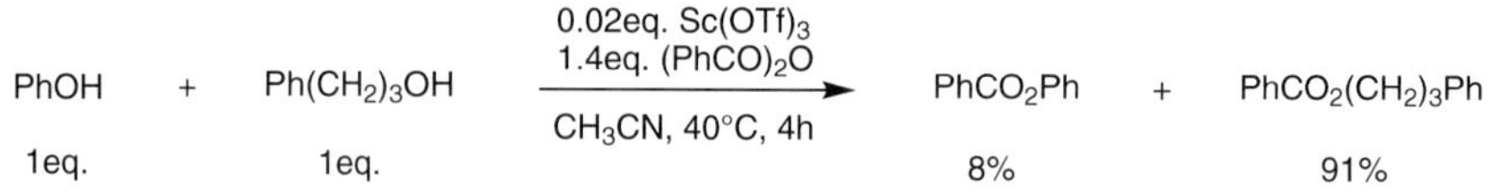

PhOH (1eq.) + $Ph(CH_2)_3OH$ (1eq.) $\xrightarrow[CH_3CN,\ 40°C,\ 4h]{0.02eq.\ Sc(OTf)_3,\ 1.4eq.\ (PhCO)_2O}$ $PhCO_2Ph$ (8%) + $PhCO_2(CH_2)_3Ph$ (91%)

Scheme 1.170

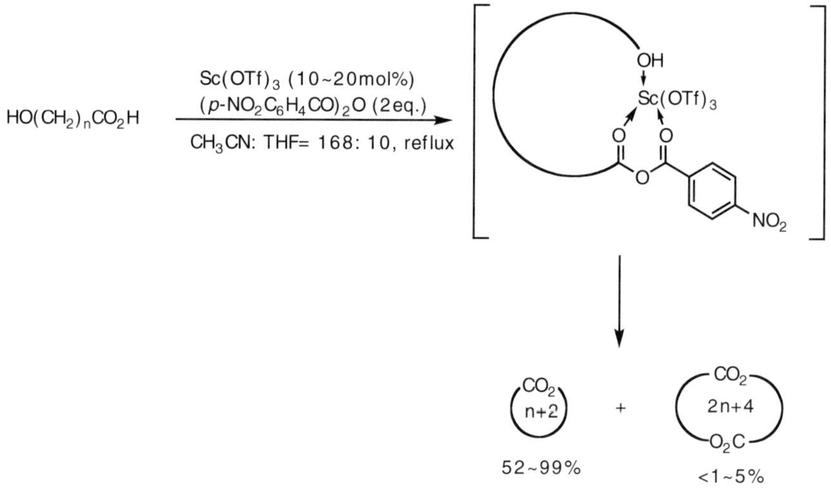

$HO(CH_2)_nCO_2H$ $\xrightarrow[CH_3CN:\ THF=\ 168:\ 10,\ reflux]{Sc(OTf)_3\ (10\sim20mol\%),\ (p\text{-}NO_2C_6H_4CO)_2O\ (2eq.)}$ [intermediate: OH → $Sc(OTf)_3$ ← O, O; NO_2] → CO_2 (n+2) 52~99% + CO_2 2n+4 O_2C <1~5%

Scheme 1.171

The direct transformation of chiral *O*-trimethylsilyl cyanohydrins into the corresponding *O*-acyl cyanohydrins is feasible with acid anhydrides or acid halides with $Sc(OTf)_3$ catalysis (Scheme 1.172) [504]. The reaction occurs with full retention of stereochemistry.

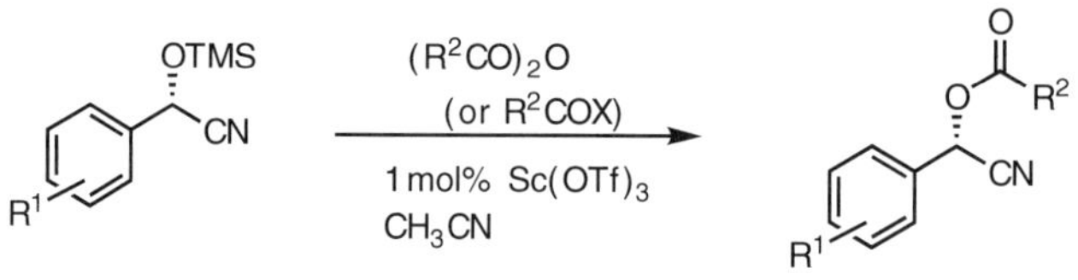

OTMS, CN, R^1 $\xrightarrow[1\,mol\%\ Sc(OTf)_3,\ CH_3CN]{(R^2CO)_2O\ (or\ R^2COX)}$ O, R^2, CN, R^1

Scheme 1.172

The slow acylation rate occasionally encountered with $Sc(OTf)_3$ is accelerated by use of $Sc(NTf_2)_3$, which is believed to result from $Sc(OAc)_3$ and Tf_2NH [505]. Acetylation of a tertiary alcohol and benzoylation of menthol, for example, proceed much more rapidly (Scheme 1.173).

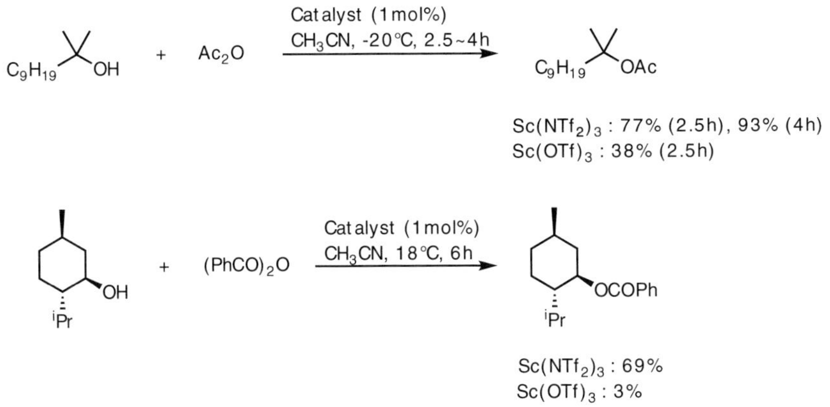

C_9H_{19} OH + Ac_2O → Catalyst (1 mol%), CH_3CN, -20 °C, 2.5~4 h → C_9H_{19} OAc

$Sc(NTf_2)_3$: 77% (2.5h), 93% (4h)
$Sc(OTf)_3$: 38% (2.5h)

OH, iPr + $(PhCO)_2O$ → Catalyst (1 mol%), CH_3CN, 18 °C, 6 h → OCOPh, iPr

$Sc(NTf_2)_3$: 69%
$Sc(OTf)_3$: 3%

Scheme 1.173

TMSOTf is also a useful catalyst for acylation of alcohols and phenols [506, 507]. It is claimed that this catalyst is more powerful and cheaper than $Sc(OTf)_3$. The reaction is extremely fast, selective, and mild. The following functional groups are tolerated: acetylene, allylic ester, aromatic ring, carbamate, diene, enone, ester, α,β-unsaturated ester, ether, halide, ketal, ketone, nitrile, sulfonate ester, thioester, and triene. A plausible mechanism is shown in Scheme 1.174. The alcohol reacts with TMSOTf to form the TMS ether and TfOH. The latter reacts with the anhydride to form the active acylating species, which could be a mixed anhydride or an acylium ion. This species then reacts with the TMS ether to form the ester and regenerate TMSOTf, which reenters the catalytic cycle. It may be said that TMSOTf is not a genuine Lewis acid catalyst if this mechanism is valid. The more popular TMSCl also catalyzes acetylation, but to somewhat limited extent [508].

Experimental Procedure Scheme 1.174 [507] A solution of alcohol (1 mmol) in CH_2Cl_2 (2 mL) is treated at 0 °C with the acid anhydride (1.5 mmol), followed by a CH_2Cl_2 solution of TMSOTf (1 M, 20 μl). Upon completion (TLC or HPLC), the reaction mixture is treated with saturated aqueous $NaHCO_3$, and the two phases are separated. In cases in which an excess of acid anhydride has been used, the reaction is quenched with stoichiometric quantities of methanol, followed by washing with $NaHCO_3$. The organic extracts are washed with water and dried, and the solvent is evaporated. The products are generally very clean and do not require any further purification.

$$R^1OH + TMSOTf \longrightarrow R^1OTMS + TfOH$$

$$(R^2CO)_2O + TfOH \longrightarrow R^2COOTf + R^2COOH$$

$$R^1OTMS + R^2COOTf \longrightarrow R^2COOR^1 + TMSOTf$$

Scheme 1.174

$Bi(OTf)_3$, which can conveniently be prepared from Ph_3Bi (Scheme 1.175), is more synthetically useful than $Sc(OTf)_3$ and TMSOTf [509, 510]. The catalytic activity is so high that the catalyst concentration can be lowered to a 0.005 mol% level for practical uses. The lack of a need for anhydrous reaction conditions is a great operational advantage. Less reactive acylation reagents such as benzoic and pivalic anhydrides are employable. The elimination reaction of tertiary alcohols is virtually suppressed, and, more remarkably, acid-labile THP- or TBS-protected alcohols, furfuryl alcohol, geraniol, and linalool can be acylated, as well as base-labile alcohols. Although a slight excess amount of acylating reagent should be used for complete conversion of the alcohol in question, this sometimes necessitates tedious separation of the acylation product from the remaining acylating reagent. The methanolysis technique is useful to overcome such an obstacle. Addition of methanol after completion of acylation transforms the remaining acylating reagent into the methyl ester, which is separable from the desired acylation products. This technique is particularly useful for the $Bi(OTf)_3$/anhydride technique, while addition of methanol induces decomposition of the acylated product in the cases of $Sc(OTf)_3$ and TMSOTf. Other bismuth(III) salts such as $BiCl_3$ and $Bi(OCOCF_3)_3$ also effect acylation, but not so usefully as the triflate [511].

Experimental Procedure [510] $Bi(OTf)_3$-Catalyzed Acetylation in Ac_2O: Acetic anhydride (8.4 mL, 89 mmol), 2-phenethyl alcohol (1.22 g, 10.0 mmol), and an acetic anhydride solution (1.0 mL) of $Bi(OTf)_3$ – prepared from $Bi(OTf)_3$ (7.3 mg, 0.01 mmol, calculated as the tetrahydrate) and Ac_2O (10 mL) – are placed in a round-bottomed flask, and the mixture is stirred at 25 °C for 10 min. After ethyl acetate and aqueous $NaHCO_3$ have been added, the organic and aqueous layers are separated. The aqueous layer is extracted three times with ethyl acetate, and the organic layers are combined and dried over $MgSO_4$. GC analysis of the crude mixture shows the formation of 2-phenethyl acetate in 98% yield.

$Bi(OTf)_3$-Catalyzed Acetylation in THF: THF (0.5 mL), furfuryl alcohol (98.1 mg, 1.0 mmol), Ac_2O (0.94 mL, 10.0 mmol), and $Bi(OTf)_3$ (0.36 mg, 0.05 mol%, calculated as the tetrahydrate) are placed in a flame-dried, round-bottomed flask. The mixture is stirred at 25 °C for 4 h. After ethyl acetate and aqueous $NaHCO_3$ have been added, the organic and aqueous layers are separated. The aqueous layer is extracted three times with ethyl acetate, and the organic layers are combined and dried over $MgSO_4$. GC analysis of the crude mixture shows the formation of furfuryl acetate in 93% yield.

Bi(OTf)$_3$-Catalyzed Pivalation of Menthol with (t-BuCO)$_2$O: A CH_2Cl_2 solution (3 mL, not purified, wet) of menthol (156.3 mg, 1.0 mmol) and (t-BuCO)$_2$O (279.5 mg, 1.5 mmol) is stirred at 25 °C in the presence of Bi(OTf)$_3$ (21.8 mg, 3.0 mol%, calculated as the tetrahydrate) for 4 h. MeOH (10 mL, unpurified and wet) is added, the mixture is passed through a pad of silica gel with hexane, and the filtrate is evaporated. Ethyl acetate (30 mL) is added to the crude product, and this organic layer is washed three times with aqueous $NaHCO_3$ and dried ($MgSO_4$). Evaporation furnishes the pure pivalate ester (98% yield, 235.6 mg).

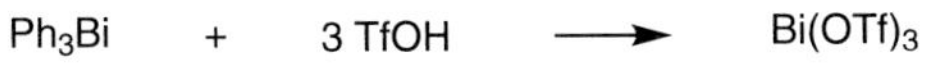

Ph_3Bi + 3 TfOH ⟶ Bi(OTf)$_3$

Scheme 1.175

In the presence of a catalytic amount of Ph_2BOTf, cyclic *meso*-dicarboxylic anhydrides are esterified stereoselectively with the diphenylboric ester of (R)-2-methoxy-1-phenylethanol to afford chiral diesters (Scheme 1.176) [512].

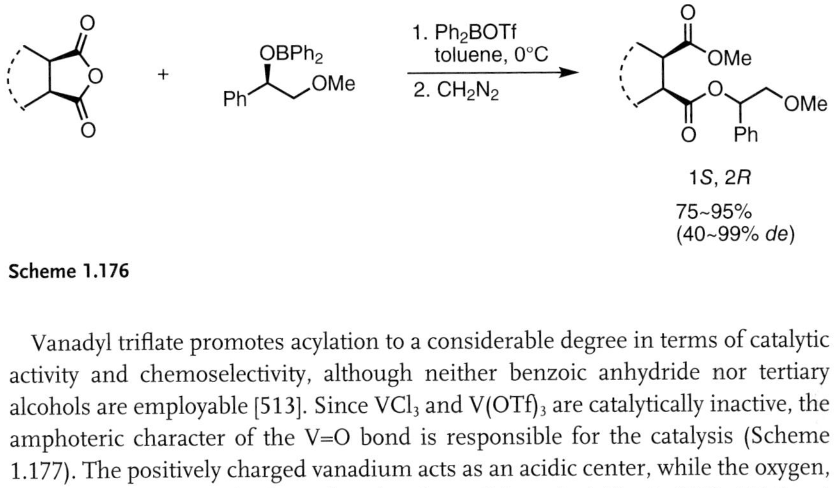

1. Ph_2BOTf toluene, 0 °C
2. CH_2N_2

OBPh$_2$, Ph, OMe, OMe, O, O, Ph

1S, 2R

75~95%
(40~99% *de*)

Scheme 1.176

Vanadyl triflate promotes acylation to a considerable degree in terms of catalytic activity and chemoselectivity, although neither benzoic anhydride nor tertiary alcohols are employable [513]. Since VCl_3 and V(OTf)$_3$ are catalytically inactive, the amphoteric character of the V=O bond is responsible for the catalysis (Scheme 1.177). The positively charged vanadium acts as an acidic center, while the oxygen, as a nucleophile, attacks the carbonyl carbon of the anhydride. In(OTf)$_3$ [514] and Cu(OTf)$_2$ [68] serve equally well for acylation.

Experimental Procedure Scheme 1.177 [513] General Procedure for the Preparation of Vanadyl Triflate, V(O)(OTf)$_2$·x(H_2O): Vanadyl sulfate trihydrate (342 mg, 2.1 mmol) is placed in a dry, 50-mL, two-necked, round-bottomed flask, and methanol (2 mL) is then added. A solution of Ba(OTf)$_2$ (872 mg, 2 mmol) in methanol (2 mL) is then added at ambient temperature. After 30 min of stirring, precipitation of barium sulfate is observed; this is filtered off over a short pad of dry Celite. The filtrate is concentrated and dried *in vacuo* at 120 °C for 4 h to furnish vanadyl triflate (622 mg, 85% yield) as a faintly blue solid, which is used directly for the acylation reactions.

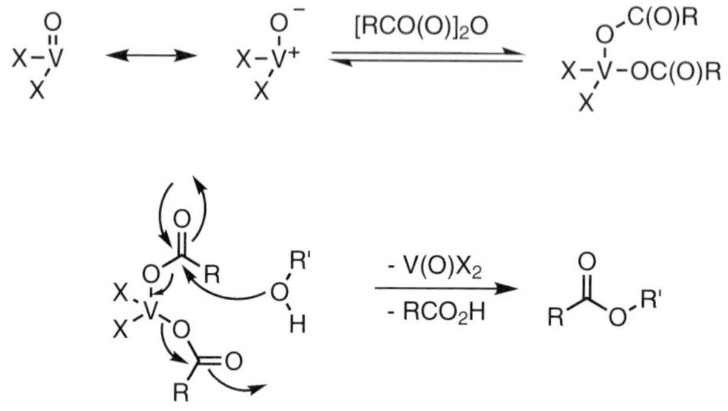

Scheme 1.177

General Procedure for Acylation Reactions: Vanadyl triflate (3.7 mg, 0.01 mmol) in 3 mL of anhydrous CH_2Cl_2 is placed in a dry 50-mL, two-necked, round-bottomed flask. The acyl anhydride (1.5 mmol) is slowly added to the above solution at ambient temperature. After 10 min, a solution of the nucleophile (1.0 mmol in CH_2Cl_2, 2 mL) is slowly added to the dark green solution, and the reaction mixture is stirred for the required period. After completion of the reaction as monitored by TLC, the reaction mixture is quenched with cold, saturated aqueous $NaHCO_3$ solution (5 mL). For the acylation of β-hydroxy ketones or esters, ice-cold water is used to quench the reaction, to prevent β-elimination. The separated organic layer is washed with brine, dried ($MgSO_4$), filtered, and evaporated. The crude product is purified by column chromatography on silica gel if required (in most of the acetylation reactions essentially pure material is obtained without further purification).

Metal bis(trifluoromethylsulfonyl)amides such as $(^iPrO)_2Ti(NTf_2)_2$ and $Yb(NTf_2)_3$ are powerful acylation catalysts [515], and $Yb[N(SO_2C_4F_9)_2]_3$ is more active than $Yb(OTf)_3$ for the reaction between ethanol and benzoic anhydride [516]. $M[C(SO_2C_8F_{17})_3]_3$ (M = Sc and Yb) work as catalysts for acetylation with acetic anhydride both in mono- and biphase systems containing fluorocarbon and non-fluorous organic solvents [517, 518]. $M[N(SO_3CH(CF_3)_2)_2]_3$ (M = La, Sm, Ga, Yb) also act as catalysts for acylation of alcohols and phenols [519].

Experimental Procedure [518] Cyclohexanol (0.21 mL, 0.20 g, 2 mmol) and acetic anhydride (0.19 mL, 0.20 g, 2 mmol) are added to a mixture of perfluoromethylcyclohexane (5 mL) and toluene (5 mL). Scandium tris(perfluorooctanesulfonyl) methide (1 mol%, 89 mg, 0.02 mmol) is added to the resultant mixture. The solution is stirred at 30 °C for 20 min, and the resultant mixture is then allowed to stand at room temperature (20 °C), so that the reaction mixture separates into the upper phase of toluene and the lower phase of perfluoromethylcyclohexane. The upper phase and the lower phase are individually analyzed by GC. Cyclohexyl acetate is obtained from the upper phase after evaporation under reduced pressure and silica gel chromatography (0.279 g, 98% isolated yield).

To the lower phase, containing the catalyst, toluene (5 mL), cyclohexanol (0.21 mL, 0.20 g, 2 mmol), and acetic anhydride (0.19 mL, 0.20 g, 2 mmol) are again added, followed by stirring at 30 °C for 20 min. The resultant mixture is again allowed to stand at room temperature (20 °C), so that the reaction mixture separates into the upper toluene phase and the lower perfluoromethylcyclohexane phase. Each phase is once more individually analyzed by GC. The overall yield of cyclohexyl acetate in the upper phase and lower phase is 100%. Essentially, the same procedure can be repeated a further three times. The overall yields of cyclohexyl acetate in the three repeated reactions are 99%, 99%, and 100%, respectively.

Metal perchlorates are another class of highly active catalyst, although they are potentially explosive. $LiClO_4$ [520], $Mg(ClO_4)_2$ [521], and $Zn(ClO_4)_2 \cdot 6H_2O$ [522] catalyze reaction of various types of alcohols including tertiary alcohols and acetic anhydride (1.05 equiv.) in the absence of solvent with low catalyst concentrations (0.1–1.0 mol%). $Bi(O)ClO_4$ is effective for acetylation of alcohols and phenols with one equivalent of acetic anhydride [523].

A variety of metal halides are also employable. $CoCl_2$ catalyzes acylation with aliphatic acid anhydride [524]. With this catalyst, β-hydroxy carbonyl compounds do not undergo elimination to afford α,β-unsaturated carbonyl compounds (Scheme 1.178). LiCl [525] and $MgBr_2$ [526] are other Lewis acids effective for acylation. Notably, addition of a tertiary amine dramatically enhances the activity [527], so the combination of $MgBr_2$ (2 equiv.), acid anhydride (2 equiv.), and tertiary amine (1.5–3 equiv.) effects benzoylation or pivaloylation very quickly. The high reactivity is explained on the basis of the dual activation arising from magnesium alkoxide formation and formation of $MgBr_2$-anhydride complex, which may fragment into acyl bromide and magnesium carboxylate (Scheme 1.179). MgI_2 also acts as a promoter [528].

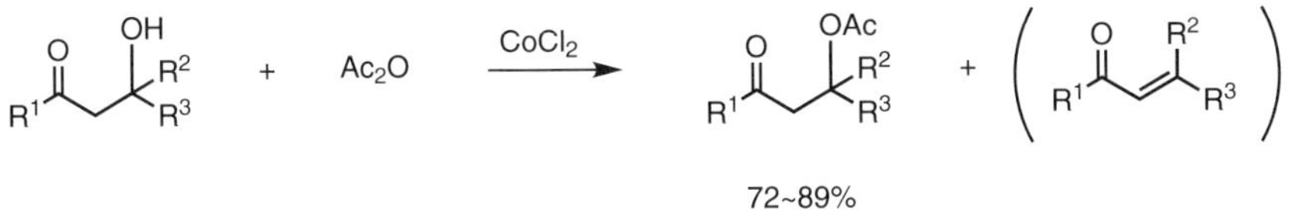

OH R² R³ R¹ O + Ac₂O —CoCl₂→ OAc R² R³ R¹ O + (O R² R¹ R³)

72~89%

Scheme 1.178

ROH + MgBr₂ ⟶ ROMgBr

Br₂Mg⁻ O⁺ O R' O R' ⇌ O R' OMgBr + O Br R'

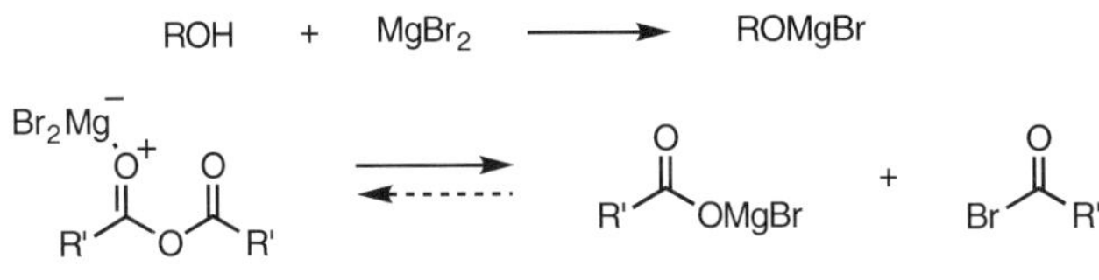

Scheme 1.179

$InCl_3$ is a powerful catalyst for acylation of alcohols and phenols [529]. Clay-supported metal halides such as $InCl_3$, $GaCl_3$, $FeCl_3$, $ZnCl_2$, etc. [530] or alumina-supported $InCl_3$, $GaCl_3$, $FeCl_3$, $ZnCl_2$, $CuCl_2$, $NiCl_2$, $CoCl_2$, etc. [531] effect acetylation of *tert*-butyl alcohol without troublesome elimination. $RuCl_3$ [532] and $ErCl_3$ [533] also catalyze acylation of alcohols and phenols.

Tris(2-methoxyphenyl)bismuthane acts as a good template for macrocyclic ester synthesis [534]. Treatment of an anhydride (1 equiv.), polyethylene glycol (1 equiv.), and the bismuthane (0.4 equiv.) in refluxing toluene furnishes the macrocyclic diester as the major product (Scheme 1.180). A small amount of tetraester is also formed. A unique template effect of Bi(III) is proposed.

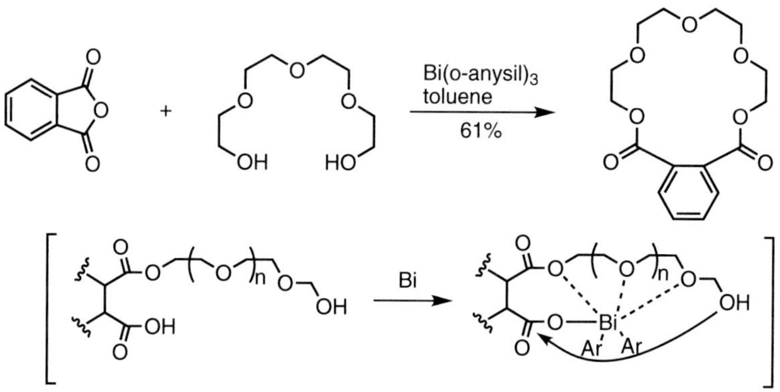

Scheme 1.180

Various organotin Lewis acids are useful catalysts. Treament of an alcohol with acetic anhydride (1.1–3.0 equiv.) in the presence of 1,3-dichlorotetrabutyldistannoxane results in quantitative formation of the corresponding acetates [350]. The mildness of this reaction is apparent from the successful use of THP- or TBS-containing substrates. The selectivity between primary and secondary alcohols is high, and this technique is applied to monoacetylation of chiral 3-chloropropane-1,2-diol at the primary position (Scheme 1.181) [535]. The resulting monoacetate is transformed into optically pure epichlorohydrin via 2-bromide.

Experimental Procedure [350] An acetic anhydride solution (1.4 mL, 15 mmol) of 2-phenylethanol (611 mg, 5.0 mmol) and 1,3-dichlorotetrabutyldistannoxane (27.6 mg, 0.25 mmol) is stirred at 30 °C for 24 h. After conventional workup with aqueous sodium hydrogen carbonate solution and ethyl acetate, the organic layer is dried over magnesium sulfate and evaporated. The crude mixture is subjected to column chromatography to give phenethyl acetate (788 mg, 96%).

The catalytic activity of organotin Lewis acids is dependent on the amount of positive charge on the tin [536, 537]. Thus, the activity for acid anhydride acetylation decreases in the order: $[Bu_2Sn(OH)(H_2O)]_2^{2+}\ 2OTf^- > [(BuSn)_{12}O_{14}(OH)_6]^{2+}$

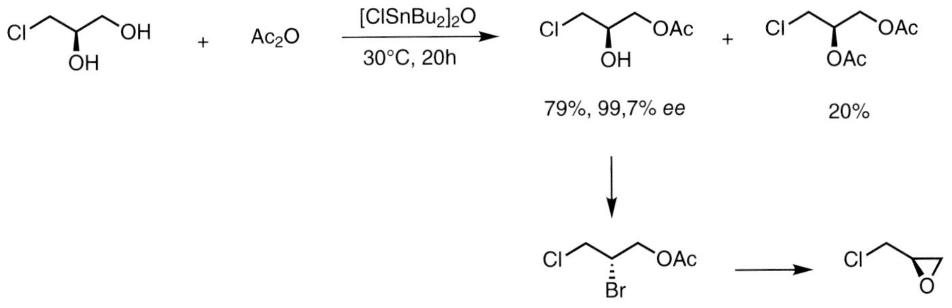

[ClSnBu2]2O
30°C, 20h

79%, 99,7% ee

20%

Scheme 1.181

$2Cl^- > (ClBu_2SnOSnBu_2Cl)_2$. The dicationic character is shared by the two tin atoms in the first compound but by 12 tin atoms in the second, while the third compound is neutral. Cationic tin(IV) porphyrin perchlorate and triflate [538, 539] catalyze acetylation of alcohols and phenols.

Iodine mediates acetylation of sugars [540]. Addition of I_2 to a suspension of sugars in Ac_2O followed by stirring affords the corresponding per-*O*-acetates (Scheme 1.182). It is postulated that iodine acts as a Lewis acid that activates the carbonyl group of the anhydride.

Experimental Procedure Scheme 1.182 [540] The sugar is suspended in acetic anhydride (5 mL per g of sugar for acetylation, 10 mL per g of sugar for acetolysis) and stirred. Iodine (10 to 500 mg per g sugar) is added, and stirring is continued until TLC shows the reaction to be complete. In small-scale reactions the reaction mixture is diluted with CH_2Cl_2 and washed successively with dilute aqueous sodium thiosulfate and aqueous sodium carbonate solutions. The organic layer is then dried (Na_2SO_4) and concentrated to give the product. The products obtained are in most cases pure enough for use elsewhere directly, or they are otherwise purified by column chromatography. In large-scale acetylations the reaction mixture is poured with stirring into ice-cold, dilute sodium thiosulfate solution. The products are allowed to crystallize in the refrigerator, and are separated by filtration.

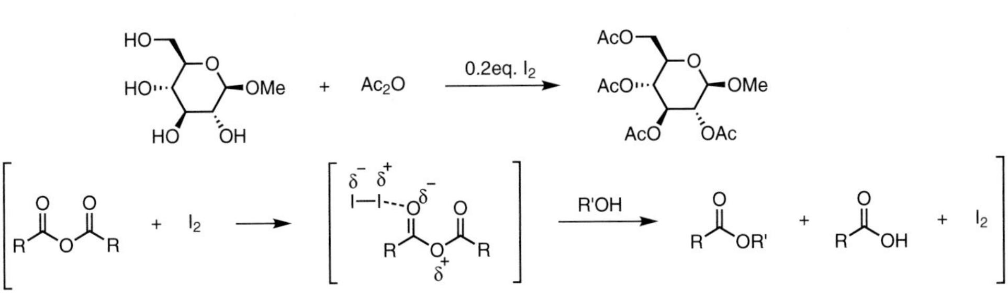

0.2eq. I_2

R'OH

Scheme 1.182

1.3.2.3 Solid Acids

Various types of zeolites are employable for acylation. With zeolite HSZ-360, alcohols and phenols are acetylated at 60 °C [541]. Chiral alcohols are acetylated with complete retention of optical purity, and β-nitroalcohols experience no dehydration to nitroolefins. Mono- and disaccharides undergo per-*O*-acetylation on stirring in acetic anhydride in the presence of H-beta zeolite at room temperature [542]. This zeolite is better than HY, H-EMT, H-ZSM-5, H-ZSM-12, and H-ZSM-22. Hβ-zeolite is effective for synthesis of dimethyl maleate from maleic anhydride and methanol, while monomethyl maleate is produced in the absence of the catalyst [543]. Al-MCM-41 molecular sieves promote the reaction between phthalic anhydride and methanol [544]. Silicoaluminophosphate molecular sieves selectively catalyze the esterification reaction between phthalic anhydride and 2-ethylhexanol [545]. A novel acylative cyclization reaction between phenol and acetic anhydride occurs over CeNaY zeolite to give 4-methylcoumarin (>70%) (Scheme 1.183) [546]. Mesoporous molecular sieves HMCM-41 effect acetylation of linalool [547].

Experimental Procedure [541] In a typical procedure, the alcohol or phenol (10 mmol) and acetic anhydride (20 mmol) are placed with stirring in a two-necked flask. After 5 min, zeolite HSZ-360 (0.2 g) is added, and the mixture is heated at 60 °C for the appropriate time. After cooling, the mixture is extracted with Et_2O and the catalyst is filtered off. After evaporation of the solvent, the acetate is purified by distillation or flash chromatography.

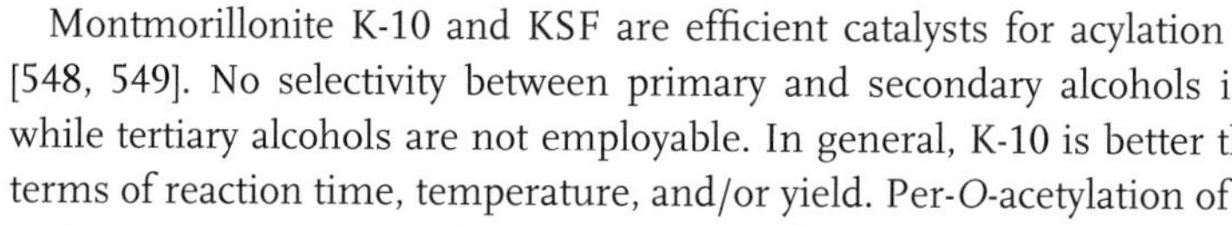

OH + O O O —zeolite (CeNaY) (cat.), 75%→ O O

Scheme 1.183

Montmorillonite K-10 and KSF are efficient catalysts for acylation of alcohols [548, 549]. No selectivity between primary and secondary alcohols is observed, while tertiary alcohols are not employable. In general, K-10 is better than KSF in terms of reaction time, temperature, and/or yield. Per-*O*-acetylation of mono-, di-, and trisaccharides is feasible with Montmorillonite K-10 [550]. Formation of α,β-furanose per-*O*-acetates takes place mostly in the monosaccharide cases, partial anomerization and acetolysis occurring under these conditions. Most disaccharides are smoothly converted into mixtures of pyranose per-*O*-acetates. The reaction of succinic anhydride with *p*-cresol to yield di-(*p*-cresyl)succinate is catalyzed by montmorillonite clay exchanged with various cations such as Al^{3+}, Cr^{3+}, Fe^{3+}, Cu^{2+}, Zn^{2+}, Ni^{2+}, Mn^{2+}, and H^+ [551].

Experimental Procedure [549] Acetic anhydride (10.0 mmol) is added to a mixture of alcohol (5.00 mmol), K-10 (100 mg) and CH_2Cl_2 (10 mL). After the mixture has been stirred at room temperature for 2 h, the catalyst is removed by filtration and washed with CH_2Cl_2 (10 mL). The solvent is evaporated under reduced pressure. The residue is pure enough for general purposes, while further purification may be achieved by column chromatography on silica gel, to give acetate in 98% yield.

Acidic resins such as Nafion H [552] and Amberlyst-15 [553] are also useful. In particular, the former resin effects acylation of a variety of alcohols such as primary, secondary, and allylic alcohols, as well as monosaccharides. Interestingly, acid-sensitive groups (ketals, acetals, THP ethers, cyclopropanes, etc.) are tolerated.

Experimental Procedure [552] Nafion-H (30 mg) is added to a solution of acetic anhydride (4 mmol) and an alcohol (2 mmol) in CH_2Cl_2 (3 mL). The resulting heterogeneous mixture is stirred at room temperature for 3–20 h. The CH_2Cl_2 layer is decanted into another flask, and the residual Nafion-H is washed with CH_2Cl_2 (2 × 2 mL). Evaporation of the solvent, followed by purification, affords pure acetates.

Yttria-zirconia based Lewis acid efficiently catalyzes acylation of structurally diverse alcohols [554]. The catalytic activity of $TaCl_5$ is enhanced when it is supported on silica gel [555]. Primary, secondary, allylic, and benzyl alcohols, and also phenols are smoothly acetylated, but tertiary alcohols are not employable.

Zirconium sulfophenyl phosphonate (of formula $\alpha\text{-}Zr(O_3PCH_3)_{1.2}(O_3PC_6H_4SO_3H)_{0.8}$) is an efficient heterogeneous catalyst for acetylation of primary, secondary, tertiary, allylic, benzylic, and acetylenic alcohols, and also phenols [556]. Good selectivity between primary and secondary alcohols is attained (pri > sec), but phenols cannot be discriminated from alcohols.

Polyoxometalates, $K_2CoW_{12}O_{40}\cdot 3H_2O$ [557] and $(NH_4)_8[CeW_{10}O_{36}]\cdot 20H_2O$ [558], are good catalysts for esterification of alcohols and phenols. Metal(IV) tungstates (M = Sn, Ti, Zr) catalyze reaction between phthalic anhydride and 2-ethylhexanol [140]. For this reaction, sulfated ZrO_2 exhibits higher activity and efficiency than zeolites [559].

1.3.3 Base Activators

1.3.3.1 Metal Salts

Acetylation with acetic anhydride is often accelerated by addition of sodium acetate. Acetolysis of methyl glycosides of *N*-acetylneuraminic acid is effected by treatment with acetic anhydride and sulfuric acid in glacial acetic acid (Scheme 1.184) [560]. The yields of the corresponding 2,3-unsaturated derivatives are improved by increasing the pH of the reaction mixture to 5 during workup by addition of sodium acetate.

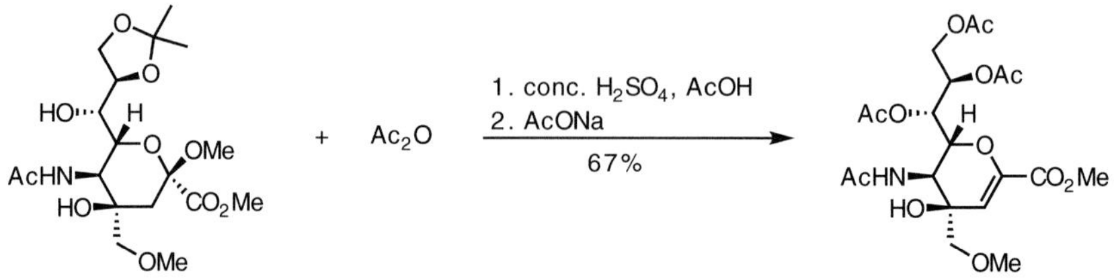

O O HO H O OMe AcHN HO CO_2Me OMe + Ac_2O → 1. conc. H_2SO_4, AcOH 2. AcONa 67% → OAc OAc AcO H O AcHN HO CO_2Me OMe

Scheme 1.184

Acetylation of phenols with acetic anhydride is smoothly conducted in the presence of sodium acetate (Schemes 1.185–1.188) [561–564]. Similarly, enols are converted into vinyl esters (Scheme 1.189) [565–567].

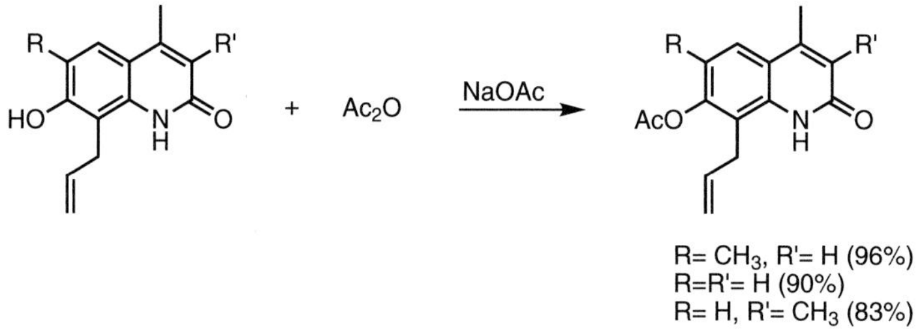

R R' HO N H O + Ac_2O → NaOAc → R R' AcO N H O

R= CH_3, R'= H (96%)
R=R'= H (90%)
R= H, R'= CH_3 (83%)

Scheme 1.185

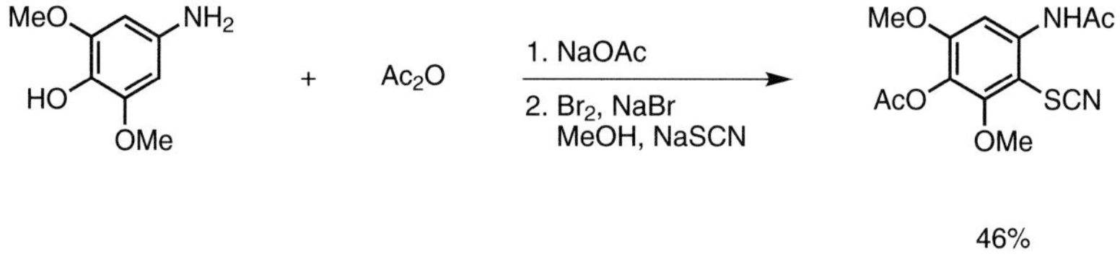

MeO NH_2 HO OMe + Ac_2O → 1. NaOAc 2. Br_2, NaBr MeOH, NaSCN → MeO NHAc AcO SCN OMe

46%

Scheme 1.186

OH HO O O + Ac_2O → NaOAc → OAc AcO O O

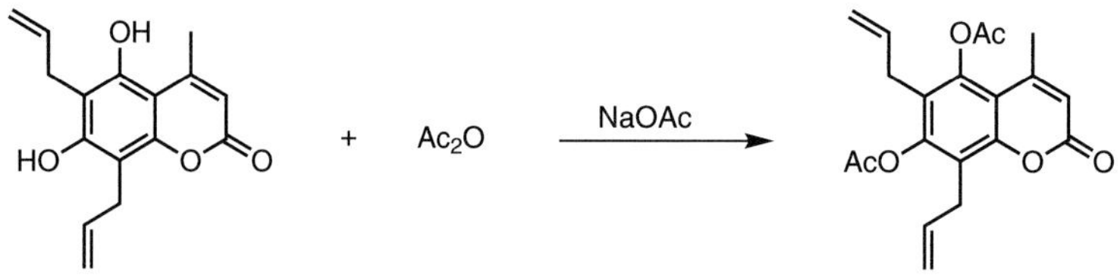

Scheme 1.187

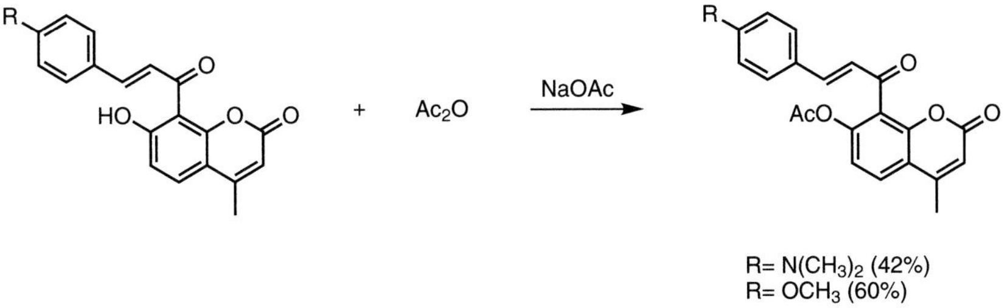

R= $N(CH_3)_2$ (42%)
R= OCH_3 (60%)

Scheme 1.188

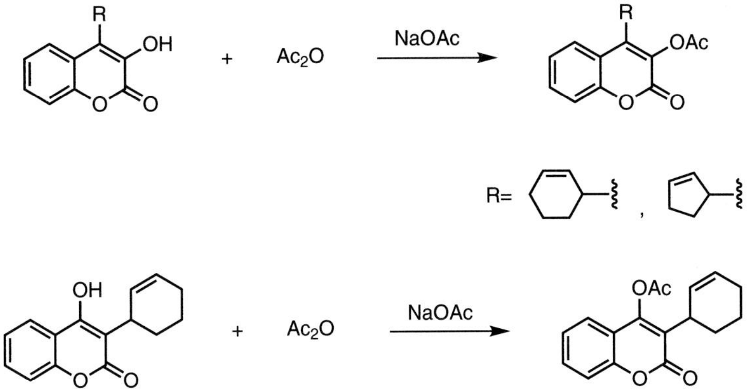

R=

Scheme 1.189

Potassium acetate works equally well for the acetylation of phenol (Scheme 1.190) [568] and benzyl alcohol (Scheme 1.191) [569].

Experimental Procedure Scheme 1.190 [568] KOAc (0.55 g, 5.57 mmol) is added to methyl-4-hydroxy-6-methylbenzofuran-5-carboxylate (0.750 g, 5.06 mmol) in Ac_2O (20 mL). The mixture is stirred at 120 °C for 12 h, cooled to room temperature, diluted with NaOH (2 N, 50 mL), and extracted with ether (3 × 60 mL). The organic layer is dried over Na_2SO_4, concentrated *in vacuo*, and purified by silica gel flash chromatography with hexane/ethyl acetate (10 : 1) to give the acetate of methyl-4-hydroxy-6-methylbenzofuran-5-carboxylate (1.24 g, 100% yield).

Experimental Procedure Scheme 1.191 [569] A mixture of 4-(hydroxymethyl)-2-methylaniline (16.58 g, 0.12 mol), acetic anhydride (34.0 mL, 0.36 mol) and potassium acetate (23.71 g, 0.24 mol) in $CHCl_3$ (240 mL) is stirred at room temperature for 3 h, heated at reflux for 2 h, and stirred at room temperature overnight. *n*-Amyl nitrite (32 g, 0.27 mol) and 18-crown-6 (1.59 g, 6.0 mmol) are then added, and the mixture is heated at reflux for 28 h. After being cooled to room temperature, the reaction mixture is added to acetic anhydride (10 mL) and stirred at room temperature overnight. The reaction mixture is diluted with CH_2Cl_2 (400 mL), washed with saturated $NaHCO_3$ (200 mL), water, and brine, and dried (Na_2SO_4), and the solvent is evaporated to give a dark brown solid. Chromatography (silica gel, 15% EtOAc/hexane) gives 1-acetyl-5-(acetoxymethyl) indazole as a yellow solid (16.98 g, 58%).

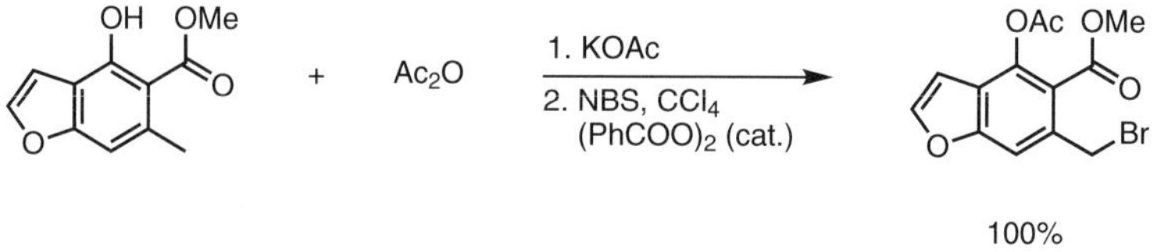
OH OMe
O
O
+
Ac2O
1. KOAc
2. NBS, CCl4
(PhCOO)2 (cat.)
OAc OMe
O
O
Br
100%

Scheme 1.190

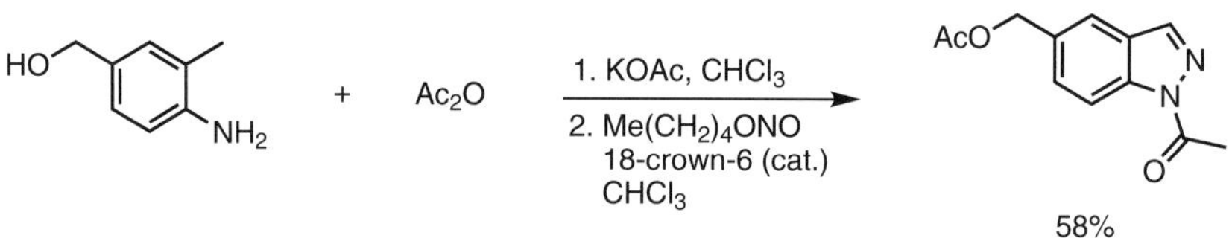
HO
NH2
+
Ac2O
1. KOAc, CHCl3
2. Me(CH2)4ONO
18-crown-6 (cat.)
CHCl3
AcO
N
N
O
58%

Scheme 1.191

Acetylation of phenols is also effected by the use of a suspension of potassium carbonate (Schemes 1.192–1.194) [570–572].

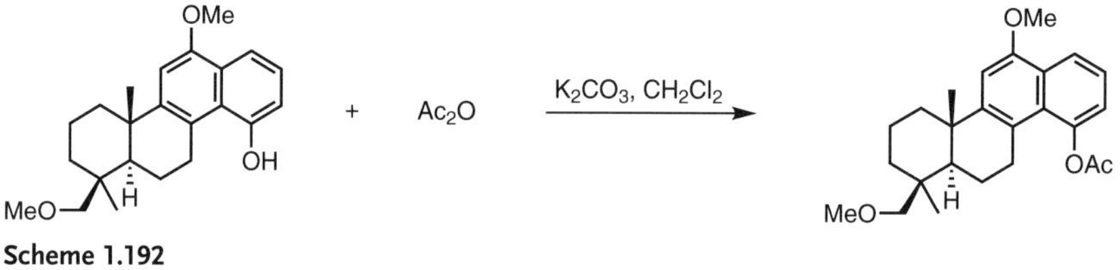
OMe
OH
MeO
H
+
Ac2O
K2CO3, CH2Cl2
OMe
OAc
MeO
H

Scheme 1.192

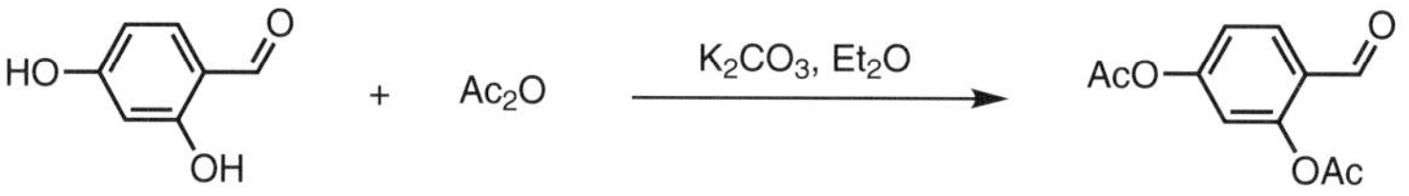
HO
O
OH
+
Ac2O
K2CO3, Et2O
AcO
O
OAc

Scheme 1.193

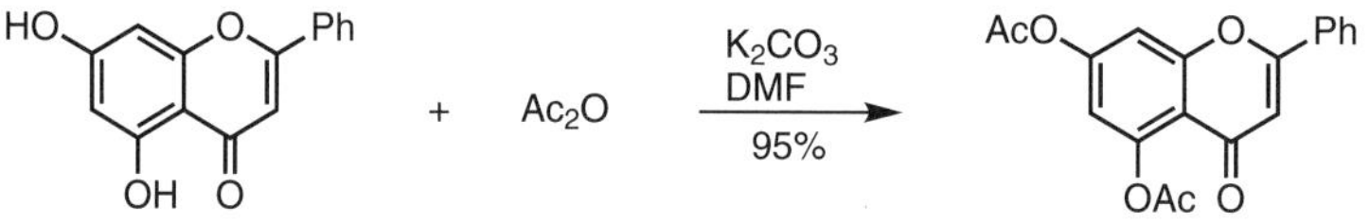
HO
O
Ph
OH O
+
Ac2O
K2CO3
DMF
95%
AcO
O
Ph
OAc O

Scheme 1.194

In contrast to the above reactions in suspension, use of an aqueous solution of sodium bicarbonate allows selective acetylation of phenol, leaving aliphatic alcohol intact (Scheme 1.195) [573]. The analogous selective acetylation is also feasible in a liquid-liquid two-phase system consisting of aqueous sodium hydroxide and isopropanol (Scheme 1.196) [553].

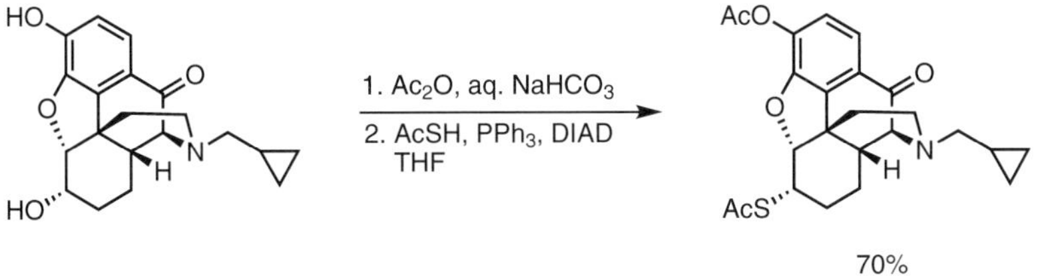

1. Ac_2O, aq. $NaHCO_3$
2. AcSH, PPh_3, DIAD THF

70%

Scheme 1.195

Chiral

+ Ac_2O — 2.7 eq. NaOH, ^iPrOH, H_2O — 95%

Chiral

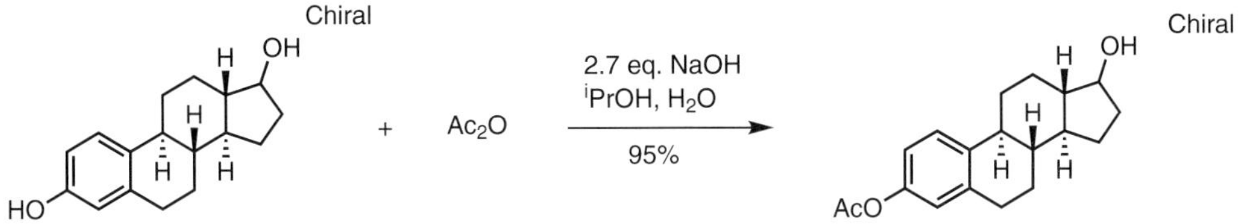

Scheme 1.196

A unique mechanistic switch is seen with $NaCo(CO)_4$ catalyst (Scheme 1.197) [574]. The reaction between alcohol and acetic anhydride proceeds via an acylcobalt intermediate in acetonitrile, since the ionization of the catalyst is induced in polar solvent, while in nonpolar toluene the sodium cation acts as a Lewis acid.

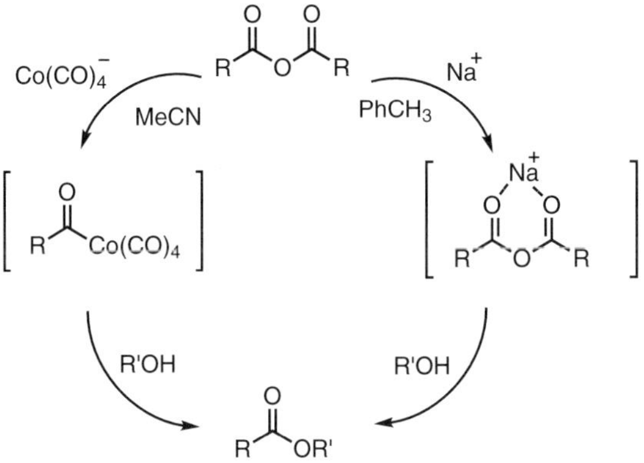

$Co(CO)_4^-$ MeCN PhCH$_3$ Na^+

R'OH R'OH

Scheme 1.197

In addition to acylation of alcohols, the acid anhydride technique also serves for the synthesis of half-esters of dicarboxylic acids. Enantioselective cleavage of cyclic meso anhydride is achievable by treatment with lithium salt of chiral mandelate ester (Scheme 1.198) [575]. The obtained half-ester can be used as a starting material for the synthesis of S-1452, an orally active potent thromboxane A_2 receptor antagonist.

Experimental Procedure Scheme 1.198 [575]
A solution of benzyl (*R*)-mandelate (5.33 g, 22.0 mmol) in THF (50 mL) is cooled to −78 °C, butyllithium in hexane 1.6 M, 13.13 mL, 21.0 mmol) is added dropwise, and the mixture is stirred for 15 min. A solution of bicyclo[2.2.1]hept-5-ene-2,3-*endo*-dicarboxylic anhydride (3.32 g, 20.0 mmol) in THF (20 mL) is added to the reaction mixture, and the resulting mixture is stirred at −78 °C for 1 h. The reaction mixture is acidified with HCl (2 N), and the product is extracted with ethyl acetate. The organic layer is washed with water and an aqueous solution of sodium chloride, and is then concentrated to yield a mixture of (1*S*, 2*R*, 3*S*, 4*R*)-bicyclo[2.2.1]hept-5-ene-2,3-dicarboxylic acid 2-(benzyl (R)-mandelate) and its diastereomer (9.33 g, 99%).

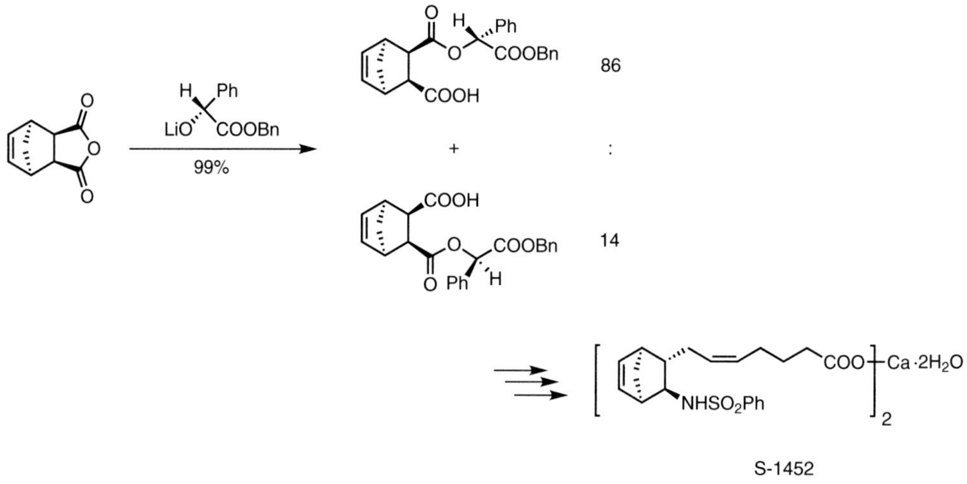

Scheme 1.198

Highly diastereoselective alcoholysis of σ-symmetric dicarboxylic acid anhydrides can be performed by use of a sodium salt of 1-phenyl-3,3-bis(trifluoromethyl) propane-1,3-diol (Scheme 1.199) [576].

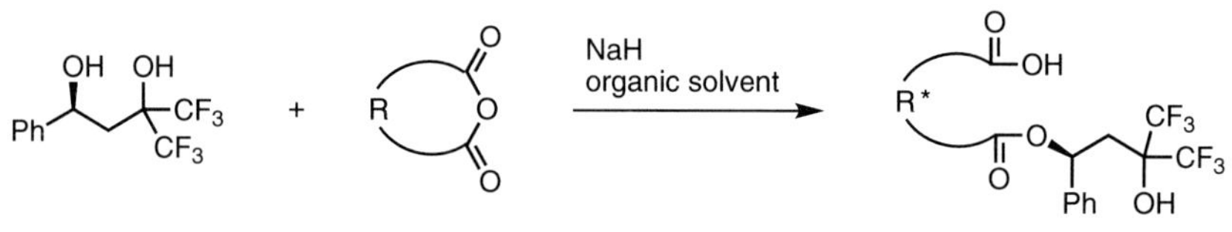

Scheme 1.199

1.3.3.2 Amines

The acylation of alcohols by acid anhydrides can most conveniently be carried out with the aid of amines. Although simple amines are not always employable, because of their relatively low basicity, they are the reagents of choice when the alcohol function is reactive enough, the substrates are too labile toward strong bases, or the acid anhydrides are reactive. The following are some representative examples.

Acetylation of azabicyclo[3.2.1]octanone is realized by stirring with Ac_2O in Et_3N at room temperature (Scheme 1.200) [577].

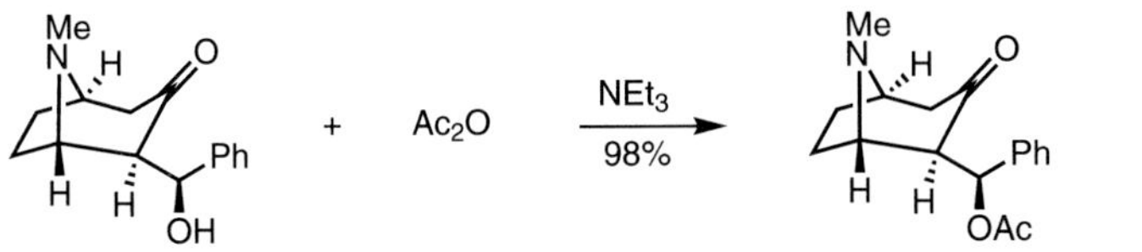

NEt_3, 98%

Scheme 1.200

π-Allyltungsten complexes undergo acetylation by Ac_2O/Et_3N in high yields (Scheme 1.201) [578].

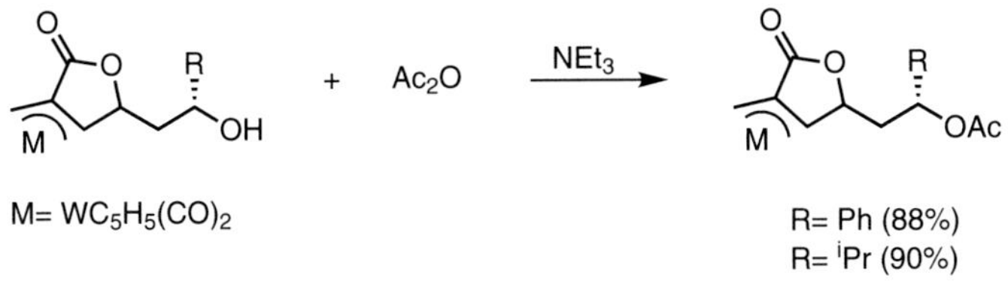

M= $WC_5H_5(CO)_2$

R= Ph (88%)
R= ^{i}Pr (90%)

Scheme 1.201

Trifluoroacetylation can normally be carried out by use of triethylamine. Propargylic alcohols [579] and silylmethanols [580] are smoothly converted into the corresponding trifluoroactates (Scheme 1.202).

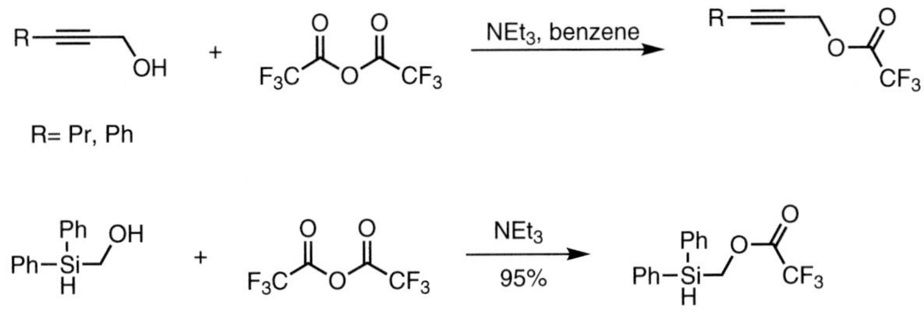

NEt_3, benzene

R= Pr, Ph

NEt_3, 95%

Scheme 1.202

The reaction between two equivalents of maleic anhydride and a benzene dimethanol in the presence of *N,N*-diisopropylethylamine affords dicarboxylic acid salts, which are further converted into cyclophanes upon treatment with α,α-dibromoxylenes (Scheme 1.203) [581].

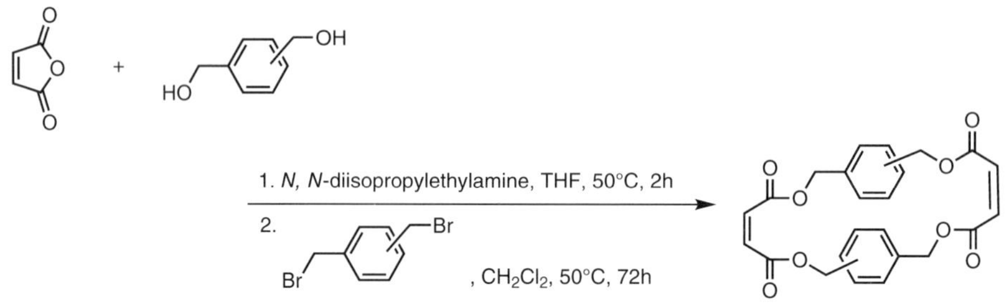

Scheme 1.203

The limitations frequently encountered in the simple amine protocol can be overcome to a considerable degree by use of 4-dialkylaminopyridines, usually DMAP or PPY (4-pyrrolidinopyridine) [582–584]. The use of these reagents in either stoichiometric or catalytic amount is effective: in the latter case, excess triethylamine or pyridine should be added to trap the acid formed. A wide range of alcohols, including highly sterically demanding tertiary alcohols and phenols, are acylated. It is postulated that *N*-acylpyridinium carboxylates are key intermediates, undergoing nucleophilic attack by alcohols (Scheme 1.204) [585]. The kinetic study revealed that a DMAP-catalyzed reaction proceeds in parallel with an uncatalyzed reaction [586].

Experimental Procedure [582]
A mixture of 1-chloro-1-methylhexanol (11.4 g, 0.1 mole), acetic anhydride (20 mL, 0.21 mol), triethylamine (20 mL, 0.15 mol), and DMAP (0.5 g, 4.1 mmol) is stirred at room temperature. After 14 h at room temperature, an 86% yield of the corresponding acetate is obtained. The amounts of acetate produced by use of pyridine and/or triethylamine only are less than 5%.

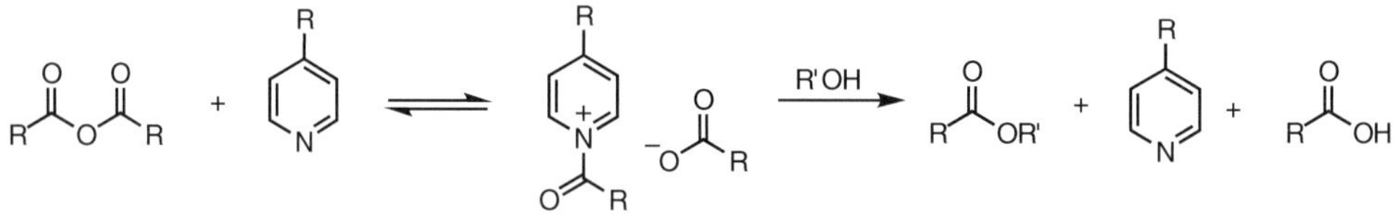

Scheme 1.204

The DMAP/pyridine technique can be further elaborated by use of high-pressure (1.5 GPa) to effect acylation of cyclic anhydrides with sterically demanding alcohols (Scheme 1.205) [587]. The monoesters are obtained selectively in 80–90% yields.

A planar-chiral DMAP derivative catalyzes kinetic resolution of arylalkylcarbinols [588] and secondary propargylic alcohols (Scheme 1.206) [589]. The use of *tert*-amyl alcohol as solvent is crucial for attaining high selectivity.

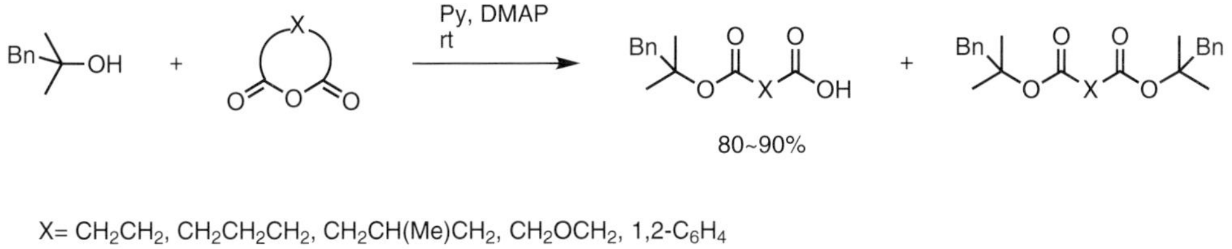

X= CH_2CH_2, $CH_2CH_2CH_2$, $CH_2CH(Me)CH_2$, CH_2OCH_2, 1,2-C_6H_4

Scheme 1.205

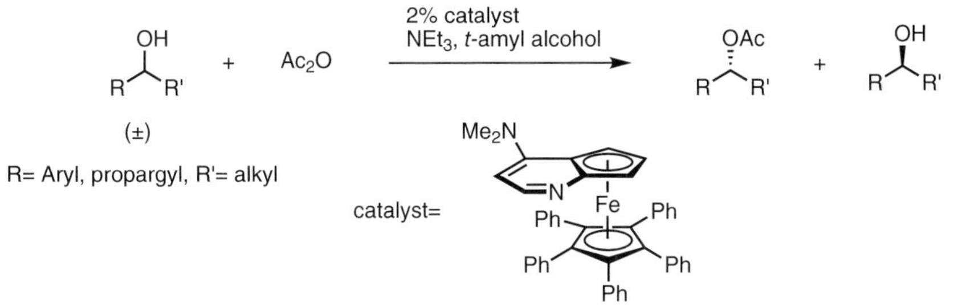

R= Aryl, propargyl, R'= alkyl

Scheme 1.206

Another chiral analog of DMAP is also effective for kinetic resolution of arylalkylcarbinols (Scheme 1.207) [590].

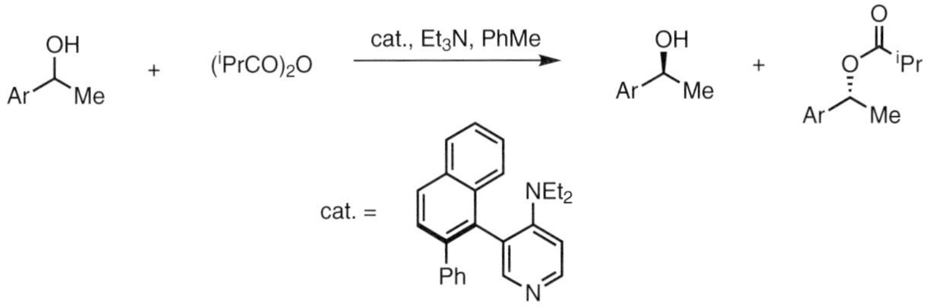

Scheme 1.207

A chiral catalyst derived from PPY is also available (Scheme 1.208) [591]. Kinetic resolution of various half-esters of cyclic 1,2-diols can be achieved. An acyliminium ion is proposed as a key species responsible for asymmetric induction.

A peptide bearing 3-(1-imidazoyl)-(*S*)-alanine as the *N*-terminal amino acid catalyzes the kinetic resolution of *trans*-2-(*N*-acetylamino)cyclohexan-1-ol (Scheme 1.209) [592]. An acyl imidazolium intermediate is postulated. This catalyst exhibits catalyst specificity, no resolution being feasible with methylnaphthylcarbinol. The second-generation catalysts involve variation of the amino acid at the (i + 1) and (i + 3) positions (Scheme 1.210) [593]. Alteration of the amino acid unit at the (i + 3) position has little influence on the selectivity of *trans*-2-(*N*-acetylamino) cyclohexan-1-ol in kinetic resolution, while substitution of L-proline at the (i + 1) site with D-proline results in a tenfold increase in the selectivity.

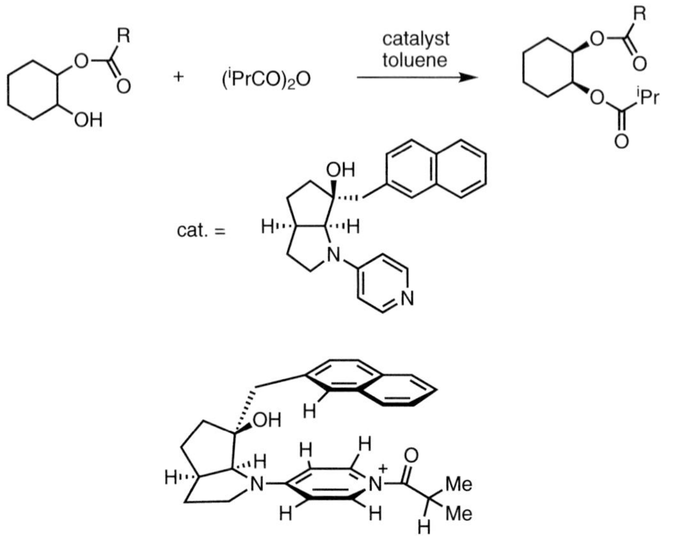

catalyst
toluene

$(^iPrCO)_2O$

cat. =

Scheme 1.208

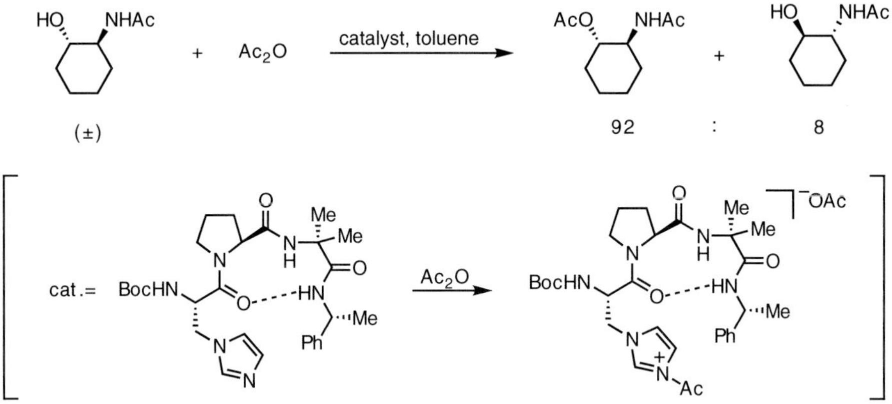

catalyst, toluene

Ac_2O

(±)

92 : 8

cat.=

Ac_2O

Scheme 1.209

Xaa= L-, D-amino acid
(i+1)= L-Pro or D-Pro

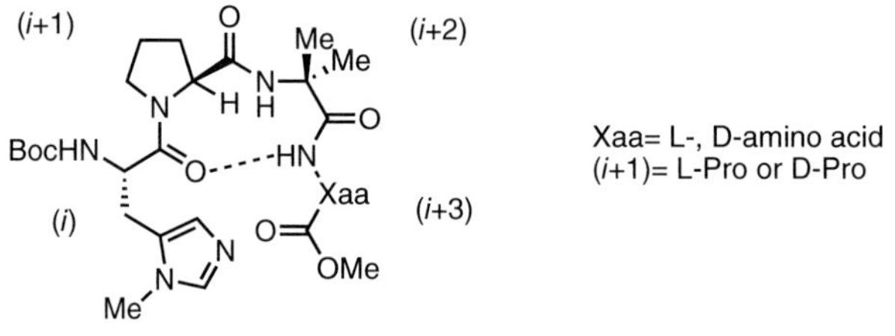

Scheme 1.210

Asymmetric ring-opening of prochiral cyclic acid anhydrides is a useful method for obtaining optically active compounds. Naturally occurring cinchona alkaloids mediate the ring-opening of meso cyclic anhydrides with methanol to furnish the corresponding half-esters, which can be transformed into lactones with up to 70% *ee* (Scheme 1.211) [594, 595]. Interestingly, the diastereomers obtained by inversion of the C-9 hydroxy group exhibit low activity as well as % *ee*s.

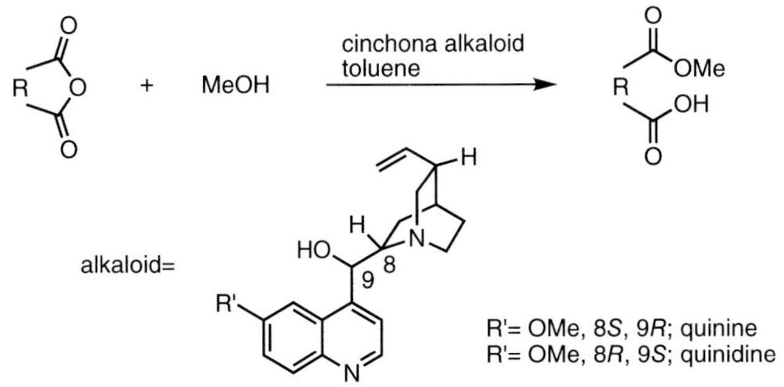

R'= OMe, 8*S*, 9*R*; quinine
R'= OMe, 8*R*, 9*S*; quinidine

Scheme 1.211

Cinchona alkaloids are also effective for ring-opening of tri- and tetracyclic anhydrides to afford the corresponding half-esters with 35–67% *ee*s (Scheme 1.212) [596, 597]. The absolute configurations of the products are tunable by selection of quinine and quinidine.

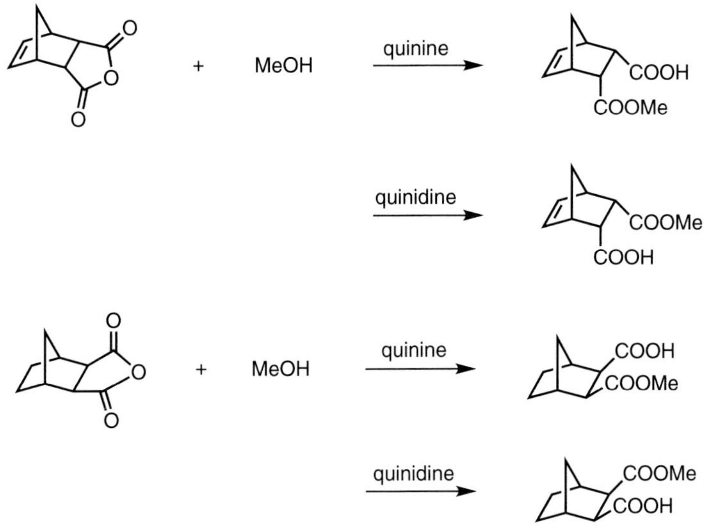

Scheme 1.212

Optimization of the cinchona alkaloid procedure improves the selectivity up to 99% *ee* in the presence of a stoichiometric amount of the promoter [598]. The catalytic version (0. 1 equiv.) is feasible with co-use of pempidine (1.0 equiv.) resulting in a 100% yield with 90% *ee*.

Ethers of cinchona alkaloids – $(DHQD)_2AQN$ and $(DHQ)_2AQN$ – are more efficient catalysts (Scheme 1.213) [599]. High yields (77–90%) and *ee*s (82–98%) are attained in desymmetrization of cyclic carboxylic anhydrides with methanol, while the absolute configurations can be switched by the selection of the two catalysts above (Scheme 1.214).

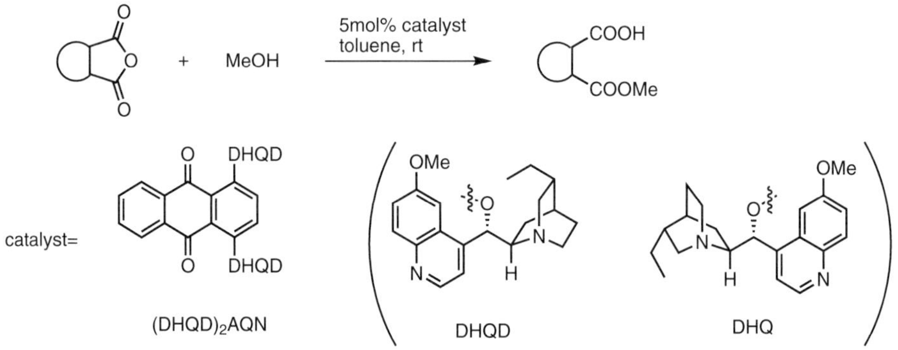

+ MeOH → 5mol% catalyst, toluene, rt → COOH / COOMe

catalyst= $(DHQD)_2AQN$ (DHQD, DHQ)

Scheme 1.213

$HOCH_3/(DHQ)_2AQN$

$HOCH_3/(DHQD)_2AQN$

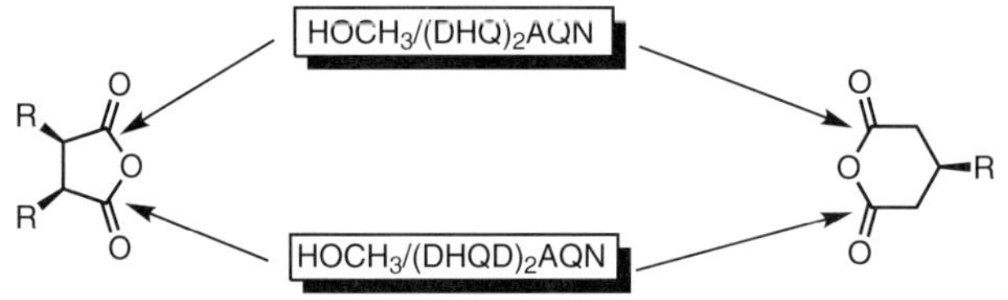

Scheme 1.214

These ethers are also useful for parallel kinetic resolution with a single catalyst (Scheme 1.215) [600]. Alcoholysis of 2-alkyl or 2-aryl succinic anhydrides in the presence of $(DHQD)_2AQN$ provides the two corresponding stereoisomers. The stereocenters at the C-2 position of the substrates are fully recognized, followed by regioselective alcoholysis. Quite naturally, $(DHQ)_2AQN$ induces the opposite selectivities.

Diethylzinc complexes with amino alcohols such as ephedrine, cinchonine, cinchonidine, and quinidine also effect desymmetrization of cyclic carboxylic anhydrides [601].

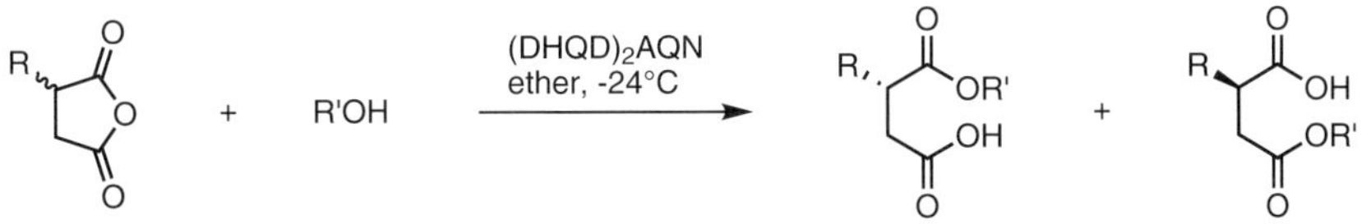

R'OH

$(DHQD)_2AQN$
ether, -24°C

R= alkyl, aryl

Scheme 1.215

1,3-Dibromo-5,5-dimethylhydantoin (DBDMH) and trichloroisocyanuric acid (TCCA) act as catalysts (0.05–0.1 mol%) for acylation of various alcohols (Scheme 1.216) [602].

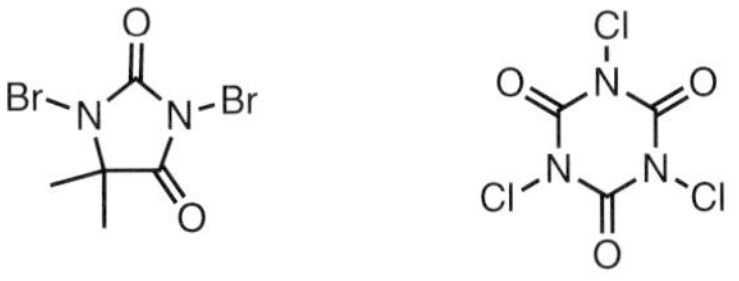

DBDMH TCCA

Scheme 1.216

1.3.3.3 **Phosphines**

Phosphines possess a considerable degree of basicity, and so can be used for promoting acylation. Bu_3P is a weak base in organic solvent but exhibits activities nearly comparable to those of DMAP [603, 604]. For benzoylation, the Bu_3P-catalyzed reaction is faster than its DMAP counterpart. DMAP is more versatile, however, since it shows catalytic activity in reactions between alcohols and a larger variety of electrophiles than Bu_3P does. The phosphine also appears to be somewhat more sensitive than DMAP to counter-ion effects. Nevertheless, use of Bu_3P is advantageous in that it is cheaper and less toxic than DMAP, is not easily deactivated by the carboxylic acid generated in reactions using acid anhydride, and can be used under nearly neutral conditions. Enantiopure 2,3-dialkyl-1-phenylphosphapentane is employable for desymmetrization of racemic 1,2-diols and kinetic resolution of secondary alcohols through acid anhydride acylation [605], but the *ee*s of the product are modest. The enatioselectivity is improved by the use of 2-aryl-4,4,8-trimethyl-2-phosphabicyclo[3,3,0] octanes (Scheme 1.217) [606]. These phosphines can be stored as borane adducts and released by warming with pyrrolidine. Aryl alkyl carbinols are resolved with isobutyric anhydride, an acylphosphonium intermediate being postulated.

The nonionic superbase $P[MeNCH_2CH_2]_3N$ acts as a catalyst for acylation with acid anhydride for hindered alcohols and also for acid-sensitive alcohols [607]. The reaction is likely to proceed with assistance of transannular coordination, as shown in Scheme 1.218.

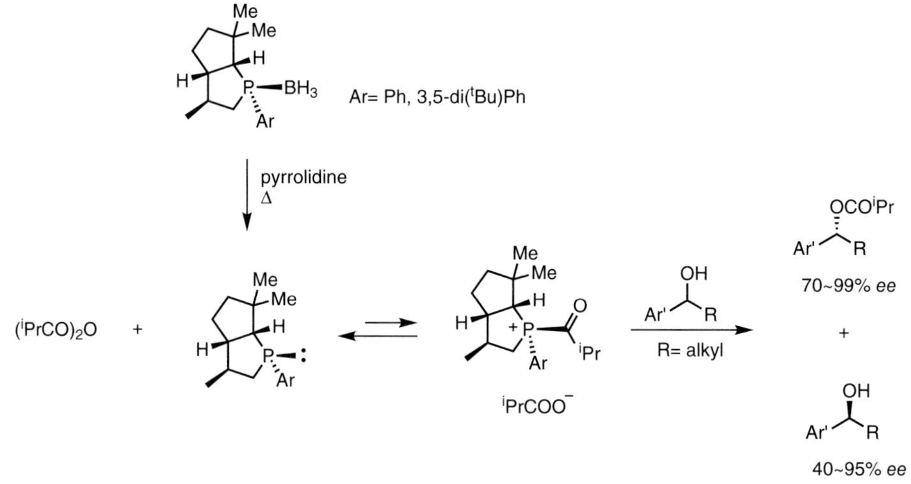

Ar= Ph, 3,5-di(^{t}Bu)Ph

pyrrolidine
Δ

(^{i}PrCO)$_2$O +

^{i}PrCOO$^-$

R= alkyl

70~99% *ee*

40~95% *ee*

Scheme 1.217

Experimental Procedure Scheme 1.218 [607] After the superbase P[MeNCH$_2$CH$_2$]$_3$N (1.1 g, 5.1 mmol) has been dissolved in solvent (30 mL) at 24 °C under N$_2$, the appropriate acid anhydride (5.1 mmol) is added and the mixture is stirred for 5 min. The alcohol (4.8 mmol) is then added with stirring. After ~1.5 h, water (0.05 mL) is added with continued stirring. This is followed by the addition of ether (80 mL), and stirring is continued for 5 min more. (When acetonitrile or benzene is used as the solvent, ~95% of the solvent is evaporated before addition of the ether.) The mixture is then filtered, and the residue is washed with ether (20 mL). The organic layer is dried with anhydrous sodium sulfate, followed by concentration under vacuum, to afford the crude ester, which is purified by chromatography on silica gel. The residue obtained from the reaction is treated with KO-*t*-Bu to recover P[MeNCH$_2$CH$_2$]$_3$N.

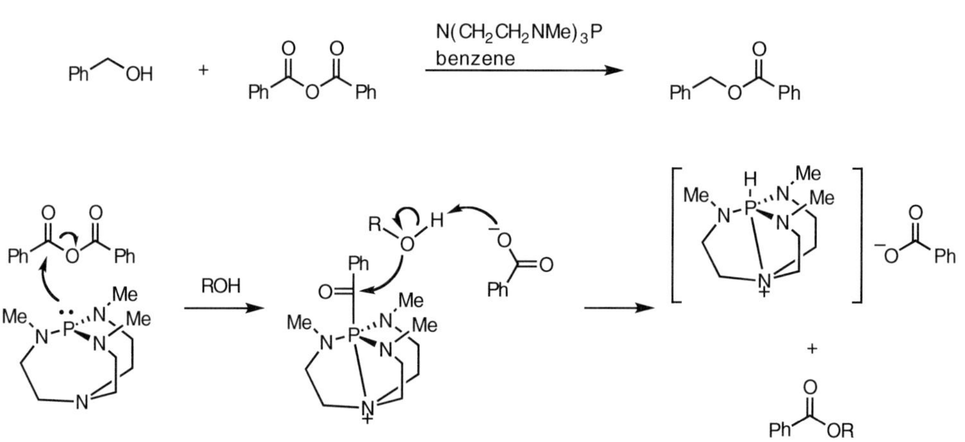

N(CH$_2$CH$_2$NMe)$_3$P
benzene

ROH

Scheme 1.218

1.3.4 Enzymes

The acid anhydride technique with enzymes is in some respects more advantageous than esterification and transesterification. Since no water is formed, the enzyme is protected from hydrolysis, and undesired side reactions requiring water (such as racemization) are avoided. The equilibrium is more readily shifted to the product ester side than it is in the case of transesterification, because the acid formed is completely removed from the equilibrium system. Lipase Amano P from *Pseudomonas fluorescens* adsorbed on Celite 577 can be used for kinetic resolution of primary and secondary alcohols [608]. Analogously, lipase AY-30 from *Candida cylindacea* adsorbed on celite can bring about kinetic resolution of highly lipophilic substrates (Scheme 1.219) [609]. The adsorption or the addition of the organic base 2,6-lutidine or the inorganic base $KHCO_3$ improves the selectivity.

Experimental Procedure [608] Amano P lipase (0.12 g, 3600 units) supported on Celite 577 (0.48 g) is added to a magnetically stirred solution of racemic 1-phenylethanol (2 g, 16.3 mmol) and acetic anhydride (1.66 g, 16.3 mmol) in benzene (40 mL), and the reaction mixture is stirred at room temperature. Periodically 1-μl aliquots of the liquid phase are withdrawn and analyzed by GC. After 24 h, approximately 50% conversion has been reached, and the reaction is stopped. The solid enzyme preparation is filtered off, and the filtrate is washed with aqueous Na_2CO_3 (5%, 40 mL), dried over sodium sulfate, and evaporated to dryness. Chromatography on silica gel with hexane/ether (95:5) as eluent affords 0.86 g (43%) of (*S*)-(−)-1-phenylethanol.

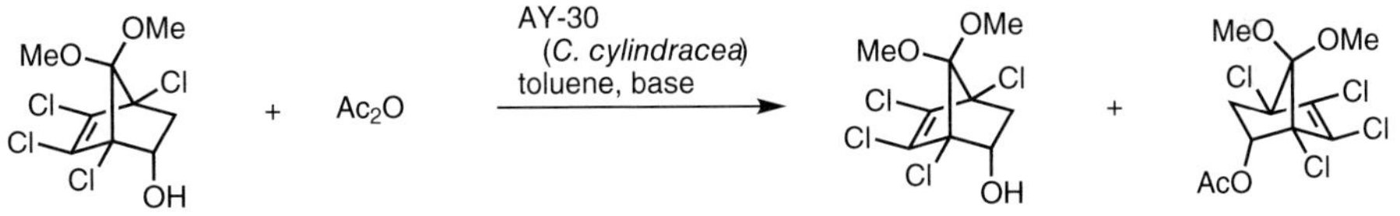

AY-30 (*C. cylindracea*) toluene, base

Scheme 1.219

Selective acylation of primary alcohols in preference to secondary ones in diols and triols is achieved by use of PPL in organic solvents (Scheme 1.220) [610]. Propanoic anhydride gives better outcomes than acetic anhydride.

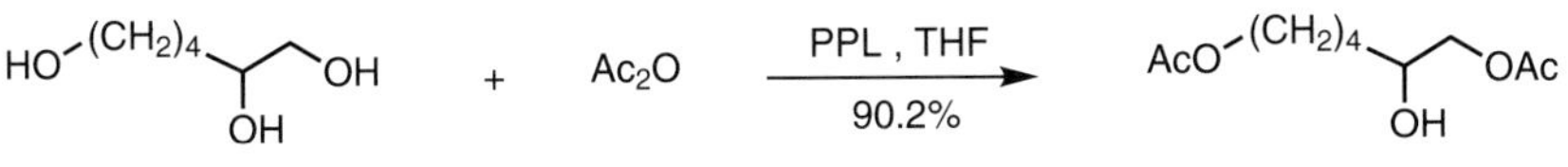

PPL, THF 90.2%

Scheme 1.220

1.3.5
Mixed Anhydrides

The reactivities of anhydrides toward nucleophiles can be enhanced by combining different carboxylic acid partners. Benzoic acid derivatives with electron-withdrawing groups are most commonly employed to activate the counterpart. Mixed anhydrides with a 2,4,6-trichlorobenzoic acid moiety (Yamaguchi procedure) are extremely versatile (Scheme 1.221) [611]. The mixed anhydride is prepared by treatment of a carboxylic acid with trichlorobenzoyl chloride (1 equiv.) in the presence of Et_3N (1 equiv.) and then treated with alcohol (1–2 equiv.) in the presence of DMAP (2–4 equiv.) to provide the desired esters. Both secondary and tertiary alcohols are smoothly esterified at room temperature. Of great significance is the applicability to lactonization of ω-hydroxy acids. This procedure is now recognized as the most versatile means in macrolactone synthesis.

Experimental Procedure Scheme 1.221 [611] Relative Rates of Alcoholysis of Mixed Anhydrides: The acid chloride to be examined (0.3 mmol) is added to a mixture of 2-methylpentanoic acid acid (37 μL, 0.3 mmol) and triethylamine (42 μL, 0.3 mmol) in THF (2 mL), after which the mixture is stirred at room temperature for 20 min. After the removal of triethylamine hydrochloride by filtration, the filtrate is evaporated under nitrogen and the residue is dissolved in dichloromethane (1 mL). To this solution is added a mixture of 2-methyl-2-propanol (56 μl, 0.6 mmol) and 4-dimethylaminopyridine (73 mg, 0.6 mmol) in dichloromethane (1 mL), and the resulting mixture is stirred at room temperature. The formation of the ester is followed by GLC by addition of bromobenzene (50 μL) as an internal standard.

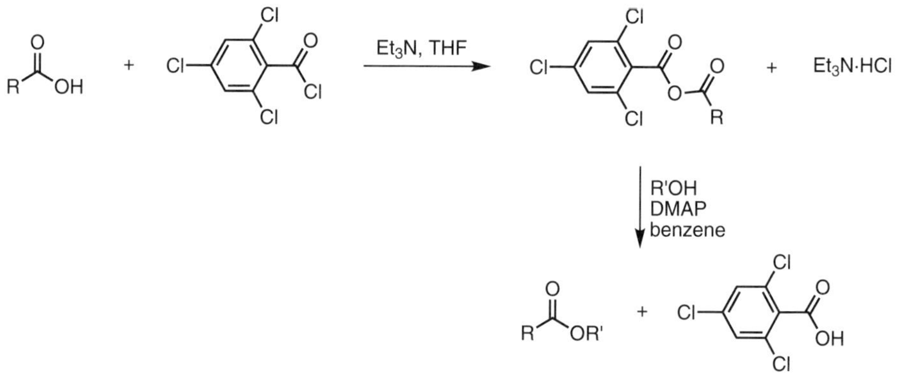

Scheme 1.221

Preparation of Lactones: 2,4,6-Trichlorobenzoyl (or 2,3,6-trimethyl-4,5-dinitrobenzoyl) chloride (1.0 mmol) is added to a mixture of a hydroxy acid (1.0 mmol) and triethylamine (1.1 mmol) in THF (10 mL), after which the mixture is stirred at room temperature for 1–2 h (or 12 h in the case of 2,36-trimethyl-

4,5-dinitrobenzoyl chloride). After removal of triethylamine hydrochloride, the filtrate is diluted with toluene (500 mL) and added under the high-dilution conditions, over a period of 1.5–8 h, to a refluxing solution of DMAP (3–6 mmol) in toluene (100 mL). The reaction mixture is worked up and purified by preparative TLC (silica gel G, Merck). The crude products are purified by distillation or recrystallization.

Subjection of enantiopure 3-hydroxybutanoic acid to the Yamaguchi procedure gives three cyclic oligomers (Scheme 1.222) [612]. Only the pentamer, hexamer, and heptamer are formed, without tetramer.

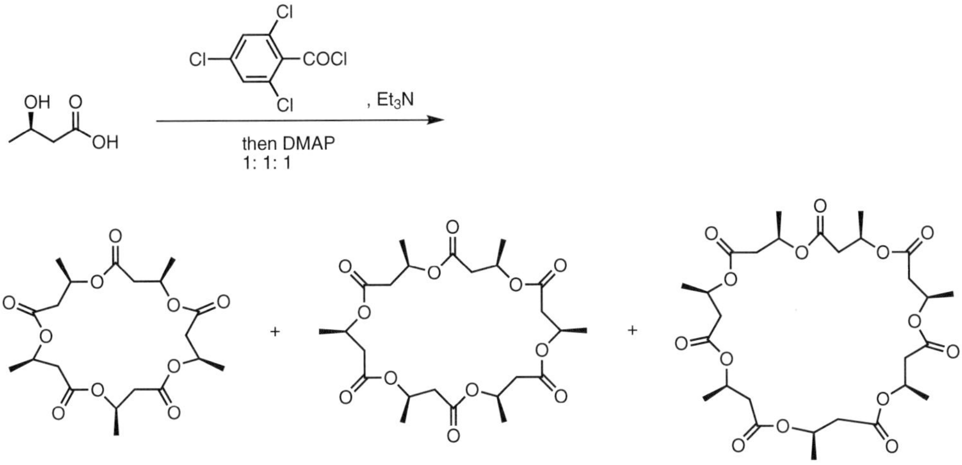

Scheme 1.222

A modified Yamaguchi procedure, with use of 2,6-dichlorobenzoyl chloride also serves for both inter- and intramolecular esterification (Schemes 1.223 and 1.224) [613–615].

Experimental Procedure Scheme 1.223 [613] 2,6-Dichlorobenzoyl chloride (577 mg, 2.76 mmol) is added under Ar to a mixture of (2*S*, 3*S*)-3-*tert*-butyldimethylsilyloxy-2-methylpentanoic acid (678 mg, 2.76 mmol) and Et_3N (306 mg, 3.03 mmol) in dry THF (14 mL). The mixture is stirred overnight at room temperature. After the removal of $Et_3N \cdot HCl$ by filtration, the filtrate is concentrated under N_2, and the residue is dissolved in dry C_6H_6 (10 mL). Solutions of hydroxy ketone (358 mg, 2.76 mmol) in dry C_6H_6 (3 mL) and DMAP (370 mg, 3.03 mmol) in dry C_6H_6 (3 mL) are added to this solution at 0° under Ar. The resulting mixture is stirred for 5 h at 0°. It is then diluted with ether (15 mL), washed with HCl (1 N), water, sat. aq. $NaHCO_3$, and brine, dried ($MgSO_4$), and concentrated *in vacuo*. The residue is purified by chromatography over SiO_2 (Fuji Davison BW-820 MH, 30 g). Elution with hexane/ether (20:1) gives (1′*S*, 2′*R*)-1′,2′-dimethyl-3′-oxopentyl (2*S*, 3*S*)-3-*tert*-butyldimethylsilyloxy-2-methylpentanoate (906 mg, 91.9%).

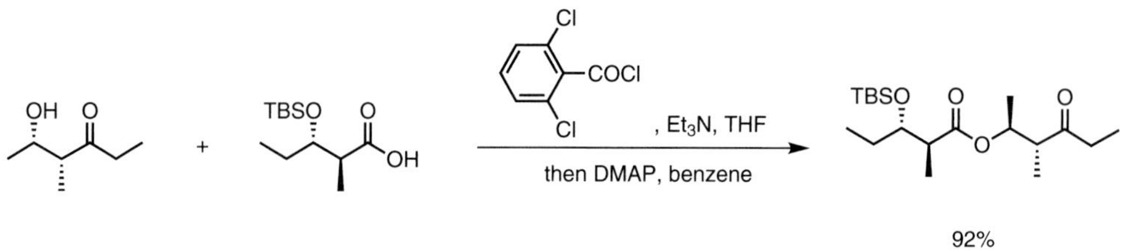

Cl
COCl
Cl , Et_3N, THF
then DMAP, benzene

92%

Scheme 1.223

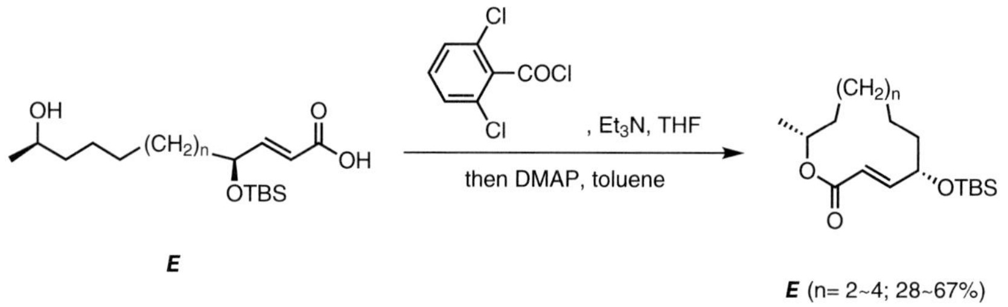

Cl
COCl
Cl , Et_3N, THF
then DMAP, toluene

E

E (n= 2~4; 28~67%)

Scheme 1.224

4-Trifluoromethylbenzoic acid anhydrides, when activated by $TiCl_4/AgClO_4$, enable nearly equimolar amounts of free carboxylic acids and alcohols to be employed, affording quantitative yields of the desired esters (Scheme 1.225) [616].

Experimental Procedure Scheme 1.225 [616] A mixture of 4-(trifluoromethyl) benzoic anhydride (740 mg, 2.04 mmol) and 3-phenylpropanoic acid (307 mg, 2.04 mmol) in dichloromethane (7.5 mL) and a solution of 1-methyl-3-phenylpropanol (278 mg, 1.85 mmol) in dichloromethane (2.5 mL) are added successively to a suspension of $AgClO_4$ (7.7 mg, 0.037 mmol), $TiCl_4$ (3.5 mg, 0.0185 mmol), and chlorotrimethylsilane (101 mg, 0.93 mmol) in dichloromethane (10 mL). The reaction mixture is kept stirring for 13 h at room temperature, and is then quenched with sat. aq. $NaHCO_3$. After conventional workup, the crude product is purified by preparative TLC on silica gel to afford 1-methyl-3-phenylpropyl 3-phenylpropanoate (518 mg, 1.83 mmol) in 99% yield.

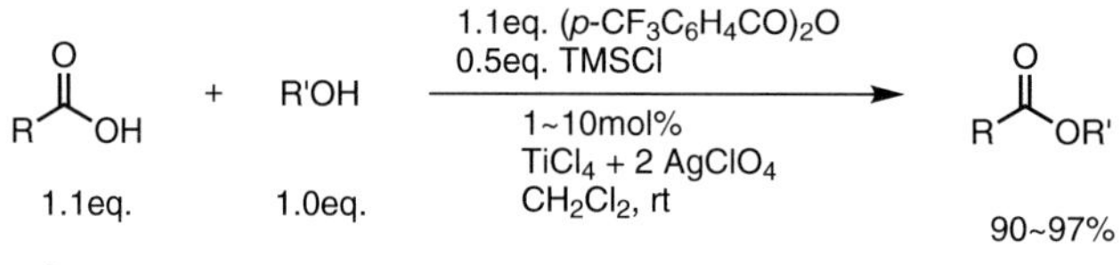

1.1eq. $(p\text{-}CF_3C_6H_4CO)_2O$
0.5eq. TMSCl
1~10mol%
$TiCl_4$ + 2 $AgClO_4$
CH_2Cl_2, rt

1.1eq. 1.0eq.

90~97%

Scheme 1.225

The employment of nearly equimolar quantities of reactants is also feasible with 2-methyl-6-nitrobenzoic anhydride (Scheme 1.226) [617–619]. The utility of 4-nitrobenzoic anhydride in the presence of scandium catalysts was described in Section 1.3.2.2.

Experimental Procedure Scheme 1.226 [617] DMAP (2.5 mg, 0.020 mmol), 2-methyl-6-nitrobenzoic anhydride (82.9 mg, 0.24 mmol) and 3-phenylpropanoic acid (36.3 mg, 0.24 mmol) are added to a solution of triethylamine (66.1 mg, 0.65 mmol) in dichloromethane (1.5 mL). After the mixture has been stirred for 10 min, a solution of 4-phenyl-2-butanol (30.1 mg, 0.20 mmol) in dichloromethane (2.0 mL) is added. The reaction mixture is stirred for 20 h at room temperature, and saturated aqueous ammonium chloride is then added. Conventional workup and purification of the mixture by TLC on silica gel afforded 1-methyl-3-phenylpropyl 3-phenylpropanoate (53.9 mg, 95%).

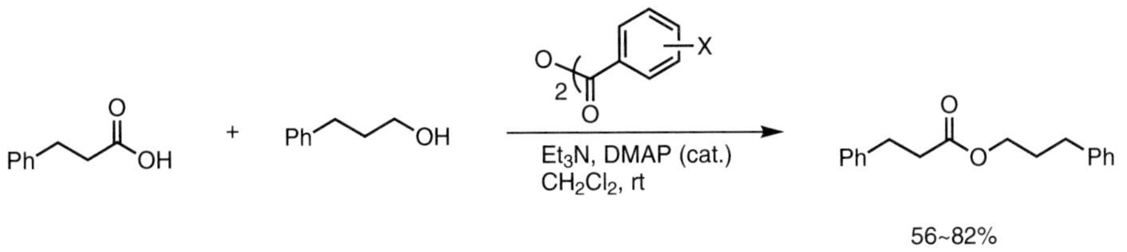

Ph OH (O) + Ph OH → [$(XC_6H_4CO)_2O$] Et_3N, DMAP (cat.) CH_2Cl_2, rt → Ph O Ph (O)

56~82%

Scheme 1.226

3,5-Bis(trifluoromethyl)benzoic acid is a useful acylating reagent when used in the presence of $TiCl_2(ClO_4)_2$ (20 mol%) and TMSCl (2 equiv.) [620]. The trifluoroacetic acid moiety acts as an activating group for acid anhydrides. Treatment of acid and alcohol in the presence of trifluoroacetic acid provides various sterically demanding esters [621]. A sterically bulky pivalic component increases the preference of mixed anhydrides for a primary hydroxy group over a secondary one. Accordingly, acetic pivalic or benzoic pivalic mixed anhydride is useful for selective acetylation or benzoylation of various sugar molecules [622].

Experimental Procedure [621] Method A: 9-Anthroic acid (2.00 g., 9.0 mmol) is suspended in benzene (40 mL), and trifluoroacetic anhydride (5.0 mL, 36 mmol) is added. The acid dissolves on gentle warming after 10 min. Methanol (5 mL) is then added. After a short period, aqueous sodium hydroxide (10%) is added to extract the acids. The benzene layer is washed with water and dried. Methyl 9-anthroate is isolated by removal of the solvent *in vacuo* and by recrystallization from ethanol or hexane.

Method B: Mesitoic acid (1.00 g., 6.1 mmol) and mesitol (0.83 g., 6.1 mmol) are treated with trifluoroacetic anhydride (5.0 mL, 36 mmol). The resulting solution is stirred at room temperature for 20 min. Benzene (20 mL) is added, and mesityl mesitoate is isolated as in method A.

Mixed carboxylic carbonic anhydrides are another useful type of acyl transfer reagent. Treatment of carboxylic acids with alkyl chloroformate affords alkoxycarbonyl esters, which can undergo facile transesterification. An ethoxycarbonyl derivative is employable for a key coupling process in the synthesis of a steroidal spin label compound (Scheme 1.227) [623].

Experimental Procedure Scheme 1.227 [623] 1-Carboxy-2,2,5,5-tetramethylpyrrolidine-*N*-oxyl (0.17 g, 0.9 mmol) is dissolved in dry tetrahydrofuran (3.5 mL), and a solution of freshly distilled ethyl chloroformate (0.08 mL, 0.9 mmol) and triethylamine (0.14 mL, 1.0 mmol) is added to it. The resultant mixture is stirred for 1 h, during which precipitation is observed. A solution of the protected alcohol (0.26 g, 0.6 mmol) in THF (5.0 mL) is added to this mixture, which is stirred at room temperature for 48 h. The solution is then filtered to remove the precipitated salt and the filtrate is concentrated *in vacuo* to give a yellow, viscous product. The crude product shows two close spots on TLC. Repeated column chromatographic purification on silica gel with ethyl acetate in petroleum ether as eluent affords the unreacted mixed anhydride (0.012 g, 5%) and the nitroxide ester as a yellow, viscous liquid (0.13 g, 37%).

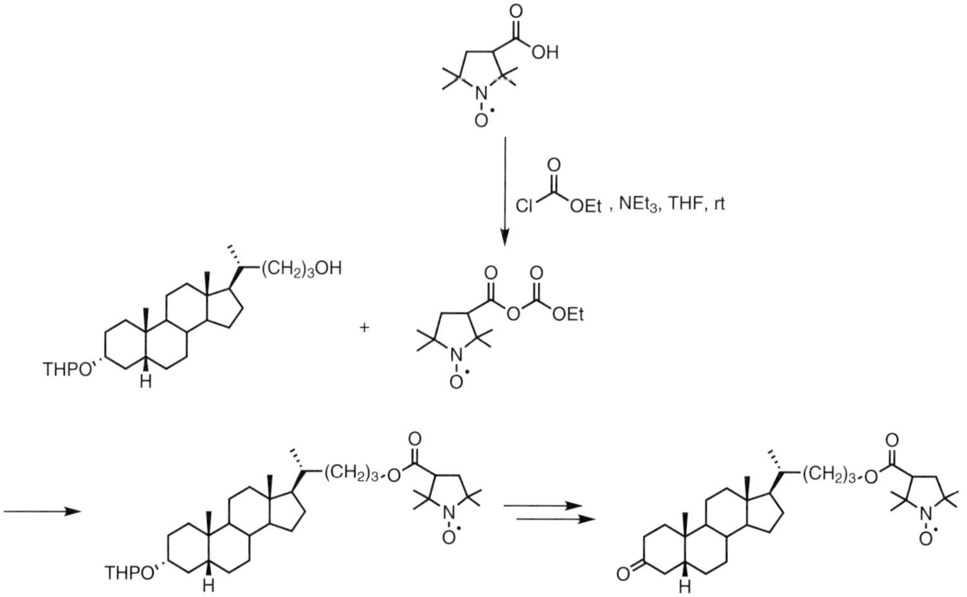

OH

Cl-C(O)-OEt, NEt_3, THF, rt

$(CH_2)_3OH$

THPO

H

+

OEt

$(CH_2)_3$-O

THPO

H

$(CH_2)_3$-O

O

H

Scheme 1.227

This procedure is effective for simultaneous *O*-esterification and *N*-carbalkoxylation (Scheme 1.228) [624].

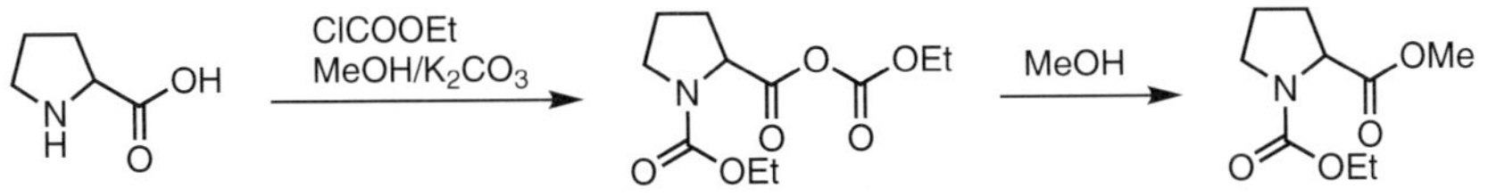

ClCOOEt
MeOH/K_2CO_3
MeOH

Scheme 1.228

Isopropenyl chloroformate is a more versatile reagent and is employable for esterification of various *N*-protected amino acids (Scheme 1.229) [625]. This methodology is applied to the synthesis of cyclic depsipeptide leualacin (Scheme 1.230) [626].

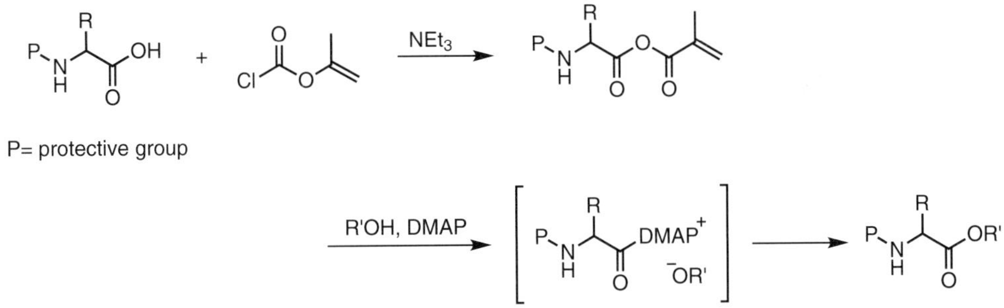

NEt_3

P= protective group

R'OH, DMAP

Scheme 1.229

, NEt_3

, DMAP

73%

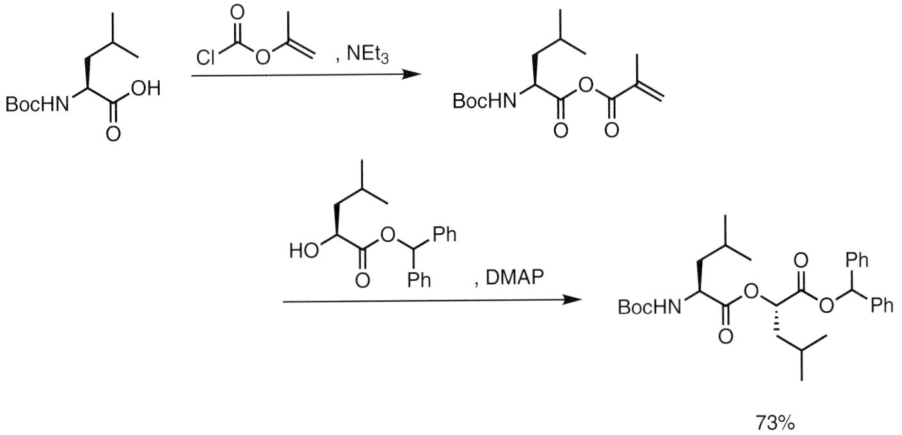

Scheme 1.230

Similar mixed anhydrides are obtained by treating carboxylic acids with di-*tert*-butyl dicarbonate (*t*-BOC) [627]. This method is applied to lactonization of ω-hydroxy carboxylic acids as shown in Scheme 1.231 [628].

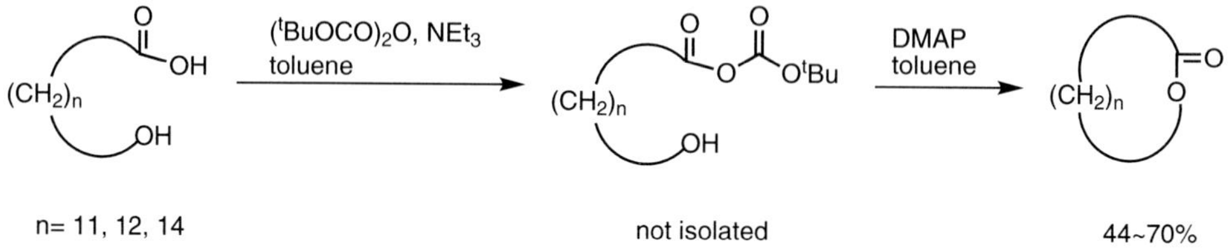

Scheme 1.231

Treatment of acid with 2-ethoxy-1-(ethoxycarbonyl)-1,2-dihydroquinoline in alcohol affords the ethoxylcarbonyl ester, which is spontaneously transformed into the corresponding ester (Scheme 1.232) [629].

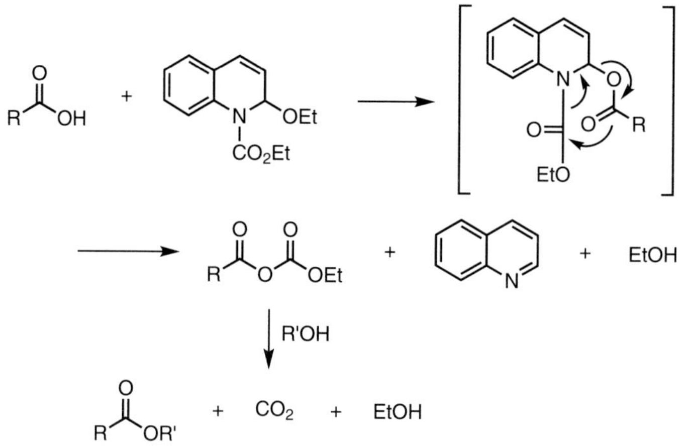

Scheme 1.232

Carboxylic carbonic anhydrides can be employed in lipase-catalyzed kinetic resolution of chiral α-substituted carboxylic acids (Scheme 1.233) [630].

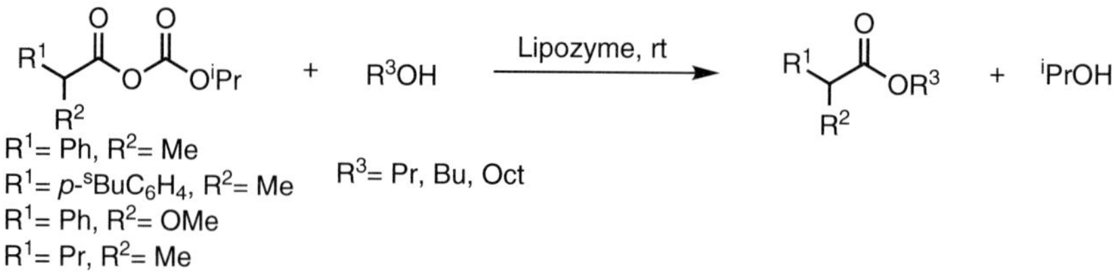

Scheme 1.233

Esterification of 1,4-dihydroxypyridine carboxylic acid is induced by formation of mixed acetic acid anhydride (Scheme 1.234) [631].

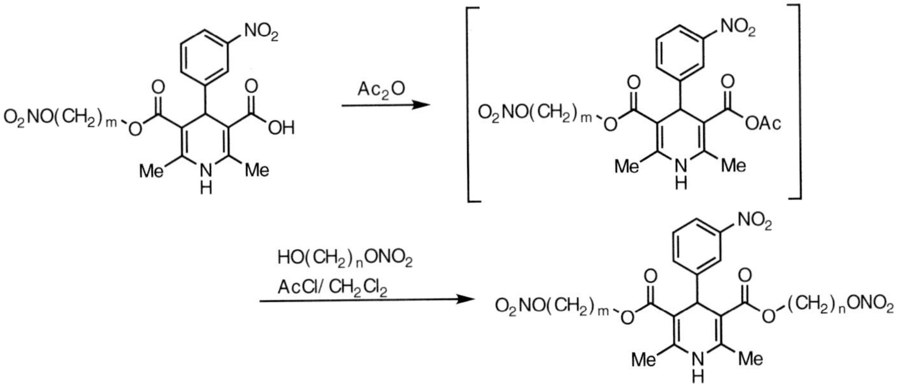

Scheme 1.234

On treatment with 1-*tert*-butoxy-2-*tert*-butoxycarbonyl-1,2-dihydroisoquinoline, *N*-protected amino acids are *in situ* converted into mixed anhydrides, which are transformed into the corresponding esters by treatment with alcohols (Scheme 1.235) [632].

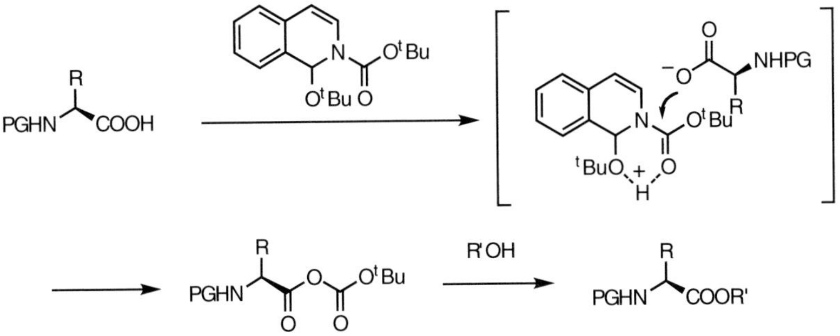

Scheme 1.235

Acylation of cellulose is effected in DMSO/tetrabutylammonium fluoride solvent with use of mixed anhydrides derived from *N,N′*-carbonyldiimidazole and carboxylic acid (Scheme 1.236) [633].

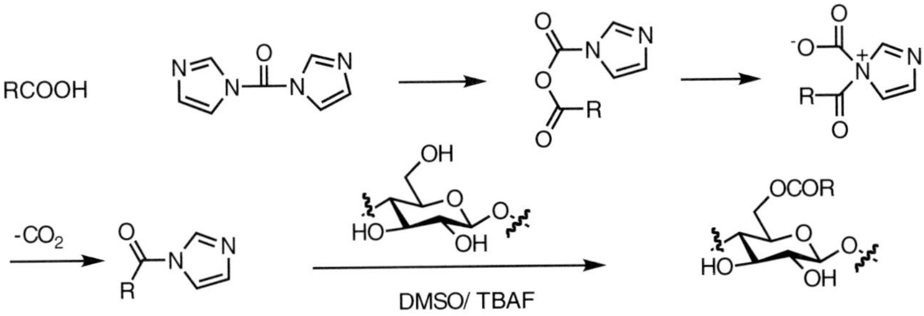

Scheme 1.236

1.4
Reaction with Acid Halides and Other Acyl Derivatives

1.4.1
Without Activator

Acid halides are acylating reagents just as classical as acid anhydrides. Acid halides are more reactive than acid anhydrides, so reactions with alcohols usually proceed spontaneously at room temperatures. Sterically demanding alcohols require higher reaction temperatures. Despite such high reactivity, acid halides are less popular than acid anhydrides, because of their labile nature. It is also not convenient that their reactions with alcohols produce hydrogen halides. Nevertheless, if the vigor of the reaction is controlled and the hydrogen halides are trapped effectively, a useful acylation methodology is attainable. One such example is seen in the synthesis of bis(2-cyanoacrylates), used as cross-linking agents for cold-setting glues (Scheme 1.237) [634]. Treatment of 2-cyanoacryloyl chloride with 2-butene- or 2-butyne-1,4-diol at 20 °C provides unsaturated bis(2-cyanoacrylates). It should be noted that direct esterification with 2-cyanoacrylic acid by *p*-toluenesulfonic acid catalysis does not work for unsaturated diols.

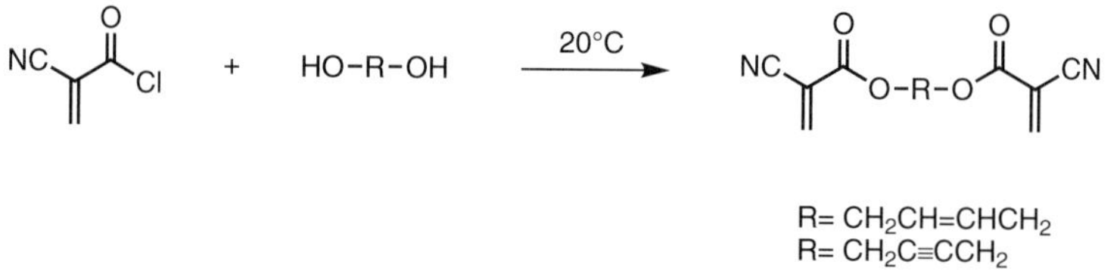

NC, Cl, O + HO–R–OH —20°C→ NC, O–R–O, CN

R= $CH_2CH{=}CHCH_2$
R= $CH_2C{\equiv}CCH_2$

Scheme 1.237

N-Phthaloylglutamic acid is transformed into the corresponding 4-brominated methyl or ethyl ester by treatment with bromine in the presence of PBr_5 under irradiation followed by direct alcoholysis (Scheme 1.238) [635].

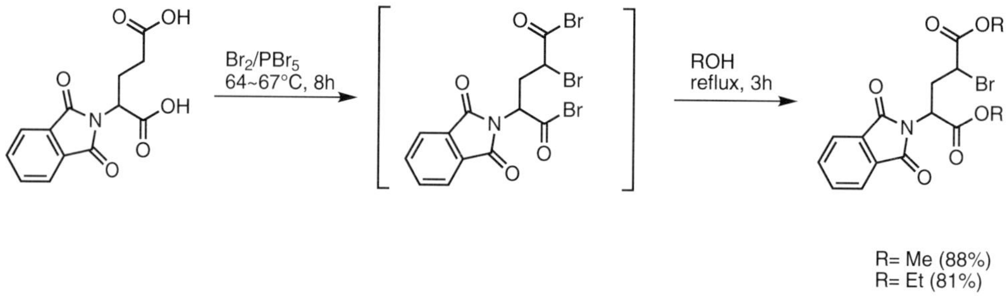

Br_2/PBr_5, 64~67°C, 8h; ROH, reflux, 3h

R= Me (88%)
R= Et (81%)

Scheme 1.238

Other acyl derivatives such as amides and thioesters are stable but less reactive, and so are not easily transformed into esters unless activated in some manner.

The inertness of amides is attributable to the resonance of the nitrogen lone pair electrons with the carbonyl group. Twisted amides, which lose the resonance energy through their non-planarity, function as acylating reagents without activator [636]. 3-Pivaloyl-1,3-thiazolidine-2-thione, for example, selectively pivaloylates primary alcohols in preference to secondary ones and phenols (Scheme 1.239), the bulky pivaloyl group not allowing the substrates to adopt a planar conformation. This situation brings about a unique reversal of reactivity in terms of steric bulk: less bulky acetyl and benzoyl analogs, normally more reactive than the pivalates, exhibit no acylation ability.

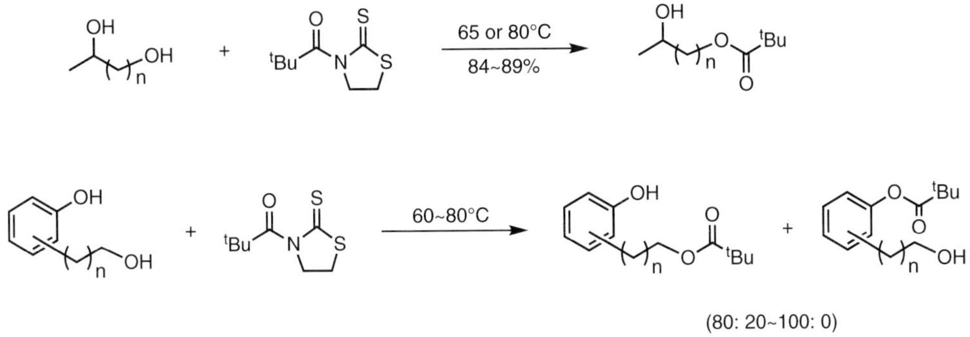

65 or 80°C
84~89%

60~80°C

(80: 20~100: 0)

Scheme 1.239

Trifluoroacetylations that cannot usually be readily done are feasible with great ease through the use of (trifluoroacetyl)benzotriazole, prepared from trifluoroacetic anhydride and benzotriazole (Scheme 1.240) [637]. A variety of trifluoroacetates are obtained simply by heating with alcohol and phenol. Notably, *p*-nitrophenol, which is difficult to trifluoroacetylate by other means, undergoes the desired transformation smoothly.

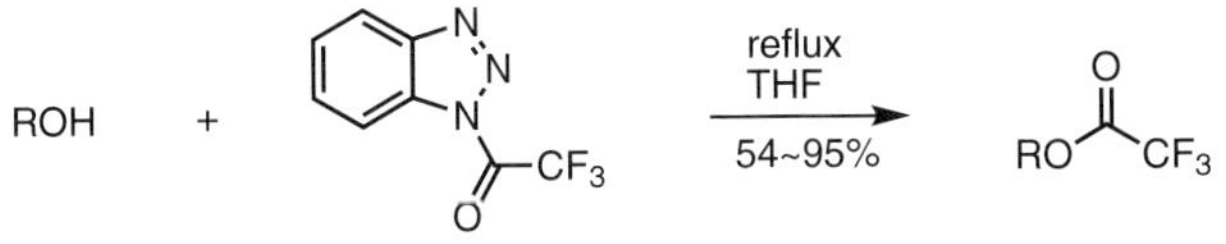

ROH + ... reflux THF, 54~95% → RO(CO)CF_3

Scheme 1.240

1.4.2
Acid Catalysts

1.4.2.1 Bønsted Acids

While acid catalysts are useless for the reactions between acid halides and alcohols, amides can be transformed into esters with catalysis by various acids.

Some Brønsted acids serve for this purpose. Lactonization of hydroxy amide by aqueous HCl is utilized for the penultimate step in the synthesis of fusarentin methyl ethers (Scheme 1.241) [638].

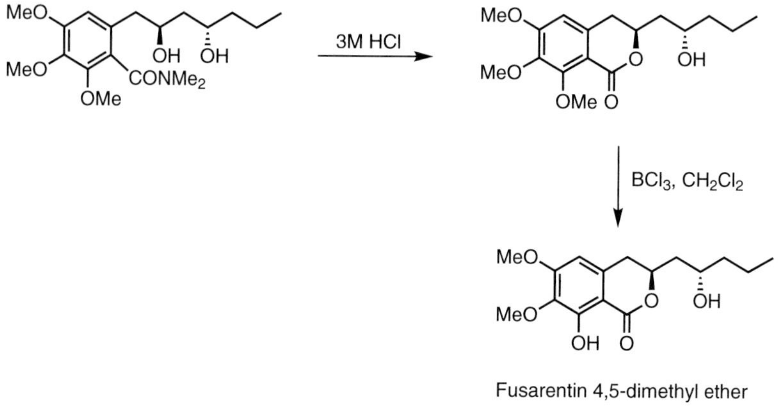

3M HCl

BCl_3, CH_2Cl_2

Fusarentin 4,5-dimethyl ether

Scheme 1.241

Alcoholysis of lactams, which is of great synthetic use for the synthesis of amino acid esters, can be achieved with Brønsted acids. α-Hydroxy β-lactams undergo methanolysis by action of dry HCl produced *in situ* from TMSCl and methanol (Scheme 1.242) [639]. The α-hydroxy β-amino acid moiety occurs in taxol and bestatin.

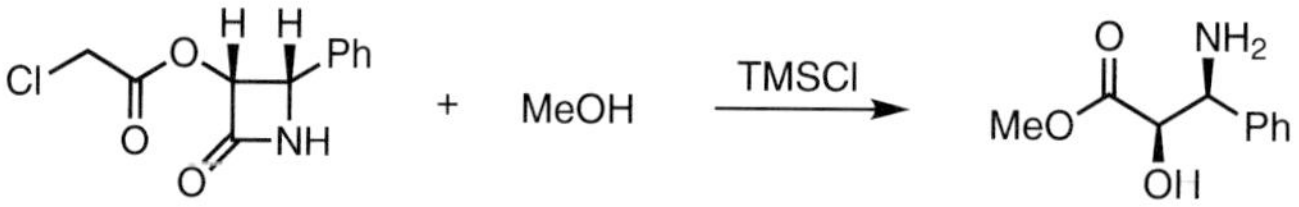

\+ MeOH —TMSCl→

Scheme 1.242

Alkoxide ions cannot be employed for alcoholysis of lactams bearing *N*-alkoxycarbonyl or -arylsulfonyl groups, because the nucleophilic attack may occur on the *N*-substituent as well as the desired lactam carbonyl. Lewis acid-mediated reactions avoid such problems. *p*-Toluenesulfonic acid catalyzes alcoholysis of lactams possessing *N*-COOMe, -CSSMe, or -SO_2Ar groups without cleavage of these groups (Scheme 1.243) [640].

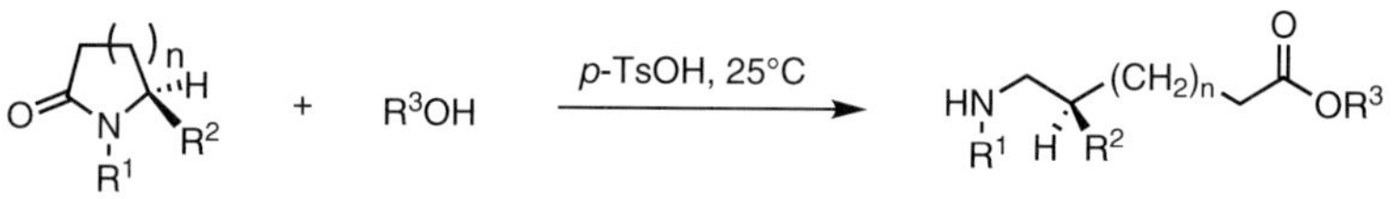

\+ R^3OH —*p*-TsOH, 25°C→

R^1= COOMe, CSSMe, Ts

Scheme 1.243

1.4.2.2 **Lewis Acids**

Reaction between alcohols and acid halides on chiral templates can result in kinetic resolution of racemic secondary alcohols. Racemic secondary alcohols are resolved by their reaction with benzoyl halides in the presence of tin(II) dihalide/chiral diamine /molecular sieves (Scheme 1.244) [641].

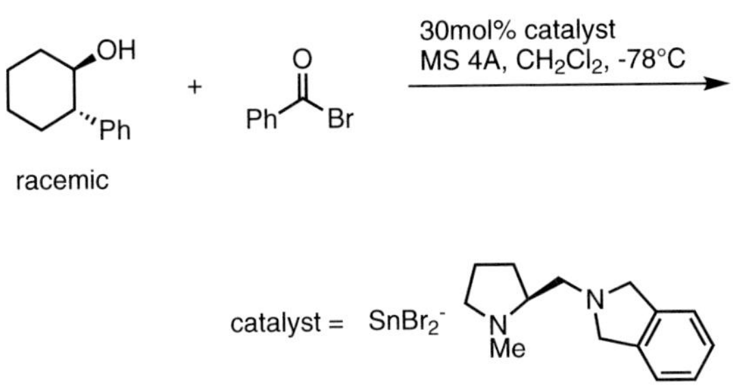

30mol% catalyst
MS 4A, CH_2Cl_2, -78°C

racemic

97% *ee*   84% *ee*

catalyst = $SnBr_2$ · N-Me

Scheme 1.244

Crown ether esters are obtained by the acid halide approach when polyether diols are treated with dicarboxylic acid dichlorides in the presence of Ph_3M (M = Sb, Bi) (Scheme 1.245) [642]. A unique template effect is responsible for this reaction

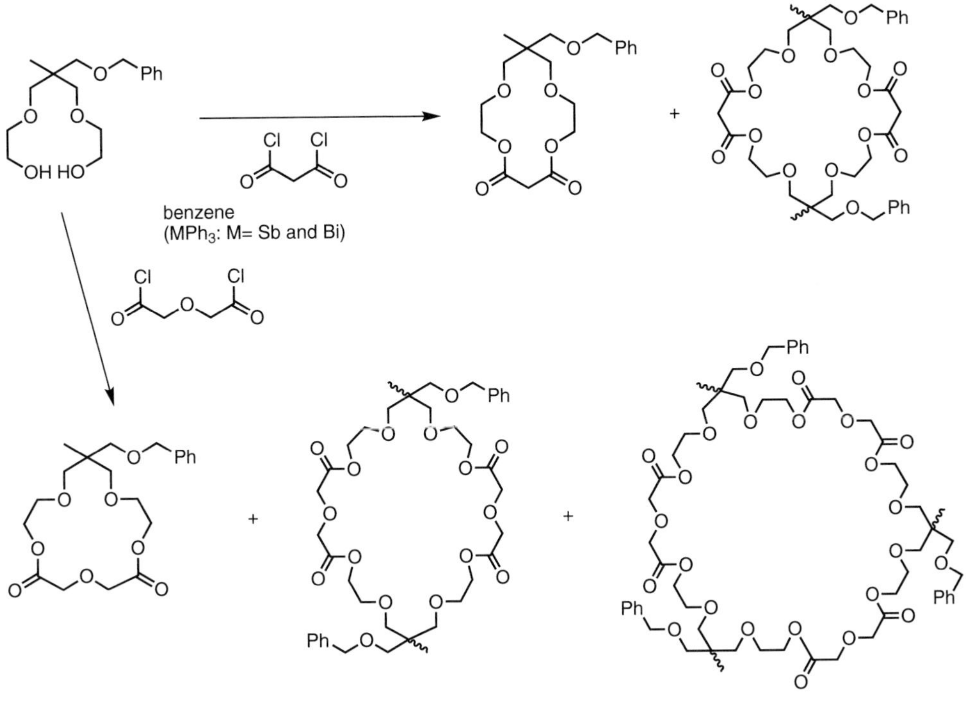

benzene
(MPh_3: M= Sb and Bi)

Scheme 1.245

(Scheme 1.246). $La(NO_3)_3 \cdot 6H_2O$ is employable for pivaloylation of primary alcohols [643] as well as acetylation of phenols [644]. $CeCl_3$ catalyzes acylation of primary and secondary alcohols upon treatment with acyl chlorides, leaving phenolic hydroxy groups intact [645].

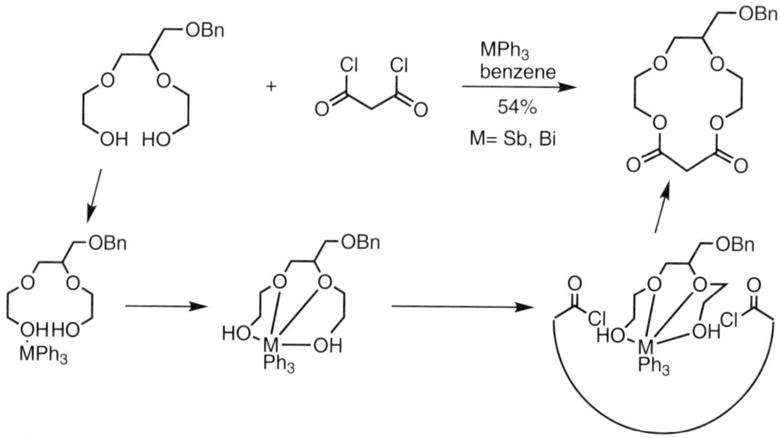
OBn
OH HO
Cl Cl
MPh3
benzene
54%
M= Sb, Bi
OHHO
MPh3
HO
M
OH
Ph3
Cl
Cl

Scheme 1.246

Alcoholysis of *N*-unsubstituted amides and *O*-methylhydroxamates proceeds smoothly with a catalytic amount of $TiCl_4$ and one equivalent of aqueous HCl in alcohol solvent (Scheme 1.247) [646]. However, *N*-alkyl and *N,N*-dialkyl carboxamides are not affected under these conditions. Although the reaction occurs in the absence of HCl, addition of one molar equivalent of aqueous HCl increases the rate considerably. $Ti(OR)_4$ species also catalyze the reaction under similar conditions. These results suggest that the Ti(IV) species involved in the alcoholysis reaction is not $TiCl_4$, but rather some intermediate $TiCl_x(OR)_{4-x}$ complex. Boron trifluoride etherate, tin tetrachloride, and silicon tetrachloride also catalyze the reaction, but with less efficiency.

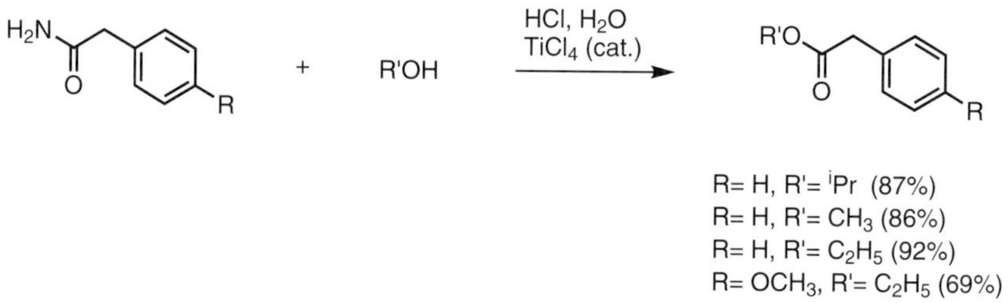
H2N
R
R'OH
HCl, H2O
TiCl4 (cat.)
R'O
R
R= H, R'= iPr (87%)
R= H, R'= CH3 (86%)
R= H, R'= C2H5 (92%)
R= OCH3, R'= C2H5 (69%)

Scheme 1.247

Conversion of acyloxazolidinones into the corresponding methyl or benzyl esters is feasible with LaI_3 catalyst (Scheme 1.248) [647].

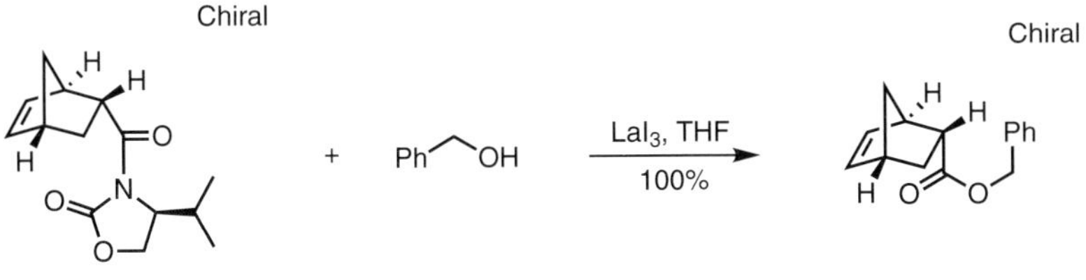
Chiral
H
H
O
H
N
O
O
+
Ph
OH
LaI3, THF
100%
Chiral
H
H
Ph
H
O
O

Scheme 1.248

Hydrazides are transformed into the corresponding esters upon treatment with thallium(III) nitrate trihydrate (2 equiv.) in alcohol (Scheme 1.249) [648]. The reaction is triggered by oxidation of the hydrazides to acyl diimides.

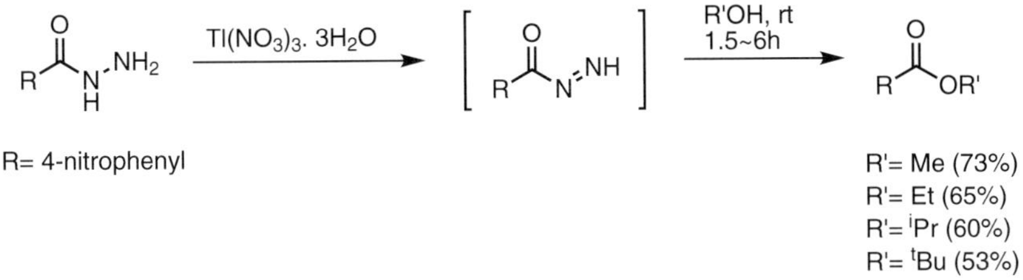
O
R
N
NH2
H
Tl(NO3)3. 3H2O
O
R
N
NH
R'OH, rt
1.5~6h
O
R
OR'
R= 4-nitrophenyl
R'= Me (73%)
R'= Et (65%)
R'= iPr (60%)
R'= tBu (53%)

Scheme 1.249

Thioesters, though stable, can be converted into the corresponding esters. Reaction between thioesters and alcohols is activated by thiophilic metal cations (Scheme 1.250) [649, 650]. Various sterically demanding thioesters and alcohols, even tertiary alcohols, can be employed. Soft metal salts of Hg(II), Ag(I), Cu(I), and Cu(II) are effective. The synthetic utility of this protocol is apparent from the successful lactonization as shown in Scheme 1.251.

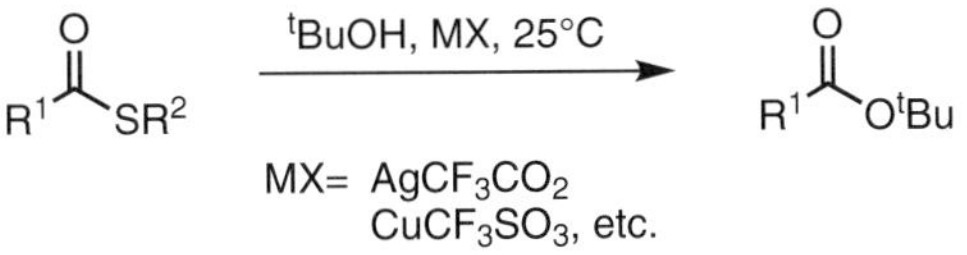
O
R1
SR2
tBuOH, MX, 25°C
MX= AgCF3CO2
CuCF3SO3, etc.
O
R1
OtBu

Scheme 1.250

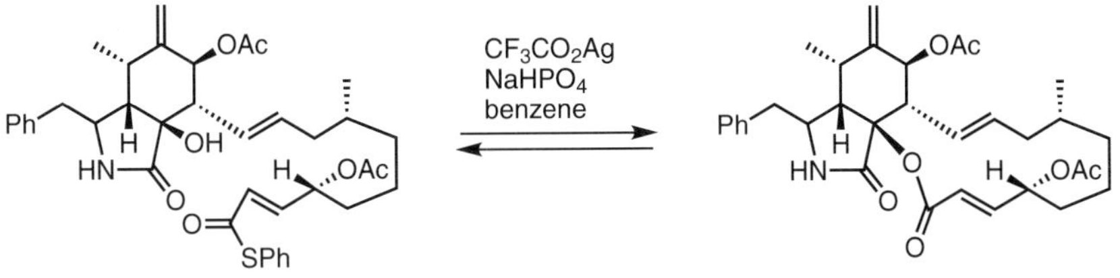
OAc
Ph
H
OH
HN
O
H
OAc
O
SPh
CF3CO2Ag
NaHPO4
benzene
OAc
Ph
H
HN
O
O
H
OAc
O

Scheme 1.251

The combined use of $Hg(OCOCF_3)_2$ and $BF_3 \cdot OEt_2$ brings about the S-to-O ester conversion (Scheme 1.252) [651].

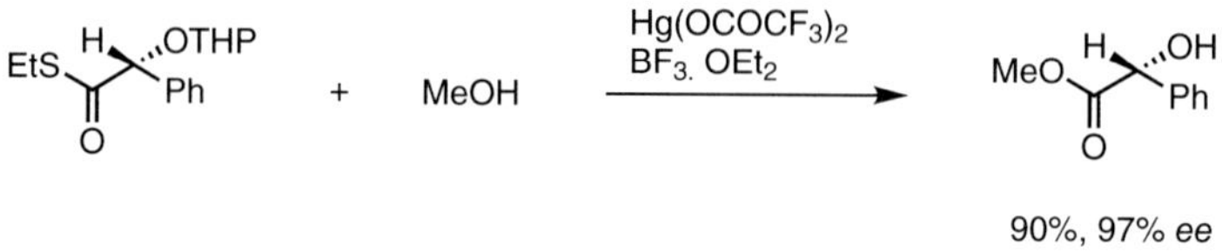

EtS, H, OTHP, Ph, O + MeOH → ($Hg(OCOCF_3)_2$, $BF_3 \cdot OEt_2$) MeO, H, OH, Ph, O

90%, 97% *ee*

Scheme 1.252

A thioethyl ester of 2-hydroperfluoro acid is transformed into the corresponding ethyl ester by use of $Ti(O^iPr)_4$ (Scheme 1.253) [652]. It should be noted that the acidic procedure is crucial for this transformation, since base-induced transesterification is not available, because of the acidic α-proton. However, this procedure is employable only with the thioethyl ester, the volatility of the ethanethiol being able to favor the equilibrium toward the *O*-ester.

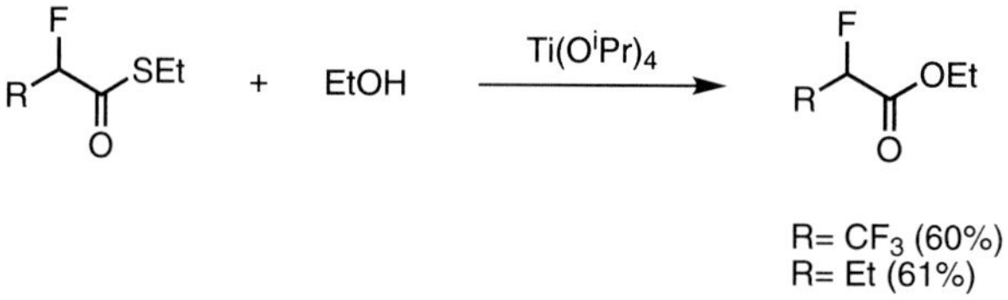

R, F, SEt, O + EtOH → ($Ti(O^iPr)_4$) R, F, OEt, O

R= CF_3 (60%)
R= Et (61%)

Scheme 1.253

1.4.2.3 **Solid Acids**

Montmorillonite K-10 and KSF are efficient catalysts for acylation of alcohols with acid chlorides [549]. Microwave irradiation of a mixture of naphthol, acyl bromide, and a copper/cupric chloride composite induces solventless reaction. Both α- and β-naphthols undergo acylation [653]. Reactions between acyl chlorides and *tert*-butyl alcohol to afford *tert*-butyl esters are promoted by activated alumina [654]. Notably, the reaction does not involve a ketene intermediate, often encountered in the chemistry of acid halides, and so optically active Naproxen, which is quite susceptible to ketene formation from its acid chloride, can be transformed into the corresponding *tert*-butyl ester without racemization (Scheme 1.254).

Alumina is also used for selective acetylation of unsymmetrical diols [655]. The primary alcohol moieties in 2-substituted 1,5-diols are selectively acetylated over the secondary alcohols upon treatment with acetyl chloride, although the selectivity is not particularly high (monoacetate : diacetate $< 60:20$) (Scheme 1.255). ZnO acts as a reusable catalyst for reactions between alcohols or phenols with acid halides [656].

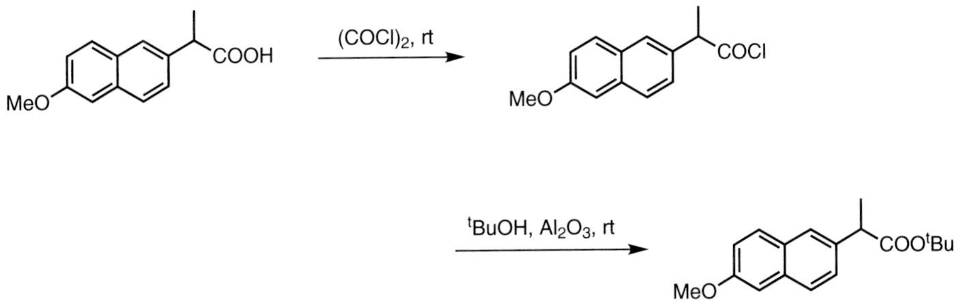

$(COCl)_2$, rt

^tBuOH, Al_2O_3, rt

Scheme 1.254

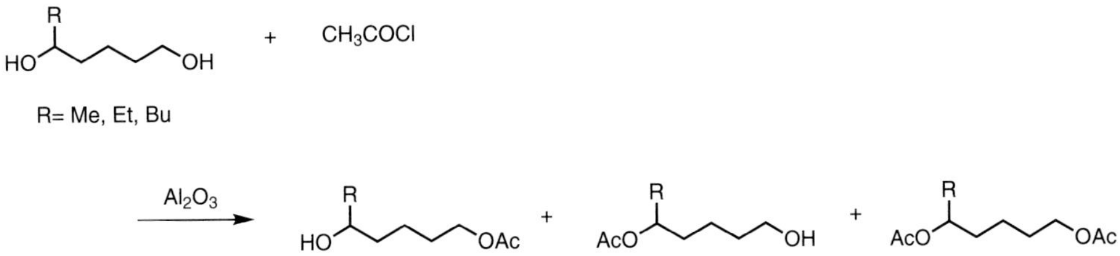

+ CH_3COCl

R= Me, Et, Bu

Al_2O_3

Scheme 1.255

A mild and selective conversion of carboxamides and carboxylic acid hydrazides into the corresponding esters is brought about by the use of acidic resins (Scheme 1.256) [657]. Heating of the substrates with Amberlyst 15 or XN-10101 or Amberlite 120 in refluxing methanol or ethanol provides the corresponding esters. The process is specific for unsubstituted carboxamides, with even an *N*-methyl substituent preventing the reaction from proceeding.

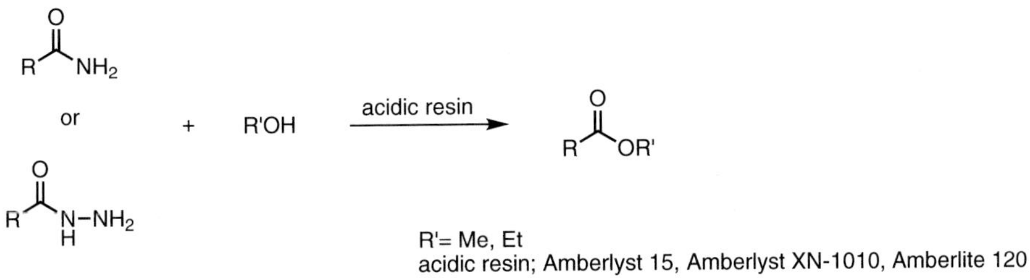

or + R'OH

acidic resin

R'= Me, Et
acidic resin; Amberlyst 15, Amberlyst XN-1010, Amberlite 120

Scheme 1.256

1.4.3
Base Activators

1.4.3.1 Metal Salts

Treatment of alkoxide ions with acid halides and other acyl derivatives is a common way to arrive at esters. Treatment of acid chlorides with lithium alkoxides prepared from BuLi and alcohols affords esters (Scheme 1.257) [658]. This reaction allows the use of sterically demanding alcohols, such as tertiary alcohols, as well as acid-sensitive alcohols, but is limited to acid chlorides that do not possess labile α-hydrogens or at least to those halides that produce ketenes that either are not volatile or do not dimerize.

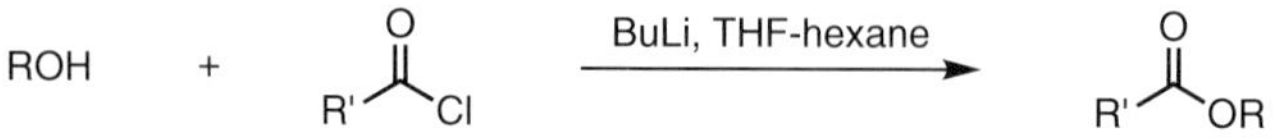

ROH + R'C(O)Cl —(BuLi, THF-hexane)→ R'C(O)OR

Scheme 1.257

Treatment of the dipotassium salt of *p-tert*-butylcalix[4]arene with diacid halides results in bridging at the lower rim (Scheme 1.258) [659]. The reaction is dependent on the shape of the bridging unit: ring-closure reaction takes place only when fitting occurs between the calix[4]arene and the capping reagent.

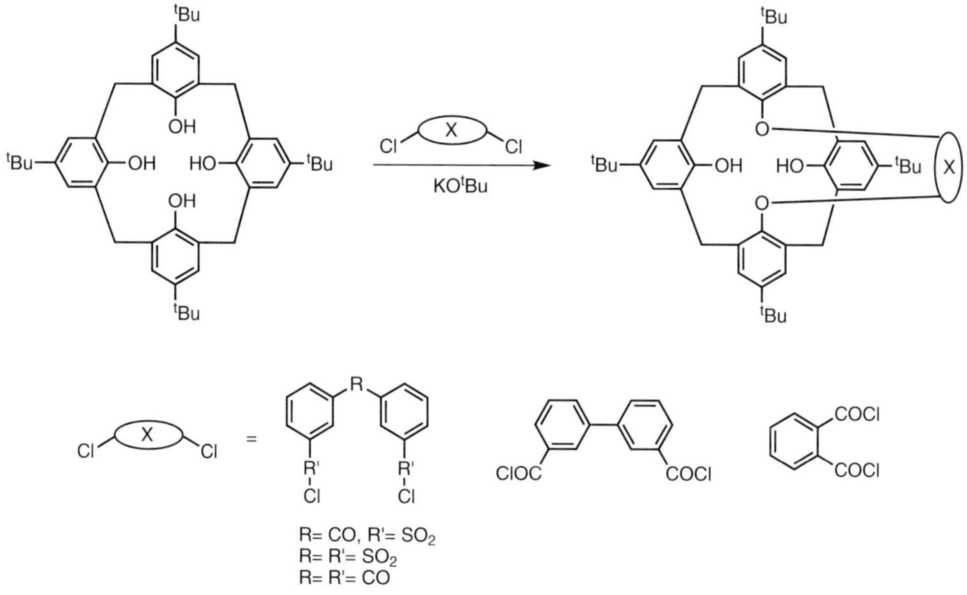

R= CO, R'= SO_2
R= R'= SO_2
R= R'= CO

Scheme 1.258

The dialkoxide/diacid halide procedure is applicable to kinetic resolution. Treatment of methyl 4,6-*O*-benzylidene-α-D-glucopyranoside with hexamethoxydiphenic acid dichloride provides only *RD* cyclic diester (Scheme 1.259) [660]. The reaction is highly stereoselective, and so the *R* isomer is obtained in 38% yield (76% theoretical yield) and the *RD*/*SD* ratio of the isomers is >1500:1. The *SD* isomer-selective reaction is best performed with Et_3N, but the selectivity is only 4.4:1. The enantiopure diphenic acid is used for synthesis of trideca-*O*-methyl-α-pedunculagin [661] and ferrocenyl allagitannins [662].

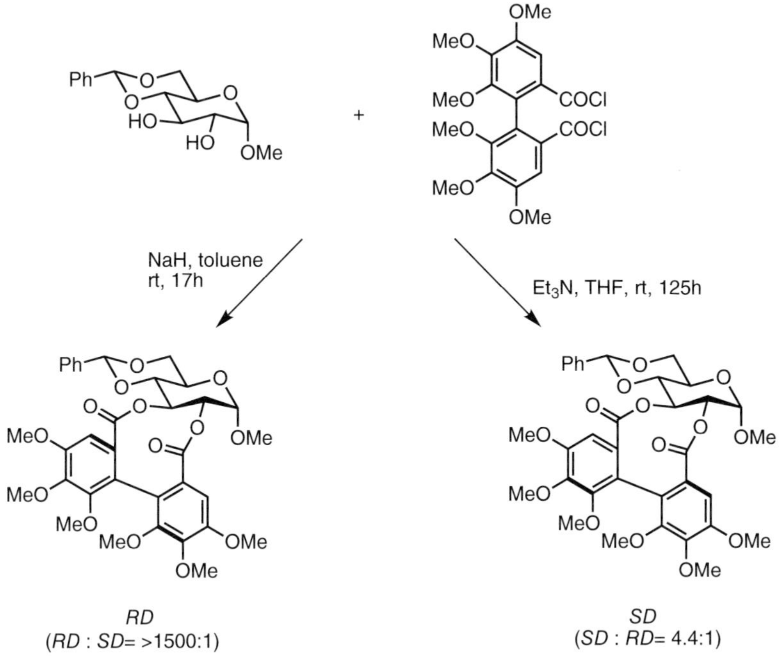

RD
(*RD* : *SD*= >1500:1)

SD
(*SD* : *RD*= 4.4:1)

Scheme 1.259

In the synthesis of the BCD carbohydrate domain of calicheamicin, the coupling between acyl chloride and a cyclohexanol unit shown in Scheme 1.260 is best performed by use of NaH [663]. Use of Et_3N/DMAP or BuLi results in poor yields.

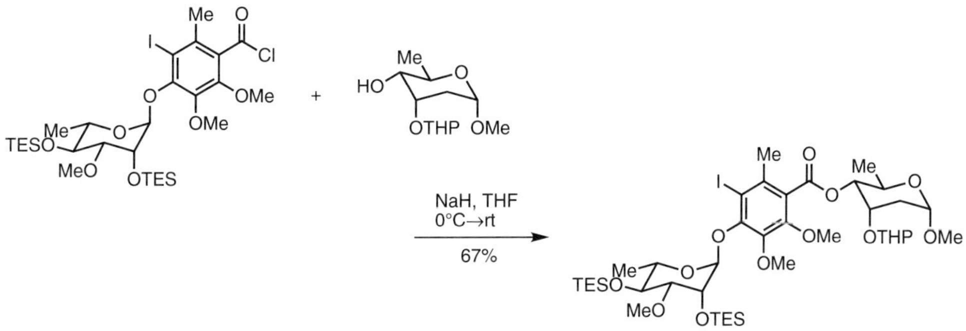

Scheme 1.260

Acyl fluorides are more reactive than the chlorides, and so the acylation of a sterically demanding alcohol is achievable, as shown in Scheme 1.261 [664]. A 70% yield of the desired ester is obtained by use of sodium hexamethylsilazane or BuLi.

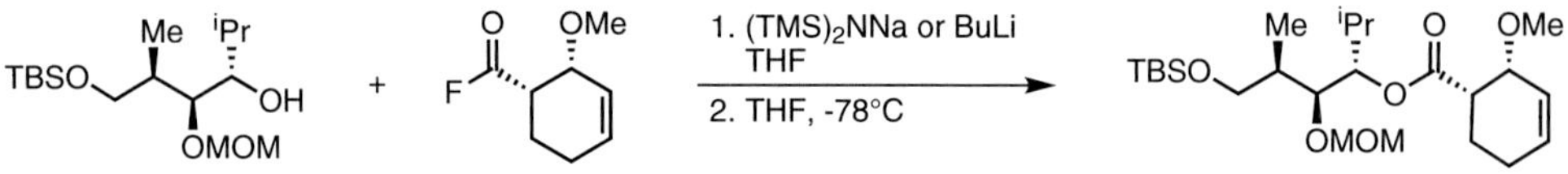

Scheme 1.261

This method is successfully applied to the synthesis of glucuronide conjugates of retinoid carboxylic acids [665]. Retinoic acid is converted into the corresponding acyl fluoride upon treatment with diethylaminosulfur trifluoride (Scheme 1.262). This product, after purification, is treated with the sodium salt of glucuronic acid in the presence of $NaHCO_3$ in acetone/water to provide the desired ester.

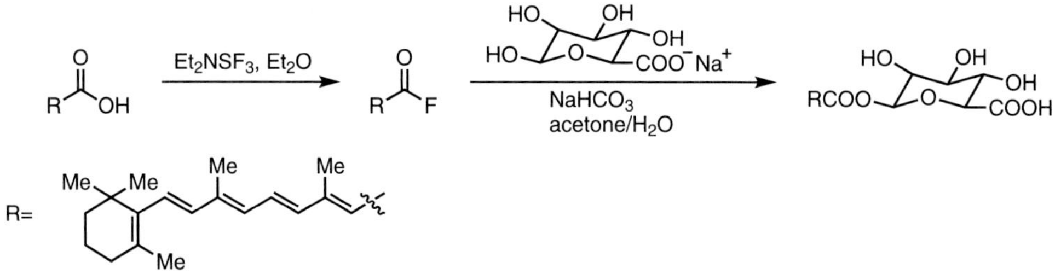

Scheme 1.262

Ring-opening of lactams is frequently required in organic synthesis. *N*-Boc-Lactams can be converted into amino esters by treatment with sodium methoxide in methanol (Scheme 1.263) [666]. This methodology is also applied to secondary amide substrates.

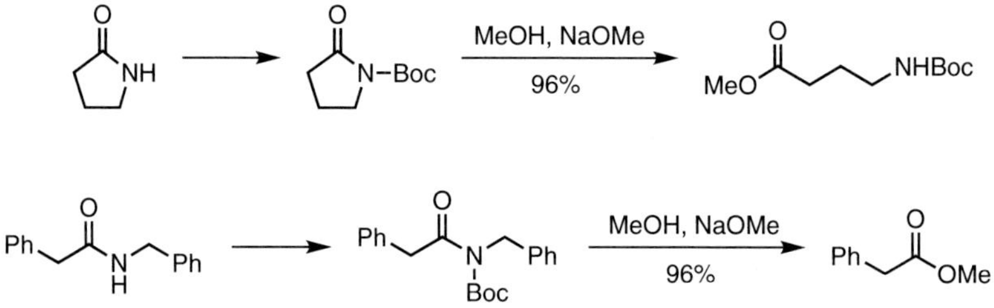

Scheme 1.263

The 2-azetidinone ring undergoes alcoholysis by catalytic NaOH (Scheme 1.264) [667]. The facile reactivity is attributable to the presence of the imino group attached to N1, which labilizes the 2-azetidinone ring toward nucleophiles, probably reducing the amide resonance and hence the strength of the N1—C2 bond.

1-Acetyl-*v*-triazolo/4,5-*b*/pyridine works for selective acetylation of phenol in preference to primary alcohol (Scheme 1.265) [668]. ω-Hydroxyalkylphenols undergo monoacetylation on the phenolic alcohol when treated with this reagent in aqueous NaOH/THF.

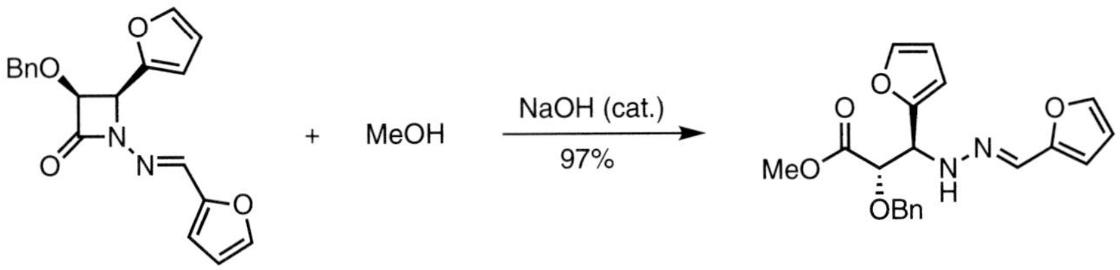

+ MeOH —NaOH (cat.), 97%→

Scheme 1.264

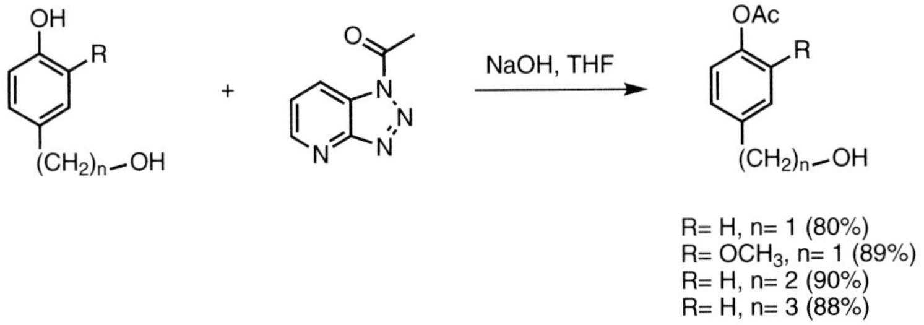

+ —NaOH, THF→

R= H, n= 1 (80%)
R= OCH_3, n= 1 (89%)
R= H, n= 2 (90%)
R= H, n= 3 (88%)

Scheme 1.265

Silver cyanide works as a more active promoter than the conventional alternatives for reactions between acid chlorides and alcohols [669]. Various sterically hindered acid chlorides such as pivaloyl, 2,2-diethylbutyryl, mesitoyl, and 2-methylpentanoyl chlorides are therefore employable.

1.4.3.2 **Amines**

Acylation of alcohols by acid halides is most conveniently achieved in the presence of amine(s). Pyridine is among the most popular, and triethylamine is the second most popular. DMAP follows the above amines, but is much less common, because the *N*-acylpyridinium halide intermediate emerging from interaction between DMAP and acid halide is a very tightly bound ion pair, and so its reactivity is low [585]. Nevertheless, the combined use of catalytic DMAP/tertiary amine is very powerful. Since these procedures are now well known, the standard procedures are not discussed in this section, but more specialized examples are the subjects of choice.

The tertiary amine/acid halide procedure is relatively mild, so the selective acylation can be performed. Treatment of calix[4]arenes with various acid halides in the presence of Et_3N affords only the distal isomers of the diesters, no proximal counterparts being detected (Scheme 1.266) [670]. Spiro[4,4]nonane-1,6-diol undergoes stepwise acylation on subjection to 2-naphthoyl chloride and methacroyl chloride in succession in the presence of Et_3N (Scheme 1.267) [671].

An α,β-dihydroxy ester is selectively acylated at the α-position upon treatment with benzoyl chloride in pyridine (Scheme 1.268) [672]. The β-hydroxy group is much less reactive, particularly when a bulky substituent is present at the γ-position. Benzoylation of the diol shown in Scheme 1.269 with benzoyl chloride/DMAP gives selective reaction at the C2 site [673]. The regioselective acylation of unsymmetrical diol is also seen in Scheme 1.270 [674].

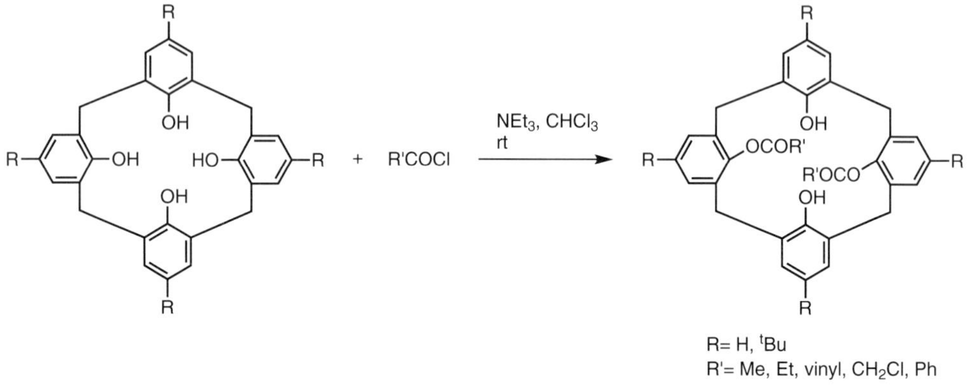

NEt$_3$, CHCl$_3$ rt

R'COCl

R= H, ^{t}Bu
R'= Me, Et, vinyl, CH$_2$Cl, Ph

Scheme 1.266

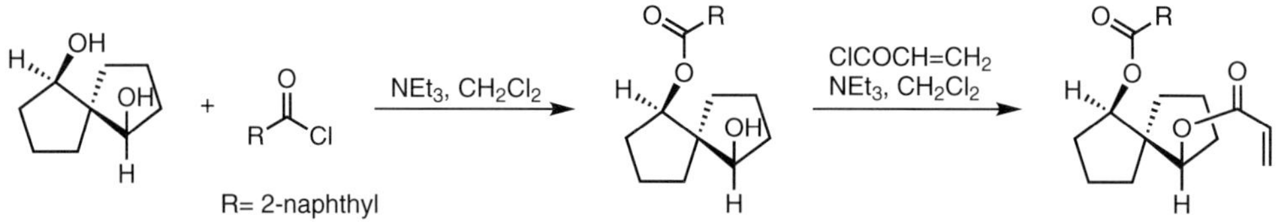

NEt$_3$, CH$_2$Cl$_2$

ClCOCH=CH$_2$
NEt$_3$, CH$_2$Cl$_2$

R= 2-naphthyl

Scheme 1.267

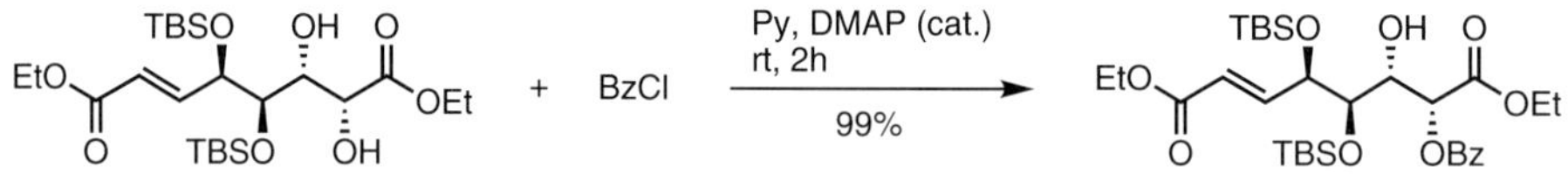

Py, DMAP (cat.)
rt, 2h
99%

Scheme 1.268

BzCl, DMAP, Py

Scheme 1.269

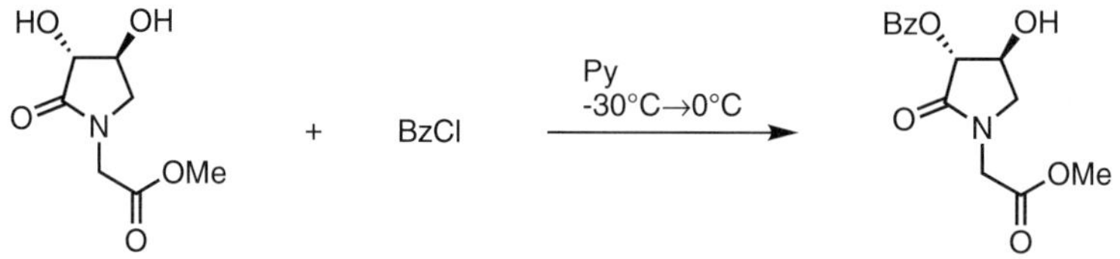

Py
-30°C→0°C

Scheme 1.270

When 2-acetoxypurpurin is treated with benzoyl chloride/pyridine, the 1-acetoxy-2-benzoyloxy derivative is obtained (Scheme 1.271) [675]. Migration of the acetyl group from position 2 to position 1 occurs, probably under the influence of pyridine, and the benzoyl group enters in place of the acetate.

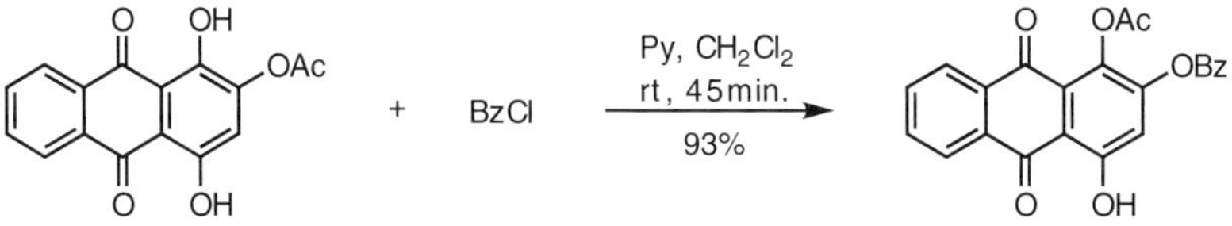

Py, CH_2Cl_2, rt, 45 min., 93%

Scheme 1.271

Asymmetric acylation of meso diols are feasible with benzoyl chloride in the presence of a chiral diamine derived from (*S*)-proline to give monobenzoates in high optical yield (Scheme 1.272) [676].

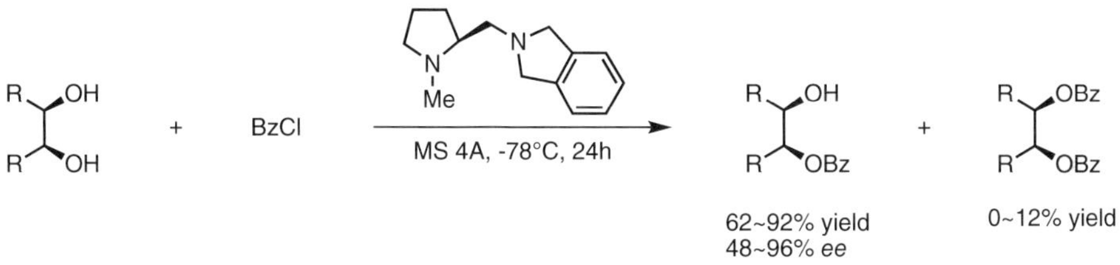

MS 4A, -78°C, 24h

62~92% yield
48~96% *ee*

0~12% yield

Scheme 1.272

myo-Inositol undergoes selective benzoylation upon treatment with benzoyl chloride, to give the symmetrical 1,3,5-tri-*O*-benzoyl-*myo*-inositol (Scheme 1.273) [677]. Methyl 6-*O*-(*tert*-butyldiphenyl)silyl-α-D-mannopyranoside is benzoylated selectively at the 3-position (Scheme 1.274) [678]. The use of 1.0 equiv. of benzoyl chloride is crucial for monobenzoylation; when more than 1 equiv. of this reagent is used, a considerable amount of 2,3-dibenzoate is produced.

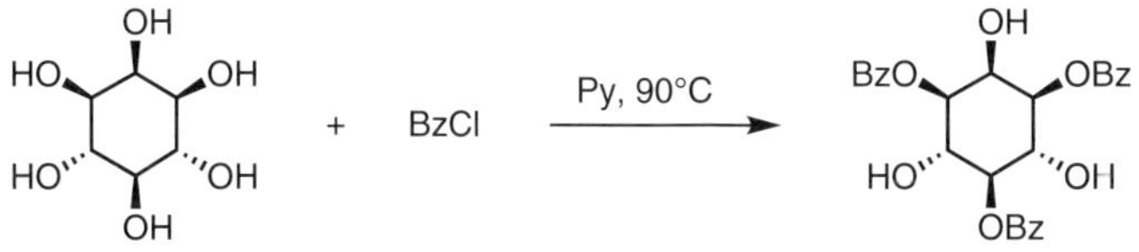

Py, 90°C

Scheme 1.273

Py, CH_2Cl_2

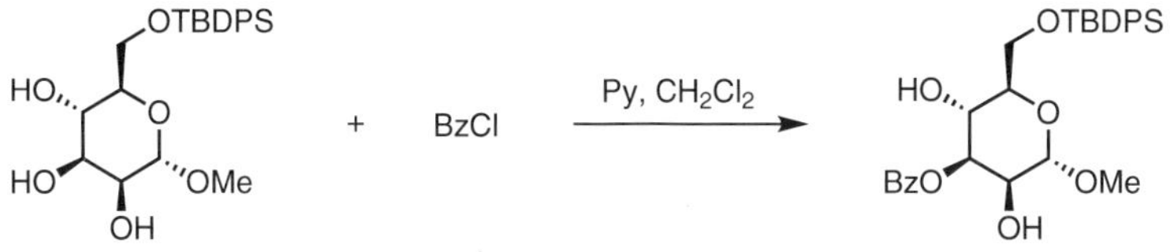

Scheme 1.274

Acetylation of cellulose with acetyl chloride is achieved by use of crosslinked polyvinylpyridine in *N*-methyl-2-pyrrolidinone [679]. Grinding acid anhydrides with alcohols in the presence of diazabicyclo[2,2,2]octane (1 equiv.) affords the corresponding esters [680].

As illustrated in Scheme 1.275, when salicyl alcohol is treated with a substituted benzoyl chloride in the presence of Et_3N/DMAP at −20 to −30 °C, the phenolic hydroxy group reacts first [681, 682]. Upon warming the reaction mixture to room temperature, the monoester thus formed rearranges completely to the benzyl ester. Without isolation of this intermediate, the second acylation by addition of another aroyl chloride to the reaction mixture at −20 to −30 °C is feasible, to provide the diester in good yield.

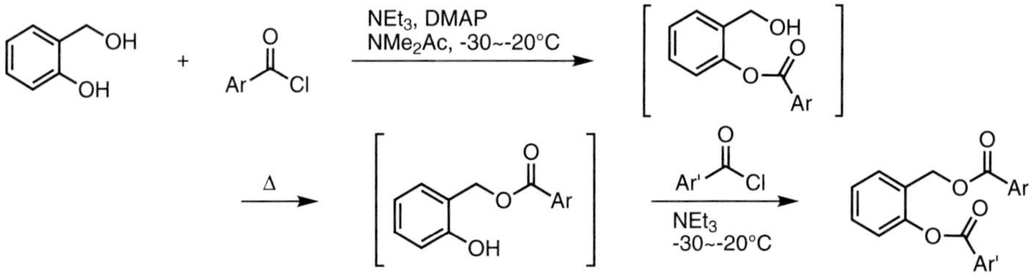

Scheme 1.275

Discrimination between primary and secondary alcohols can be achieved by this procedure. A high preference for the primary alcohol is obtained by use of 2,4,6-collidine, *N,N*-diisopropylethylamine, or 1,2,2,6,6-pentamethylpiperidine [683]. Quite naturally, there are too many examples of selective acylation of primary hydroxy groups in preference to secondary ones in sugar chemistry to be described here. The summary is given in Section 7.1.1.

The diphenylacetoxy group is a useful protecting group because high selectivity for primary hydroxyl groups is to be expected due to its steric bulk and, moreover, they can be deprotected under neutral conditions by free-radical bromination with *N*-bromosuccinimide on the benzylic position, followed by treatment of the resulting bromo esters with thiourea or hydrazinedithiocarbonate (Scheme 1.276) [684]. The

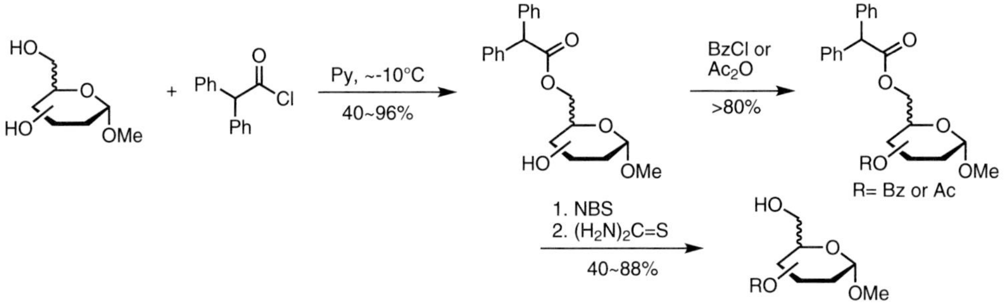

Scheme 1.276

acylation is carried out at low temperature (~−10 °C) by addition of diphenylacetyl chloride in anhydrous pyridine to the corresponding sugar in anhydrous pyridine.

Oxalyl chloride reacts with a wide range of acyclic 1,2-glycols in the presence of Et_3N to produce 1,3-dioxolan-2-ones, together with 1,4-dioxolane-2,3-diones (Scheme 1.277) [685, 686]. Ethylene glycol, monosubstituted ethylene glycols, and *erythro*-1,2-disubstituted ethylene glycols provide the cyclic carbonates, while *threo*-1,2-disubstituted ethylene glycols afford the 1,4-dioxane-2,3-diones.

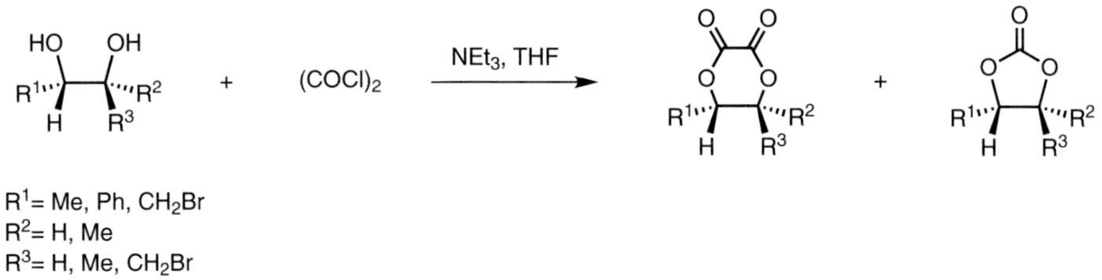

HO OH R1 R2 H R3 + $(COCl)_2$ —(NEt_3, THF)→ [1,4-dioxane-2,3-dione] + [1,3-dioxolan-2-one]

R^1= Me, Ph, CH_2Br
R^2= H, Me
R^3= H, Me, CH_2Br

Scheme 1.277

Reactions between alcohols and benzoyl chloride are accelerated by TMEDA (*N,N,N′,N′*-tetramethylethylenediamine) (Scheme 1.278) [687]. Even secondary alcohols are normally acylated at −78 °C within 20 min.

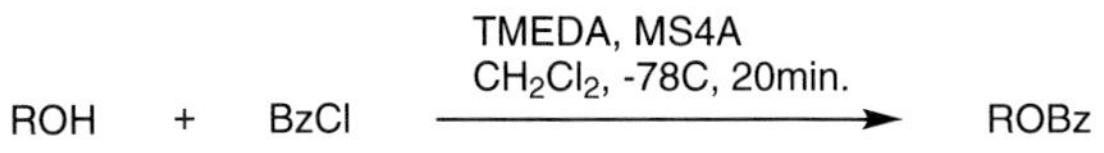

ROH + BzCl —(TMEDA, MS4A, CH_2Cl_2, -78C, 20min.)→ ROBz

Scheme 1.278

Combination of *N*-methylimidazole (0.1 equiv.) and *N,N,N′,N′*-tetramethylethylenediamine (TMEDA) (0.1 equiv.) can be used for reactions between acid chlorides and alcohols in water/EtOAc (Scheme 1.279) [688]. In these reactions, *N*-methylimidazole forms highly reactive ammonium intermediates, and TMEDA acts as an HCl captor.

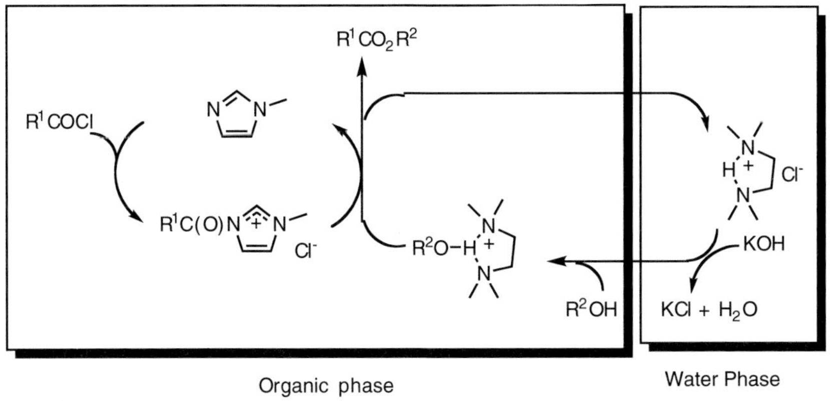

$R^1CO_2R^2$
R^1COCl
$R^1C(O)N^+$ Cl^-
R^2O–H+
R^2OH
Cl^-
KOH
KCl + H_2O
Organic phase
Water Phase

Scheme 1.279

1.4.3.3 Phase Transfer Catalyst

Phase transfer technology is successfully employed in a variety of alkali-catalyzed reactions. However, it should be taken into account that this technology cannot be invoked for esterification with acid halides because these compounds are readily hydrolyzed upon contact with the aqueous phase. The use of a solid/liquid two-phase system overcomes this predisposition. Stirring a mixture of alcohol, $Bu_4N^+HSO_4^-$, powdered NaOH, and acetyl chloride provides the desired esters in good yield [689]. More interestingly, even aqueous NaOH is employable. Phenyl acetate is obtained in 84% yield by treatment of phenol and acetyl chloride in a two-phase system consisting of CH_2Cl_2 and aqueous NaOH containing $Bu_4N^+Cl^-$ [690]. In the absence of the ammonium salt, the yield is only 15%. The use of NaOH is crucial for a good yield, but the use of this reagent in high concentrations (over 30%) decreases the yield, probably because of hydrolysis of the resulting ester. Both variants of this technology are applied to the synthesis of diaryl 1,4-phenylene dioxydiacetate (Scheme 1.280) [691]. Hydroquinone is acylated with chloroacetic acid in the presence of PEG-600 (polyethylene glycol-600) and KI with powdered K_2CO_3 in toluene to give dipotassium 1,4-phenylenedioxydiacetate, which is converted into the diacid by treatment with aqueous HCl. The diacid is transformed into the acyl chloride, which is then converted into the final products by treatment with phenols under liquid-liquid phase-transfer catalysis conditions, with 3% PEG-400 as phase transfer catalyst and NaOH in H_2O/CH_2Cl_2.

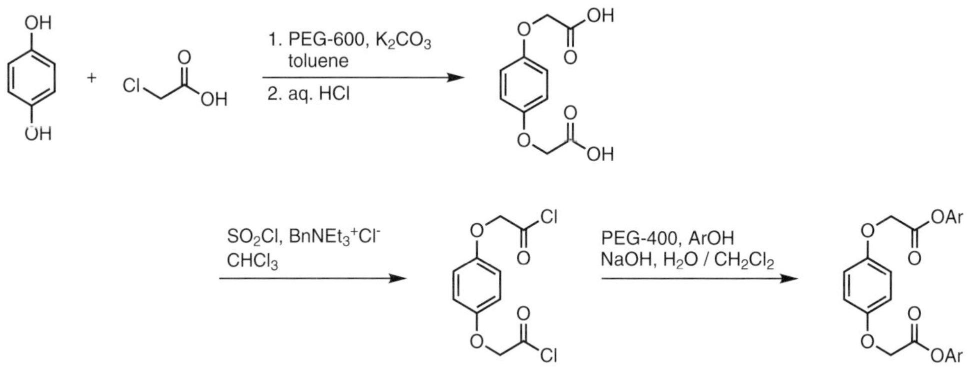

Scheme 1.280

Esters of (*Z*)-1,4-but-2-enediols are obtainable only under phase-transfer conditions (Scheme 1.281) [692]. Treatment of the diol with acyl chlorides in the presence of benzyltriethylammonium bromide in H_2O/CH_2Cl_2 at –5 °C furnishes the desired esters. No other conventional acylation methods give satisfactory yields.

Phase-transfer technology is the method of choice in the acylation shown in Scheme 1.282 [693]. Attempted regioselective benzoylation of the primary alcohol

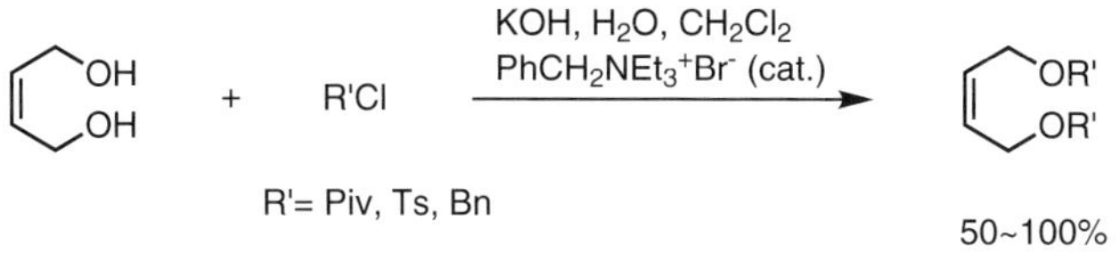

KOH, H_2O, CH_2Cl_2
$PhCH_2NEt_3^+Br^-$ (cat.)
R'Cl
R'= Piv, Ts, Bn
50~100%

Scheme 1.281

by use of benzoyl chloride in pyridine results in only 45% yield. On the other hand, phase-transfer catalysis with $Bu_4^+HSO_4^-$ in aqueous $NaOH/CH_2Cl_2$ gives the desired mono-*O*-benzoylated compound in 79% yield.

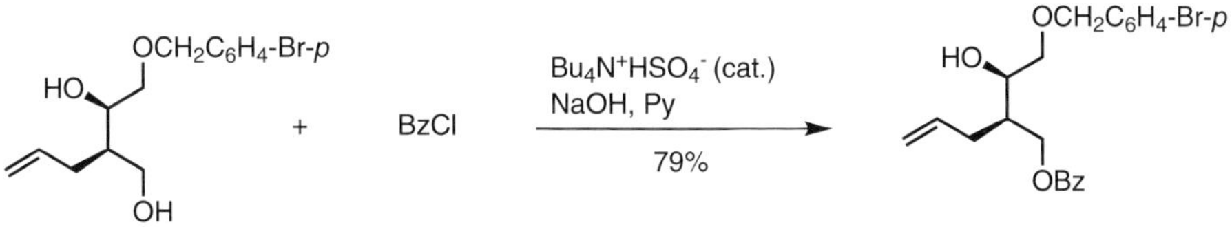

$Bu_4N^+HSO_4^-$ (cat.)
NaOH, Py
BzCl
79%

Scheme 1.282

1.4.4
Other Activators

Reactions between acid halides and alcohols are promoted by metallic samarium (0.67 equiv.) [694] or CsF/celite (1.5 equiv.) [695]. Template effects of heterocyclic compounds are often invoked for activation of carboxylic acids. 2-Pyridinethiol esters of ω-hydroxy carboxylic acids, which can be prepared from 2,2′-pyridyldisulfide and the acids, undergo lactonization upon heating in refluxing xylene under high-dilution conditions (Scheme 1.283) [696]. Various macrolides, including medium-sized ones, are obtained, although with the exception of nine-membered rings.

Experimental Procedure Scheme 1.283 [696] The hydroxy acid (0.5 mmol), 2,2′-dipyridyl disulfide (165 mg, 0.75 mmol), and triphenylphosphine (197 mg, 0.75 mmol) are dissolved under argon in dry, oxygen-free xylene and stirred at 25 °C for 5 h. The reaction mixture containing the 2-pyridinethiol ester is diluted with dry, oxygen-free xylene (10 mL), and the resulting solution is added slowly from a mechanically driven syringe, over 15 h, to dry xylene (100 mL) at reflux under argon. Refluxing is continued for an additional 10 h. Quantitative GLC analysis is performed directly on the obtained solution of product (10 ft, 10% silicone SE-30 column), by using the next higher homologous lactone as standard. The solvent is removed under reduced pressure and the ether-soluble part of the residue is subjected to preparative thin layer chromatography on silica gel (10% ether in pentane for development) to afford pure lactones and diolides.

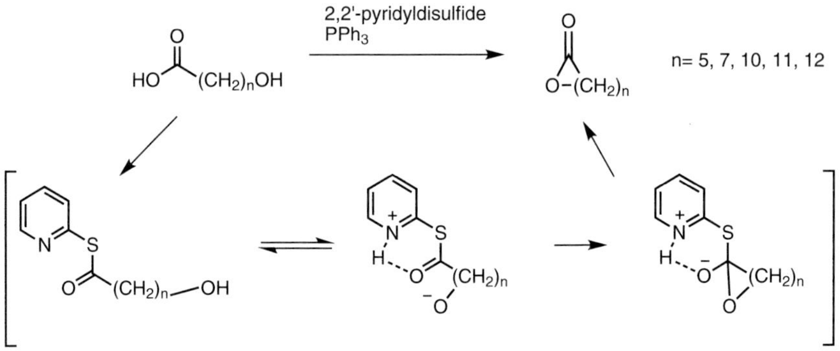
2,2'-pyridyldisulfide
PPh3
n= 5, 7, 10, 11, 12
HO
(CH2)nOH
O-(CH2)n
N
S
(CH2)n
OH
H
(CH2)n

Scheme 1.283

N-Acyl oxazolidinones are useful chiral building blocks, but are not easy to convert into esters or carboxylic acids. Esters are produced by treatment with the magnesium alkoxides, obtainable either by alcohol deprotonation with MeMgBr or through deprotonation with the Lewis acid-base combination of $MgBr_2/R_3N$ (Scheme 1.284) [697]. Selective benzoylation of racemic carbinols with kinetic selectvities of 20–30 : 1 for the *R* enantiomer can be attained. This method is applied to enantioselective acylation of aryl alkyl carbinols with *N*-benzoyl-4(*S*)-*tert*-butyl-2-oxazolidinone (Scheme 1.285) [698, 699]. Lewis acids are also effective for methyl ester synthesis from 4-substituted and unsubstituted *N*-acyl oxazolidinones [700].

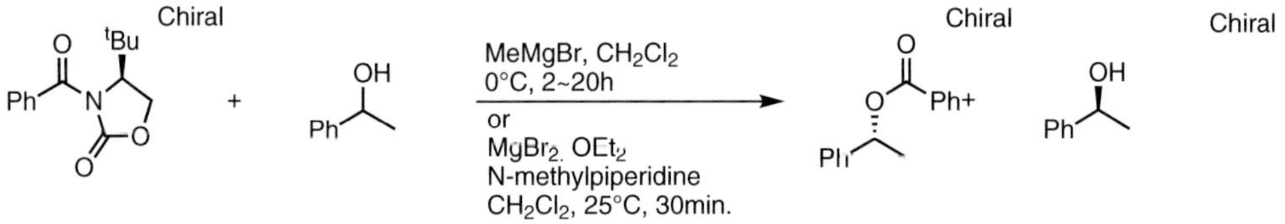
Chiral
tBu
Ph
OH
MeMgBr, CH2Cl2
0°C, 2~20h
or
MgBr2. OEt2
N-methylpiperidine
CH2Cl2, 25°C, 30min.
Chiral
Ph+
Chiral

Scheme 1.284

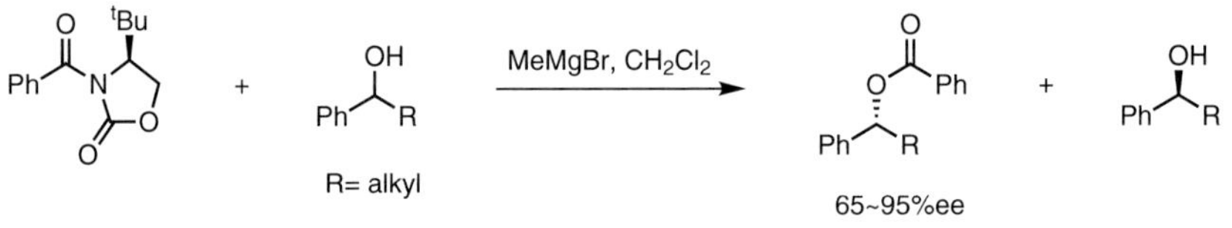
tBu
Ph
OH
R
R= alkyl
MeMgBr, CH2Cl2
Ph
65~95%ee

Scheme 1.285

Hydrazides readily undergo alcoholysis through the action of iodobenzene diacetate [701]. The reaction proceeds via a key acyldiimide intermediate (Scheme 1.286). A similar oxidative process is feasible with CAN (ceric(IV) ammonium nitrate) [702].

Primary carboxamides are converted into the corresponding esters by treatment with dimethylformamide dimethylacetal (2 equiv.) in methanol (Scheme 1.287)

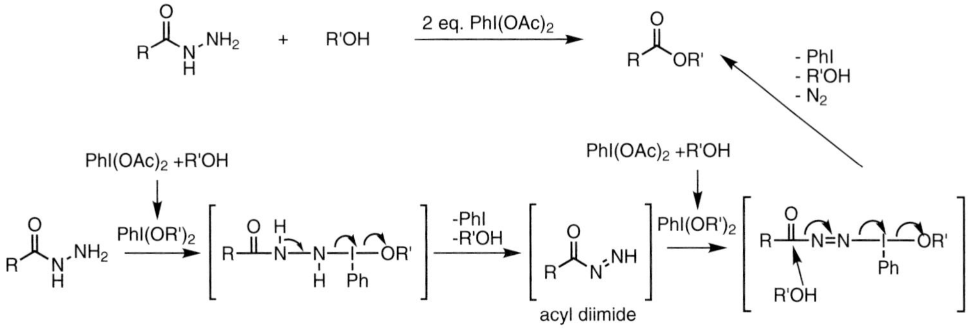

Scheme 1.286

[703]. The hydroxy group is tolerant in this reaction, but some functional groups such as NH_2, NHR, and COOH are not employable. The transient *N*-acylformamidine can be observed by monitoring the reaction course by HPLC or GC.

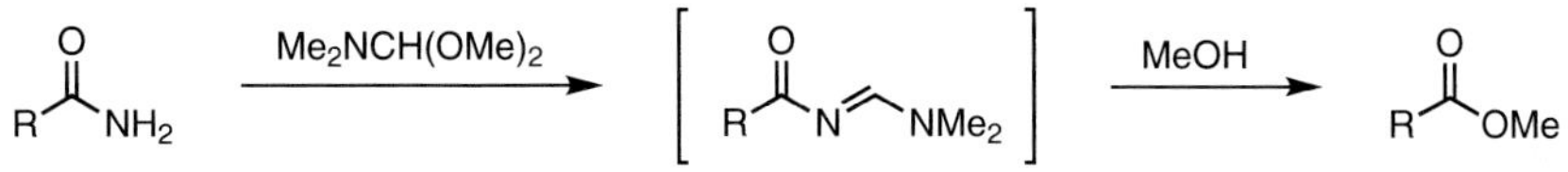

Scheme 1.287

Direct conversion of silyl ethers into esters is feasible (Scheme 1.288) [704]. Treatment of silyl ethers with acid chlorides at room temperature affords the corresponding esters.

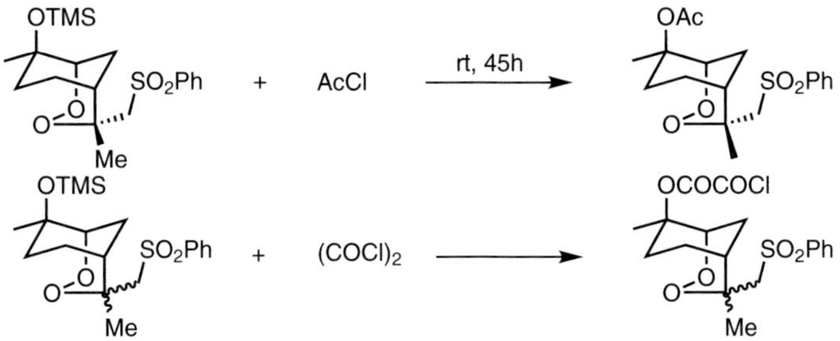

Scheme 1.288

Thioesters are usually more stable than normal esters. Nevertheless, conversion into the corresponding esters can be effected by electrochemical means (Scheme 1.289) [705]. The reaction requires an electrolyte: $Bu_4N^+I^-$ or $Li^+BF_4^-$ in acetonitrile.

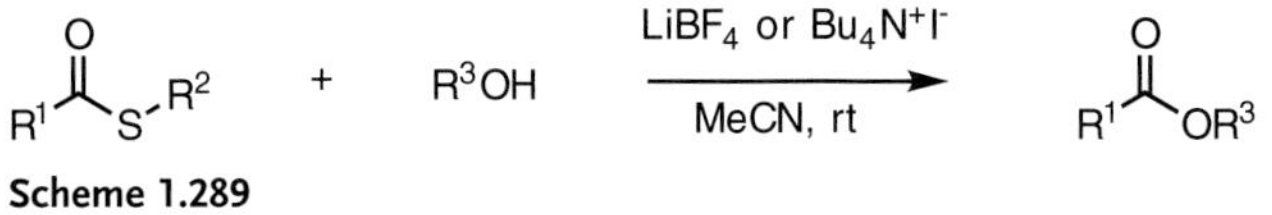

Scheme 1.289

The reaction proceeds differently depending on the electrolyte, and a wider range of substrates are employable with use of $Bu_4N^+I^-$.

Acid fluorides are generated by treatment of carboxylic acids with tetramethylfluoroformamidinium hexafluorophosphate (TFFH) (Scheme 1.290) [706]. Treatment of TFFH thus formed with alcohols *in situ* in the presence of DMAP provides the corresponding esters.

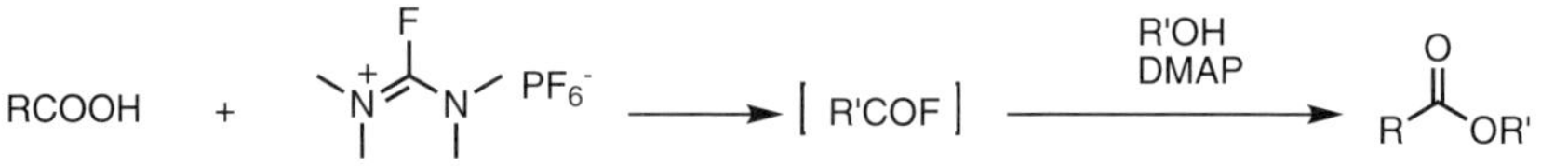

RCOOH + TFFH (PF_6^-) → [R'COF] —(R'OH, DMAP)→ R(C=O)OR'

Scheme 1.290

1.4.5 Enzymes

Thiol esters function as acyl donors in enzymatic acylation. 2,3-Butanediol, 2,4-pentenediol, and 2,5-hexanediol are resolved by a lipase derived from *Candida antarctica* (Scheme 1.291) [707]. The (*R,R*) isomers react more rapidly than their (*S,S*) counterparts, both enantiomers being available with >99% *ees* for the hexane- and pentanediols, but the selectivity being lower for the butanediol derivative.

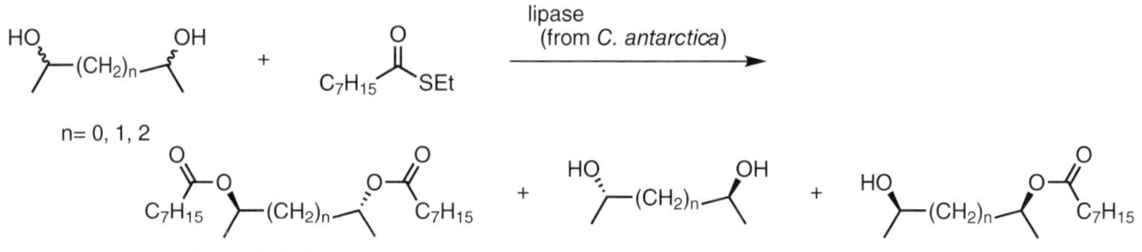

HO–CH(CH3)–$(CH_2)_n$–CH(CH3)–OH + C_7H_{15}COSEt —lipase (from *C. antarctica*)→
n= 0, 1, 2
diester + diol + monoester (C_7H_{15})

Scheme 1.291

Racemic indolmycenic ester is acetylated with lipase OF-360 (Scheme 1.292) [708]. Phenyl thioacetate is a better acyl donor than isopropyl acetate and nitrophenyl acetate in terms of chemical yield.

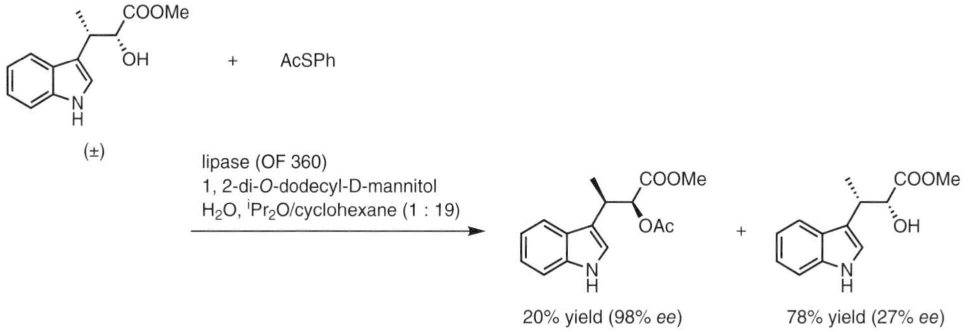

(±) COOMe, OH indole + AcSPh —lipase (OF 360), 1, 2-di-*O*-dodecyl-D-mannitol, H_2O, $^iPr_2O$/cyclohexane (1 : 19)→ COOMe/OAc product + COOMe/OH product

20% yield (98% *ee*)   78% yield (27% *ee*)

Scheme 1.292

Oxime acetates and acrylates are efficient irreversible acyl transfer agents for lipase-catalyzed transesterification in organic media (Scheme 1.293) [709].

N-Pivaloyl imidazole serves as an acyl donor in pivaloylation of monosaccharides and lactose [710]. The primary hydroxy group is selectively acylated in preference to the secondary hydroxy groups, but the selectivity is not always high.

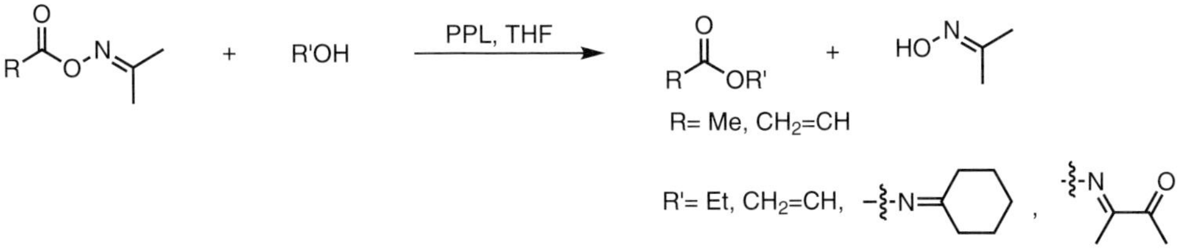

R(C=O)O–N=C(Me)₂ + R'OH —(PPL, THF)→ R(C=O)OR' + HO–N=C(Me)₂

R= Me, CH_2=CH

R'= Et, CH_2=CH, –N=(cyclohexylidene), –N=C(Me)C(=O)Me

Scheme 1.293

2
Use of Tin and Other Metal Alkoxides

The oxygen atom of metal alkoxides exhibits enhanced nucleophilicity due to the electropositive character of the metal. If the metal-oxygen bond is chemically labile, then the alkoxide should function well as a nucleophile. Tin alkoxides, especially organotin(IV) alkoxides, are extremely useful because the tin-oxygen bond is readily formed, thanks to its thermodynamic stability, while this bond is reactive enough to attack electrophiles. Because of its mildness and high selectivity, this procedure has found a wide range of applications, particularly in sugar chemistry, and this subject is therefore addressed in this chapter.

Bu_3SnOMe undergoes exothermic reaction with acid anhydrides and halides to give the corresponding esters [711]. This reactivity of Bu_2SnO is utilized for the selective acylation of nucleosides (Scheme 2.1) [712]. The 2′,3′-*O*-(dibutylstannylene) nucleosides are obtained by heating a methanolic suspension of the nucleosides and an equimolar amount of Bu_2SnO. It is postulated that $Bu_2Sn(OMe)_2$ is initially formed under these conditions. Treatment of the stannylene nucleosides with acetic or benzoic anhydride or chloride results in selective acylation at the 3′-position, no acylation taking place on the primary alcohol. The stannylene intermediate does not need to be isolated, resulting in a one-pot acylation: the addition of 5–10 equiv. of Ac_2O and Et_3N to a solution of the stannylene prepared in methanol effects selective monoacetylation.

Experimental Procedure [712] Scheme 2.1 2′,3′-*O*-(Dibutylstannylene)uridine: A suspension of uridine (488 mg, 2 mmol) and dibutyltin oxide (500 mg, 2 mmol) in methanol (100 mL) is heated under reflux for 30 min, and the resulting clear solution is then evaporated to dryness and dried *in vacuo*. The resulting crystalline residue (915 mg, 96%) is analytically pure and has m.p. 232–234°.

3′-*O*-Benzoyluridine: A solution of 2′, 3′-*O*-(dibutylstannylene)uridine (2 mmol) in methanol (100 mL) is prepared *in situ* as above. Triethylamine (1.4 mL, 10 mmol) and benzoyl chloride (1.2 mL, 10 mmol) are added, and the mixture is stirred at room temperature for 10 min, at which point TLC (ethyl acetate/acetone 1 : 1) shows no remaining uridine. The solvent is evaporated *in vacuo* and the residue is partitioned between ether (100 mL) and water, and filtered. The aqueous phase

Esterification. Methods, Reactions, and Applications. 2nd Ed. J. Otera and J. Nishikido

ISBN: 978-3-527-32289-3

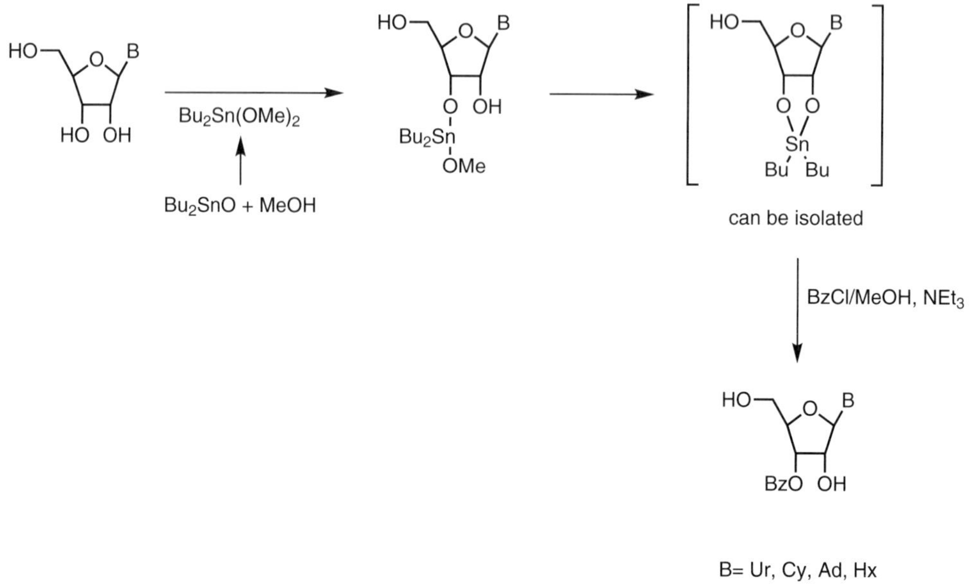

Scheme 2.1

is concentrated to about 30 mL and allowed to crystallize. Recrystallization from aqueous ethanol gives pure (NMR and TLC) 3′-*O*-benzoyluridine (570 mg, 78%) as the dihydrate.

Methyl α-D-hexopyranosides undergo selective acylation at the 2-position by the stannylene method [713]. While 4,6-*O*-benzylidene D-hexopyranosides are converted into the corresponding stannylene derivatives, which undergo acylation at the 2-position upon treatment with acyl halides in dioxane in the presence of Et_3N, the unprotected sugars are also successfully transformed into the C2 monoesters in one-pot fashion without isolation of the stannylene intermediates (Scheme 2.2). The C2 esters of methyl α-D-gluco-, α-D-allo-, and α-D-galactopyranosides are obtained by this procedure.

Experimental Procedure Scheme 2.2 [713] Methyl 2,3-*O*-Dibutylstannylene-α-D-glucopyranoside: Dibutyltin oxide (12.50 g, 50 mmol) is added to a solution of methyl α-D-glucopyranoside (9.7 g, 50 mmol) in methanol (200 mL), and the resulting milky solution is heated at reflux until it becomes homogeneous and clear (45 min). After further heating at reflux for an additional 0.5 h, the solvents are evaporated *in vacuo* to leave a white solid, m.p. range 105–115 °C.

Methyl 2-*O*-Benzoyl-α-D-glucopyranoside: Triethylamine (1.54 mL, 11 mmol) is added to a magnetically stirred, slightly cloudy solution of methyl 2,3-*O*-

dibutylstannylene-α-D-glucopyranoside (4.25 g, 10 mmol) in dioxane (75 mL), followed by slow addition of benzoyl chloride (1.32 mL, 11 mmol). The solution becomes clear upon addition of the benzoyl chloride, but a white precipitate starts to form after about 2 min. TLC examination of the solution (ethyl acetate, silica gel G) after 1 h shows the presence of a major spot at R_f 0.50 and a minor spot at R_f 0.70. The salts are filtered off and washed with dioxane (20 mL), and the combined filtrates are evaporated *in vacuo* to leave a syrup. This is fractionated on a column of silica gel G (120 g) with ethyl acetate as eluent. The first compound eluted from the column is methyl 2,6-di-*O*-benzoyl-α-D-glucopyranoside (0.08 g, ~2%). The second compound eluted from the column is the desired material (2.05 g, 70%).

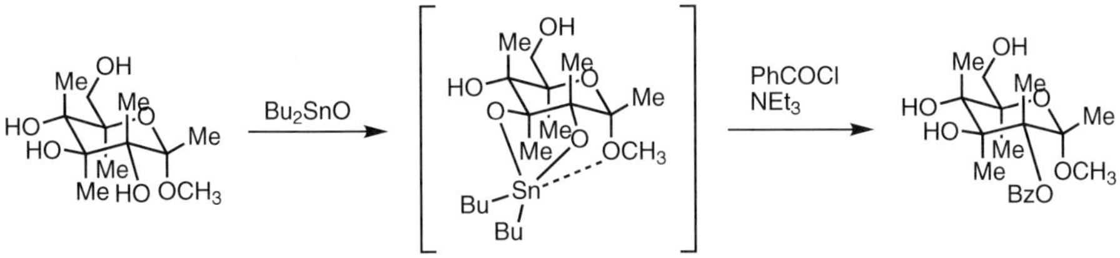

Scheme 2.2

Regioselective acylation of sugars through the use of $(Bu_3Sn)_2O$ is also feasible [714]. In this case, the selectivity is governed by a monoalkoxytin intermediate coordinated by the proximate hydroxy group. For example, stannylation of methyl α-D-glucopyranoside with two equiv. of $(Bu_3Sn)_2O$ followed by treatment with benzoyl chloride at 20 °C provides the 2,6-di-*O*-benzoyl ester (81%) and the 2,3,6-tri-*O*-benzoyl ester (18%) (Scheme 2.3). When the reaction is conducted at –10° to –5 °C, the dibenzoate is obtained in 95% yield.

Experimental Procedure Scheme 2.3 [714] Methyl α-D-glucopyranoside (283 mg, 1.46 mmol) is stannylated with $(Bu_3Sn)_2O$ (1.3 g, 2.19 mmol) in toluene (23 mL). A solution of benzoyl chloride (600 mg, 4.4 mmol) in toluene (5 mL) is then added dropwise to the cooled solution over 5 min at –10 °C. The mixture is stirred for 4 h at –10 °C and then left for 17 h at –5 °C. Acetic acid (0.2 mL) is added and the solvent is evaporated *in vacuo* to give an oily residue, which is triturated with diisopropyl ether to afford the crystalline product (588 mg, 95%).

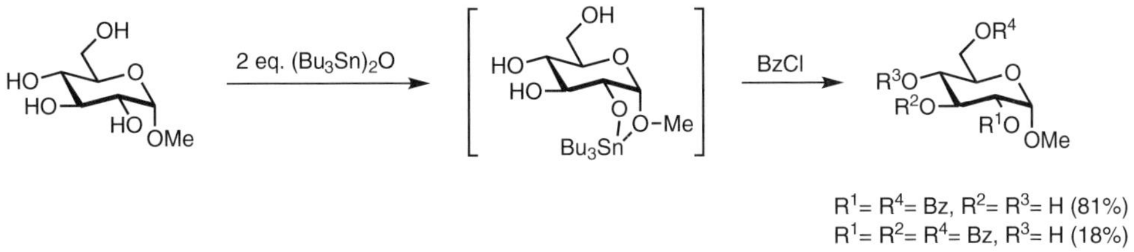

$R^1 = R^4 = Bz$, $R^2 = R^3 = H$ (81%)
$R^1 = R^2 = R^4 = Bz$, $R^3 = H$ (18%)

Scheme 2.3

The regioselectivity in monoacylation of secondary hydroxy groups in monosaccharides can be controlled by the stannylene technique (Scheme 2.4) [715]. Use of one equiv. of strong base such as *N*-methylimidazole results in the formation of the equatorial 3-benzoate in a yield of more than 90%.

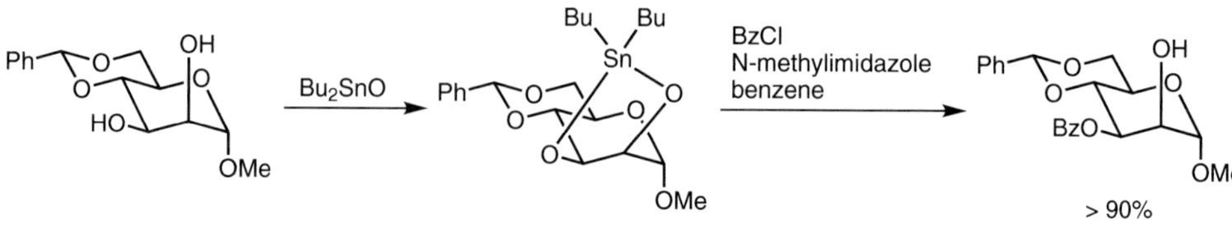

Scheme 2.4

Borylated carbohydrates are quantitatively *O*-stannylated by transmetalation with tributyltin acetylacetonate. This methodology is applicable to *O*-acylation of *myo*-inositol (Scheme 2.5) [716]. The standard stannylene procedure is not effective for acylation because of the poor solubility of the stannylene intermediate, whereas the hexa-*O*-diethylboryl derivative prepared by treatment with excess Et_3B is soluble in hexane, and treatment of this compound with tributyltin acetylacetonate furnishes a partially borylated-stannylated intermediate. Treatment of this compound with benzoyl chloride in the presence of *N*-methylimidazole affords 1-*O*-benzoyl-*myo*-inositol.

Experimental Procedure Scheme 2.5 [716] A solution of hexa-*O*-diethylboryl derivative (234 mg, 0.4 mmol) in toluene is treated with tributyltin acetylacetonate (0.5 mmol), and *N*-methylimidazole and benzoyl chloride are then added at −5 °C. The reaction mixture is stirred at 5 °C for 10 h. After conventional workup the mixture is purified by column chromatography on silica gel (chloroform/methanol, 3 : 1).

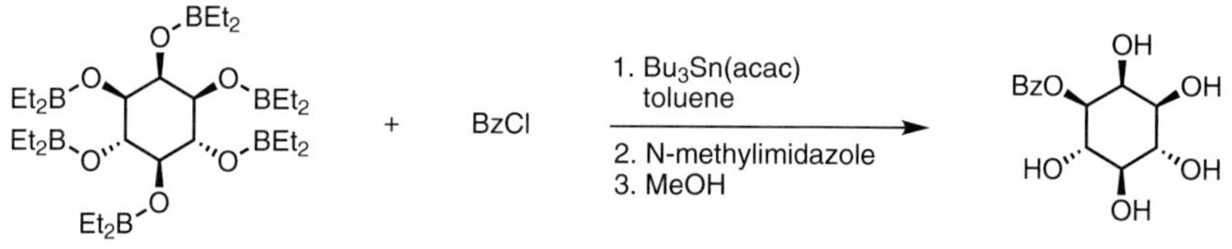

Scheme 2.5

Selective *O*-acylation and -alkylation are feasible, as shown in Scheme 2.6 [717] and Scheme 2.7 [718].

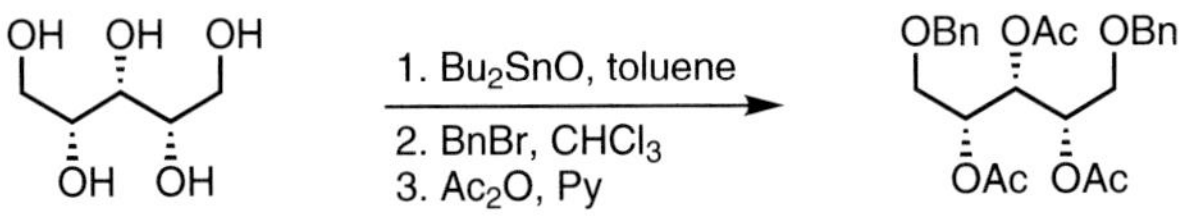

Scheme 2.6

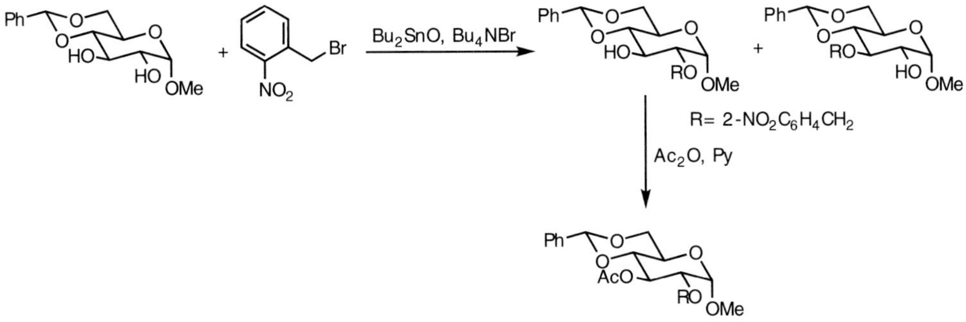

Scheme 2.7

The use of chiral acid halides in the stannylene approach allows asymmetric acylation of *meso*-1,2-diols. Thus, treatment of *meso*-dimethyl tartrate with Bu_2SnO followed by (1*S*)-ketopinic acid chloride exclusively affords the monoacylation product in high yield and with high diastereoselectivity (Scheme 2.8) [719].

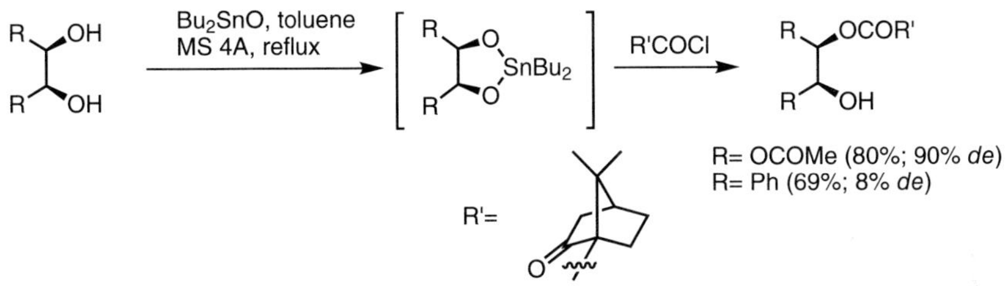

Scheme 2.8

Optically active glycerol acetonide can be prepared analogously from achiral glycerol (Scheme 2.9) [720].

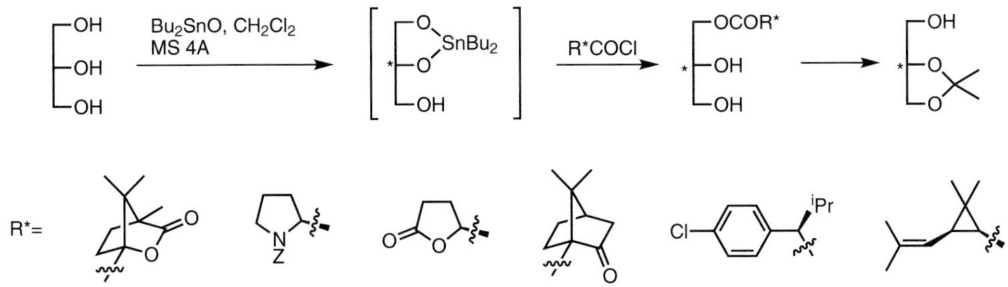

Scheme 2.9

Kinetic resolution of chiral 1,2-diols is effected by use of an organotin catalyst with a binaphthyl moiety as a chiral source (Scheme 2.10) [721]. Addition of sodium carbonate and a small amount of water improves the selectivity.

Experimental Procedure Scheme 2.10 [721] Sodium carbonate (1.5 mmol) base is suspended in HF (5 mL) containing racemic alcohol (1 mmol), water (100 μl, 5.5 mmol), and organotin catalyst (0.25 mol%). Benzoyl chloride (0.5 mmol) is then added to the suspension at −10 °C, and the resulting solution is stirred at that temperature for 14 h. After conventional workup, a mixture of primary monobenzoate (86% *ee*, 38% yield) and recovered racemic alcohol (46% *ee*, 58% yield) is obtained, with a trace amount of secondary monobenzoate.

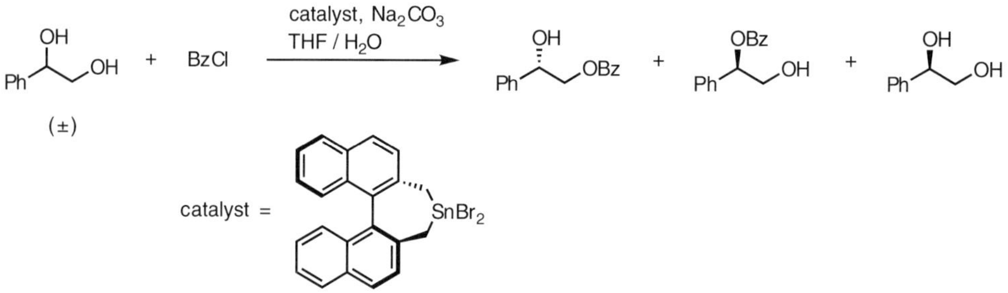

catalyst, Na_2CO_3
THF / H_2O

Ph–CH(OH)–CH₂OH (±) + BzCl → Ph–CH(OH)–CH₂OBz + Ph–CH(OBz)–CH₂OH + Ph–CH(OH)–CH₂OH

catalyst = $SnBr_2$ (binaphthyl)

Scheme 2.10

When the treatment of unsymmetrical diols by the stannylene procedure is accompanied by *in situ* quenching with chlorosilane or oxalic acid, acylation takes place on the more substituted hydroxy group (Scheme 2.11) [722, 723]. This method can be applied to 1,2-, 1,3-, and 1,4-diols of primary-secondary, primary-tertiary, and secondary-tertiary natures. Benzoyl chloride and other acid chlorides are employable, but acid anhydrides are of no use.

Experimental Procedure Scheme 2.11 [723] General Acylation Procedure: A portion of the diol (1.5 mmol) is dissolved or suspended in toluene (30 mL), and, after the addition of dibutyltin oxide in 5% molar excess, water is separated by azeotropic distillation in a Dean-Stark apparatus for a variable length of time, depending on the substrate. After evaporation of the solvent, the residue is dried under vacuum, dissolved under nitrogen in anhydrous $CHCl_3$ (30 mL), and cooled to 0–5 °C. An equimolar amount of the appropriate acylating reagent in the same solvent (1 mL) is added dropwise to the stirred solution by syringe through a septum cap, and the reaction mixture is allowed to react at room temperature for 1 h, and then quenched by one of two different methods.

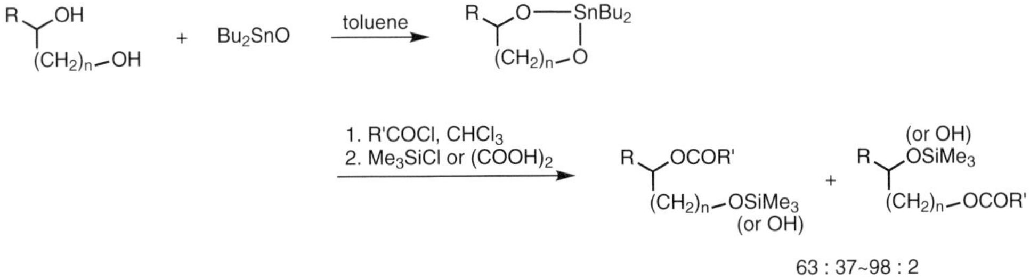

Scheme 2.11

Method A – Quenching with Trialkylsilyl Chlorides: A solution of the appropriate silyl chloride (5% molar excess) in anhydrous $CHCl_3$ (1 mL) is added dropwise by syringe to the cooled solution (0–5 °C), and the mixture is then allowed to react at room temperature for 1–2 h. The mixture obtained is then analyzed by ^1H NMR spectroscopy.

Method B – Quenching with Oxalic Acid: The solvent is evaporated under vacuum, and the residue is dissolved in anhydrous CH_3CN (8 mL). The solution is cooled to 0–5 °C, oxalic acid (0.75 mmol) in CH_3CN (3.5 mL) is added, and the mixture is stirred at room temperature for 20 h. The resulting suspension is filtered under nitrogen, and the solid is washed several times with CH_3CN. The residue obtained after evaporation of the solvent under vacuum is dissolved in $CDCl_3$, and the mixture is analyzed by ^1H NMR spectroscopy.

Microwave irradiation is useful not only for shortening the time required to prepare the stannylene intermediates but also for rendering the procedure catalytic. As shown in Scheme 2.12, if $Bu_2Sn(OH)Cl$ is successfully transformed to

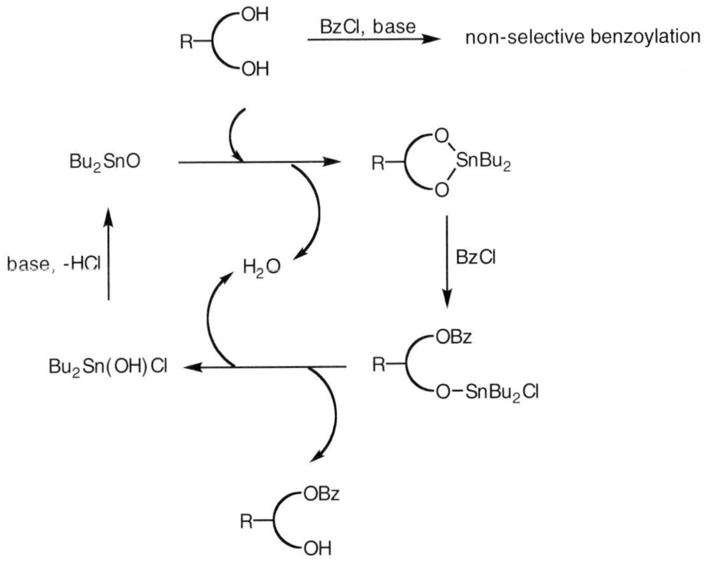

Scheme 2.12

Bu_2SnO in the presence of a base, the organotin species can be recycled. Under normal conditions, however, in which heating is necessary to prepare stannylate diols, the base promotes the reaction between benzoyl chloride and diols. However, no such direct reaction occurs under microwave conditions, thus allowing the catalytic cycle to complete.

The catalytic process is also achievable through the use of Me_2SnCl_2 in the presence of K_2CO_3, although the mechanism is not clear (Scheme 2.13) [724]. A variety of cyclic and acyclic diols, 1,2-diols in particular, are selectively monobenzoylated in good yield.

Experimental Procedure Scheme 2.13 [724] A catalytic amount of dimethyltin dichloride (0.01 mmol), solid K_2CO_3 (2.0 mmol), and benzoyl chloride (1.2 mmol) are successively added at room temperature to a THF (5 mL) solution of *trans*-1,2-cyclohexanediol (1 mmol). After the mixture has been stirred at room temperature until *trans*-1,2-cyclohexanediol has disappeared (checked by thin layer chromatography), the mixture is poured into water and the organic portion is extracted with CH_2Cl_2. After evaporation of the solvent, a residue is obtained, and this is confirmed by NMR to be pure monobenzoylated product (>99%).

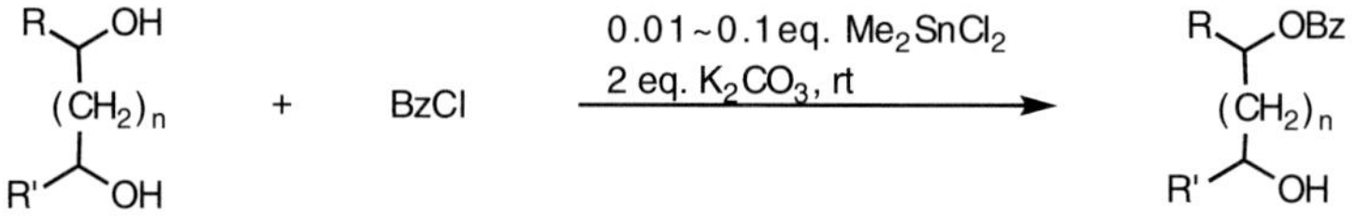

Scheme 2.13

Cyclic stannoxanes made up of dibutylstannylene and 1, n-dialkoxy units function as covalent templates in reaction with acid dihalides to give macrocycles (Scheme 2.14) [725, 726]. Acid anhydrides are also employable, and the use of chiral tin templates affords diastereomeric macrocyles (Scheme 2.15) [727, 728]. *N*-(Trifluoroacetyl)-glutamic and -aspartic anhydrides provide macrocylces with an

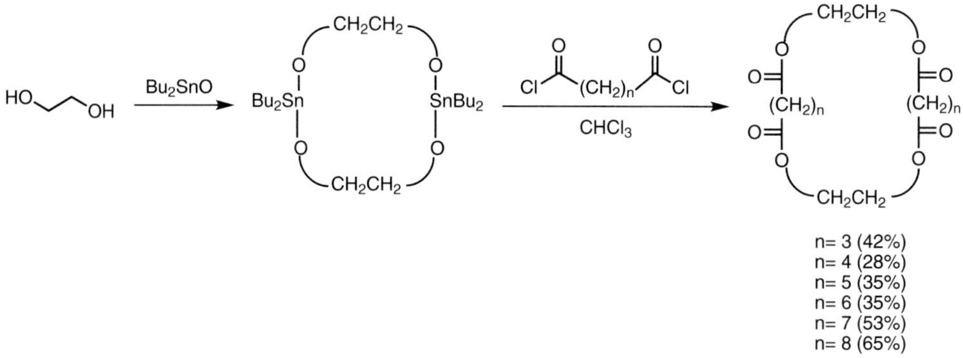

n= 3 (42%)
n= 4 (28%)
n= 5 (35%)
n= 6 (35%)
n= 7 (53%)
n= 8 (65%)

Scheme 2.14

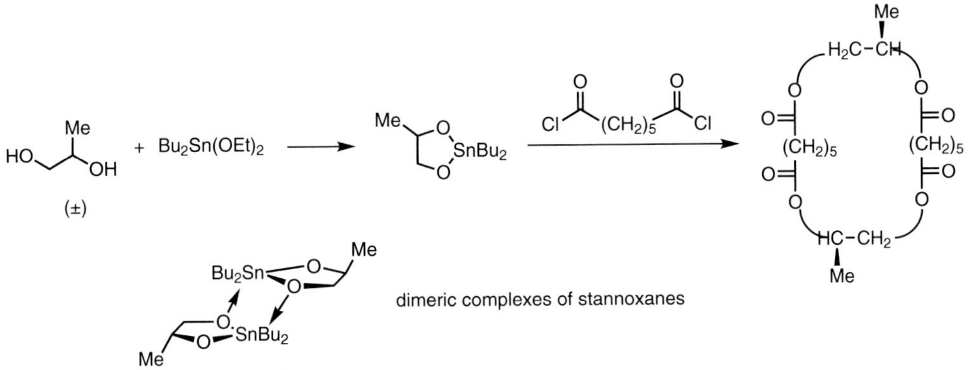

dimeric complexes of stannoxanes

Scheme 2.15

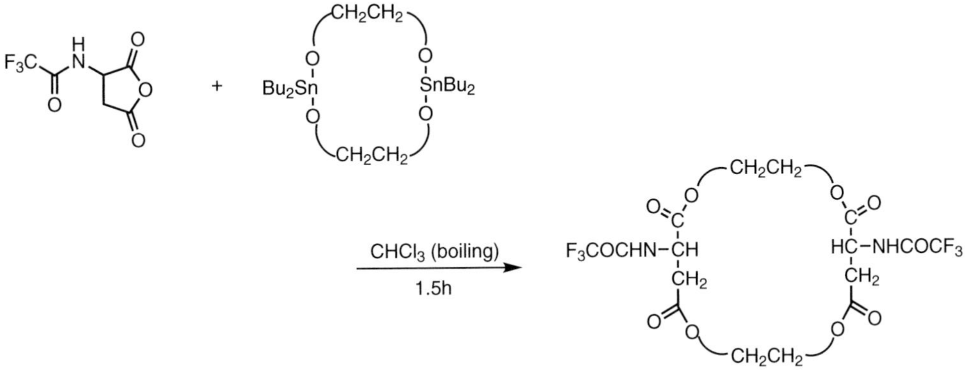

CHCl3 (boiling)
1.5h

Scheme 2.16

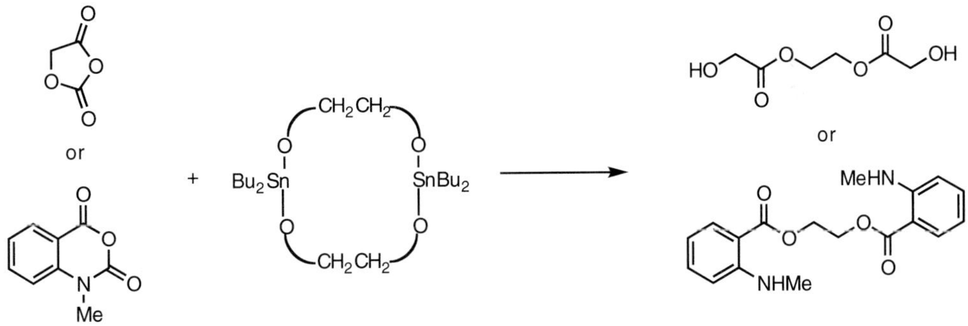

Scheme 2.17

amino residue on the ring (Scheme 2.16). Treatment of the distannoxane with a cyclic carboxy-carbonate derived from glycolic acid or isatoic anhydride results in ring opening (Scheme 2.17) [729].

Tin(II) alkoxides are also effective. Reactions between 1,1′-dimethylstannocene and alcohols readily occur at room temperature, and the resulting tin(II) alkoxides provide esters upon treatment with acid halides (Scheme 2.18) [730]. When this reaction is carried out in the presence of a chiral amine ligand, asymmetric desymmetrization of 2-*O*-protected glycerols is feasible, furnishing monoesters with up to 84% *ee* (Scheme 2.19) [731].

Experimental Procedure Scheme 2.18 [730] 3-Phenylpropanol (125 mg, 0.921 mmol) in toluene (1.5 mL) is added at room temperature, under an argon atmosphere, to a toluene solution (2 mL) of 1,1′-dimethylstannocene (153 mg, 0.552 mmol). After the mixture has been stirred for 30 min, hexamethylphosphoric triamide (1 mL) and benzoyl chloride (155 mg, 1.11 mmol) in toluene (1.5 mL) are successively added. The mixture is kept stirring at room temperature for 2 h and quenched with pH 7 phosphate buffer. The aqueous phase is extracted three times with ether and the combined extracts are washed with brine and dried over anhydrous Na_2SO_4. After evaporation of the solvent, the resulting crude product is purified by silica-gel thin layer chromatography to afford 3-phenylpropyl benzoate (200 mg, 90%).

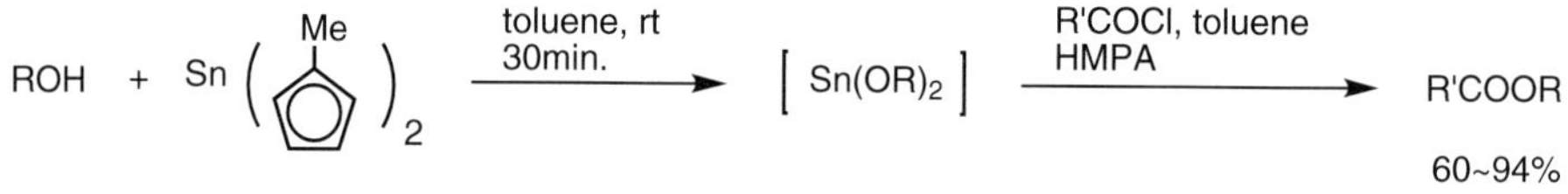

ROH + Sn(MeC5H4)2 —(toluene, rt, 30min.)→ [Sn(OR)2] —(R'COCl, toluene, HMPA)→ R'COOR 60~94%

Scheme 2.18

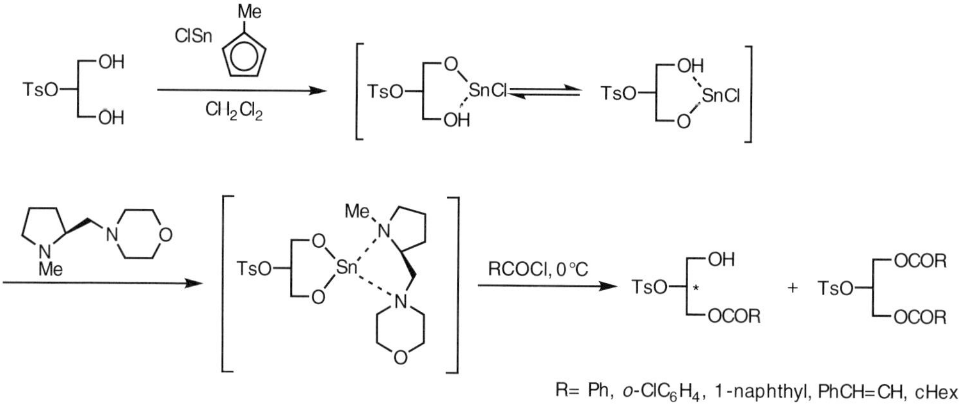

R= Ph, *o*-ClC_6H_4, 1-naphthyl, PhCH=CH, cHex

Scheme 2.19

The stannylene technique has been successfully applied to synthesis of (–)-integerrimine (Scheme 2.20) [732] and (–)-senecionine (Scheme 2.21) [733].

Metallacycles with metals other than tin also react with acid halides. When an arsole or a stibole is treated with one equiv. of acid halides with subsequent hydrolysis, a monoester is produced (Scheme 2.22) [734]. On the other hand, treatment with two equiv. of acid halides affords the corresponding diesters.

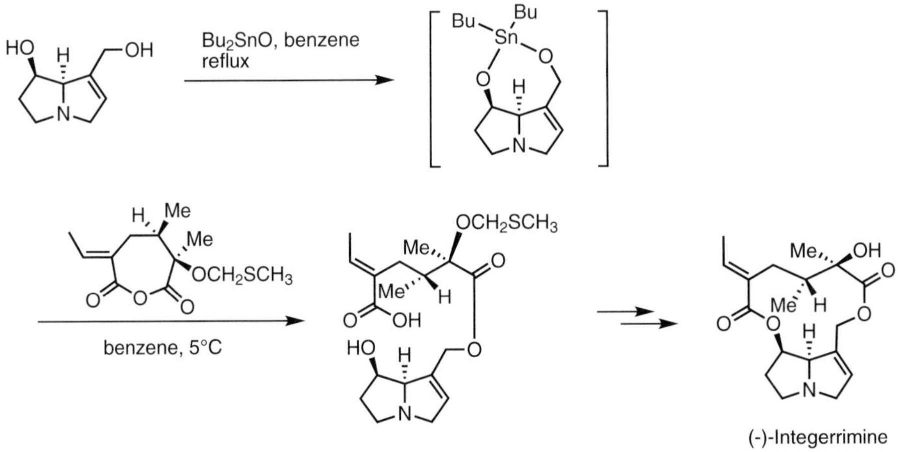

Scheme 2.20

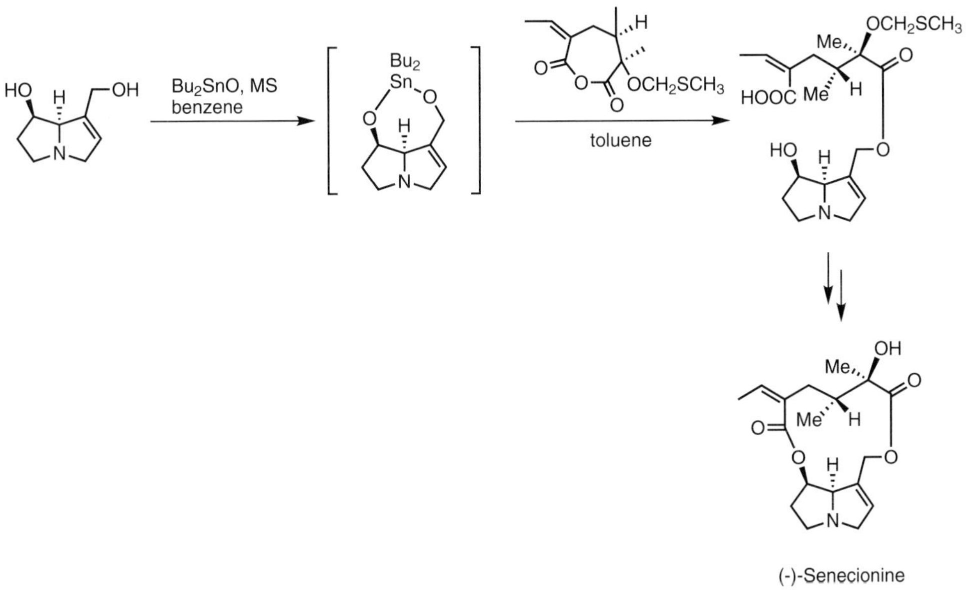

Scheme 2.21

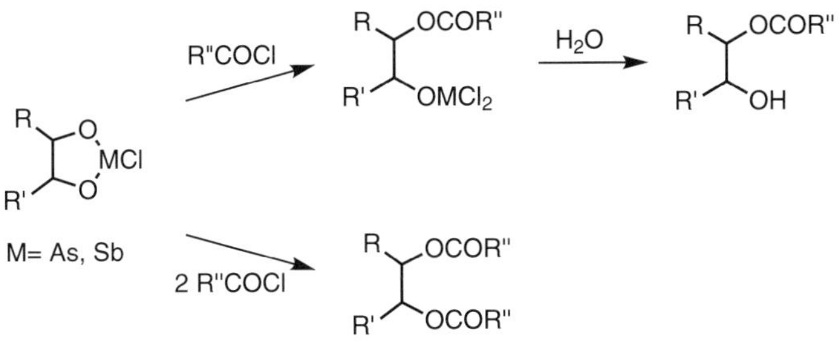

Scheme 2.22

Macrocyles can be obtained by the treatment of stiboles with diacid dihalides (Scheme 2.23) [735].

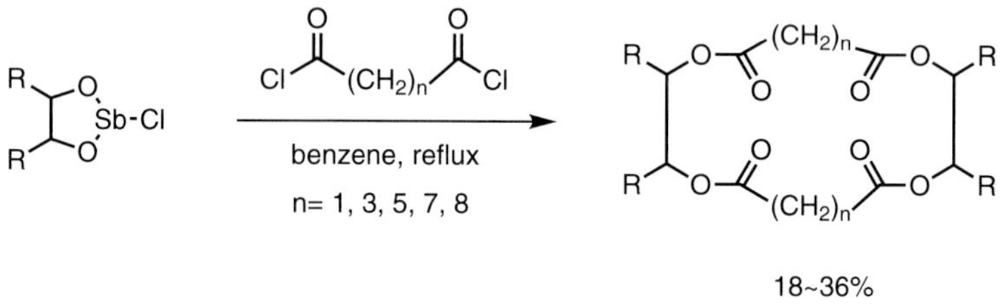

benzene, reflux

n= 1, 3, 5, 7, 8

18~36%

Scheme 2.23

Copper(II) also functions as a template. Regioselective C3 O-acylation is achieved by treatment of sodium salts of 4,6-*O*-benzylidene-glucopyranosides (prepared by addition of NaH) with $CuCl_2$ followed by acid halides (Scheme 2.24) [736].

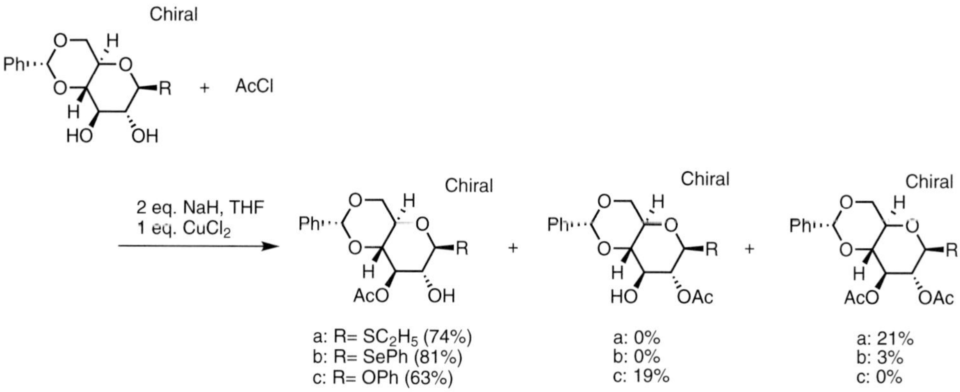

2 eq. NaH, THF
1 eq. $CuCl_2$

a: R= SC_2H_5 (74%)
b: R= SePh (81%)
c: R= OPh (63%)

a: 0%
b: 0%
c: 19%

a: 21%
b: 3%
c: 0%

Scheme 2.24

$Si(OMe)_4$ acts as a catalyst/reagent in the selective methylation of 2-hydroxycarboxylic acids (Scheme 2.25) [737]. 2-Hydroxy acids play a crucial role: the hydroxy acid attaches to silicon through the alkoxy group and subsequently through the carboxy group in an intramolecular rearrangement to form an unstable and reactive cyclic intermediate. This intermediate may accelerate methylation of the carboxylic acid via nucleophilic attack of MeOH at the carbonyl group. $B(OBu)_3$ acts similarly for butylation of dicarboxylic acids (Scheme 2.26) [738]. Heating of a mixture of dicarboxylic acid and $B(OBu)_3$ in a 3:2 ratio affords the desired diester in good yield.

Scheme 2.25

$$3\ HOOC{-}Z{-}COOH + 2\ B(OBu)_3 \longrightarrow 3\ BuOOC{-}Z{-}COOBu + 2\ H_3BO_3$$

Scheme 2.26

In transesterification, a tetranuclear zinc cluster, $Zn_4(OCOCF_3)_6O$, acylates alcohols in the presence of free amine [739], in contrast to the conventional reactivity, by which amines are more reactive than alcohols due to their stronger nucleophilicity. In this sense, the zinc cluster is similar to lipase. With this catalyst, various amino alcohols are transformed into the corresponding esters without violating the amine function (Scheme 2.27).

$$PhCOOMe + H_2N{-}Z{-}OH \xrightarrow[i\text{-}Pr_2O,\ reflux]{Zn_4(OCOCF_3)_6O} H_2N{-}Z{-}OCOPh\ (>82) + PhCONH{-}Z{-}OCOPh\ (<18)$$

Scheme 2.27

3
Conversion of Carboxylic Acids into Esters without Use of Alcohols

3.1
Reaction with Diazomethane

Diazomethane chemistry was pioneered by von Pechmann [740, 741]. Since then, the use of this reagent–despite its explosive and toxic nature–has proven to be one of the most convenient ways by which to convert carboxylic acids into their methyl esters (Scheme 3.1) [742]. The reaction proceeds at room temperature under mild conditions, and even sterically hindered carboxylic acids are employable. Thanks to the simplicity of the preparation of the reagent and its operation in esterification, numerous papers on the use of this reagent are available. Notwithstanding, since no special technique is involved, there is no need to describe this method in more detail here, save for the following special cases.

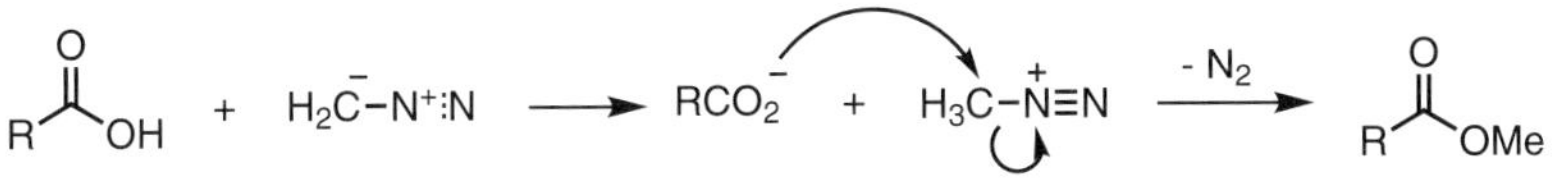

$$RCOOH + H_2\bar{C}{-}N^+{:}N \longrightarrow RCO_2^- + H_3C{-}\overset{+}{N}{\equiv}N \xrightarrow{-N_2} RCOOMe$$

Scheme 3.1

The hazards associated with diazomethane can be circumvented by the use of trimethylsilyldiazomethane, which is stable, nonexplosive, and nontoxic (Scheme 3.2) [743, 744]. Various carboxylic acids, including aromatic, heteroaromatic, alicyclic and aliphatic ones, smoothly undergo methylation.

Experimental Procedure Scheme 3.2 [743] $TMSCHN_2$ (1.3 mmol) in benzene (1 mL) is added at room temperature to a stirred carboxylic acid (1 mmol) in methanol (2 mL)/benzene (7 mL). The mixture is stirred at room temperature for 30 min and concentrated to give corresponding methyl esters.

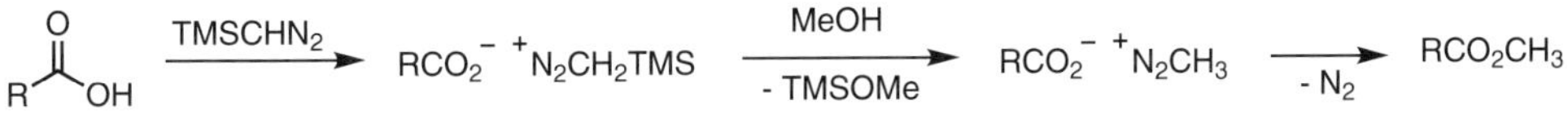

$$RCOOH \xrightarrow{TMSCHN_2} RCO_2^-\ {}^+N_2CH_2TMS \xrightarrow[-\ TMSOMe]{MeOH} RCO_2^-\ {}^+N_2CH_3 \xrightarrow[-\ N_2]{} RCO_2CH_3$$

Scheme 3.2

Esterification. Methods, Reactions, and Applications. 2nd Ed. J. Otera and J. Nishikido

ISBN: 978-3-527-32289-3

Diphenyldiazomethane is employed to transform a carboxylic acid into the corresponding diphenylacetate (Scheme 3.3) [745]. After oxidation of a terminal diol to the carboxlic acid via aldehyde, the treatment of the carboxylic acid with diphenyldiazomethane provides the desired ester.

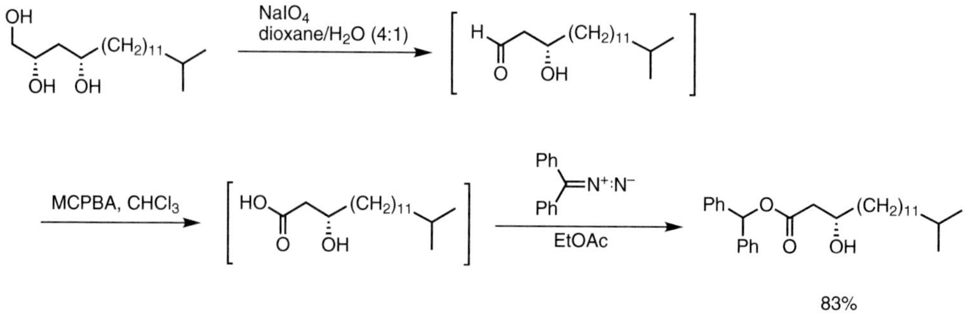
OH
(CH2)11
OH OH
NaIO4
dioxane/H2O (4:1)
H
(CH2)11
O OH
MCPBA, CHCl3
HO
(CH2)11
O OH
Ph
=N+:N−
Ph
EtOAc
Ph
O
(CH2)11
Ph O OH
83%

Scheme 3.3

A terminal acetylene moiety is converted into a benzyl ester by use of phenyldiazoacetate (Scheme 3.4) [746]. After conversion of the acetylene into the carboxylic acid, the crude product is treated with phenyldiazomethane to furnish the benzyl ester.

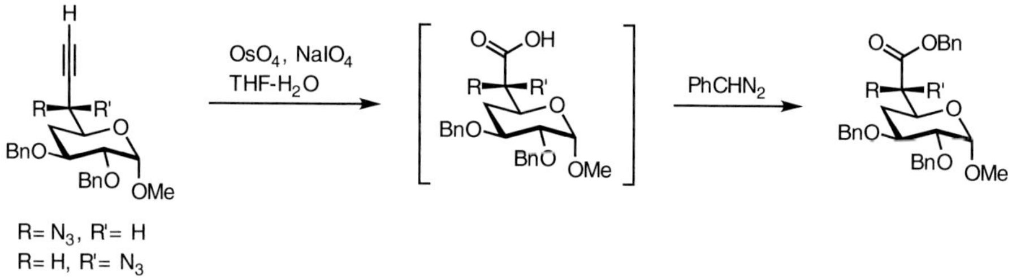
H
R
R'
O
BnO
BnO
OMe
OsO4, NaIO4
THF-H2O
O
OH
R
R'
O
BnO
BnO
OMe
PhCHN2
O
OBn
R
R'
O
BnO
BnO
OMe
R= N3, R'= H
R= H, R'= N3

Scheme 3.4

Selective monomethylation of dicarboxylic acids with diazomethane is effected with the aid of alumina (Scheme 3.5) [747]. Adsorption of one of the carboxylic acids permits selective esterification of the remaining free carboxylic acid function. High selectivity for the monoesters can be achieved.

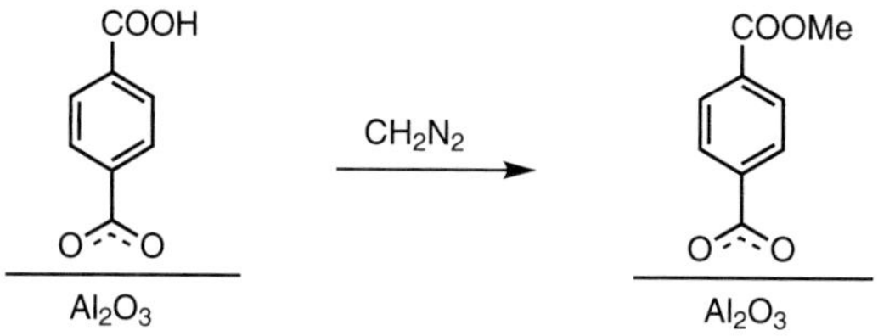
COOH
O O
Al2O3
CH2N2
COOMe
O O
Al2O3

Scheme 3.5

There is also a more general procedure to generate diazoalkanes, which provide a variety of esters upon *in situ* treatment with carboxylic acids (Scheme 3.6) [748].

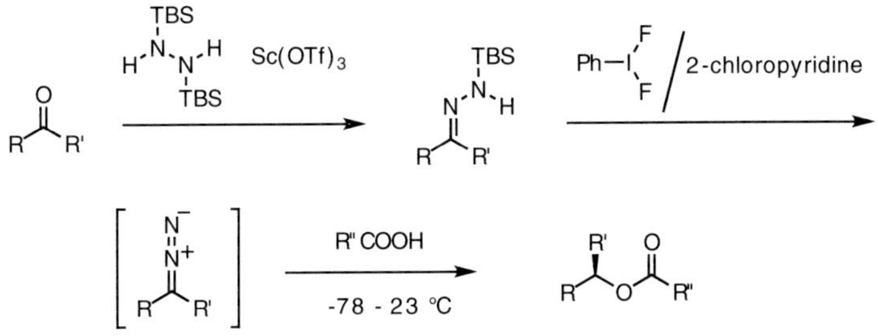

Scheme 3.6

3.2
Reaction with Alkyl Halides

The reaction between carboxy anions and alkyl halides is a simple way to arrive at esters, but not particularly easy to achieve because of the weak nucleophilicity of the anion. The most classical way is to use silver salts of carboxylic acids, taking advantage of the insolubility of the silver halides formed. However, use of silver halides may be impractical because of the expense involved. More common alkali metal salts react with alkyl halides in polar solvents [749]. The use of HMPA (hexamethylphosphoramide) as solvent or cosolvent is useful to promote the alkylation [750, 751].

Benzylation of the sodium salts of dicarboxylic acid is smoothly carried out by reaction with benzyl bromide in DMF (Scheme 3.7) [752]. *Z*-protected L-amino acids react smoothly with alkyl halides at room temperature in DMF in the presence of sodium hydrogen carbonate [753]. Reaction between sodium salicylate and benzyl chloride in DMF proceeds more smoothly than the reaction with a phase-transfer catalyst [754]. Phenolic acids undergo selective esterification on the carboxy function leaving a phenolic hydroxy group intact upon treatment with $KHCO_3$/alkyl halide in DMF (Scheme 3.8) [755]. Similarly, *O*-methylation of carboxylic acids can be performed with MeI/KOH in DMSO [756].

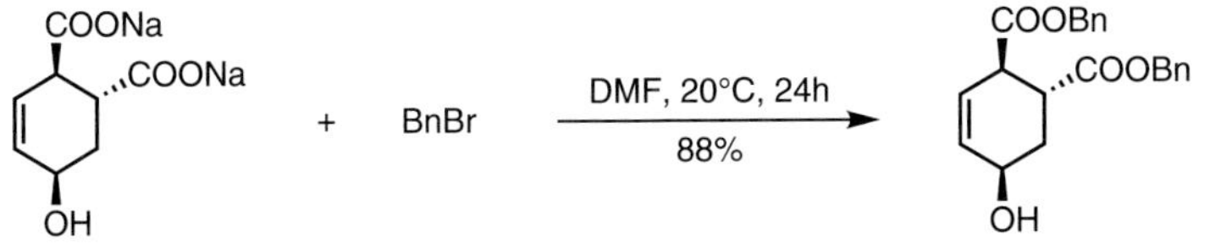

Scheme 3.7

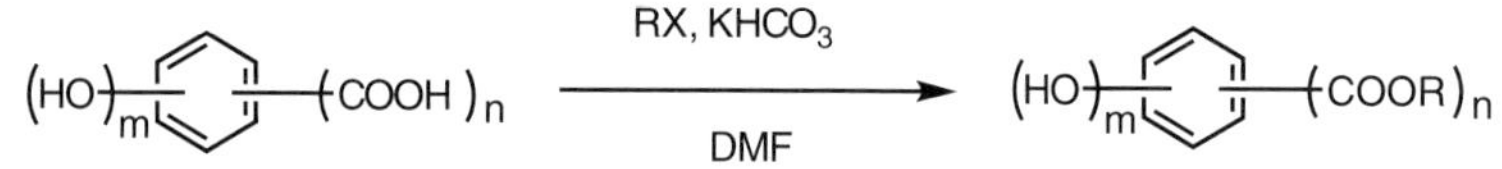

$(HO)_m$–C6H4–$(COOH)_n$ —RX, $KHCO_3$ / DMF→ $(HO)_m$–C6H4–$(COOR)_n$

Scheme 3.8

Another feasible mode of activation of carboxylic acids is by the use of KF, which promotes the reaction through hydrogen bonding between the hydroxyl and the fluorine, resulting in increased nucleophilicity of the hydroxy oxygen [757]. Since KF is very soluble in acetic acid, heating of chloroacetic acids or benzyl chloroacetate in the presence of KF in acetic acid provides the corresponding alkylation products (Scheme 3.9) [758]. Similar treatment of dichloro- and dibromo-alkanes in carboxylic acids affords the corresponding diesters (Scheme 3.10) [759]. It should be noted that no such reaction occurs in the absence of KF. Phenacyl esters that are difficult to prepare by other conventional methods are readily obtained by treatment of α-bromoacetophenone and carboxylic acid with KF in DMF [760].

Experimental Procedure Scheme 3.9 [758] Potassium fluoride (58 g, 1.0 mol) and $ClCH_2CO_2H$ (47 g, 0.5 mol) are heated at 150 °C, and the reaction is monitored by ^1H NMR spectroscopy. This shows the disappearance of $ClCH_2CO_2H$ and the appearance of (chloroacetoxy)acetic acid, $ClCH_2CO_2CH_2CO_2H$, and later the formation of higher polymers. These appear before the complete disappearance of the starting material. At no time are any resonances attributable to fluoroacetic acid present in the spectrum. After 7 h the reaction mixture is cooled, extracted with diethyl ether, filtered, dried, and evaporated. The product is distilled and (chloroacetoxy)acetic acid (10.2 g, 0.066 mol, 27%) is obtained.

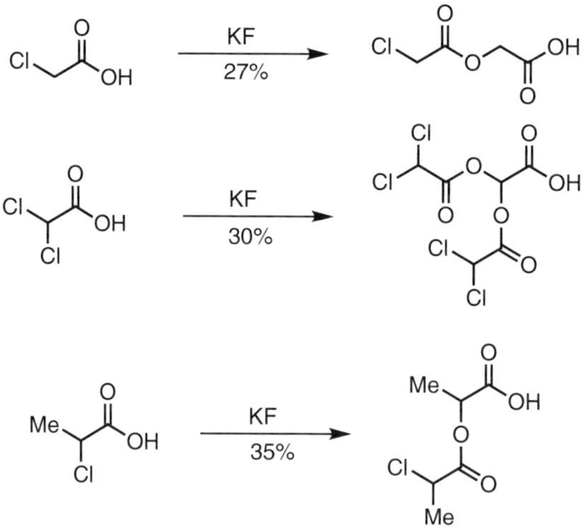

KF, 27%

KF, 30%

KF, 35%

Scheme 3.9

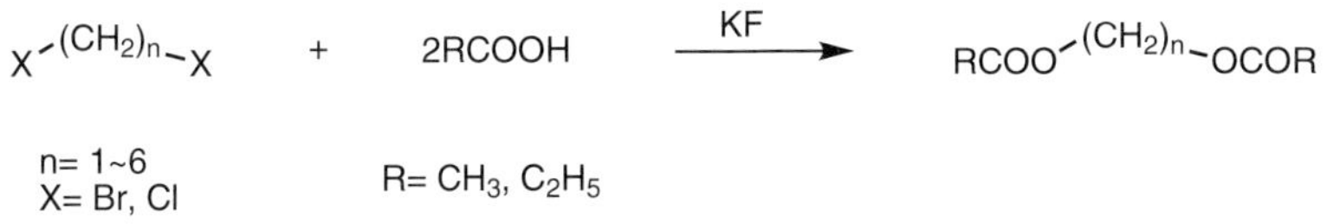

$$X-(CH_2)_n-X \quad + \quad 2RCOOH \xrightarrow{KF} RCOO-(CH_2)_n-OCOR$$

n= 1~6
X= Br, Cl

R= CH_3, C_2H_5

Scheme 3.10

KF-promoted alkylation protocol has been applied to the synthesis of Boc-aminoacyloxymethylphenylacetic acid, useful in solid phase peptide synthesis (Scheme 3.11) [761]. The yields of the first and second are larger than those obtained from reactions with amine base because the competing reaction between the bromomethyl group and the amine base is avoided.

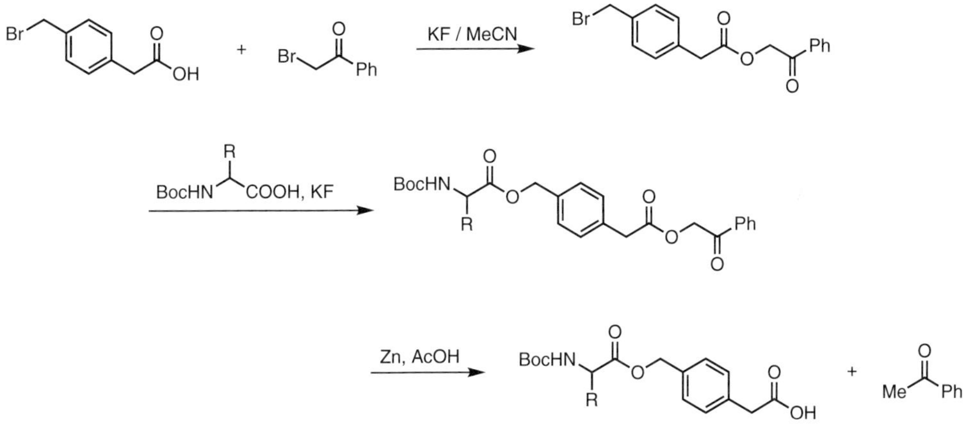

KF / MeCN

BocHN COOH, KF

Zn, AcOH

Scheme 3.11

Tetrabutylammonium fluoride is a good fluoride anion source, but its utility is limited by its hygroscopic nature. This drawback can be circumvented by the use of a silica gel support. Silica gel-supported Bu_4NF effects reaction between α-bromoacetophenone and acetic acid to give phenacetyl acetate [762].

Cesium salts also promote the alkylation of carboxylic acids. Merrifield resins are prepared without side reactions by treatment of chloromethylated polystyrene/1% divinylbenzene resin with cesium salts of *N*-protected amino acids [763]. A similar method is utilized for the synthesis of a wide variety of esters derived from protected amino acids and peptides [764]. The carboxylic acid to be esterified is first titrated to pH 7 with cesium carbonate or cesium bicarbonate, and the obtained neutral salt is then allowed to react with different alkyl halides to form the corresponding esters. The reaction is simple, easily scaled up, and proceeds without observable racemization. Many amino acid and peptide esters that might be difficult to prepare by other methods are obtainable. The synthesis of enkephalin is illustrated in Scheme 3.12 as a representative example.

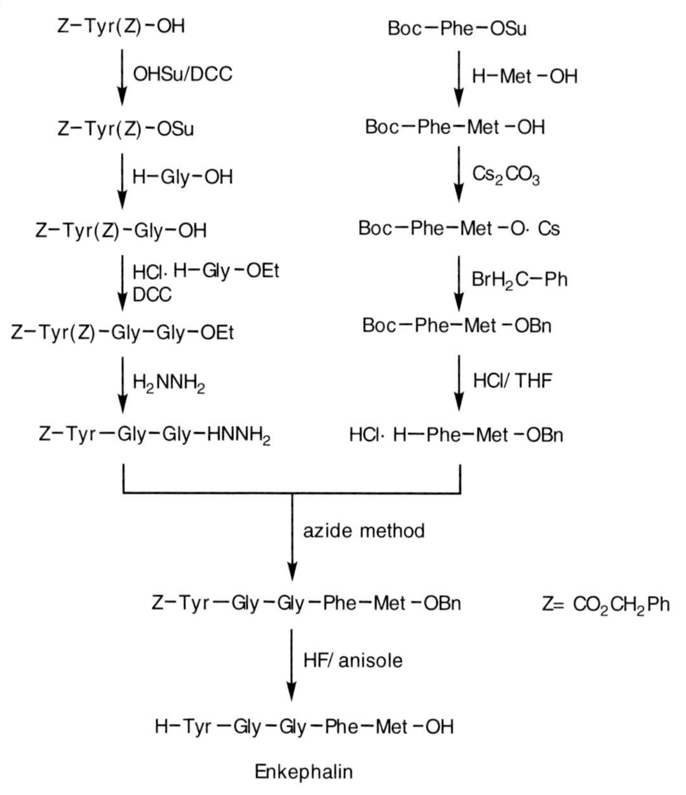

Scheme 3.12

Experimental Procedure [764] Boc-Asn-OBzl: Boc-Asn-OH (11.0 g, 47.5 mmol) is dissolved in MeOH (200 mL), and water (20 mL) is added. The solution is titrated to pH 7.0 (pH paper) with a 20% aqueous solution of Cs_2CO_3 (~55 mL). The mixture is evaporated to dryness and the residue is re-evaporated twice from DMF (120 mL, 45 °C). The obtained white, solid cesium salt is stirred with benzyl bromide (8.9 g, 52 mmol) in DMF (120 mL) for 6 h. On evaporation to dryness and treatment with a large volume of water (500 mL) the product solidifies. It is taken into ethyl acetate, washed with water, dried over Na_2SO_4, evaporated to a solid mass, and crystallized from ethyl acetate with petroleum ether: yield 13.8 g (90.3%); mp 120–122 °C; $[\alpha]^{25}$D −17.29° (c = 1, DMF).

Depsipeptides containing D-α-hydroxycarboxylic acids are efficiently synthesized in an analogous manner (Scheme 3.13) [765]. Moreover, optically active D-α-hydroxy carboxylic acids are obtained from L-amino acids via L-α-halocarboxylic acids and their stereoselective reaction with cesium *p*-nitrobenzoate (Scheme 3.14) [766].

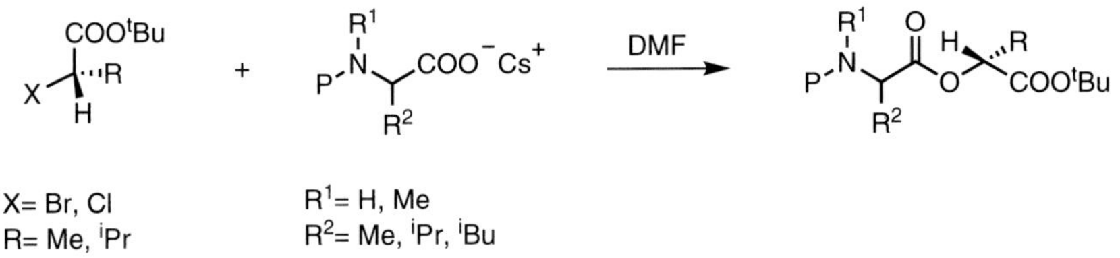

X= Br, Cl
R= Me, ^{i}Pr

R^{1}= H, Me
R^{2}= Me, ^{i}Pr, ^{i}Bu

Scheme 3.13

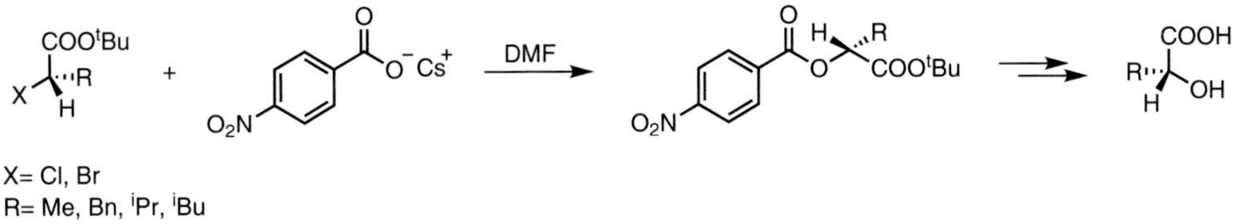

X= Cl, Br
R= Me, Bn, ^{i}Pr, ^{i}Bu

Scheme 3.14

Treatment of ω-halo carboxylic acids with an equivalent amount of dry Cs_2CO_3 in DMF affords the corresponding lactones (Scheme 3.15) [767]. The reaction proceeds through intramolecular nucleophilic attack of cesium carboxylate. Cesium carboxylates undergo ring-closure more readily and in far better yield than the carboxylates of lithium, sodium, potassium, rubidium, silver, thallium, magnesium, strontium, or barium. The origin of the unique effect induced by the cesium ion is an issue of controversy [768, 769].

Experimental Procedure Scheme 3.15 [767] Cs_2CO_3 (180 mg, 1.1 equiv.) is added to 16-iodohexadecanoic acid (190 mg, 0.5 mmol) dissolved in dry DMF (50 mL) in a single-necked flask containing a magnetic stirring bar. The reaction mixture is stirred magnetically at 40 °C for 24 h, during which time all the solid Cs_2CO_3 goes into solution. A precipitate of cesium halide usually forms slowly during the course of the reactions. The DMF is removed under vacuum (2–3 Torr) on a rotary evaporator. A saturated aqueous NaCl solution is added to the residue, and extraction with diethyl ether is carried out (three times). A mixture of macrolide and diolide (127 mg, 0.5 mmol in total) is obtained: GLC with use of an internal standard shows this to consist of at least 99% of macrolide and diolide, present in yields of 85% and 15%, respectively.

In a reaction carried out with 16-iodohexadecanoic acid (1 mmol), the cyclization products are isolated by preparative layer chromatography on silica gel, with a 9 : 1 light petroleum (40–60 °C)/ether eluent. The macrolide is isolated pure in 68% yield and the diolide in 4% yield. Both are identified from their mass spectra and by comparison with authentic materials.

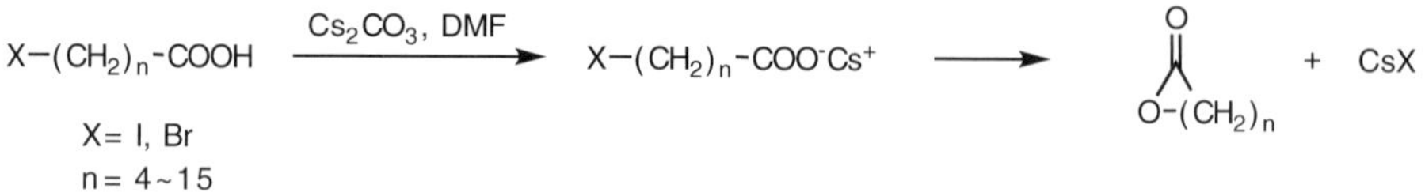

$X-(CH_2)_n-COOH \xrightarrow{Cs_2CO_3,\ DMF} X-(CH_2)_n-COO^-Cs^+ \longrightarrow$ lactone $O-(CH_2)_n$ + CsX

X= I, Br
n= 4~15

Scheme 3.15

The cesium carboxylate technique can also be used for alkylation of carboxylic acid on sugar, as shown in Scheme 3.16 [770]. The pentaacetate is converted into the corresponding methyl ester by treatment with cesium carbonate and iodomethane in DMF.

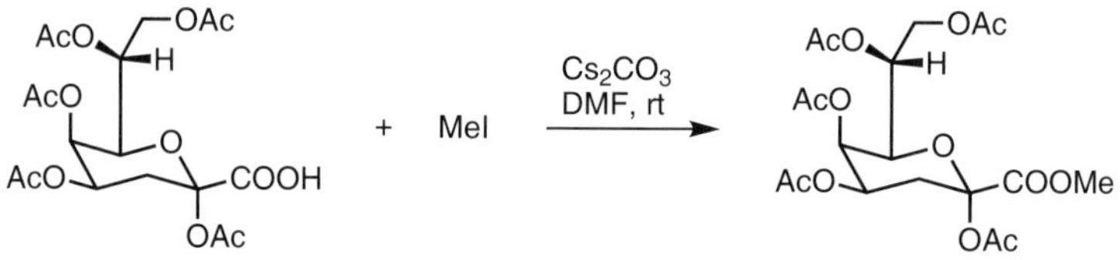

AcO, OAc, H, AcO, O, AcO, COOH, OAc + MeI $\xrightarrow{Cs_2CO_3,\ DMF,\ rt}$ AcO, OAc, H, AcO, O, AcO, COOMe, OAc

Scheme 3.16

Cesium fluoride is another reagent of choice for reaction between carboxylic acids and alkyl halides [771]. Alkyl bromides and iodides (except for *tert*-BuI) are employable, and a wide range of functional groups are tolerated since CsF is a good scavenger of hydrogen halides (Scheme 3.17). As a result, this technique is able to replace diazomethane-based esterification in many cases. Organotin carboxylates are protected carboxylic acids frequently used in peptide synthesis, and these compounds also undergo alkylation with RX/CsF. No epimerization occurs with α-amino acids. The DMF solvent is crucial in the above procedure, but CH_3CN is employable when the CsF is supported [772].

Experimental Procedure Scheme 3.17 [771] A mixture of (*S*)-*N*-acetylphenylalanine (207 mg, 1 mmol), CsF (227 mg, 1.5 mmol), EtI (234 mg, 1.5 mmol), and DMF (3 mL) is stirred at 15 °C for 24 h. The reaction mixture is combined with aqueous $NaHCO_3$ (50 mL) and extracted with EtOAc (100 mL). The organic layer is dried (Na_2SO_4) and evaporated. Column chromatography on silica gel (50 : 50 hexane/EtOAc) gives ethyl (*S*)-*N*-acetylphenylalanine (211 mg, 90%, ≥98% *ee*).

Large-Scale Preparation of Methyl 7-(Tetrahydropyranyloxy)-5-heptynoate from 7-(Tetrahydropyranyloxy)-5-heptynoic Acid and Methyl Iodide: A mixture of 7-(tetrahydropyranyloxy)-5-heptynoic acid (6.0 g, 26.4 mmol), CsF (8.8 g, 58 mmol), MeI (8.2 g, 58 mmol), and DMF (100 mL) is stirred at 30 °C for 18 h. The reaction mixture is extracted with EtOAc and washed with saturated aqueous $NaHCO_3$ (100 mL × 3). The organic layer is dried (Na_2SO_4) and evaporated to give an oil. Column chromatography on silica gel (95 : 5 hexane/EtOAc) provides methyl 7-(tetrahydropyranyloxy)-5-heptynoate (5.0 g, 79%).

General Procedure for the Preparation of Esters from Tributyltin Carboxylates and Alkyl Halides in the Presence of Cesium Fluoride: A mixture of hexabutyldistannoxane (656 mg, 1.1 mmol), (*S*)-*N*-acetylphenylalanine (414 mg, 2 mmol), and benzene (30 mL) is heated at reflux in a Dean-Stark apparatus for 3 h. The benzene is removed under reduced pressure, and DMF (6 mL) is added. CsF (456 mg, 3 mmol) and EtI (468 mg, 3 mmol) are added to this solution, and the reaction mixture is stirred at 30 °C for 30 h. Aqueous workup as described for the reaction between (*S*)-*N*-acetylphenylalanine and ethyl iodide and column chromatography on silica gel (50:50 benzene/EtOAc) affords ethyl (*S*)-*N*-acetylphenylalanine (430 mg, 91%, ≥98% *ee* determined by ^1H NMR in the presence of Eu(hfc)$_3$.

$$RCOOH + R'X \xrightarrow{CsF,\ DMF} RCOOR'$$

$$RCOOSnBu_3 + R'X \xrightarrow{CsF,\ DMF} RCOOR'$$

Scheme 3.17

Tetraalkylammonim fluorides are also effective for the alkylation of carboxylic acids [773]. Bu_4NF generated *in situ* from Bu_4NHSO_4 and KF promotes alkylation even of sterically demanding carboxylic acids such as adamantanecarboxylic acid and triphenylacetic acid. Use of a chiral ammonium fluoride derivative can effect kinetic resolution of *sec*-alkyl halides (Scheme 3.18).

Experimental Procedure [773] Tetrabutylammonium hydrogen sulfate (8.7 mg, 0.025 mmol) and potassium fluoride dihydrate (240 mg, 2.5 mmol) are placed under argon in a two-necked flask fitted with a stirring bar and dried *in vacuo* (0.6 mmHg) for 10 min. After replacement of the argon atmosphere, freshly distilled THF (1 mL) is introduced. 3-Phenylpropanoic acid (76.6 mg, 0.5 mmol) and benzyl bromide (66.8 μl, 0.55 mmol) are then added sequentially at room temperature. After having been stirred for 3 h at room temperature, the resulting reaction mixture is poured into water (5 mL) and extracted with ether (2 × 10 mL). The combined organic extracts are washed with brine and dried over Na_2SO_4. Evaporation of solvents and purification of the residual oil by column chromatography on silica gel (hexane/ether = 10:1 as eluent) gives pure benzyl ester (120 mg, 99% yield) as a colorless oil.

Strongly basic amines also serve for esterification. DBU (1,8-diazabicyclo[5.4.0] undec-7-ene) is one such amine, in the presence of which a variety of carboxylic acids including simple acids, sterically hindered acids, thermally unstable acids, and *N*-protected amino acids react with alkyl halides [774]. A novel base can be prepared by electroreduction of 2-pyrrolidone in the presence of a tetraalkylammonium salt in DMF. By use of this base, sterically demanding acids and amino

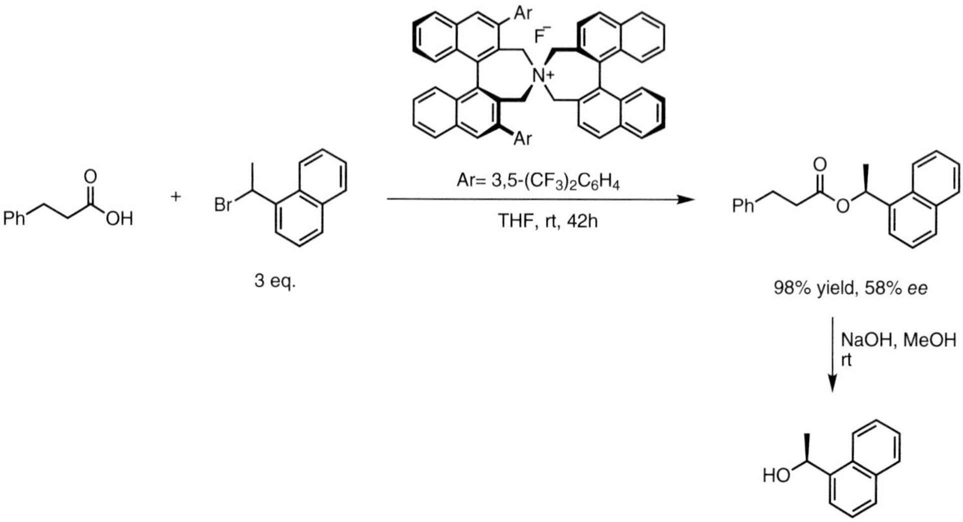

Scheme 3.18

acids can be alkylated to give the corresponding esters (Scheme 3.19) [775]. In addition, ω-bromo carboxylic acids are transformed into lactones.

Experimental Procedure [774] A solution of ethyl iodide (1.56 g, 0.01 mol) in benzene (5 mL) is added to a solution of 2,4,6-trimethylbenzoic acid (1.64 g, 0.01 mol) and DBU (1.52 g, 0.01 mol) in benzene (15 mL), and the mixture is stirred at room temperature for 2 h. The reaction mixture is then washed with water, dried over anhydrous magnesium sulfate, and distilled. Ethyl 2,4,6-trimethylbenzoate is obtained (1.53 g, 80% yield).

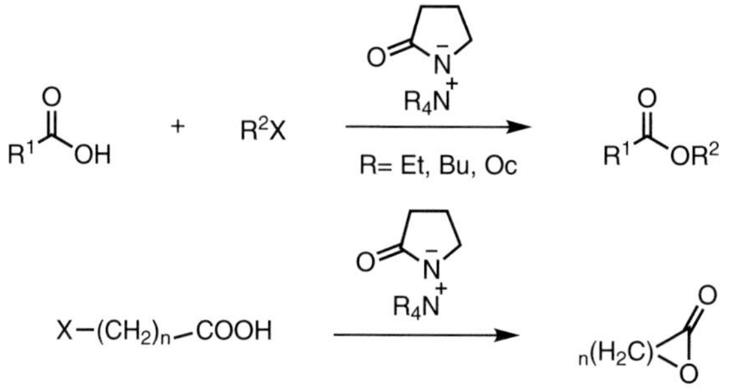

Scheme 3.19

A wide variety of suitably protected amino acids and peptides are alkylated with alkyl halides under phase-transfer conditions (Scheme 3.20) [776, 777]. The method is particularly convenient for acid-sensitive amino acids. Didecyldimethylammonium bromide has been claimed to be a more powerful phase-transfer catalyst than others for reaction between benzyl chloride and sodium formate [778].

Experimental Procedure Scheme 3.20 [776] A mixture of trioctylmethylammonium chloride (Adogen-464; 0.404 g, 1 mmol) and organic halide (1.2 mmol) in dichloromethane (1 mL) is added at room temperature to a solution of Cbz- or Boc-*N*-protected L-amino acid or dipeptide (1 mmol) in saturated aqueous sodium hydrogen carbonate solution (1 mL). After the reaction is complete (3–24 h) the mixture is extracted twice with dichloromethane. The organic phase is washed with water, dried with sodium sulfate, and concentrated to a small volume under vacuum at room temperature. The residue is purified by percolation on a silica gel column by elution with hexane/ethyl acetate (8 : 2).

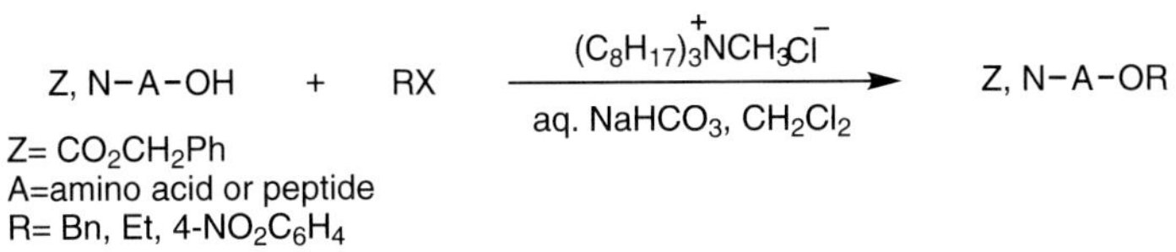

$$\text{Z, N–A–OH} + \text{RX} \xrightarrow[\text{aq. NaHCO}_3,\ \text{CH}_2\text{Cl}_2]{(\text{C}_8\text{H}_{17})_3\overset{+}{\text{N}}\text{CH}_3\overset{-}{\text{Cl}}} \text{Z, N–A–OR}$$

Z= CO_2CH_2Ph
A=amino acid or peptide
R= Bn, Et, 4-$NO_2C_6H_4$

Scheme 3.20

3.3 Treatment with Other Electrophiles

An alkyl group on an onium center undergoes facile attack even by a weak electrophile such as carboxylate ions. Trialkyloxonium tetrafluoroborates are onium centers of this kind, and they react with carboxylic acids in the presence of diisopropylethylamine to furnish methyl or ethyl esters (Scheme 3.21) [779, 780]. A wide variety of sterically hindered and polyfunctional carboxylic acids are employable.

Experimental Procedure Scheme 3.21 [780] Carboxylic acid (0.010 mol) is added to a suspension of trimethyloxonium tetrafluoroborate (1.63 g, 0.011 mol) in dichloromethane (75 mL). The resulting suspension is stirred magnetically while *N,N*-diisopropylethylamine (1.42 g, 0.011 mol) is introduced by syringe. For di- and triprotic acids, a corresponding increase in the number of equivalents of oxonium salt and amine is used. During the addition of the amine, warming of the reaction mixture is observed; larger-scale reactions require a dropping funnel and a reflux condenser. The flask is stoppered after the addition of the amine, and the suspension is allowed to stir for 1–24 h. After approximately 1 h, virtually all of the originally undissolved oxonium salt has gone into solution. After the end of the reaction time, the organic solution is extracted with hydrochloric acid (1 N, 3 × 50 mL), potassium bicarbonate (1 N, 3 × 50 mL), and saturated sodium chloride. With maleic acid, fumaric acid, 3-butenoic acid, and citric acid, the HCl and the $KHCO_3$ solutions are back-extracted, because of the water solubility of the corresponding esters. The organic solution is then dried over anhydrous sodium sulfate and the solvent is removed by evaporation under reduced pressure. The residue is purified by short-path (Kugelrohr) distillation or by recrystallization to afford 70–97% of the desired esters with purities of 95% or greater.

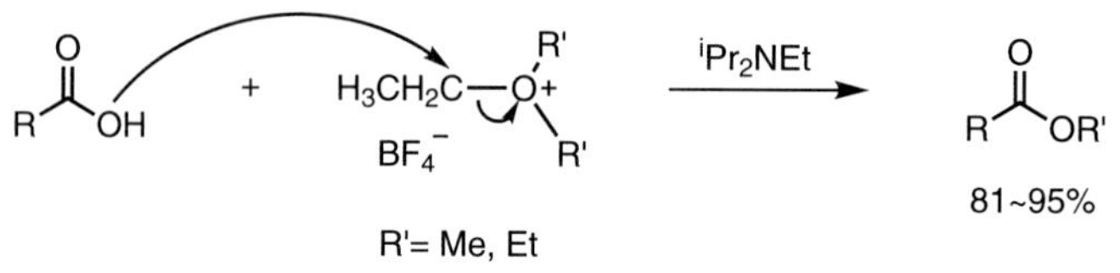

$RCOOH + H_3CH_2C{-}O^+R'_2\ BF_4^- \xrightarrow{^iPr_2NEt} RCOOR'$ (81~95%)

R'= Me, Et

Scheme 3.21

Isobutylene is readily protonated to give a *tert*-butyl cation. Treatment of carboxylic acids with *tert*-butyl acetoacetate in the presence of catalytic H_2SO_4 thus gives the corresponding *tert*-butyl esters through *in situ* formation of isobutylene from the *tert*-butyl acetoacetate (Scheme 3.22) [781].

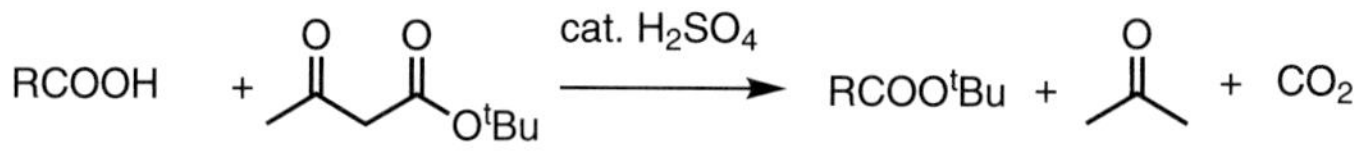

$RCOOH + CH_3COCH_2COO^tBu \xrightarrow{\text{cat. } H_2SO_4} RCOO^tBu + CH_3COCH_3 + CO_2$

Scheme 3.22

Sulfonium salts are also powerful alkylating agents. Heating of a mixture of trimethlsulfonium hydroxide and carboxylic acids affords the corresponding methyl esters [782]. In the presence of copper(I) salts, reactions of allyl sulfonium salts are highly accelerated [783]. Trimethylselenonium hydroxide works similarly [784]. Pentamethylphosphorane $(CH_3O)_5P$ reacts with carboxylic acids such as benzoic acid and salicylic acid to afford the corresponding methyl esters [785]. *O*-Methylcaprolactam effects methylation of carboxylic acids just on being heated with one equivalent of the acids (Scheme 3.23) [786].

Experimental Procedure [785] $(CH_3O)_5P$ (4.0 g, 0.0215 mol) is added dropwise over a period of 10 min and under an atmosphere of nitrogen to a solution of benzoic acid (2.44 g, 0.02 mol) in methylene chloride (20 mL). After the addition, the methylene chloride solution is washed with water to remove trimethyl phosphate. The methylene chloride solution is evaporated to give essentially pure methyl benzoate (2.48 g, 90%).

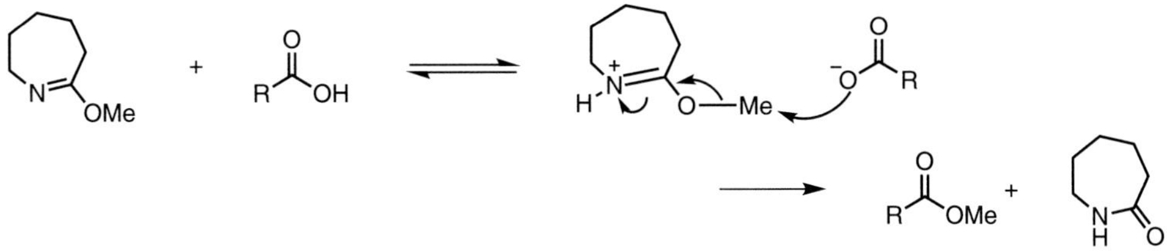

Scheme 3.23

Dimethyl sulfate is used for various *O*-methylations. The reaction takes place with carboxylic acids in the presence of a base (Scheme 3.24) [787]. This technique has been applied to the preparation of bile acid methyl esters (Scheme 3.25) [788]. In these reactions, however, the use of aqueous alkali induces side reactions such as hydrolysis of dimethyl sulfate and the ester formed. This drawback can be overcome by employing lithium carboxylates (Scheme 3.26) [789].

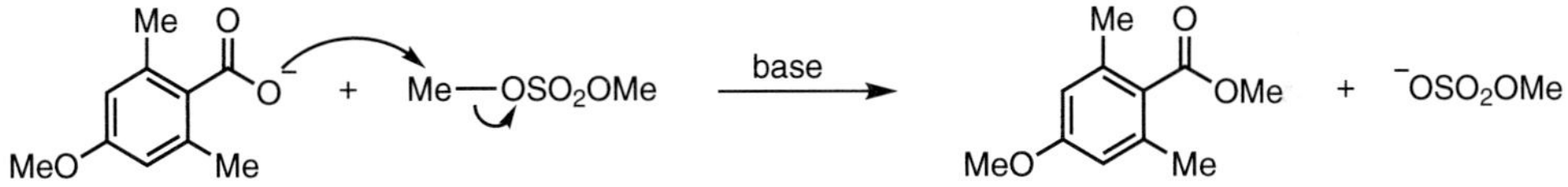

Scheme 3.24

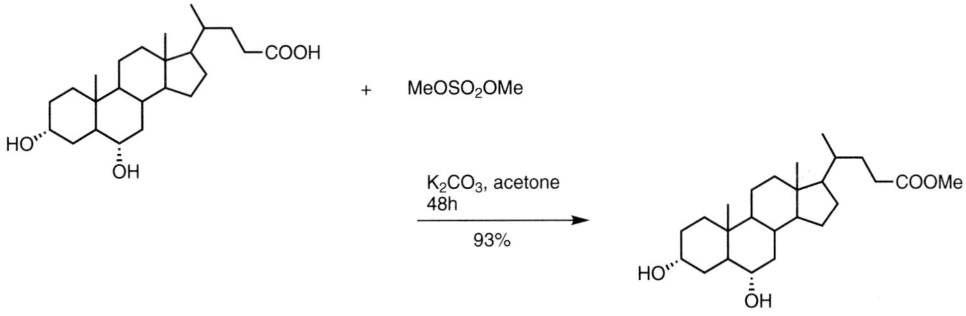

Scheme 3.25

Experimental Procedure Scheme 3.26 [789] 2-Hydroxy-1-naphthoic acid (470.5 mg, 2.5 mmol) in dry THF (2.5 mL) is treated with $LiOH \cdot H_2O$ (104.9 mg, 2.5 mmol) at room temperature for 30 min. Me_2SO_4 (0.12 mL, 1.25 mmol) is then added, and the mixture is heated under reflux for 3 h. Solvent is distilled off, and the mixture is diluted with saturated aqueous $NaHCO_3$ and extracted with Et_2O to afford the ester (white solid, 404.3 mg, 80%). The yield may be improved to 96% by the use of 2.5 mmol of Me_2SO_4.

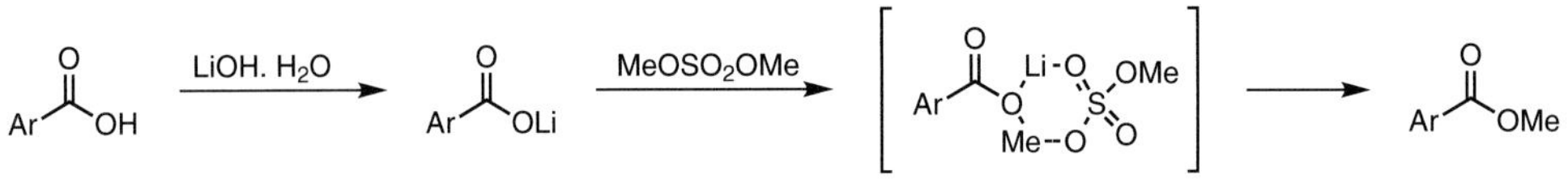

Scheme 3.26

Orthoesters are reagents which have long been used for alkylation of carboxylic acids; alkyl orthoformates are most popular. The reaction is catalyzed by ammonium chloride or nitrate, ferric chloride, and acidic ion-exchange resins. Reaction between triethyl orthoacetate and carboxylic acids proceeds without catalyst in refluxing toluene or in the absence of solvent (Scheme 3.27) [790]. The *O*-alkylation mechanism is suggested by the fact that no significant rate difference is observed between the esterification of 2,6-dimethylbenzoic acid and that of benzoic acid. The enhanced rate of esterification observed for the orthoacetate over the orthoformate is ascribed to the higher stability of the cationic intermediate. This method is employable for ethyl ester formation from amino acid (Scheme 3.28) [791]. When *vic*-diols are treated with $MeC(OMe)_3$ (3 equiv.) in the presence of $Yb(OTf)_3$ (5 mol%) followed by hydrolysis, the secondary alcohol is selectively acylated (Scheme 3.29) [792].

Experimental Procedure Scheme 3.27 [790] Triethylorthoacetate (16.0 mL, 87 mmol, 3 equiv.) is added dropwise to a solution of 1-naphthoic acid (5.0 g, 29 mmol) in toluene (35 mL). The reaction mixture is heated at reflux for 24 h. After the mixture has been cooled, HCl (2 M, 30 mL) is added. The organic extract is washed with saturated $NaHCO_3$ (1 × 30 mL) and brine (1 × 30 mL) and dried with $MgSO_4$. The solvent and excess reagent are removed *in vacuo* to give a brown oil. Kugelrohr distillation of the product at 100 °C (0.45 Torr) gives the ethyl ester (5.15 g, 89%) as a colorless liquid.

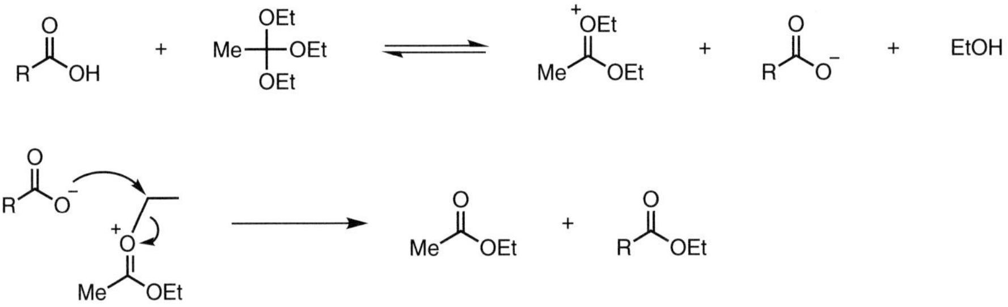

R= 1-naphthoic, 1-naphthylacetic, nicotinic, adipic, 2,6-dimethylbenzoic

Scheme 3.27

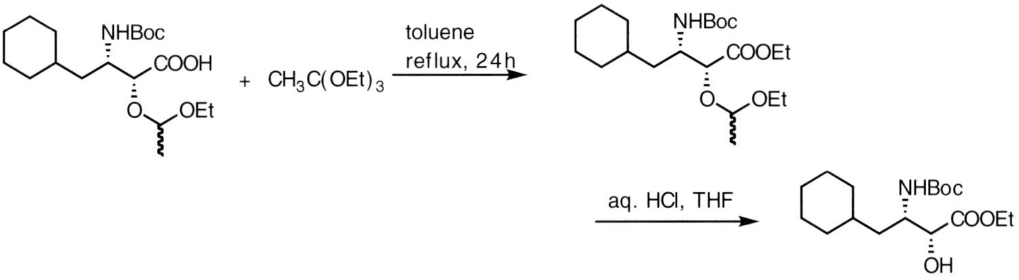

Scheme 3.28

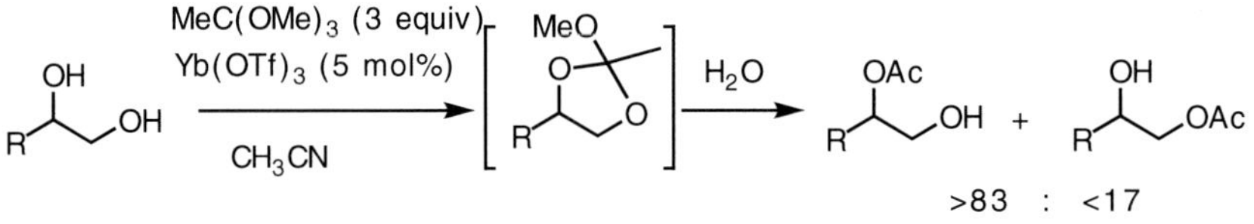

Scheme 3.29

Methyl trichloroacetate is a unique methyl donor, releasing chloroform and carbon dioxide upon treatment with carboxylic acids (Scheme 3.30) [793]. The reaction is catalyzed by potassium carbonate and 18-crown-6.

Experimental Procedure Scheme 3.30 [793] A mixture of methyl trichloroacetate (12.5 mmol), carboxylic acid (12.5 mmol), potassium carbonate (0.035 g, 0.25 mmol) and 18-crown-6 (0.066 g, 0.25 mmol) is placed in a 25 mL round-bottomed flask fitted with a distillation head and a dry ice-cooled receiver and trap. The stirred reaction mixture is gradually warmed to 90 °C, whereupon the evolution of carbon dioxide begins. The temperature is gradually raised to 150 °C and maintained there until the evolution of carbon dioxide and the distillation of chloroform (1.3–1.4 g) cease (1–2 h). The product is isolated by distillation or crystallization.

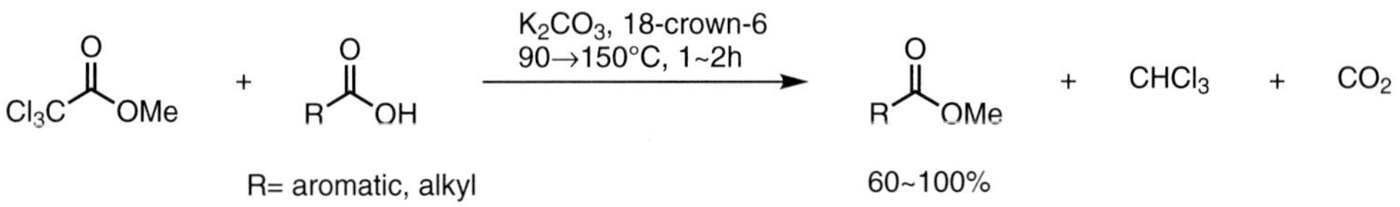

Scheme 3.30

Dimethyl carbonate reacts both with carboxylic acid and with phenols in the presence of Cs_2CO_3 (Scheme 3.31) [794]. DBU is also an effective catalyst (Scheme 3.32) [795]. A mechanism involving a carbamate intermediate has been proposed on the basis of an ^{18}O-labeling study. The dimethyl carbonate/DBU technique can be upgraded by conducting the reaction in a continuous-flow reactor under microwave irradiation conditions [796].

Experimental Procedure Scheme 3.32 [795] DBU (1 equiv.) is added to a 10% solution of carboxylic acid in dimethyl carbonate, and the resulting mixture is heated to reflux. Upon completion of the reaction, the mixture is allowed to cool to room temperature and diluted with either CH_2Cl_2 or EtOAc and H_2O. The aqueous layer is removed, and the organic layer is washed once with H_2O, twice with HCl (2 M) or aqueous citric acid (10%), twice with saturated aqueous $NaHCO_3$, and twice with H_2O. The organic layer is dried over Na_2SO_4, filtered, and concentrated under vacuum to afford the ester.

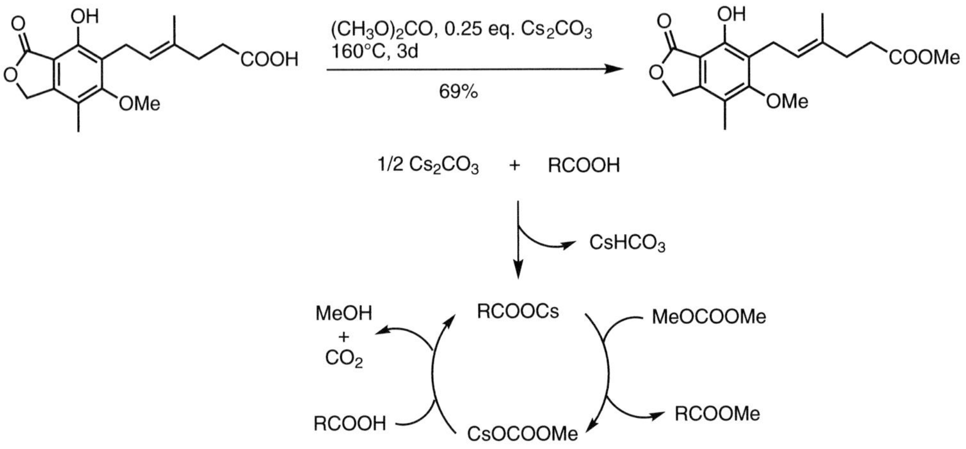

Scheme 3.31

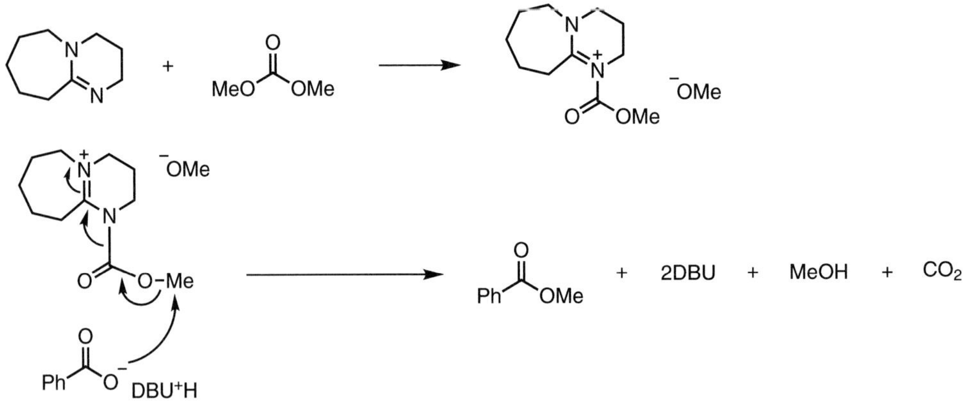

Scheme 3.32

Esterification of salicylic acid with methanol/dimethyl carbonate is catalyzed by zeolites [797] and by anion-modified metal oxides [798]. NaY faujasite effects selective esterification of phenolic carboxylic acids [799].

As briefly described in Section 1.3.5, reactions between carboxylic acids and dialkyl carbonates proceed through mixed anhydrides, and the use of dimethyl carbonate brings about instantaneous nucleophilic attack of *in situ*-formed methoxide ion on the mixed anhydride, without the possibility to obtain esters other than methyl esters. Employment of bulkier di-*tert*-butyl carbonate [$(BOC)_2O$] allows mixed anhydrides to survive the attack of less reactive *tert*-butoxide. Alternatively, other alcohols added separately can react to give the desired esters (Scheme 3.33) [627]. Triphosgene is employable for esterification of carboxylic acids and amino-acids (Scheme 3.34) [800].

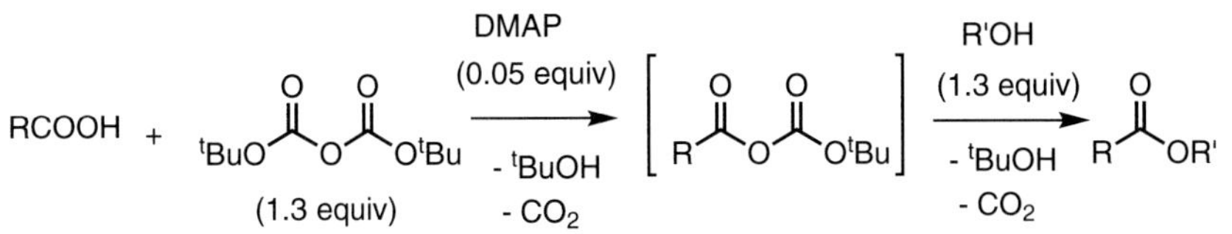

Scheme 3.33

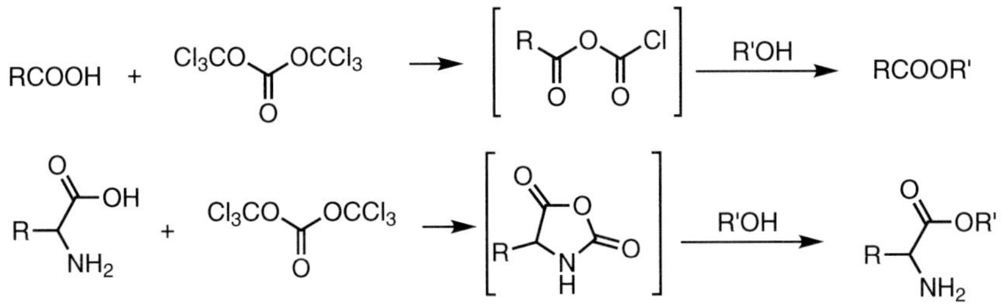

Scheme 3.34

N,N-Dimethylformamide acetals are another class of reagents used to alkylate carboxylic acids. Treatment of carboxylic acids with diethyl or dibenzyl acetal in refluxing CH_2Cl_2 or $ClCH_2CH_2Cl$ furnishes the corresponding esters [801]. A variety of acetals are useful for the preparation of esters of sterically demanding carboxylic acids [802]. The sterically bulky dineopentyl actal can be employed for reaction between carboxylic acids and *p*-dodecylbenzyl alcohol or *p*-methoxybenzyl alcohol (Scheme 3.35) [803]. The reaction involves alkylation of the carboxylate anion with an alkoxyimmonium ion. The use of *tert*-butyl acetal affords the corresponding *tert*-butyl esters [804]. Pyrrole- and indolecarboxylic acids are converted into the corresponding *tert*-butyl esters by treatment with *N,N*-dimethylformamide di-*tert*-butyl acetal [805]. The *tert*-butyl ester formation is achieved by slow addition of the dineopentyl acetal to a solution of amino acid in benzene/*tert*-butyl alcohol (2 : 1) (Scheme 3.36) [806].

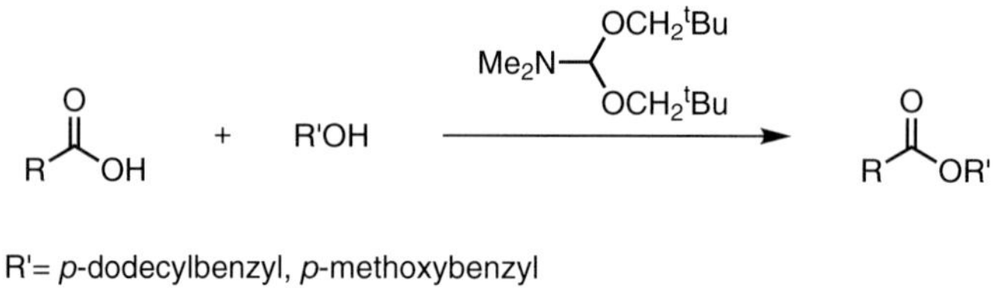

R'= p-dodecylbenzyl, p-methoxybenzyl

Scheme 3.35

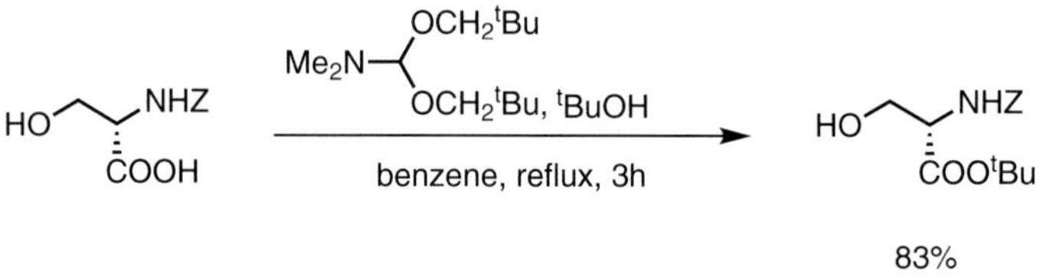

benzene, reflux, 3h

83%

Scheme 3.36

3-Alky-1-*p*-tolyltriazenes (2.1–2.5 equiv.) can be used for alkylation of cephalosporanic acids in ether/THF (Scheme 3.37) [807]. The reaction proceeds at room temperature. *O*-Alkylisoureas effect *O*-alkylation of carboxylic acids under microwave irradiation conditions (Scheme 3.38) [808, 809]. The reaction is normally complete in 5 min. An immobilized *O*-alkylisourea is also available.

Experimental Procedure Scheme 3.37 [807] 1-Alkyl-1-*p*-tolyltriazene in diethylether (2.1–2.5 equiv.) is added to a solution of cephalosporanic acid in THF, and the mixture is stirred under nitrogen at room temperature for 6 h (R_4 = Me) or overnight (R_4 = Bn). Conventional workup gives the desired Δ^3 ester in a crude state, and the product is confirmed free of the Δ^2 ester by NMR (no peak due to the vinylic proton of the Δ^2 isomer in the δ = 6 region is observed). Recrystallization or column chromatography affords analytically pure samples.

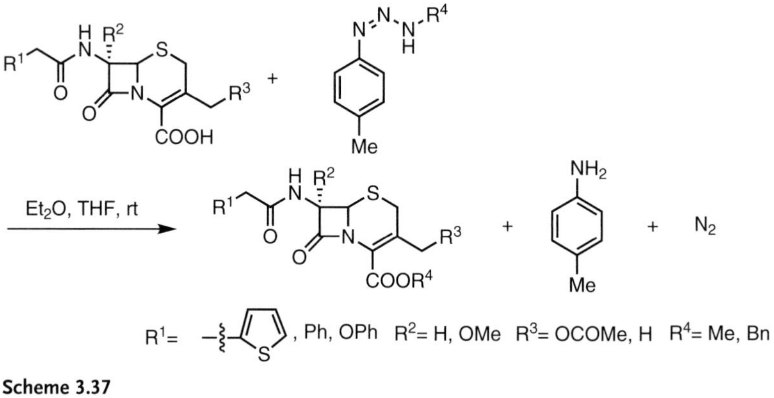

Et_2O, THF, rt

R^1= 2-thienyl, Ph, OPh R^2= H, OMe R^3= OCOMe, H R^4= Me, Bn

Scheme 3.37

Experimental Procedure Scheme 3.38 [808] A microwave vial is charged with the carboxylic acid (1.0 mmol) and *O*-alkylisourea, followed by addition of THF (2 mL). The vial is heated at the required temperature in a Smith Synthesizer for 5 min. The white solid is filtered off, and the solvent is evaporated under vacuum. The residue is then further purified by column chromatography.

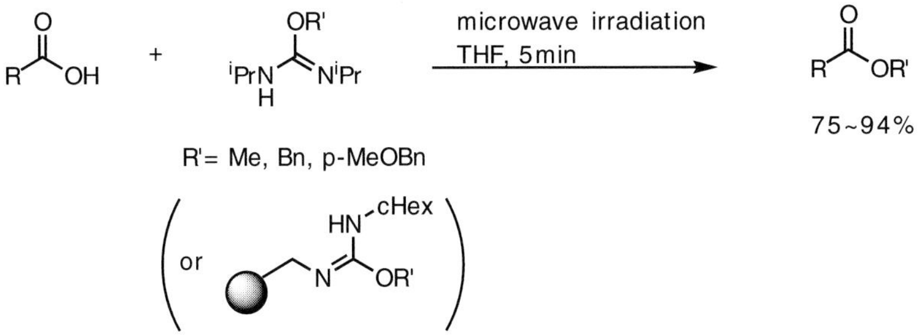

Scheme 3.38

Chiral allylic esters are produced by chiral palladium-catalyzed reaction between (*E*)-allylic trichloroacetimidates and carboxylic acids (3 equiv.) with high enantiopurity (Scheme 3.39) [810]. *tert*-Butyl esters of various heterocyclic carboxylic acids, including indole-5-carboxylic acid, are obtained by reaction between the acids and *tert*-butyl trichloroacetimidate (Scheme 3.40) [811].

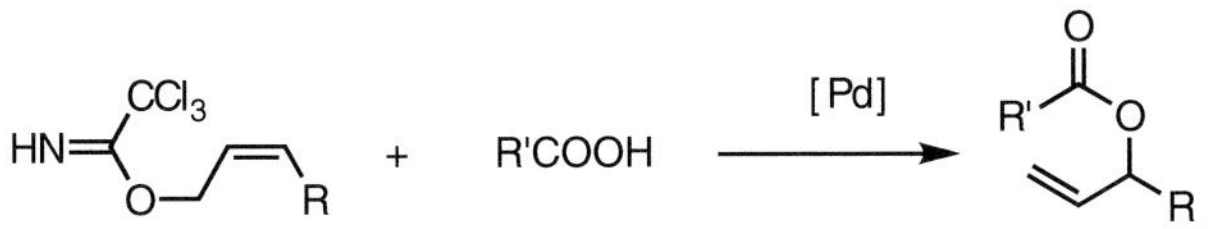

Scheme 3.39

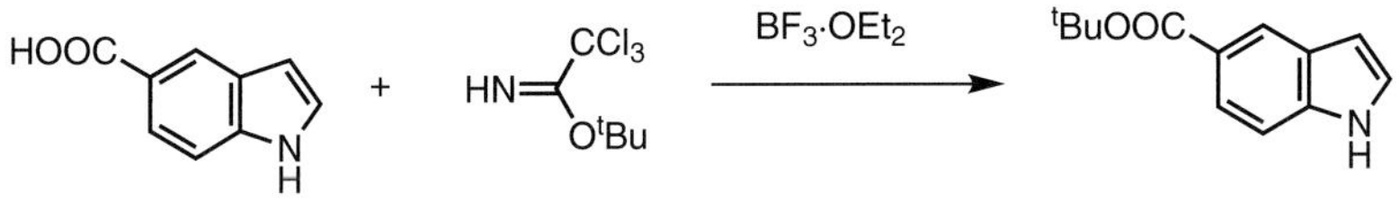

Scheme 3.40

Carboxylic acids with alkoxydiphenylphosphines, which are readily prepared from *N,N*-dimethylaminodiphenylphosphine, react with alcohols to give redox condensation products with the aid of 2,6-dimethyl-1,4-benzoquinone [812]. Chlorodiphenylphosphine is also employable as a precursor of the alkoxyphosphines [813, 814]. As shown in Scheme 3.41, inverted *tert*-alkyl carboxylates are obtained with high stereoselectivity by this method. 1,4-Benzoquinone is also effective [815].

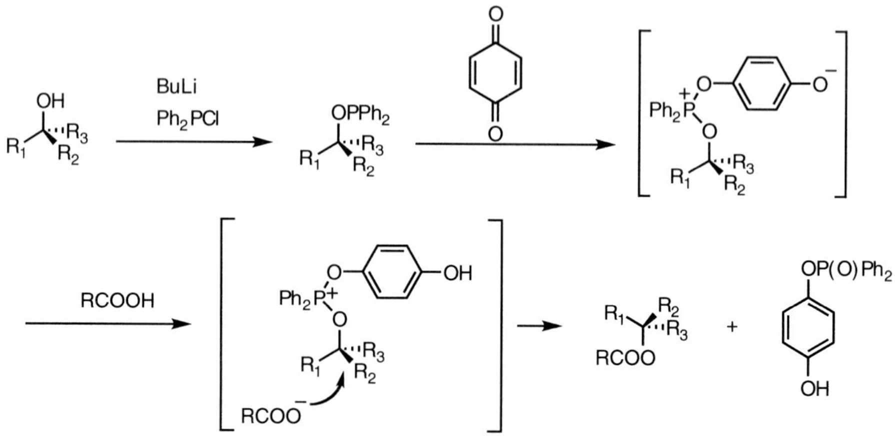

Scheme 3.41

2-(3-Methoxypyridyl)-β-D-gluco- and -D-galactopyranosides are employable for esterification of carboxylic acids, affording the corresponding α-1-esters in high yields (Scheme 3.42) [816].

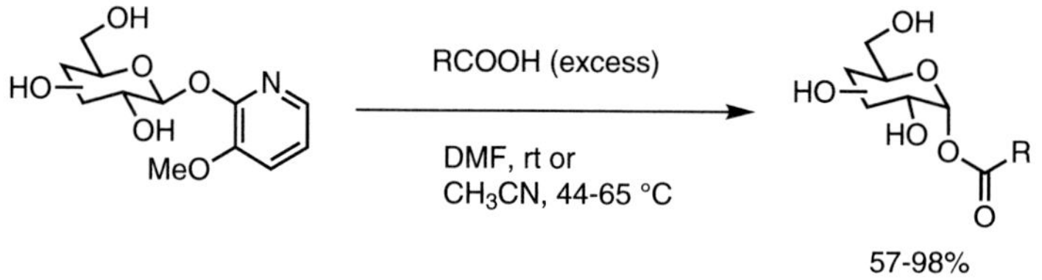

Scheme 3.42

4
Ester Interchange Reaction

Reactions between two different esters result in exchange of the alkoxy components, but it is not easy to bias the reaction in favor of the one side, because of equilibration. Tributyltin alkoxides catalyze reactions between methyl and ethyl esters, for example, but these unfortunately reach equilibrium at 30–60% conversions [345]. A similar situation applies with $[La(O^iPr)_3]_n$ [817].

These reactions are dramatically accelerated, however, by catalytic alkali metal alkoxides (Scheme 4.1) [818, 819].

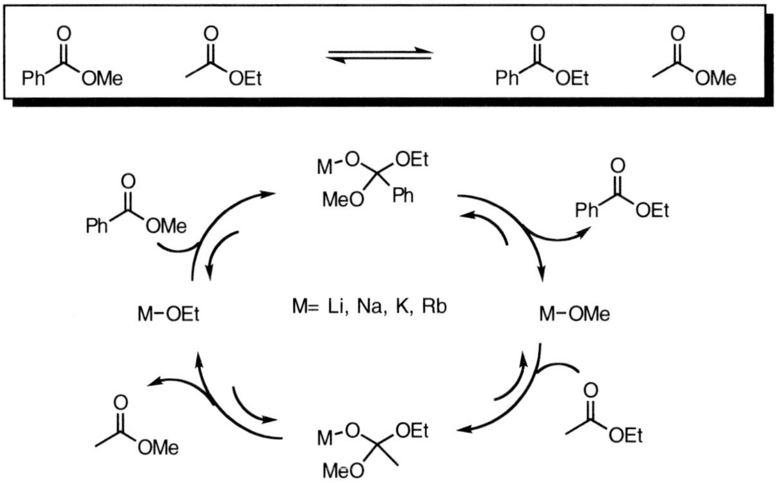

Ph–C(=O)–OMe + CH3–C(=O)–OEt ⇌ Ph–C(=O)–OEt + CH3–C(=O)–OMe

M–OEt   M= Li, Na, K, Rb   M–OMe

Scheme 4.1

Reactivities increase in the order: $Li^+ < Na^+ < K^+ < Rb^+ < Cs^+$. When methyl or ethyl esters are treated with *tert*-butyl acetate under reduced pressure (to remove the methyl or ethyl acetate formed), the *tert*-butyl esters are formed quantitatively [820]. These reactions proceed by stepwise exchange of the *tert*-butoxy group with the methoxy group in the tetrameric cluster (Scheme 4.2). When the third substitution is completed, the resulting mixed alkoxide precipitates, and so no further ester exchange takes place. The catalyst turnovers are increased by use of a mixed

Esterification. Methods, Reactions, and Applications. 2nd Ed. J. Otera and J. Nishikido

ISBN: 978-3-527-32289-3

alkoxide composed of NaO^tBu and $NaOC_6H_4$-4-^tBu in a 1:3 ratio (Scheme 4.3) [821, 822]. The aryloxy groups are never substituted by other alkoxy groups and are longer-lived than the simple metal alkoxides.

Experimental Procedure [819] A typical kinetic measurement is run as follows: methyl benzoate (5 mmol, 0.63 mL) and *tert*-butyl acetate (5 mmol, 0.67 mL) are mixed in a round-bottomed flask and diluted with the desired solvent (2.5 mL). In a separate flask, MO*t*Bu (0.025 mmol) is dissolved in the desired solvent (2.5 mL). The esters (0.8 M in each ester) are transferred to the catalyst solution by syringe. Aliquots are withdrawn by syringe, quenched in saturated aqueous NH_4Cl solution, diluted with ethyl acetate, and analyzed by GC to monitor the conversion of methyl benzoate to *tert*-butyl benzoate. Integration percentages are corrected for response factor differences between *tert*-butyl benzoate and methyl benzoate [%$PhCO_2Me$ (GC)/%$PhCO_2tBu$ (GC) = 1.46 × %$PhCO_2Me$ (actual)/%$PhCO_2tBu$ (actual)].

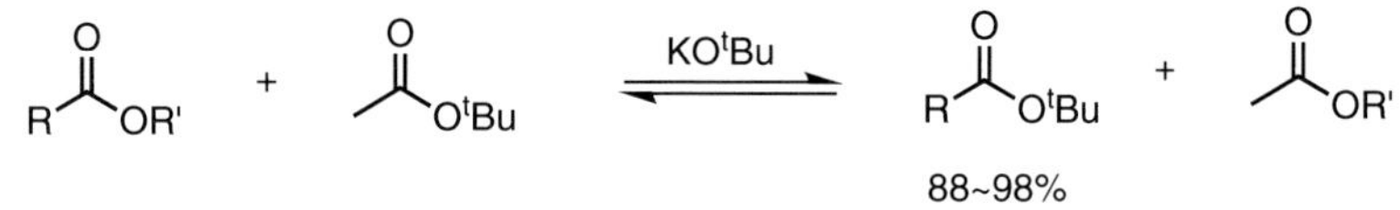

KO^tBu

88~98%

R= aromatic, aliphatic, alkyl
R' =Me, Et

Scheme 4.2

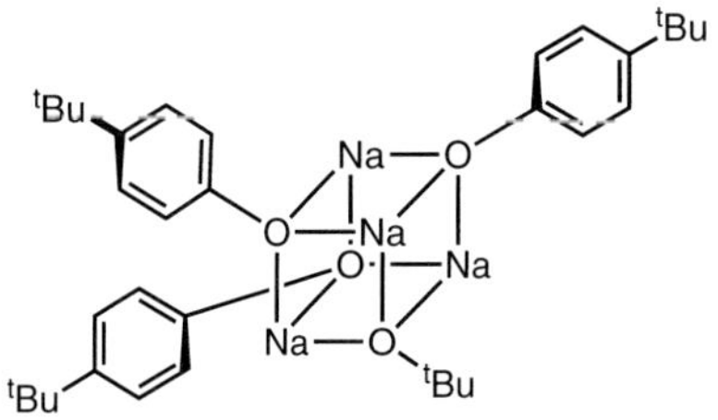

Scheme 4.3

As already described in Section 1.2.2.2, the exchange reaction between an ethyl ester and methyl propanoate in the presence of $Ti(OEt)_4$ is an effective means to prepared the corresponding methyl ester without recourse to insoluble $Ti(OMe)_4$ [341].

Part Two Synthetic Applications

Esterification. Methods, Reactions, and Applications. 2nd Ed. J. Otera and J. Nishikido

ISBN: 978-3-527-32289-3

5
Kinetic Resolution

5.1
Enzymatic Resolution

Racemic substrates can be resolved when one of the enantiomers reacts more rapidly than the other with a reagent. Esterification serves well for this end, racemic alcohols, carboxylic acids, esters, thioesters, and anhydrides successfully undergoing resolution. In this procedure, one of the enantiomers is selectively consumed while the other remains unchanged. Accordingly, the effectiveness of the reaction cannot be evaluated simply from the optical purity of the product or the remaining substrate, but the degree of conversion also has to be taken into account. In an ideal case, both product and remaining substrate are obtained in 50% yield, each with 100% *ee*. Such a degree of conversion/%*ee* relationship is measured by the stereoselectivity factor **s**, which is the relative ratio of the reaction rates of the two enantiomers [823]. For enzymatic reactions, the biochemical stereoselectivity factor *E* (which is virtually the same as **s** under steady-state conditions) is employed [824]. To obtain a 99% *ee* in the recovered material, for example, 72.1% conversion is required for **s** = 10, 52.3% conversion for **s** = 100, 51.1% conversion for **s** = 200, and 50.0% conversion for **s** = 1000.

Kinetic resolution with alcohol substrates is most frequently performed through enzymatic acylation. Representative examples for various enzymes are shown in Table 5.1.

Lipases are most popular, and acetylation with enol acetates is invoked most frequently, whereas acid anhydrides, trihaloethyl esters, and simple carboxylic acids as well as esters are employable as acyl donors. Esterases, acylase, hydrolase, and protease are also effective. The stereoselectivity factor is usually higher than that of the chemical process, as described later. Because of the specificity characteristic of eynzymatic reaction, the *E* value is dependent on substrates, acylating reagents, and reaction conditions. The acylation in most cases takes place on the *R*-alcohols, although *S*-secondary alcohols are acylated in a few cases. The use of additives such as amine often increases *E* values. Addition of metal ions is also effective to enhance the enantioselectivity [825]. The reaction is run in organic solvent, but water is occasionally added. Improved enantioselectivity may be achieved by immobilization of enzymes or by use of an ionic liquid solvent. High boiling points of the ionic

Esterification. Methods, Reactions, and Applications. 2nd Ed. J. Otera and J. Nishikido

ISBN: 978-3-527-32289-3

Table 5.1 Enzymatic kinetic resolution of racemic alcohols.

Enzyme	Acylating agent	Medium	*E*	Remarks	References
Porcine pancreatic lipase and yeast lipase	1,1,1-Trichloroethyl-alkylate	Organic solvent		PPL catalyzes transesterification. Yeast lipase catalyzes esterification. Stability of enzyme is much greater in organic solvents than in water.	[264]
Lipase from *M. miehei* (lipozyme)	*n*-Alkanoic acid	Hexane	1~50	Kinetic resolution of aliphatic alcohols by esterification.	[313]
	Vinyl acetate	Hexane or toluene		Kinetic resolution of 2-substituted cyclohexanols	[830]
	Vinyl acetate	2-Propancl		Kinetic resoluition of 3,3′-bis(hydroxymethyl)-2,2′bipyridine *N,N*-oxide	[831]
Rabbit gastric lipase	Isopropenyl acetate	Organic sɔlvent	1.1~500	Kinetic resolution of secondary benzylic and allylic alcohols using RGL.	[832]
Lipase from *P. sp.*	Vinyl acetate	Organic solvent ($^iPr_2O$, acetone)	1057	Kinetic resolution of 3-hydroxy-3-(pentafluorophenyl)propanonitrile. (*R*)- and (*S*)-optically active products can be obtained.	[833]
	Diketene	Toluene, Pr_2O	4~12	Diketene can be used as a new reagent for irreversible and enantioselective acylation of secondary alcohols.	[834]
	Ac_2O	Et_2O	2~13	Resolution of the sterically hindered *myo*-inositol derivatives is achieved. Racemic 3,4,5,6-*O*-tetrabenzyl-*myo*-inositol is resolved by this method.	[835]
Lipase from *P. cepacia*	Vinyl acetate	Organic solvent (hexane/THF, $^iPr_2O$, etc.)	2~645	Resolution of 2-acylamino-1-arylethanol is performed. *S*-specificity is obtained.	[836]

Vinyl acetate	Organic solvent ($^iPr_2O$, THF, pentane)	8~97	Kinetic resolution of *N*-hydroxymethyl γ-butyrolactams: access to optically active γ-butyrolactams, is achieved.	[837]
Vinyl acetate	CH_2Cl_2		Kinetic resolution of 5-hydroxy-4-oxa-*endo*-tricyclo[5.2.0]dec-8-en-3-ones is descibed. This reaction leads to a homochiral D-ring synthon for strigolactones.	[838]
Various vinyl esters	$^iPr_2O$	4~1057	Kinetic resolution of 2-substituted 3-cyclopentene-1-ols, and the effect of the acyl group of acylating agent are studied.	[839]
Vinyl acetate	Organic solvent and crown ether (cat. amount)	5~1200	Thiacrown ether additive enhances enantioselectivity in transesterification of allyl alcohols	[840, 841]
1-Ethoxyvinyl acetate	$^iPr_2O$ or BuOMe	640	1-Ethoxyvinyl acetate can be used as acyl donor in resolution of secondary alcohols.	[842] (see also [843])
Vinyl acetate	Et_2O, −40 °C	17~99	(*S*)-(+)-Phenyl-2H-azirine-2-methanol is obtained by using lipase-catalyzed kinetic resolution at −40 °C.	[844]
Vinyl acetate	Hexane, 30 °C	5.3~52	Resolution of N-aryl methylated *trans*-2,5-disubstituted pyrrolidines. The enantioselectivity depends on the structure of the aryl ring.	[845]
Vinyl acetate			Transesterification of *syn*-2,3-dihydroxy esters which leads to the synthesis of the taxol side chain is performed.	[846]

Table 5.1 *Continued.*

Enzyme	Acylating agent	Medium	*E*	Remarks	References
	Vinyl acetate	$^{i}Pr_2O$, 30 °C	298	The first kinetic resolution of large secondary alcohols having tetra-phenylporphyrin is performed. Lipases from *C. antarctica, R. miehei, P. aeruginosa* also give good selectivities ($E > 104$).	[847]
	2,2,2-Trifluoroethyl butanoate	$^{i}Pr_2O$	5~58	Resolution of 2-(1-hydroxyalkyl)-triazoles, important synthetic equivalents of 2-hydroxy aldehydes, is achieved.	[848]
	Vinyl 3-phenyl-butanoate	Hexane	98	*S*-preference. 2-Phenyl-1-propanol is resolved by using racemic mixture of vinyl 3-phenylbutanoate with improvement of enantioselecivity. This reaction is performed by lipase from *Alcaligenes sp.* in moderate *E* values.	[849]
	Vinyl acetate	Ionic liquid	1000	Ionic liquids can serve as the solvents for enantioselective transesterification with markedly enhanced enantioselectivity. Immobilized lipase from *C. antarctica* also gives good selectivities ($E \geq 172$).	[475]
Lipase from *P. fluorescens*	Enol ester	$^{i}Pr_2O$		Irreversible enantioselective acylation of 2-halo-1-arylethanols.	[470]
	Vinyl acetate	Benzene		Resolution of neopentyl alcohols that are produced from protected α-vinyl amine acids.	[850]
	Vinyl acetate	Organic solvent	4.5~44	Kinetic resolution of (±)-*cis*-hydroxy-methyl-2-phenyl-1,3-dioxane, an intermediate in the synthesis of pheromones and branched-chain sugars.	[851]

	Vinyl acetate	Methanol	48	Kinetic resolution of a *trans*-5,6-dihydro-1,10-phenanthroline possessing helical and central chirality	[852]
Lipase from *C. antarctica*	Isopropyl acetate	Hexane, 50 °C		Five- to seven-membered 2-iodo-2-cycloalken-1-ols are resolved by transesterificaiton.	[853]
	Vinyl acetate	Organic solvent (hexane, THF, etc.)	2~280	Resolution of 1,1,1-trifluoro-2-alkanols. *S*-preference.	[854]
	Vinyl acetate	MeCN and amine	38~43	Kinetic resolution of (±)-phenylbutan-1-ol and study of the role of the amine and acetanilide in the resolution is documented. The use of amine in the reaction enhances the enantiomeric ratio.	[855]
	Vinyl acetate	$^iPr_2O$	30	Resolution of racemic 2-hydroxymethyl-1-methylthioferrocene. Lipozyme also gives good selectivity ($E = 20$).	[856]
	Isopropenyl acetate		81	Resolution of *endo*-bicyclo[4.1.0]heptan-2-ols.	[857]
	Vinyl acetate	Isooctane	49	Kinetic resolution of tertiary 2-phenylbut-3-yn-2-ol	[858]
	Isopropenyl acetate	Toluene	>1000	Kinetic resolution of 4-*N*-Boc-amino-1-alken-3-ols	[859]
	Cyclic anhydrides	Toluene	<84	Kinetic resolution of trans-4-(4′-fluorophenyl)-3-hydroxymethyl-*N* -phenyloxycarbonylpiperidine	[860]
	Vinyl acetate	Toluene		Selective discrimination of diastereomeric 3β-hydroxy-5,6-epoxysteroids	[861]

Table 5.1 *Continued.*

Enzyme	Acylating agent	Medium	*E*	Remarks	References
	Vinyl acetate or succinic anhydride	$^iPr_2O$	<92	Kinetic resolution of nitro aldols	[862]
	Enol esters	Organic solvents		Kinetic resolution of alkyl lactates	[863]
	Vinyl acetate	Hexane	>200	Kinetic resolution of 4-hydroxy-cyclopentenone	[864]
	Vinyl acetate	Hexane	>200	Kinetic resolution of cis-fused octalols	[865]
Mutant lipase	Vinyl butyrate	Hexane	12~30	Efficient resolution of halohydrins by rational protein engineering to procedure CALB mutants is performed.	[472]
Lipase from *C. cylindracea*	Fatty acid	Apolar organic solvent, 40 °C		The resolution of α-substituted cyclo-hexanols by esterification reaction.	[312]
Lipase from *C. rugosa*	Vinyl acetate	$^iPr_2O$	72	Octahydro-3,3,8a-trimethyl-1-naphthol, a key intermediate in the total synthesis of lactaranes and marasmanes, is resolved.	[866]
	Vinyl acetate	Organic solvents	<3.4	Kinetic resolution of primary alcohols with remote stereogenic center	[867]
Lipase from *Alcaligenes* sp.	Vinyl acetate	$^iPr_2O$		Enantioselective acetylation of albicanol is performed, leading to the synthesis of several natural products. (ex. (−)-3,4-dihydroxycinnamate, etc.)	[868]

Porcine pancreatic lipase (PPL)	Vinyl acetate	$^{i}Pr_2O$		Kinetic resolution of aliphatic 2-alkanols with varying chain length is studied. A gradual decline both in the reaction rate and enantioselectivity with increase in chain length is observed.	[869]
	Vinyl acetate, palmitic acid, etc	Benzene, micro-wave irradiation	117~196	1,2,3,4-Tetrahydro-1-naphthol, 1-indanol, and menthol are resolved by microwave irradiation with enhancing of reaction rates and enantioselectivities.	[870]
Carboxylic esterase	Methyl propanoate	Biphasic aqueous-organic mixture			[479]
Acylase (from *Aspergillus* sp.)	Vinyl acetate or vinyl butylate	Organic (toluene, acetone and so on)	31~100	Several secondary alcohols, especially aryl alkanols, are resolved using acylase with high selectivities.	[489, 871, 872]
Hydrolase (papain)	5-Phenyl pentanoic acid	Isooctane, H_2O	2~28	Esterification of ethylmethylphenylsilyl-methanol.	[873]
Protease (subtilisin)	Vinyl acetate	$^{i}Pr_2O$, 30 °C	8~15	*S*-preference.	[874]
Carica papaya lipase	Methanol	Organic solvents	>200	Transesterification of 2,2,2-trifluoroethyl esters of *N*-benzyloxycarbonylate D,L-amino acids	[875]
Lipase and polymer-supported scavenger	Vinyl acetate	Toluene	3~500	This strategy features a simple two-step procedure that combines both kinetic resolution and separation of the product by solid-phase scavenging of the remaining alcohol.	[474]

P.: Pseudomonas; C.: Candida; M.: Mucor; R.: Rhizomucor

Table 5.2 Enzymatic resolution of carboxylic acids, esters, anhydrides, or thioesters by reaction with alcohols.

Substrate	Enzyme	Medium	*E*	Remarks	References
Carboxylic acids					
2-Methylalkanoic acid	Lipase from *C. cylindracea*	Heptane	1.4~70		[876]
	Lipase from *C. rugosa*		3~130	*S*-preference. Study of alcohol chain length and enantioselectivity.	[877]
2-Methylvaleric acid fluoroibuprofen	Lipase from *C. rugosa*	Organic solvents orthoformates		Orthoesters preclude accumulation of water in the reaction mixture.	[315]
Fluoroibuprofen	Lipase from *C. antarctica*	MeCN	1.0~22.2		[878]
α-Lipoic acid	Lipase from *C. rugosa*	Hexane	1.3~7.5	*S*-preference.	[879]
3,7-Dimethyl-6-octanoic acid	Lipase from *C. rugosa*	Cyclohexane H_2O	3.6–53		[880]
2-(4-Substituted phenoxy)-propanoic acid	Lipase from *C. rugosa*	$^iPr_2O$ and SDSaq.	3.8~72	SDS enhances the enantioselective lipase-catalyzed esterification of 2-(4-substituted phenoxy)-propanoic acids with BuOH.	[314]
2- to 8-Methyl decanoic acid	Lipase from *C. rugosa*		2.8~68	Racemic substrates with 1-hexadecanol. The lipase shows enantiopreference for *S*-enantiomer when the methyl group is located on even-numbered carbons.	[881]

Esters					
Methyl 3-phenyl-glycidate	Lipase from *C. antarctica*	Toluene	6.5~16.2	Kinetic resolution of 3-phenylglycidate with amino alcohols.	[471]
Vinyl esters of arylaliphatic carboxylic esters	Lipase from *C. antarctica*	Toluene Hexanol	3.5~100	Resolution of vinyl esters.	[882]
α-Methylene β-lactones	Lipase from *C. antarctica*	^{t}BuOMe	>200	Substituent of the substrate influences on the absolute configuration of the desired lactone. Lipase from *C. antarctica* gives the best result.	[883]
Ketrolac	Lipase from *C. antarctica*	Chlorinated solvent		The resolving efficiency is very high in octanol.	[884]
Cyanohydrin acetate	Lipase from *C. antarctica*	Toluene		The optically active protected cyanohydrins, important building blocks for the synthesis of drugs and agrochemicals, are prepared through enzymatic kinetic resolution.	[490]
Anhydrides					
Carboxylic-carbonic anhydride	Lipase from *M. miehei*	Organic solvent (Et_2O, ^{t}BuOMe)	7~200	Resolution of mixed carboxylic-carbonic anhydrides. This can be used for the resolution of chiral carboxylic acids.	[630, 885]
1,4-Dihydropyridine-3,5-dicarboxylates	Protease	Phosphate buffer (also aqueous solution)	<87		[886]
2-(4-Chlorophenoxy)-propanoic acid	*Carica* papaya	Cyclohexane		Best results with Me_3SiCH_2OH	[887]

M.: Mucor; P.: Pseudomonas; C.: Candida.

liquid, however, are occasionally problematic. Use of low-boiling hydrofluorocarbon solvents is more useful in terms of separation of solvent [826].

Chemical modification of lipase MY with benzyloxycarbonyl group improves the enantioselectivity in reactions between 2-(4-substituted phenoxy)propanoic acids and BuOH [827]. Combination of enzymatic kinetic resolution and the Mitsunobu reaction brings about perfect conversion of racemic alcohols. That is, initial kinetic resolution provides a 1 : 1 mixture of parent alcohol and enantiomeric ester. Subjection of the product mixture to the Mitsunobu esterification with a carboxylic acid converts the remaining alcohol to the ester with inversion, providing homochiral esters [828, 829]. As well as these advantages, there are some disadvantages: a relatively large amount of solvent is necessary and the reaction time tends to be long.

Substrates other than alcohols are also amenable to kinetic resolution. Again, lipases are the most important enzymes. As shown in Table 5.2, carboxylic acids, esters, lactones, and acid anhydrides are resolved. As in the case of alcohols, the *R* isomer is selectively consumed in many cases. Exceptions to this trend are referred to as '*S* preference' in this table.

Enzymes can be separated from reaction mixture by filtration, but immobilization enables more facile recovery of the enzymes. Immobilized enzymes frequently exhibit higher activity than parent ones. Polymer-supported lipases are used for synthesis of β-keto esters [460], and pig liver esterase is immobilized on methoxy-polyethylene glycol to effect kinetic resolution with vinyl acetate [888]. Hybrid gel-entrapped lipase on Celite shows three times higher activity than the deposited lipase on Celite for esterification of glycidol with 2-octanol [889]. A lipase from *Candida rugosa* immobilized on hydrophobic and superparamagnetic microspheres is active for reaction between carboxylic acids and alcohols, and can be recovered by magnetic separation [890]. The selectivity of kinetic resolution of primary alcohols by lipases immobilized on the porous ceramic support is improved at low temperatures (−30 to −40 °C) [891, 892]. Lipase PS-C II immobilized on the porous ceramic support is employable at high temperatures up to 120 °C, with the best results at 80–90 °C [893].

A variety of immobilized enzymes which can be used for acylation of alcohols [894–896] and for kinetic resolution of 2-(4-chlorophenoxy)propanoic acid [897] are now commercially available.

5.2 Nonenzymatic Resolution

Kinetic resolution by nonenzymatic procedures is feasible as well. Although the stereoselectivity factor is not generally so high as for the enzymatic procedures ($s < 370$ thus far), there are several synthetic merits. A variety of chemical reactions can be utilized, depending on the chiral catalysts or acyl donors. Any type of substrate is theoretically employable, thanks to the lack of the substrate specificity inevitably encountered with enzymes. The substrates to be resolved are usually alcohols, and acid anhydrides or chlorides are most popular as acyl donors. The absolute configuration can be controlled by changing the chirality of catalysts or acyl donors. The reaction time is shorter and the required amount of solvent smaller than in the enzymatic reactions. The following list covers the representative nonenzymatic kinetic resolution procedures available up to the end of 2007.

(a) **Alcohol:** secondary alcohols in general

(b) **Acylating agent:** *N*-benzoyl-*tert*-butyl-2-oxazolidinone

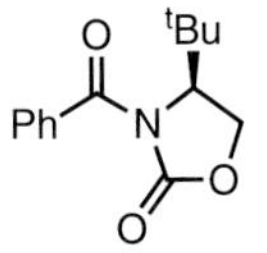

(c) **Promoter:** MeMgBr or $MgBr_2$ / NEt_3

(d) **Conditions:** 1~1.1 equiv. promoter, CH_2Cl_2

(e) 65~95% *ee*

(f) **Reference:** [697]

Remarks: The acylation process is promoted by the formation of the derived Mg alkoxides, which may be obtained either from alcohol deprotonation with MeMgBr or through deprotonation with the Lewis acid-base combination of $MgBr_2/NEt_3$.

(a) **Alcohol:** secondary alcohols in general
(b) **Acylating agent:** (*S*)-4-alkyl-3-pivaloyl-1,3-thiazolidine-2-thiones

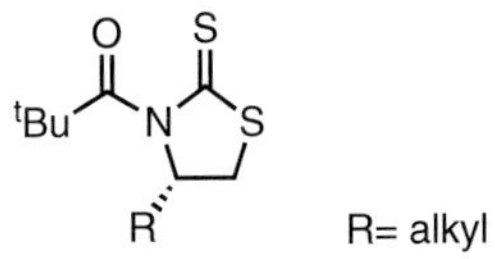

R= alkyl

(c) **Promoter:** MeMgBr
(d) **Conditions:** organic solvent (toluene, hexane, etc.)
(e) **References:** [698, 699]

Remarks: The *R* esters are produced as major products in the presence of MeMgBr. On the other hand, the *S* isomers are selectively produced under neutral conditions.

(a) **Alcohol:** secondary alcohols in general
(b) **Acylating agent:** aryl carboxylic acids
(c) **Promoter:** Chiral, non-racemic 1,3,2-dioxaphosphepanes

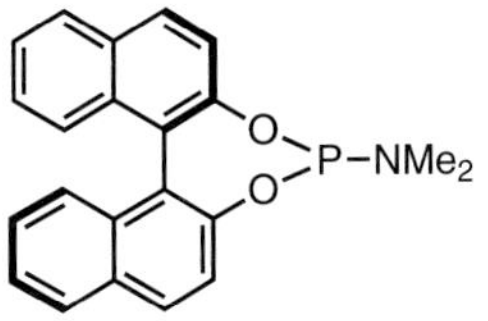

(d) **Conditions:** benzene, room temperature

(e) 14~73% conversion, 11~39% *ee*
(f) **Reference:** [214]
Remarks: The reactions are run under Mitsunobu conditions. The alcohols can be obtained in >99% *ee*.

(a) **Alcohol:** cyclic and acyclic secondary alcohols
(b) **Acylating agent:** benzoyl bromide
(c) **Catalyst:** $SnBr_2$-chiral diamine derived from (*S*)-proline complex

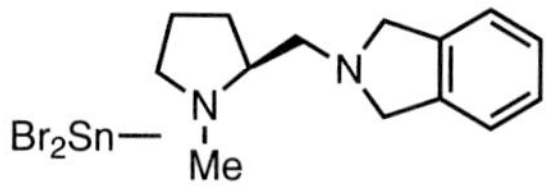

Br_2Sn— N Me N

(d) **Conditions:** 30 mol% catalyst, CH_2Cl_2, −78 °C
(e) **E:** 4.5~100
(f) **Reference:** [641]
Remarks: Acyclic racemic alcohols give lower *E* values than cyclic ones.

(a) **Alcohol:** cyclic and acyclic alcohols
(b) **Acylating agent:** benzoyl chloride
(c) **Catalyst:** chiral diamine derived from (*S*)-proline

N Me N

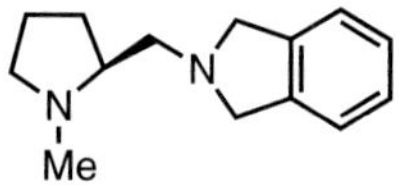

(d) **Conditions:** 0.3 mol% catalyst, CH_2Cl_2, NEt_3, −78 °C
(e) *s*: 4.5~170
(f) **Reference:** [898]
Remarks: Acyclic racemic alcohols give lower **s** values than cyclic ones.

(a) **Alcohol:** mono-substituted cyclohexane diols
(b) **Acylating agent:** $(^iPrCO)_2O$
(c) **Catalyst:** 4-(2-acylpyrrolidino)pyridine (acPPY)

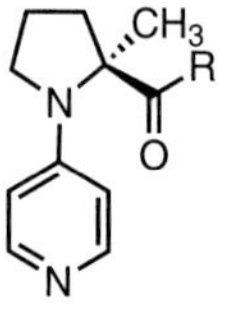

CH_3 R N O N

(d) **Conditions:** 5 mol% catalyst, toluene, room temperature
(e) *s*: ~18.8
(f) **Reference:** [899]

(a) **Alcohol:** monosubstituted 1,2-diols
(b) **Acylating agent:** $(^iPrCO)_2O$
(c) **Catalyst:** 4-pyrrolidinoacylpyridine (PacPY)

(d) **Conditions:** 1 mol% catalyst, Et_3N, CH_2Cl_2, −78 °C
(e) *s*: ~9.4
(f) **Reference:** [900]

(a) **Alcohol:** cyclic secondary alcohols
(b) **Acylating agent:** $(^iPrCO)_2O$
(c) **Catalyst:** chiral diamine derived from PPY (4-pyrrolidinopyridine)

(d) **Conditions:** 5 mol% catalyst, toluene, room temperature
(e) *s*: 4.3~12.3
(f) **Reference:** [591]

Remarks: The catalyst acts as though through an induced fit mechanism, like a natural enzyme.

(a) **Alcohol:** secondary alcohols in general
(a) **Acylating agent:** isopropenyl acetate
(b) **Catalyst:** Yttrium-salen complex

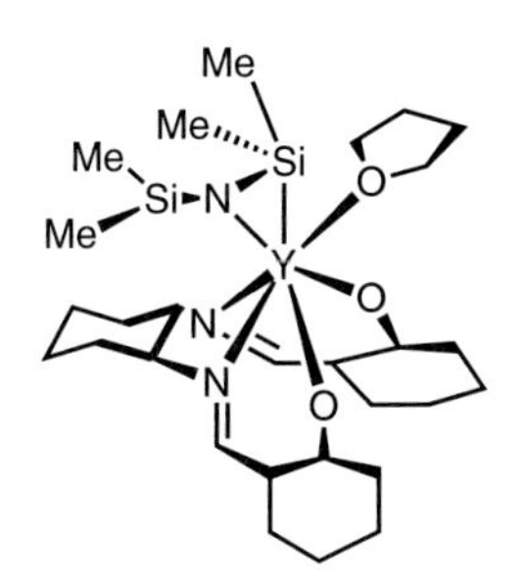

(c) **Conditions:** 1~2 mol% catalyst, toluene, −10 to −3 °C
(d) *s*: 1.5~4.8
(e) **Reference:** [367]

Remarks: This is the first example of a transition metal-catalyzed acyl transfer reaction.

(a) **Alcohol:** cyclic and acyclic secondary alcohols and PhCH(OH)R
(b) **Acylating agent:** Ac_2O or $(m\text{-}ClC_6H_5CO)_2O$
(c) **Catalyst:** chiral phosphine

Me
P–Ph
Me

(d) **Conditions:** CH_2Cl_2 or CD_2Cl_2
(e) *s*: 3~15
(f) **Reference:** [605]
Remarks: Better selectivity is obtained in the case of R = ^{t}Bu.

(a) **Alcohol:** aromatic, allylic, and benzyl alcohols
(b) **Acylating agent:** acid anhydrides
(c) **Catalyst:** 2-aryl-2-phosphabicyclo[3,3,0]octane derivatives

H H
P
Ar Ar= 3,5-(^{t}Bu)$_2$-C_6H_3

(d) **Conditions:** 2.5–6 mol% catalyst, acid anhydride 2.5 equiv., organic solvent, −40 °C to room temperature
(e) *s*: 15~390
(f) **Reference:** [606, 901, 902]

(a) **Alcohol:** arylalkylcarbinols
(b) **Acylating agent:** diketene
(c) **Catalyst:** azaferrocene derivative

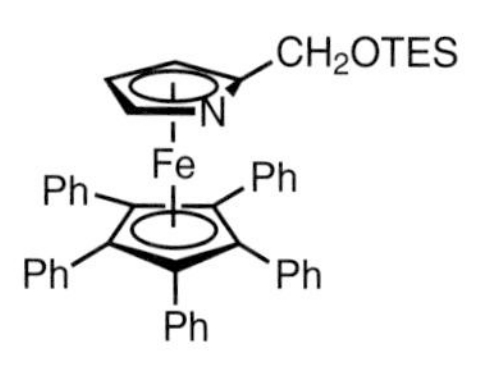

(d) **Conditions:** 10 mol% catalyst, benzene, room temperature
(e) *s*: 3.7~6.5
(f) **Reference:** [903]

(a) **Alcohsol:** propargylic and arylalkylcarbinols
(b) **Acylating agent:** Ac_2O
(c) **Catalyst:** planar chiral DMAP derivative
(d) **Conditions:** 1~2 mol% catalyst, *t*-amyl alcohol, 0 °C

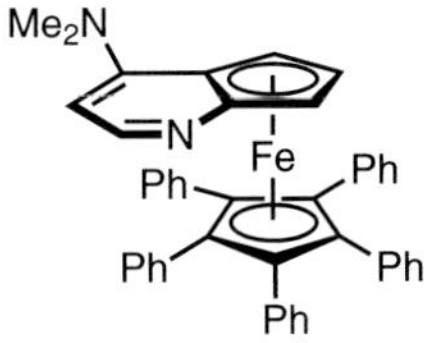

planner chiral DMAPderivative

(e) *s*: 3.8~20
(f) **References:** [588, 589, 904–906]

(a) **Alcohol:** 1-aryl ethanols
(b) **Acylating agent:** Ac_2O
(c) **Catalyst:** diethyl[3-(2-phenylnaphthyl)(4-pyridyl)]amine

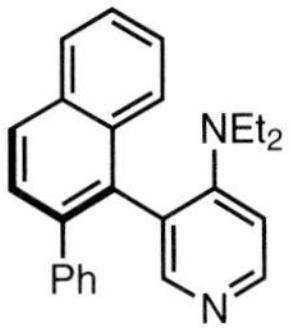

(d) **Conditions:** 0.01 mol% catalyst, organic solvent, −78 °C
(e) *s*: 8.9~29
(f) **Reference:** [590]

(a) **Alcohol:** 1-aryl ethanols
(b) **Acylating agent:** Ac_2O
(c) **Catalyst:** diethyl[3-(2-arylnaphthyl)(4-pyridyl)]amine

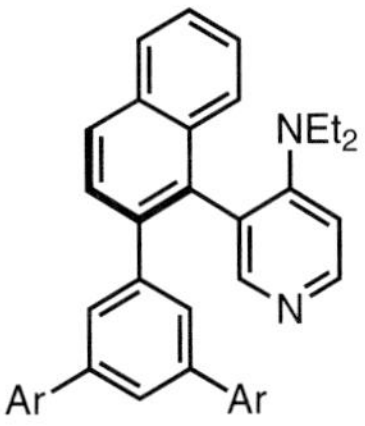

(d) **Conditions:** 0.01 mol% catalyst, toluene, −78 °C
(e) *s*: ~39
(f) **Reference:** [907]

(a) **Alcohol:** 1-arylethanols
(b) **Acylating agent:** acid anhydrides
(c) **Catalyst:** 4-(dimethylamino)pyridyl sulfoxide

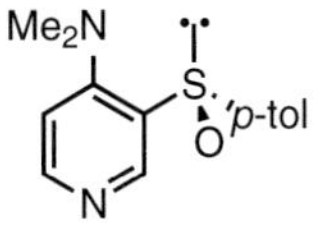

(d) **Conditions:** 5 mol% catalyst, 0.6 equiv. acid anhydride, Et_3N (0.6 equiv.), CH_2Cl_2, room temperature
(e) *s*: ~4.5
(f) **References:** [908]

(a) **Alcohol:** (±)-*trans*-2-*N*-acetamidocyclohexanol
(b) **Acylating agent:** Ac_2O
(c) **Catalyst:** chiral oligopeptide

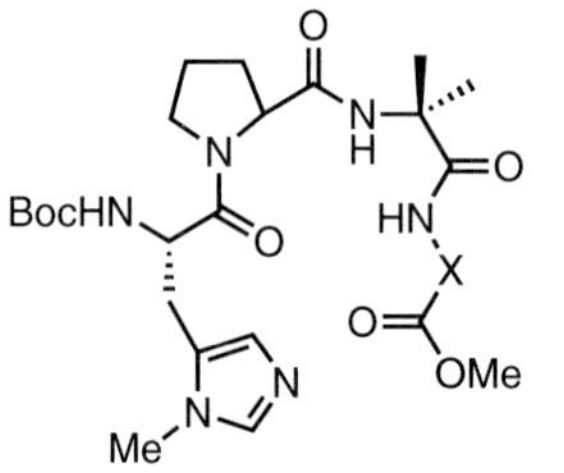

chiral terapeptide
X= L-or D-amino acid

(d) **Conditions:** 2~5 mol% or 0.25 equiv. catalyst, toluene, 25 °C
(e) *s*: 3.0~28
(f) **References:** [592, 593]

(a) **Alcohol:** (±)-*trans*-2-*N*-acetamidocyclohexanol
(b) **Acylating agent:** Ac_2O
(c) **Catalyst:** chiral oligopeptide

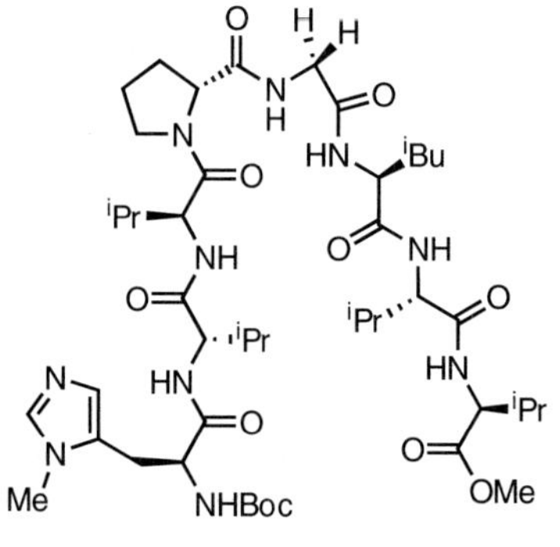

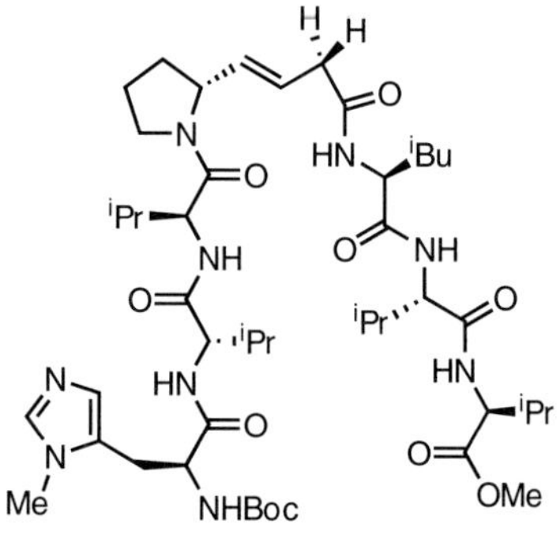

(d) **Conditions:** 1~2 mol% catalyst, toluene, 25 °C
(e) *s*: 15~51
(f) **References:** [909, 910]

(a) **Alcohol:** monosubstituted 1,2-diol
(b) **Acylating agent:** acid anhydride
(c) **Catalyst:** L-histidine-derived sulfonamide

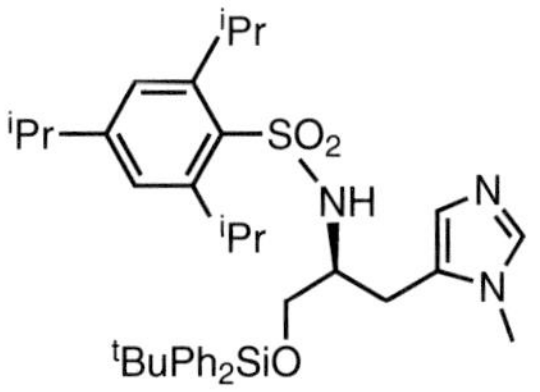

(d) **Conditions:** catalyst 5 mol%, anhydride (0.5 equiv), $^{i}Pr_2EtN$ (0.5 equiv), toluene, room temperature
(e) *s*: ~87
(f) **Reference:** [911]

(a) **Alcohol:** 1,2-diols
(b) **Acylating agent:** benzoyl chloride
(c) **Catalyst:** (*S, S*)-4,4-dibromo-4,5-dihyrdro-3*H*-dinaphtho[2,1-c:1′2-e]stannepin

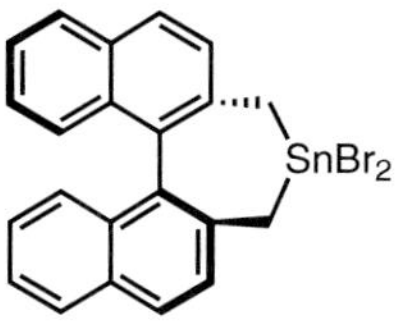

(d) **Conditions:** 0.25 mol% catalyst, H_2O, Na_2CO_3, THF, −10 °C
(e) *s*: 3.2~10.0
(f) **Reference:** [721]
Remarks: Primary hydroxy group is selectively benzoylated in the presence of secondary alcohol.

(a) **Alcohol:** 1-arylethanols
(b) **Acylating agent:** acid anhydrides
(c) **Catalyst:** 2,3-dihydroimidazo[1,2-a]pyridine

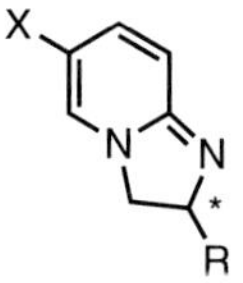

(d) **Conditions:** 20 mol% catalyst, 1 equiv. acid anhydride, $CDCl_3$, room temperature
(e) *s*: ~85
(f) **References:** [912]

(a) **Alcohol:** racemic timolol
(b) **Acylating agent:** (*R,R*)-*OO*-diacetyl or benzoyl-(*R,R*)-tartaric acid anhydride
(c) **Conditions:** acetone, room temperature, 2~20 h
(d) **Reference:** [913]
(e) **Reaction:**

Chiral

acetone

rt

R= Ph, Me

(a) **Alcohol:** propargylic alcohols
(b) **Acylating agent:** propanoic anhydride
(c) **Catalyst:** benzotetramisole

(d) **Conditions:** 4–10 mol% catalyst, 0.75 equiv. anhydride, 0 °C
(e) *s*: ~32
(f) **Reference:** [914]

(a) **Alcohol:** 1-arylethanols
(b) **Acylating agent:** vinyl acetate
(c) **Catalyst:** *N*-heterocylic carbenes

(d) **Conditions:** 3 mol% catalyst, 1.2 equiv. vinyl acetate, ether, room temperature
(e) *s*: ~6.1
(f) **Reference:** [915]

(a) **Alcohol:** 1,2:4,5-di-*O*-protected *myo*-inositols

(b) **Acylating agent:** camphanoyl chloride

(c) **Promoter:** NEt_3, DMAP (cat.)

(d) **Conditions:** CH_2Cl_2, 25 °C

(e) **References:** [916, 917]

(f) **Reaction:**

Et_3N, DMAP (cat.)
CH_2Cl_2

(±)
R= Bn, Tf

R'Cl

D (more polar)

L (less polar)

Remarks: The treatment of racemic alcohols with camphanic chloride gives the diastereomeric camphanate esters. The more polar diastereomer is obtained by three crystallizations. The second diastereomer is recovered from the mother liquors by MPLC.

(a) **Alcohol:** 4,12-dihydroxy[2.2]paracyclophane
(b) **Acylating agent:** camphanoyl chloride
(c) **Promoter:** pyridine (solevent)
(d) **Conditions:** 25 °C
(e) **References:** [918]

(a) **Alcohol:** methyl 4,6-*O*-benzylidene-α-D-glucopyranoside
(b) **Acylating agent:** racemic hexamethoxydiphenoyl chloride
(c) **Promoter:** NaH or NEt_3
(d) **Conditions:** NaH in toluene, NEt_3 in CH_2Cl_2
(e) **References:** [660–662]
(f) **Reaction:**

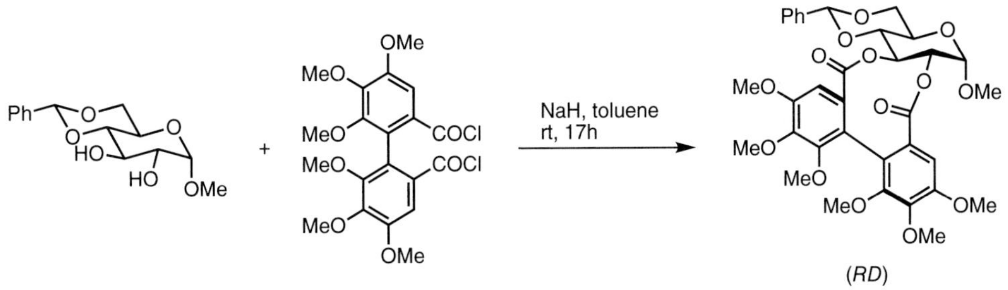

NaH, toluene
rt, 17h

(*RD*)

Remarks: The *R* isomer is obtained when the reaction is performed in the presence of NaH, while the *S* isomer is produced in the case of NEt_3.

Notably, phenyl esters are resolved through transesterification with methanol by catalysis of C_3-symmetrical zinc complexes of 1,1,1-tris{2-[(*S*)-4-isopropyl]oxazoyl} ethane $[(i\text{-Pr-triox})_2Zn_2(\mu\text{-OTf})_3]OTf$ (Scheme 5.1) [919].

5.3 Dynamic Kinetic Resolution

Kinetic resolution suffers from the intrinsic limitation that the yield of the desired enantiomer can never exceed 50% except at the cost of inclusion of the enantiomeric counterpart. This drawback can be overcome by dynamic kinetic resolution. If the substrate undergoes racemization much more rapidly than it undergoes the enzyme-catalyzed reaction or if the unreacted substrate remaining in the reaction mixture is racemized *in situ*, then the racemic substrate can be completely transformed into one enantiomer. One such case is the lipase-catalyzed methanolysis of 4-substituted oxazolin-5-ones. Treatment of 4-methyl-2-phenyloxazolin-5-one with BuOH in the presence of Lipozyme affords the corresponding amino acid

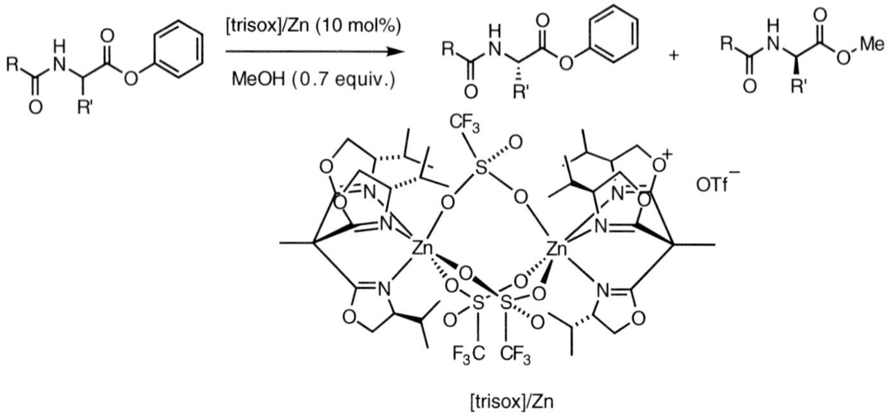

Scheme 5.1

ester in 57% *ee* at 45% conversion (Scheme 5.2) [920]. The recovered substrate exhibits no optical rotation, suggestive of *in situ* racemization. Better outcomes in terms both of yield and of optical purity are obtainable with various 4-substituted 2-phenyloxazolin-5-ones by use of *Pseudomonas capacia* lipase [921]. The *tert*-butyl derivative is transformed into (*S*)-*N*-benzoyl *tert*-leucine butyl ester in 94% yield with 99.5% *ee* in the presence of Lipozyme. The use of toluene-containing BuOH as solvent together with a catalytic amount of Et_3N is crucial (Scheme 5.3). The enantioselectivity and catalytic activity in the formation of the butyl ester in the presence of *Candida antarctica* Lipase B in organic solvent is improved by addition of Et_3N [922].

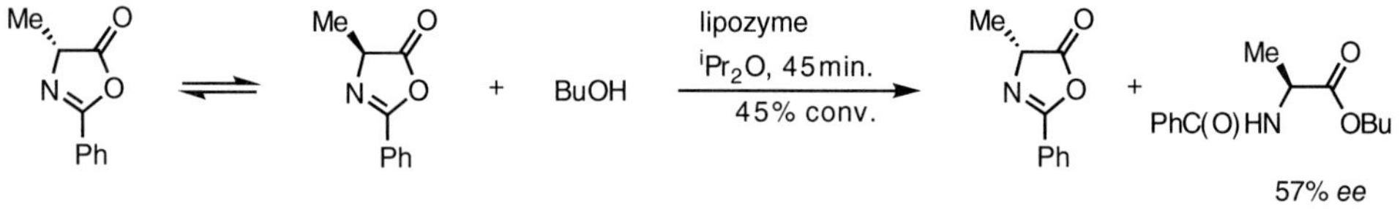

Scheme 5.2

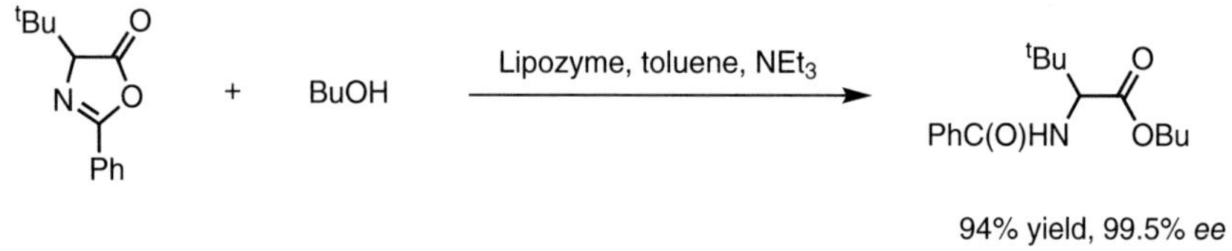

Scheme 5.3

5-Hydroxy-5*H*-furan-2-one undergoes dynamic kinetic resolution on acylation with vinyl acetate (Scheme 5.4) [923, 924]. With the aid of lipase R immobilized on Hyflo Super Cell, 100% enantiomeric excess is feasible at 90% conversion.

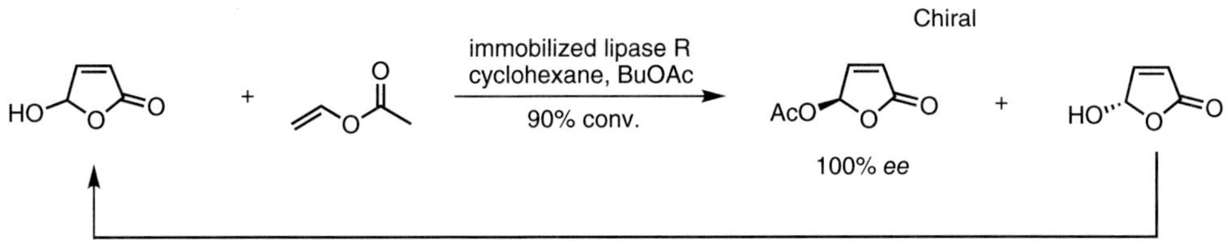

Chiral

immobilized lipase R
cyclohexane, BuOAc
90% conv.

100% *ee*

Scheme 5.4

Enzymatic hydrolysis of the trifluoroethyl thioester of 2,4-dichlorophenoxypropanoic acid is accompanied by nonenzymatic hydrolysis, resulting in racemization. On the other hand, lipase PS-30-catalyzed transesterification with BuOH furnishes the corresponding butyl ester in 75% *ee* at 98% conversion. The resulting ester is then successfully hydrolyzed to the carboxylic acid in 93% *ee* (Scheme 5.5) [925].

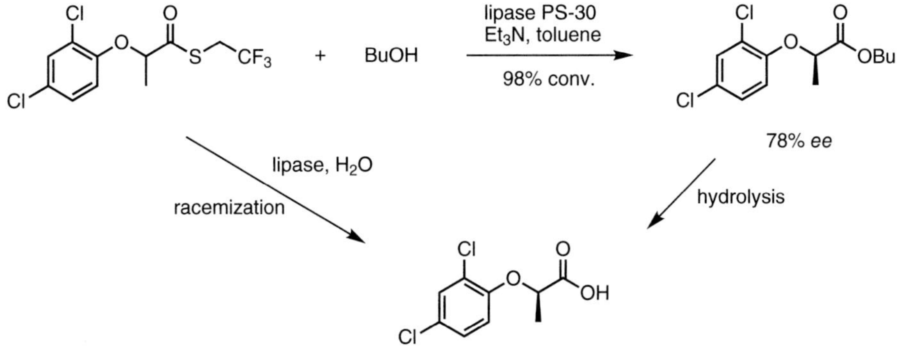

lipase PS-30
Et_3N, toluene
98% conv.

78% *ee*

lipase, H_2O
racemization

hydrolysis

Scheme 5.5

Mandelic acid is prone to racemization, so kinetic resolution with high conversion is feasible (Scheme 5.6) [926]. Racemic mandelic acid is resolved by acylation with vinyl acetate in the presence of *Pseudomonas* sp. lipase with excellent enantioselectivity ($E > 200$). The resulting 1:1 mixture of (*S*)-*O*-acetyl mandelic acid and (*R*)-mandelic acid, after filtration of the lipase, is treated with immobilized mandelate racemate to racemize the mandelic acid. Repetition of this procedure three more times provides (*S*)-*O*-acetyl mandelic acid in 80% yield and with >98% *ee*.

Integration of enzymatic kinetic resolution and transition metal-catalyzed racemization results in dynamic kinetic resolution [927]. Hydrogen-transfer reactions between secondary alcohols and ketones are useful for racemization of the

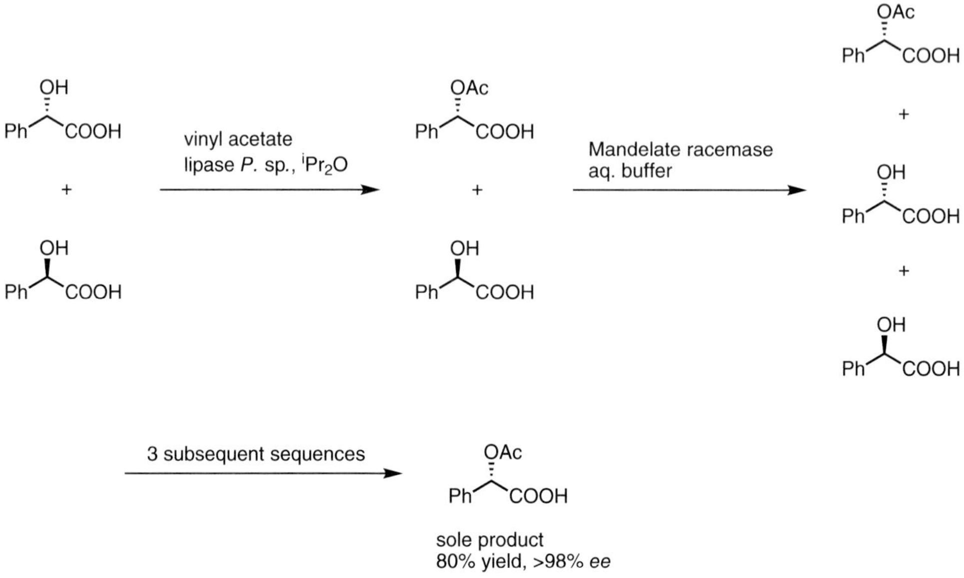

vinyl acetate
lipase *P.* sp., $^{i}Pr_2O$

Mandelate racemase
aq. buffer

3 subsequent sequences

sole product
80% yield, >98% *ee*

Scheme 5.6

alcohols. Thus, subjection of racemic 2-phenethyl alcohol to kinetic resolution with vinyl acetate in the presence of *Pseudomonas fluorescens* lipase with concurrent hydrogen-transfer reaction with acetophenone in the presence of $Rh_2(OAc)_4$ affords (*R*)-phenethyl acetate in 98% *ee* at 60% conversion (Scheme 5.7) [928].

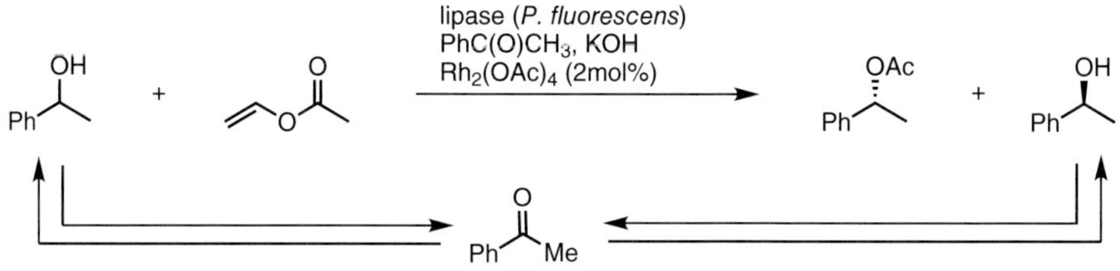

lipase (*P. fluorescens*)
$PhC(O)CH_3$, KOH
$Rh_2(OAc)_4$ (2mol%)

Scheme 5.7

The Shvo's ruthenium catalyst illustrated in Scheme 5.8 is another efficient catalyst for hydrogen transfer [929]. Various secondary alcohols are converted into the corresponding *R* acetates by use of Novozyme 435 and the ruthenium catalyst. *p*-Chlorophenyl acetate is a better acyl donor than enol acetates. The procedure is also effective even in the absence of ketone to furnish secondary alcohol acetates in reasonably high yields (~88%) and with high *ees* (~99%). In this case, the ruthenium catalyst serves for hydrogen abstraction from the alcohol, followed by re-hydrogenation of the resulting ketone. Diols are also transformed into (*R,R*) diacetates and *meso*-diacetates in ratios of 74:26~100:0 under similar conditions (Scheme 5.9) [930]. This technique can be further applied to substrates such as α-hydroxy acids [931, 932], α-hydroxy nitriles [478], β-azido alcohols [933], γ-hydroxy acid derivatives [934], δ-hydroxy esters followed by conversion into chiral δ-lactones [935], and 1,4-diols leading to γ-hydroxy ketones [936]. *p*-Chlorophenyl acetate serves best for these reactions as an acyl donor, but toxic *p*-chlorophenol is produced. This drawback can be bypassed by the use of alkyl esters in combination with a ketone [937]. Dynamic kinetic resolution of benzoins is feasible with enol acetates or trifluoroethyl ester [938].

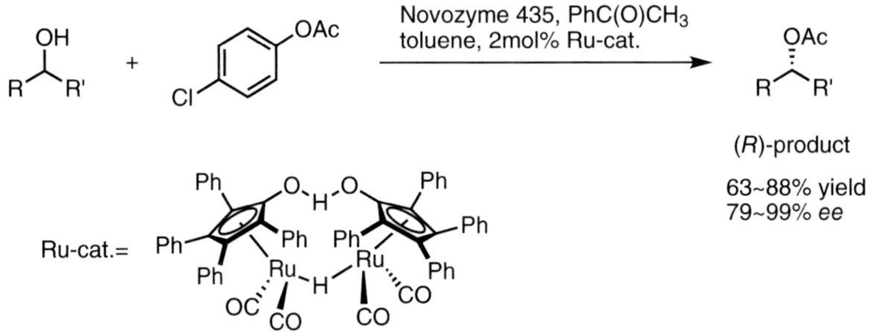

Novozyme 435, $PhC(O)CH_3$
toluene, 2mol% Ru-cat.

(*R*)-product
63~88% yield
79~99% *ee*

Ru-cat.=

Scheme 5.8

= $(CH_2)_n$ or
arylene moiety

Novozyme 435
4mol% Ru-cat., toluene

(*R*,*R*)-product
63~90% yield
96~99% *ee*

meso

R,*R*: *meso*= 86: 14~100: 0

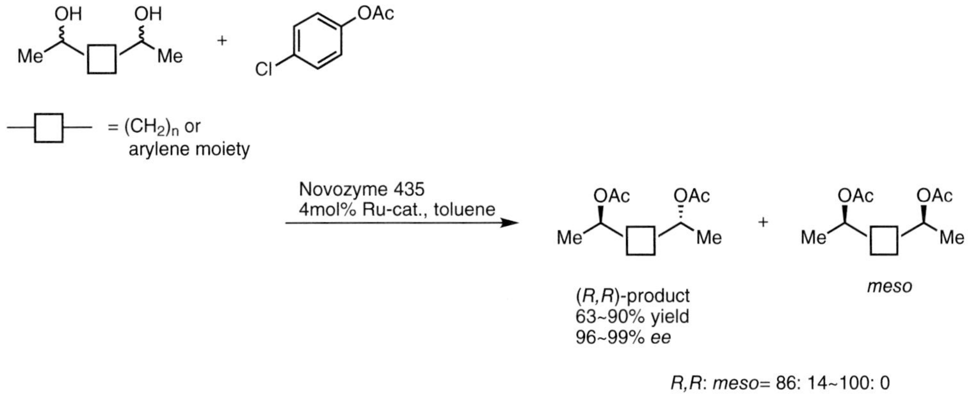

Scheme 5.9

Integration of aldol reaction and kinetic resolution gives rise to an alternative to the asymmetric aldol reaction (Scheme 5.10) [939]. In this procedure, a key role is played by racemization of the aldols by the above ruthenium catalyst. The aldol products are obtained in ca. 70% yield, and mostly with >95% *ee*.

The Shvo's catalyst, though being highly useful, needs to be activated at slightly elevated temperature, so only thermostable enzymes can be used. The catalyst shown in Scheme 5.11, in combination with CALB, effects dynamic kinetic resolution of secondary alcohols at room temperature in a very short time [940, 941]. Replacement of one of the phenyl groups with an alkoxy group improves the

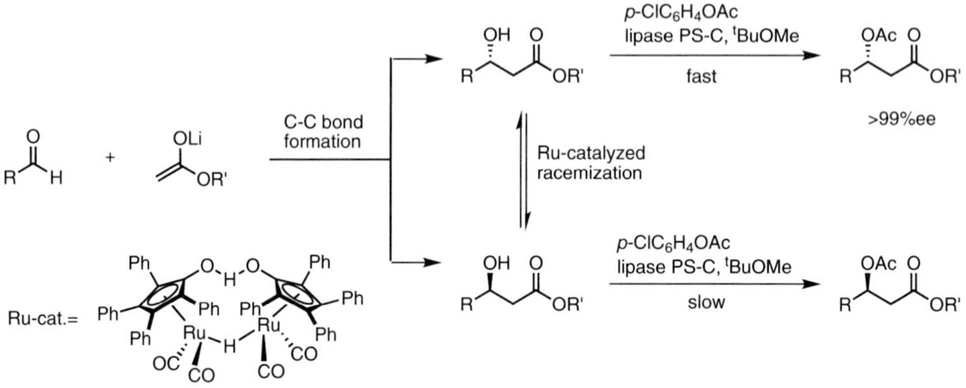

Scheme 5.10

air-sensitivity of this catalyst, allowing dynamic kinetic resolution at room temperature in air [942]. An amino group in place of the alkoxy group is also effective for conducting the resolution at room temperature, though the reaction time is longer [943, 944].

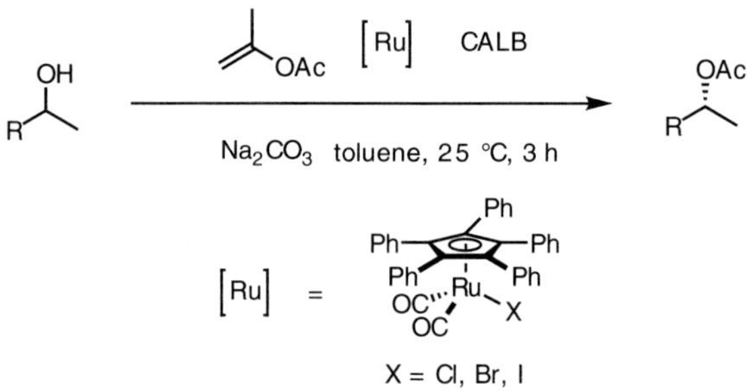

Scheme 5.11

Another ruthenium catalyst, illustrated in Scheme 5.12, is also an effective catalyst for dynamic kinetic resolution [945]. This procedure results in none of the ketone formation frequently encountered in the above technique. Various secondary alcohols including hydroxy acids, diols, and hydroxy aldehydes are successfully transformed into the corresponding esters. [Ru(cymeme)Cl_2] in combination with lipase also acts as a racemization catalyst [946].

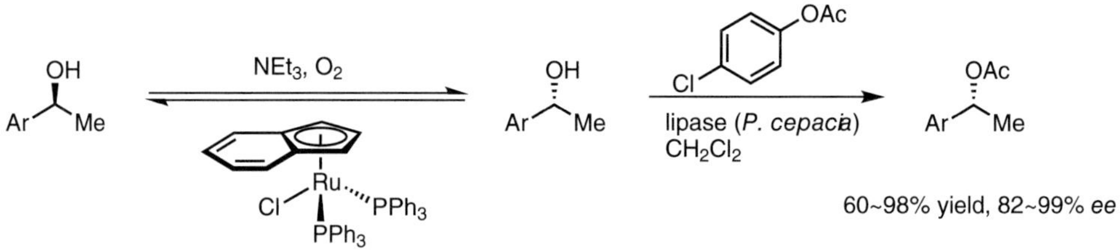

Scheme 5.12

By taking advantage of $VO(OR)_3$, which can catalyze allylic rearrangement, combination of this catalyst with CALB effects dynamic kinetic resolution of allylic alcohols (Scheme 5.13) [947].

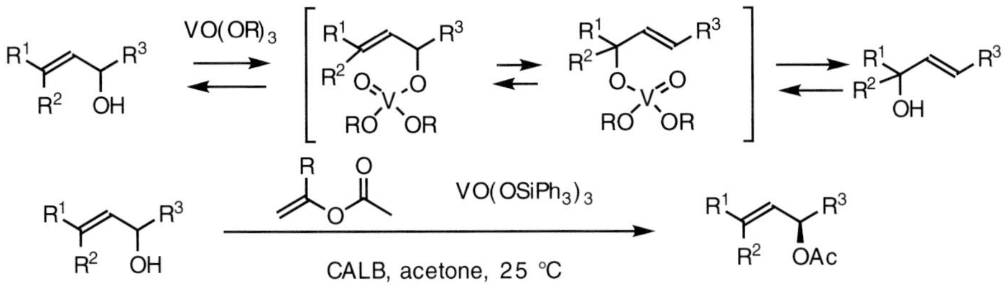

Scheme 5.13

Aliphatic cyanohydrins, which usually undergo modest dynamic kinetic resolution, can be converted into the corresponding acetates in high chemical and enantiomeric yields when NaCN is added (Scheme 5.14) [948]. Acid zeolites are also useful for dynamic kinetic resolution of secondary alcohols [949, 950].

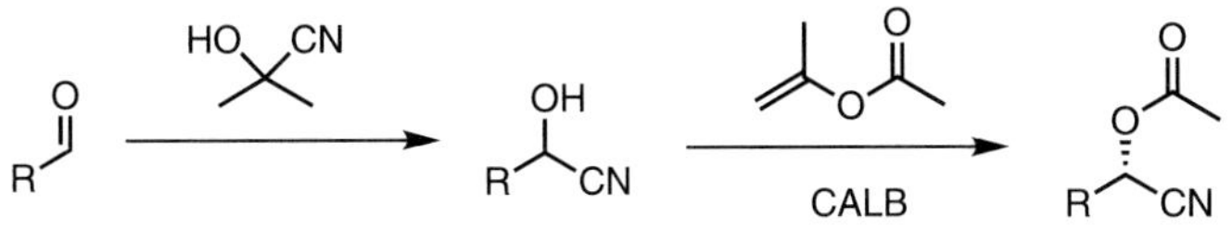

Scheme 5.14

Nonenzymatic dynamic kinetic resolution is feasible with TiTADDOLate [951] and a planar-chiral derivative of DMAP (Scheme 5.15) [952]. 4-Substituted 2-phenyloxazolin-5-ones undergo alcoholysis under catalysis by these catalysts to provide quantitative yields of α-amino acid esters with up to 78% *ee*. Racemic ibuprofen is converted into the corresponding *S,S* ester with moderate selectivity when treated with amides of (*S*)-lactic acid, DCC, and DMAP (Scheme 5.16) [953].

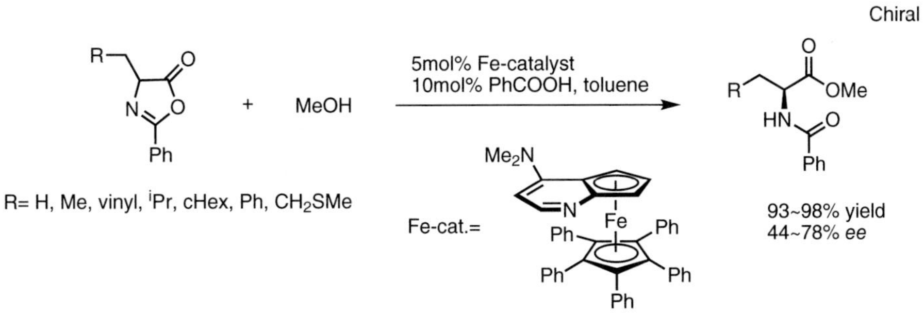

Scheme 5.15

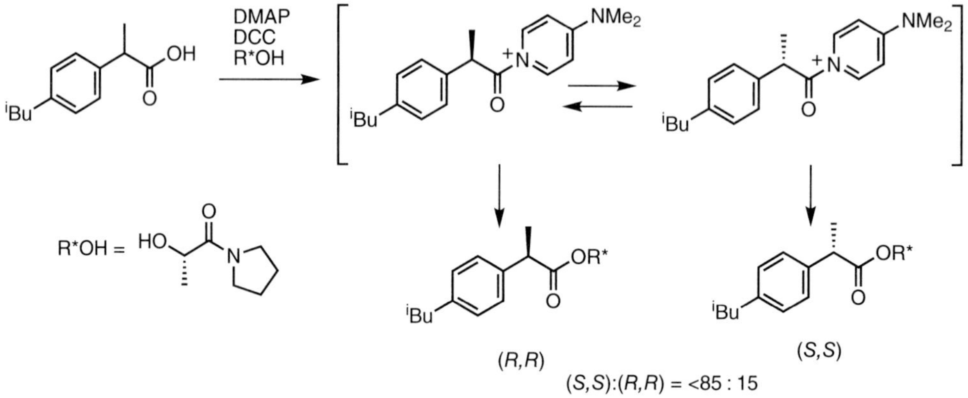

DMAP
DCC
R*OH

R*OH =

(R,R)

(S,S)

(S,S):(R,R) = <85 : 15

Scheme 5.16

5.4 Parallel Kinetic Resolution

In kinetic resolution, one enantiomer is consumed preferentially as the reaction proceeds, so the relative concentrations of the remaining substrate enantiomers deviate from the original 1 : 1 ratio with the progress of the reaction. As a result, the reaction rate of the more reactive enantiomer decreases with its lower concentration, resulting in decreasing enantioselectivity. This drawback can be offset if the both enantiomers are consumed in parallel in separate reactions, their concentrations decreasing at a comparable rate or–under ideal conditions–at the same rate.

Although being carbonation rather than esterification, the following is a nice example with which to show the effectiveness of parallel kinetic resolution. Two chiral DMAP derivatives react selectively with each of a pair of secondary alcohol enantiomers (Scheme 5.17) [954]. Treatment of a secondary alcohol with the 1 : 1 mixture of these DMAP derivatives results in preferential reaction of the trichlorobutyloxycarbonyl derivative with the *S* isomer and of the fenchyloxycarbonyl derivative with the *R* counterpart. The *R* alcohols are recovered in ~50% yields with up to 88% *ee* and the fenchyl carbonates in ~49% yields with up to 95% *ee*.

A three-phase system is useful for upgrading the selectivity (Section 1.2.5; Scheme 1.160) [477]. The system is composed of (i) insoluble cross-linked lipase acylation catalyst (ChiroCLEC), (ii) insoluble polymer-bound acyl donor, and (iii) soluble phosphine catalyst and soluble acyl donor (vinyl pivalate). The interaction between ChiroCLEC and vinyl pivalate produces the acylated lipase intermediate, which reacts with the *R* enantiomer of secondary alcohols. On the other hand, reaction between the polymer-bound acyl donor and the chiral phosphine catalyst generates the acyl phosphonium salt, which preferentially affords the *S* isomer of an ester. Satisfactory results in terms of conversion (~50%) and enantioselectivity (~98% *ee*) are obtained.

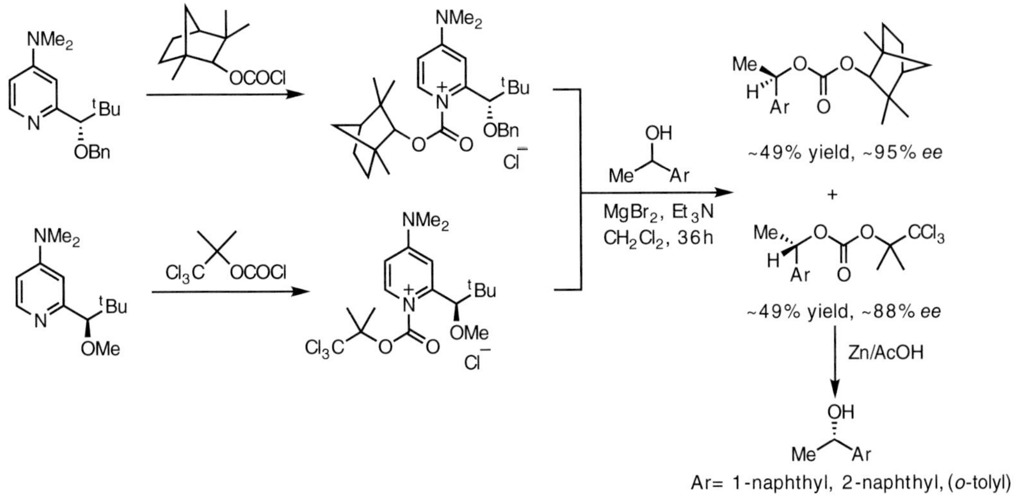

Scheme 5.17

A modified cinchona alkaloid catalyzes ring opening of methylsuccinic anhydride in which the *R* and *S* isomers undergo alcoholysis at different sites (Scheme 5.18) [600]. Thus, subjecting 2-alkyl-substituted succinic anhydrides to kinetic resolution with 1,1,1-trifluoroethanol in the presence of $(DHQD)_2AQN$ provides two regioisomers of different absolute configuration at the chiral center, giving rise to a reagent-controlled parallel kinetic resolution.

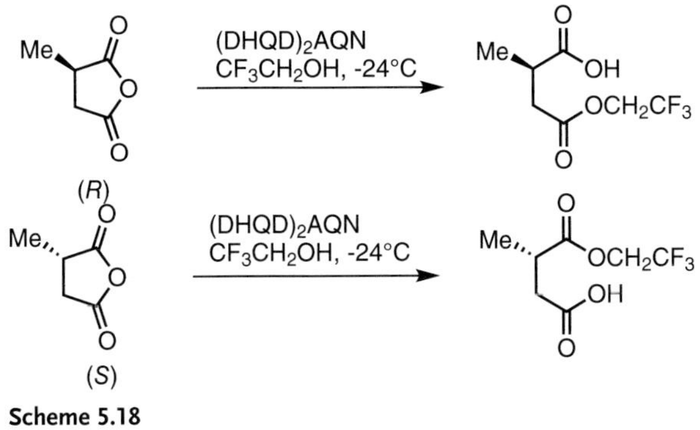

Scheme 5.18

Racemates of *tert*-butyl 3-benzyl-cyclopentene-1-carboxylate are separately converted into 1,4-addition products upon treatment with different benzyllithiums (Scheme 5.19) [955].

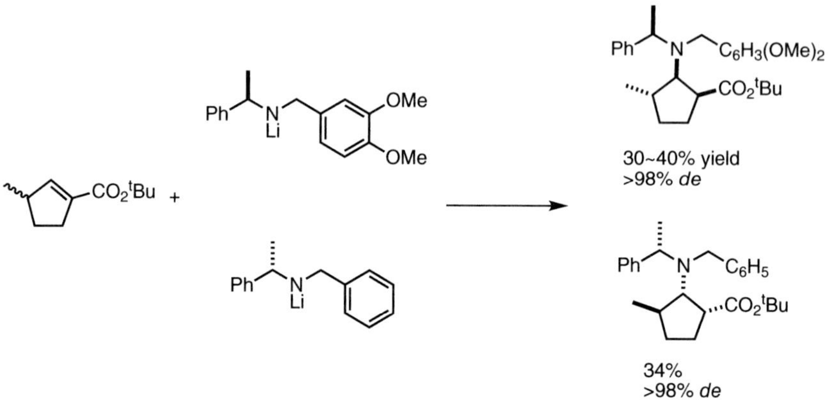

Scheme 5.19

Treatment of an equimolar mixture of D/L-thymidine with acetoxime levulinate in the presence of *Pseudomonas cepacia* lipase (PSL-C) affords a mixture of β-L-5′-*O*-levulinylthymidine and β-L-3′-*O*-levulinylthymidine in good yields (Scheme 5.20) [956].

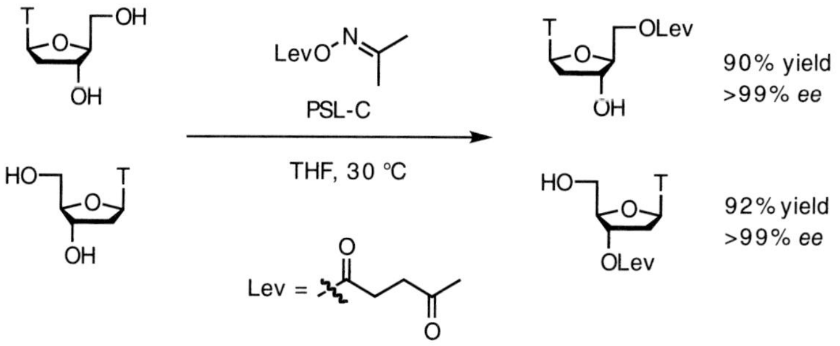

Scheme 5.20

Parallel kinetic resolution is utilized in the determination of enantiomeric excess by mass spectrometry (Scheme 5.21) [957]. The technique employs an equimolar mixture of pseudo-enantiomeric mass-tagged chiral acylating reagents possessing different substituents remote from the chiral center, so that the mass of the molecule can be correlated with its absolute configuration. A twenty-fold excess of an equimolar mixture of the mass-tagged *N*-acylprolines is treated with racemic alcohols in the presence of DCC and DMAP. The relative amounts of the enantiomers of the resulting esters are determined by mass spectrometry.

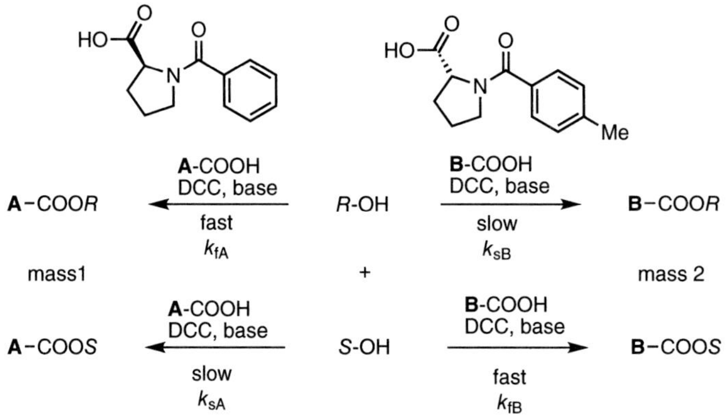

$$\textbf{A}\text{-COO}R \xleftarrow[\text{fast},\ k_{fA}]{\textbf{A}\text{-COOH, DCC, base}} R\text{-OH} \xrightarrow[\text{slow},\ k_{sB}]{\textbf{B}\text{-COOH, DCC, base}} \textbf{B}\text{-COO}R$$

mass1 + mass 2

$$\textbf{A}\text{-COO}S \xleftarrow[\text{slow},\ k_{sA}]{\textbf{A}\text{-COOH, DCC, base}} S\text{-OH} \xrightarrow[\text{fast},\ k_{fB}]{\textbf{B}\text{-COOH, DCC, base}} \textbf{B}\text{-COO}S$$

$$\frac{I_{\text{mass 1}}}{I_{\text{mass 2}}} = y \cdot q \qquad \%ee = \left[\frac{(y-1)(s+1)}{(y+1)(s-1)}\right] \cdot 100 \qquad \text{where } s = \frac{k_f}{k_s}$$

y = corrected intensity ratio
q = ionization correction factor

Scheme 5.21

6
Asymmetric Desymmetrization

Asymmetric desymmetrization is a method that allows many stereocenters to be established by enantioselective esterification of meso or prochiral compounds through breaking symmetry. The substrates that can be employed are symmetric diols and polyols, and also dicarboxylic acids and their anhydrides. The desymmetrization is performed by either enzymatic or chemical means.

PPL catalyzes desymmetrization of 2-substituted 1,3-propanediols in methyl acetate (Scheme 6.1) [958]. The enantiomeric excesses of the resulting monoacetates are dependent on the substituents at the 2-position, and range from 10% to 90%.

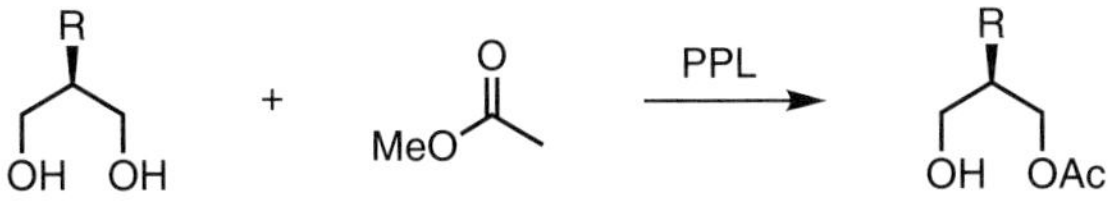

R, OH OH + MeO(C=O)CH3 —PPL→ R, OH OAc

Scheme 6.1

Isopropenyl and vinyl esters are more efficient acyl donors than alkyl esters in lipase-catalyzed desymmetrization of 2-substituted 1,3-diols such as glycerol and serinol derivatives, ferrocenylethanol, sugars, and other alcohols [959].

Similar treatment of 2-[4-(benzyloxy)-2-nitrophenyl]propane-1,3-diol with vinyl acetate as an acyl donor furnishes the corresponding monoacetate, which can be transformed into an indoline (Scheme 6.2) [960].

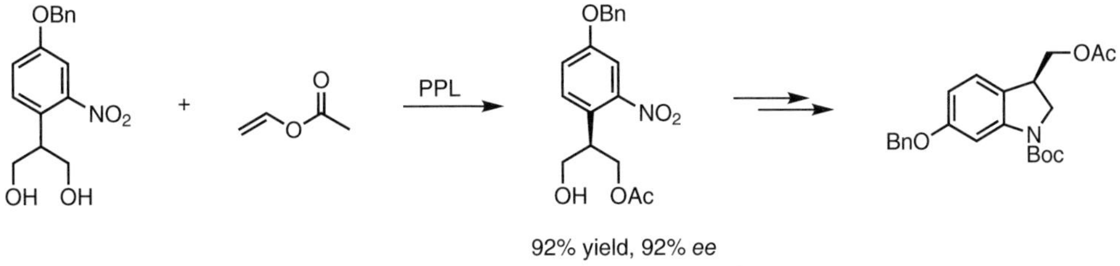

OBn, NO_2, OH OH + vinyl acetate —PPL→ OBn, NO_2, OH OAc (92% yield, 92% *ee*) ⟶ BnO, OAc, N Boc

Scheme 6.2

Esterification. Methods, Reactions, and Applications. 2nd Ed. J. Otera and J. Nishikido

ISBN: 978-3-527-32289-3

A *meso* spirodiol undergoes monoacetylation on transesterification with vinyl acetate/*Pseudomonas fluorescens* lipase in octane to afford the enantiomerically pure monoacetate in 56% yield (Scheme 6.3) [961].

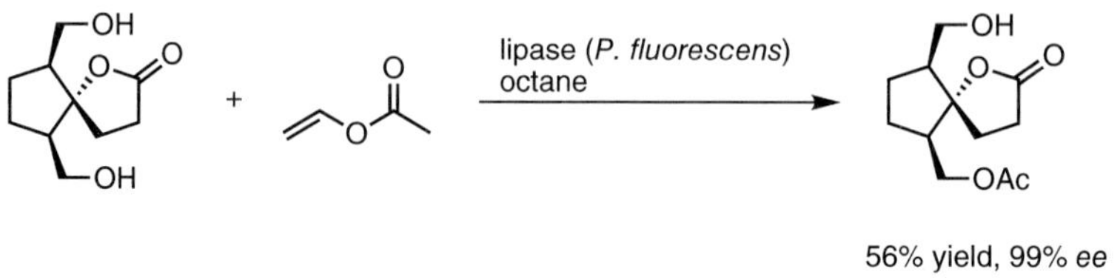

lipase (*P. fluorescens*)
octane

56% yield, 99% *ee*

Scheme 6.3

2,2-Disubstituted 1,3-propanediols are rather reluctant to undergo efficient desymmetrization because of their steric bulk and the facile racemization of the resulting monoacetate. The use of 1-ethoxyvinyl 2-furoate as an acyl donor results in successful formation of the corresponding monoesters possessing an asymmetric quaternary center with 82–99% *ee* (Scheme 6.4) [962].

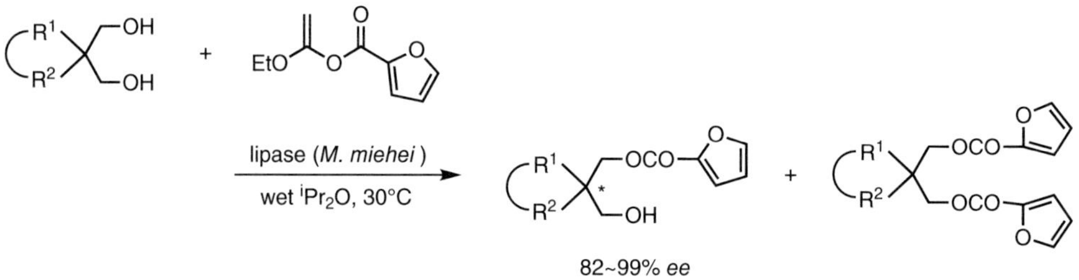

lipase (*M. miehei*)
wet $^{i}Pr_2O$, 30°C

82~99% *ee*

Scheme 6.4

Asymmetric desymmetrization by chemical means is also useful. As already described in Chapter 2, reaction of stannylenes derived from *meso*-1,2-diols with (1*S*)-ketopinic acid chloride affords the monoacylation product with high diastereoselectivity (Scheme 2.8) [719]. A similar tin(II) alkoxide derived from 2-*O*-protected glycerols coordinated with a chiral ligand catalyzes asymmetric desymmetrization (Scheme 2.19) [731]. *meso*-1,2-Diols are esterified with (1*S*,4*R*)-(–)-camphanoyl iodide in the presence of 2 equiv. of 2,6-dimethoxypyridine (Scheme 6.5) [963].

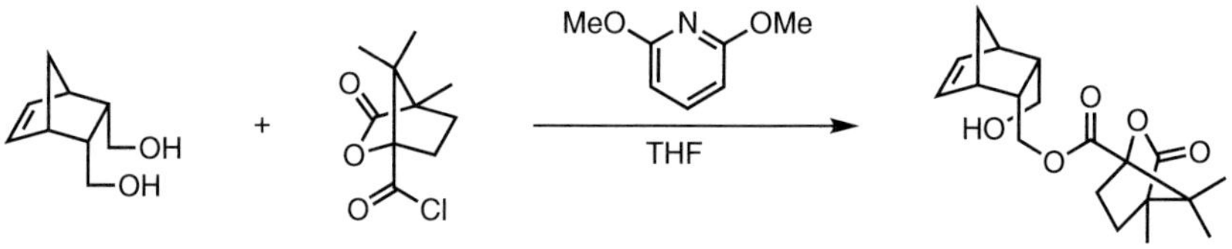

THF

Scheme 6.5

Axially chiral twisted amides effect desymmetrization of *meso*-1,2-, -1,3-, and -1,4-diols (Scheme 6.6) [964]. The corresponding monoesters are obtained, occasionally together with diester, with 33–88% *ee* by treatment of the diols with the amide at 80 °C.

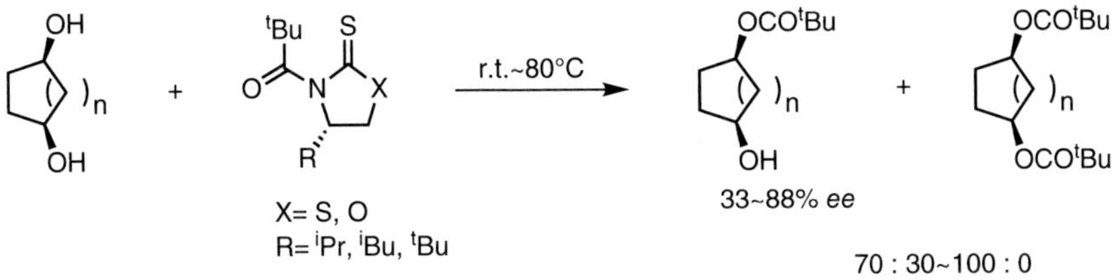

Scheme 6.6

meso-Cyclohexane-1,2-diol and *meso*-1,2-diphenylethylene glycol are transformed into the corresponding monoesters by treatment with acetic or benzoic anhydride with catalysis by a chiral phosphine, although the **s** values are modest (1.2–5.5) (Scheme 6.7) [605].

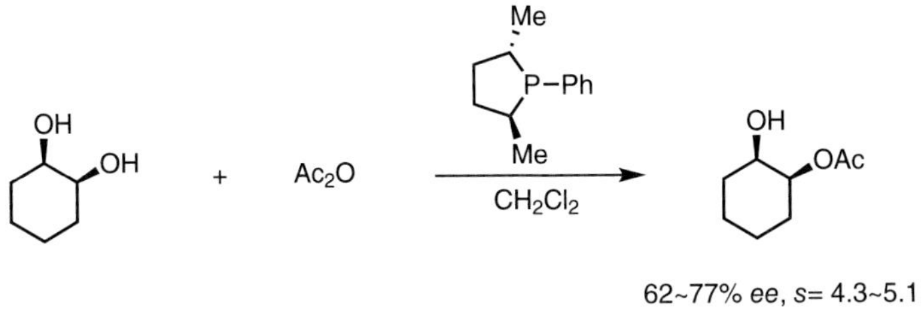

Scheme 6.7

Highly enantiomeric desymmetrization through the use of a chiral 1,2-diamine catalyst is feasible (Scheme 6.8) [965]. The reaction proceeds at −78 °C in the presence of the chiral amine catalyst (0.5 mol%) and diisopropylethylamine (1.5 equiv.) to furnish the corresponding monoesters in 85–96% *ee*.

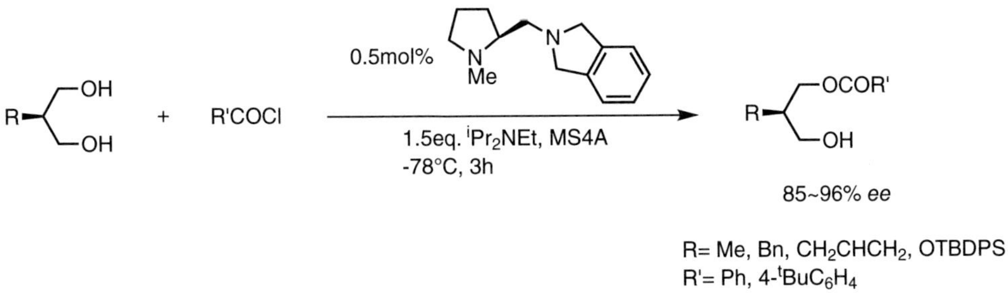

Scheme 6.8

Asymmetric desymmetrization of *meso* dicarboxylic acid anhydrides is a useful means to provide chiral half-esters. The combined use of Ph_2BOTf and (*R*)-2-methoxy-1-phenylethanol was described in Section 1.3.2.2 (Scheme 1.176) [512]. The employment of a chiral 1,3-diol has also been mentioned in Section 1.3.3.1 (Scheme 1.36) [576]. TiTADDOLates mediate ring-opening of various cyclic *meso* anhydrides with isopropanol to afford the corresponding half-esters in up to 98% *ee* (Scheme 6.9) [966, 967].

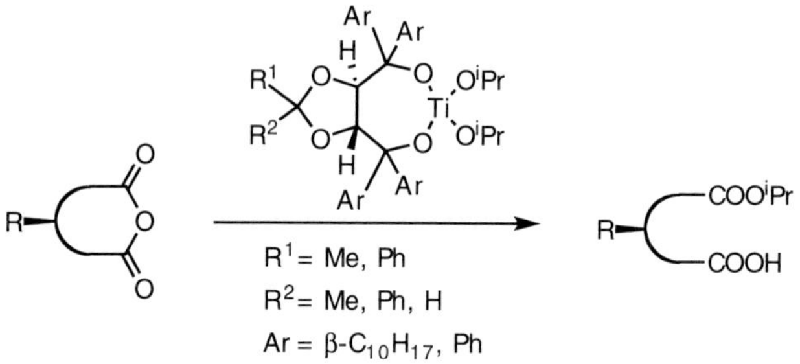

R^1 = Me, Ph
R^2 = Me, Ph, H
Ar = β-$C_{10}H_{17}$, Ph

Scheme 6.9

Some chiral amino alcohols such as ephedrine, cinchonine, cinchonidine, quinine, and quinidine are effective for the desymmetrization of 1,2-cyclohexanedicarboxylic acid anhydride [601]. Lithium salts of sterically congested chiral *N*-sulfonylaminoalcohols mediate the ring-opening of *meso* dicarboxylic acid anhydrides in a highly diastereoselective manner (up to >500 : 1 diastereomer ratio; Scheme 6.10) [968].

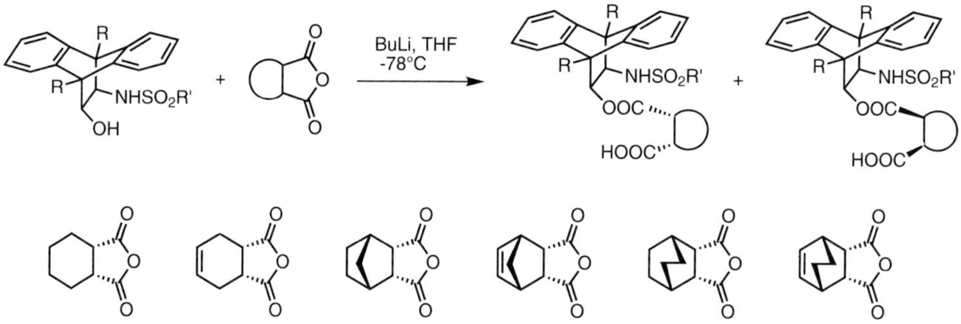

BuLi, THF
-78°C

Scheme 6.10

As described in Section 1.3.3.2, cinchona alkaloids efficiently catalyze methanolysis of *meso* dicarboxylic acid anhydrides to give half-esters (Schemes 1.48) [594–597, 599]. The enantioselectivity can be improved up to 98% *ee* by conducting the reaction at low temperature (~−50 °C) (Scheme 6.11) [969].

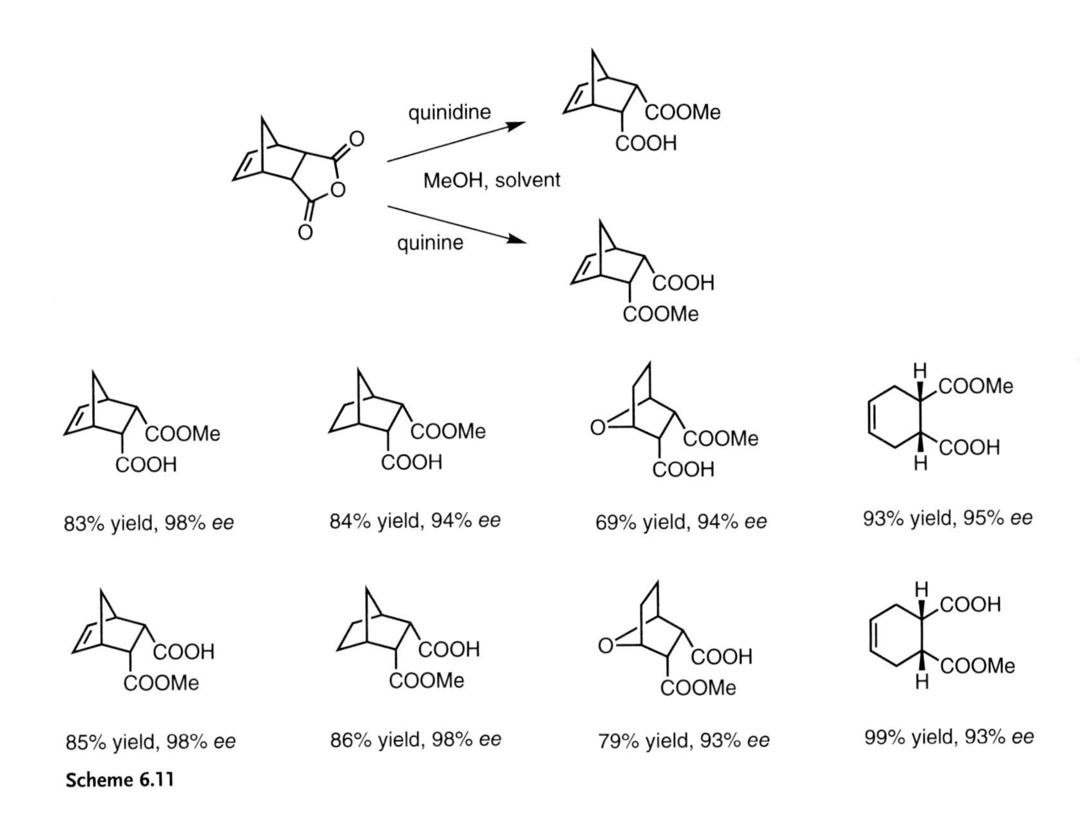

Scheme 6.11

7
Miscellaneous Topics

7.1
Selective Esterification

7.1.1
Differentiation between Primary, Secondary, and Tertiary Alcohols, and Phenols

The selective modification of one of a set of coexisting hydroxy functions is an important manipulation in organic synthesis, some examples already having been described in Part One. This chapter focuses on this subject by adding new procedures to provide an overview.

Most frequently encountered in the synthetic process is the need to discriminate between primary and secondary alcohols. Basically, a preference for a primary alcohol is generally attainable to some extent because of its innate superior reactivity, irrespective of acylating reagents. This is reflected to a degree in the simple thermal reaction of sugars (see Section 1.1.1) [2]. However, enzymes improve the selectively in a more general manner, as discussed in Sections 1.2.5 (Scheme 1.150) [455] and 1.3.4 (Scheme 1.220) [610]. The following examples constitute further addenda. Regioselective esterification on the primary alcohol moieties in chloramphenicol and thiamphenicol is successfully catalyzed by lipases (Scheme 7.1) [970]. Analogous selectivity is obtained with *Pseudomonas fluorescens* lipase (Scheme 7.2) [971]. The melanoma-associated disialoganglioside 9-*O*-acetyl GD3 can be obtained through regioselective enzymatic acetylation of GD3 with subtilisin as the biocatalyst and vinyl acetate as acetyl donor (Scheme 7.3) [972].

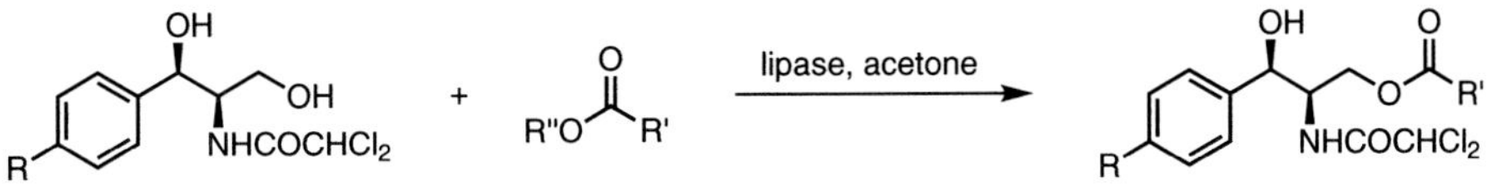

R= NO_2, SO_2CH_3
R'= $CH_2CH_2CH_3$, $(CH_2)_{14}CH_3$, CH=CHPh, CH_2CH_2COOH

Scheme 7.1

Esterification. Methods, Reactions, and Applications. 2nd Ed. J. Otera and J. Nishikido

ISBN: 978-3-527-32289-3

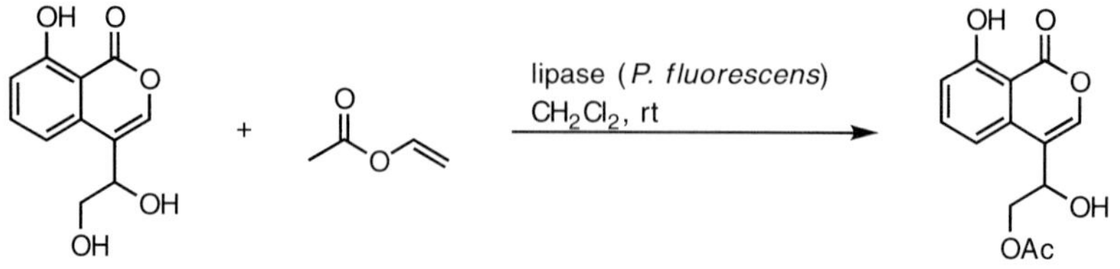

lipase (*P. fluorescens*)
CH_2Cl_2, rt

Scheme 7.2

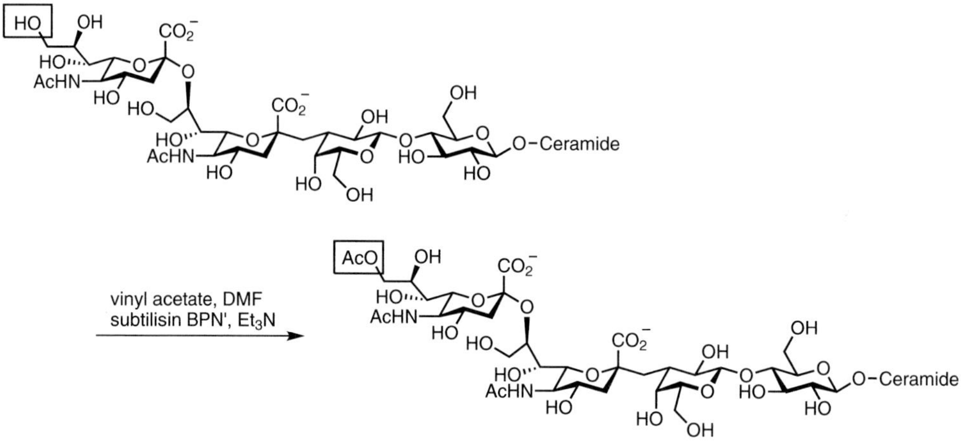

vinyl acetate, DMF
subtilisin BPN', Et_3N

Scheme 7.3

Primary/secondary alcohol differentiation finds the most important utilization in sugar and nucleoside chemistry. Selective acetylation on the 6-position of a glucal derivative with vinyl acetate occurs in high yield (>90%) with catalysis by lipase PS (Scheme 7.4) [973].

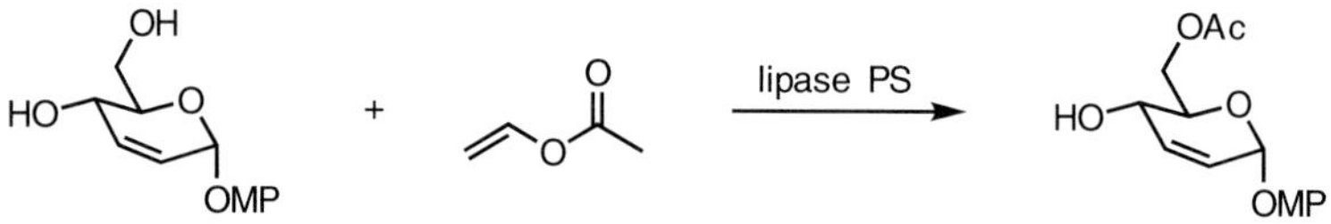

lipase PS

MP= p-$MeOC_6H_4$

Scheme 7.4

The use of oxime esters as acyl donors in lipase-catalyzed acylation of sugars effects selective acylation (Scheme 7.5) [974, 975]. The regioselectivity is dependent on the substrates, high regioselectivity on the primary alcohol being attained for hexoses such as D-galactose and D-mannose with lipase from *Pseudomonas cepacia*. Among pentoses, L-arabinose and D-ribose exhibit good selectivity, while D-xylose and D-lyxose yield complex mixtures. Notably, free pentoses are employable in this procedure.

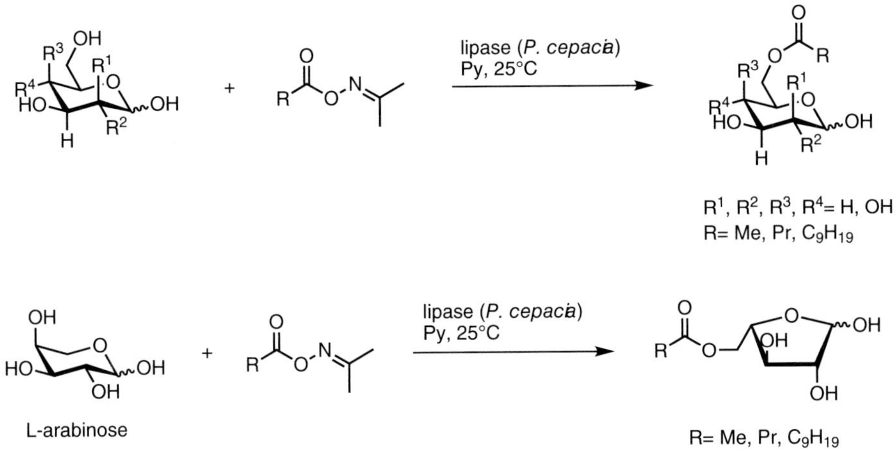

lipase (*P. cepacia*)
Py, 25°C

R^1, R^2, R^3, R^4= H, OH
R= Me, Pr, C_9H_{19}

L-arabinose

lipase (*P. cepacia*)
Py, 25°C

R= Me, Pr, C_9H_{19}

Scheme 7.5

Subjecting nucleosides to the same reaction conditions also results in preference for the primary alcohol [976]. Some 2′-deoxynucleosides, however, undergo selective acylation on the 3′-position (Scheme 7.6) [977]. On the other hand, the 2′-fluoro derivative is benzoylated on the primary hydroxy group simply on treatment with benzoyl chloride in pyridine at −20 °C (Scheme 7.7) [978].

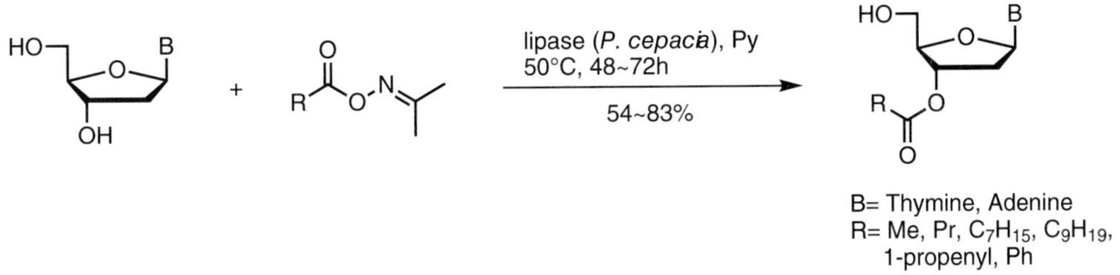

lipase (*P. cepacia*), Py
50°C, 48~72h
54~83%

B= Thymine, Adenine
R= Me, Pr, C_7H_{15}, C_9H_{19}, 1-propenyl, Ph

Scheme 7.6

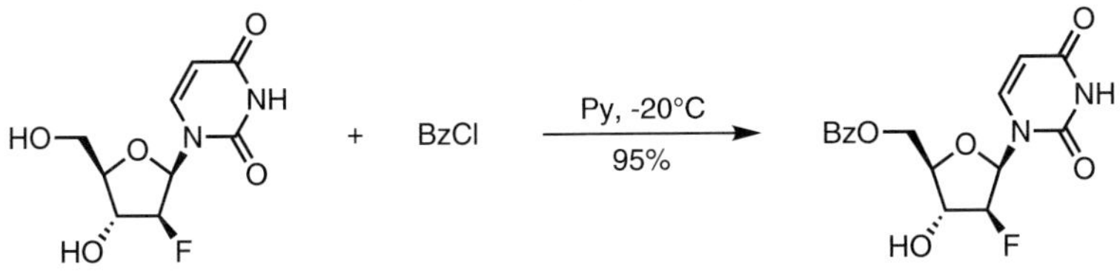

+ BzCl
Py, -20°C
95%

Scheme 7.7

A subtilisin mutant (subtilisin 8350) derived from subtilisin BPN′ through six site-specific mutations (Met50Phe, Gly169Ala, Asn76Asp, Gln206Cys, Tyr217Lys, and Asn218Ser) effects regioselective acetylation of uridine, adenosine, and cytidine with isopropenyl acetate in DMF (Scheme 7.8) [485]. 'Solvent engineering' points to pyridine as a better solvent for primary acetylation of ribonucleosides [486].

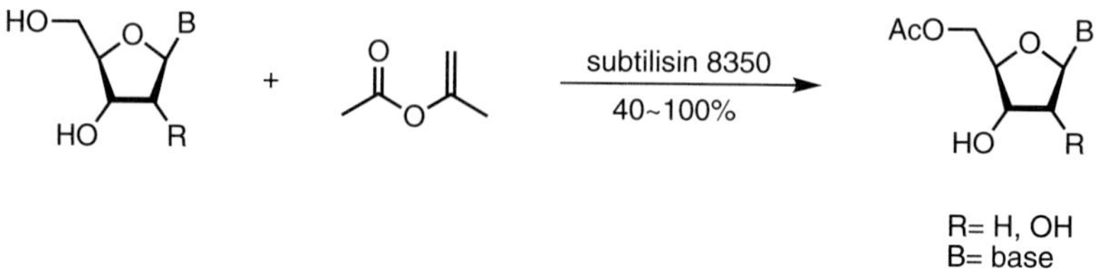

HO–(furanose ring, B, HO, R) + isopropenyl acetate —(subtilisin 8350, 40~100%)→ AcO–(furanose ring, B, HO, R)

R= H, OH
B= base

Scheme 7.8

These examples of the successful use of enzymes notwithstanding, much more attention has been paid to the use of nonenzymatic reactions. In particular, recent progress in Lewis acid chemistry has revealed that control over Lewis acidity can enable efficient discrimination. In the $HfCl_4 \cdot 2THF$-catalyzed reaction between carboxylic acids and alcohols, primary alcohols react preferentially in comparison with secondary ones (see Section 1.1.2.2; Scheme 1.18) [74]. TMSOTf is useful for selective acylation of primary alcohols (see Section 1.3.2.2) [506, 507]. Organotin Lewis acids are extremely mild, resulting in the high selectivity. 1,3-Disubstituted tetraalkyldistannoxanes $(XBu_2SnOSnBu_2Y)_2$ catalyze transesterification of primary alcohols in preference to secondary ones with various esters, including enol esters (see Section 1.2.2.2; Scheme 1.88 and Section 1.3.2.2) [52, 348–350]. This technique can be applied to the synthesis of enantiopure epichlorohydrin (see Section 1.3.2.2; Scheme 1.181) [535]. The neutral μ-hydroxy organotin dimer $[tert\text{-}Bu_2SnOH(Cl)]_2$ is particularly useful for deacetylation through transesterification, the primary alcohol being predominantly cleaved compared with the secondary one in a wide range of substrates including sugars and nucleosides (see Section 1.2.2.2; Scheme 1.89) [354, 355]. It should further be noted that solid acids are also employable. Thus, a primary alcohol is selectively consumed in competition with a secondary alcohol in alumina-catalyzed transesterification (see Section 1.2.2.3; Scheme 1.98) [375] and in zirconium sulfophenyl phosphonate-catalyzed acylation with acetic anhydride (see Section 1.3.2.3) [556].

Differentiation under basic conditions is also feasible. As described in Section 1.1.5, the Mitsunobu reaction usually provides high primary/secondary alcohol selectivity. Amines such as 2,4,6-collidine, *N,N*-diisopropylethylamine and 1,2,2,6,6-pentamethylpiperidine effect preferential acylation of primary alcohols with acid halides (see Section 1.4.3.2) [683]. Diphenylacetyl chloride is used for selective acylation of the primary alcohol in sugars, this reaction being promoted by pyridine (see Section 1.4.3.2; Scheme 1.276) [684]. Various esters of *p*-nitrothiophenol or 2,4-dinitrophenol effect regioselective acylation on the 6-position of nonprotected glycopyranosides catalyzed by DMAP in pyridine (Scheme 7.9) [1435].

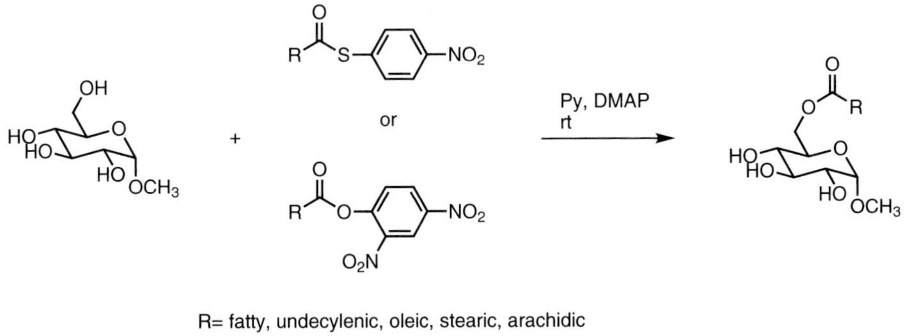

R= fatty, undecylenic, oleic, stearic, arachidic

Scheme 7.9

The use of activated esters occasionally gives high selectivity. Bipyridyl esters exhibit high preference for primary alcohols over secondary ones in the presence of a CsF promoter (see Section 1.1.6; Scheme 1.39) [222]. Benzoylation with 1-(benzoyloxy)benzotriazole in the presence of Et_3N takes place on primary alcohols in preference to secondary ones (see Section 1.2.3.2; Scheme 1.123) [416]. A phase-transfer reaction with benzoyl chloride affords the primary benzoate (see Section 1.4.3.3; Scheme 1.282) [693]. *N*-Pivaloyl imidazole is employable for pivaloylation of monosaccharides on primary hydroxy groups in preference to secondary ones (see Section 1.4.5) [710].

Differentiation between primary and secondary alcohols is quite easily achieved in some cases of practical natural product synthesis, probably because of the steric bulk of the substrates. Thus, simple treatment with acid halides in the presence of a base results in satisfactory selective primary alcohol acylation, as shown in Schemes 7.10–7.14 [979–983].

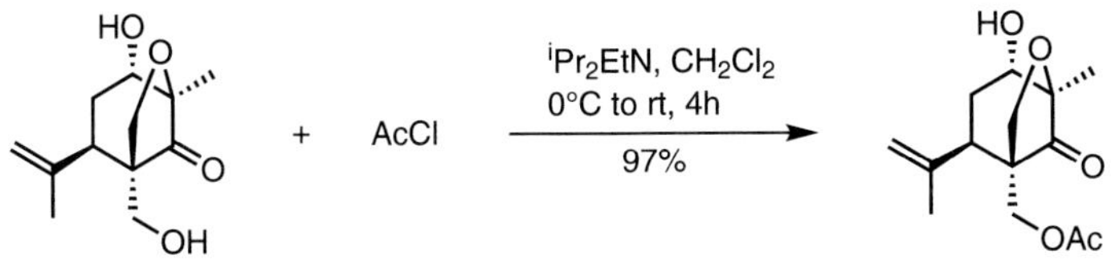

Scheme 7.10

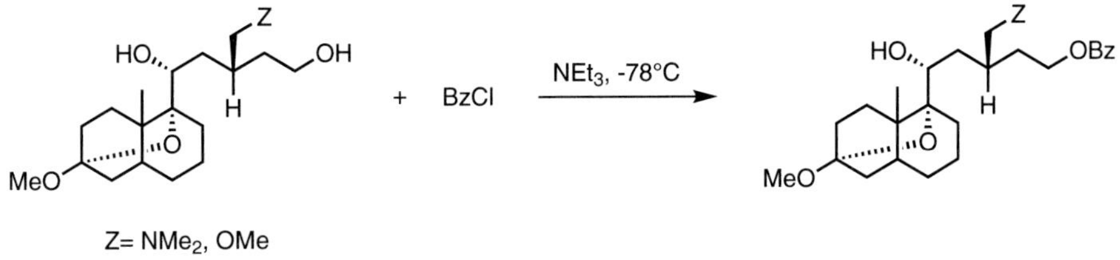

Z= NMe_2, OMe

Scheme 7.11

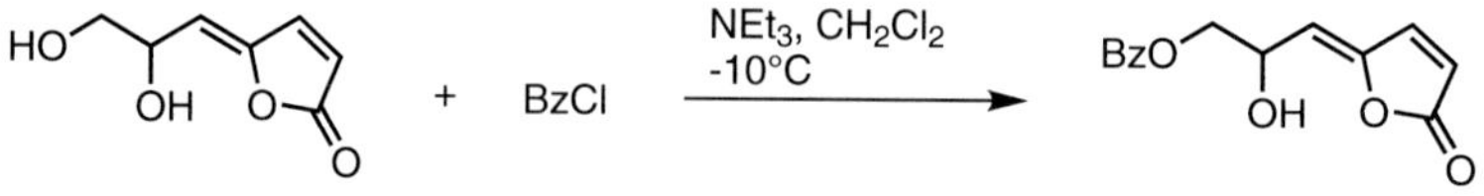

HO OH + BzCl → (NEt_3, CH_2Cl_2, -10°C) BzO OH

Scheme 7.12

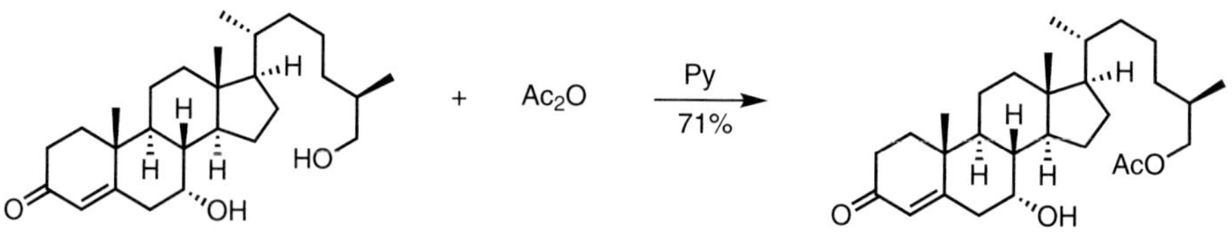

+ Ac_2O → (Py, 71%) AcO

Scheme 7.13

HO OH N–O COOMe + BzCl → (Py, DMAP (cat.), CH_2Cl_2, 82%) BzO OH N–O COOMe

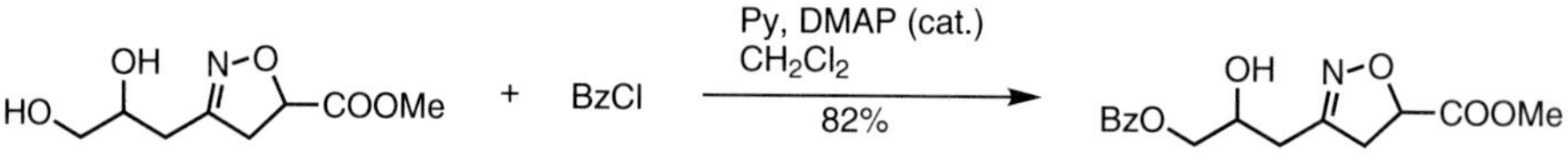

Scheme 7.14

It was mentioned in Chapter 2 that the primary alcohol in (+)-retronecine is selectively acylated via the stannylene intermediate (Schemes 2.20 and 2.21). As shown in Scheme 7.15, the sodium salt of (+)-retronecine reacts with a racemic imidazolide to afford the primary alcohol ester (1 : 1 diasteromer ratio) [984]. Treatment of (+)-retronecine with a highly reactive anhydride results in spontaneous ring opening (Scheme 7.16) [985].

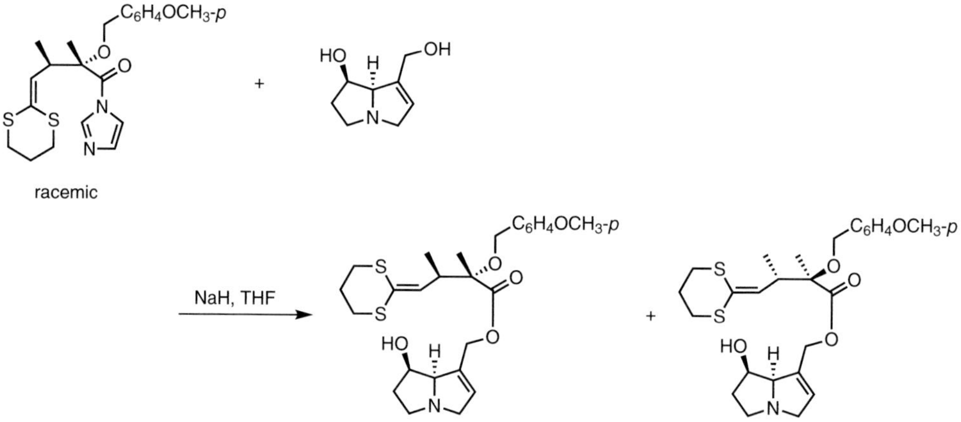

racemic + → (NaH, THF) +

Scheme 7.15

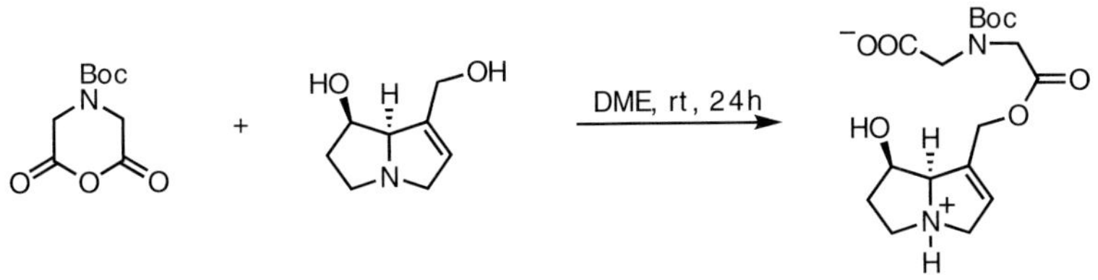

Scheme 7.16

Discrimination of primary and secondary alcohols from tertiary ones is very easily accomplishable: conventional acid anhydride or halide techniques provide satisfactory outcomes, as demonstrated in Schemes 7.17 and 7.18 [986, 987]. Primary acetates preferentially undergo methanolysis in the presence of secondary and tertiary acetates (see Section 1.2.3.1; Scheme 1.105) [396].

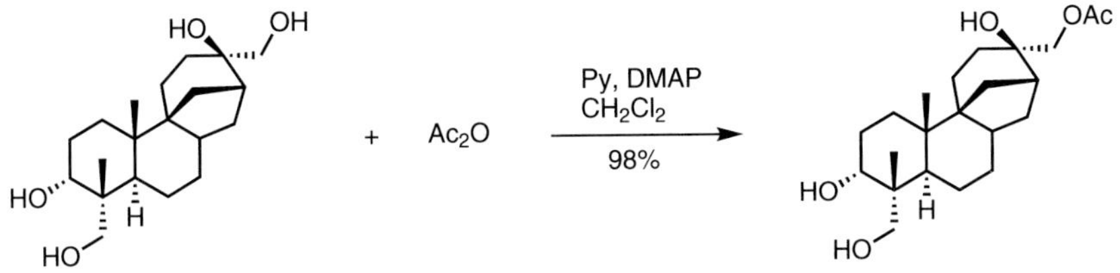

Scheme 7.17

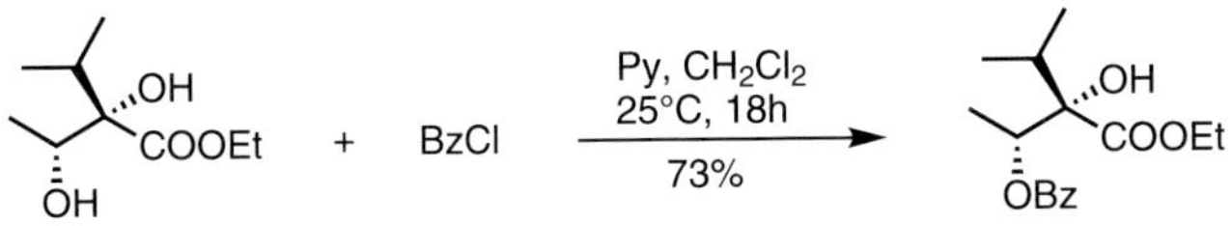

Scheme 7.18

Differentiation between aliphatic alcohols and phenols is also important in organic synthesis. Lipase catalysis induces selective acylation of the aliphatic alcohol. *Pseudomonas capacia* PS lipase adsorbed on Celite mediates selective acetylation of the aliphatic alcohol with vinyl acetate (Scheme 7.19) [988]. The same selectivity holds with immobilized lipase, even a secondary alcohol being preferentially acylated compared with phenol (see Section 1.2.5; Scheme 1.154) [460].

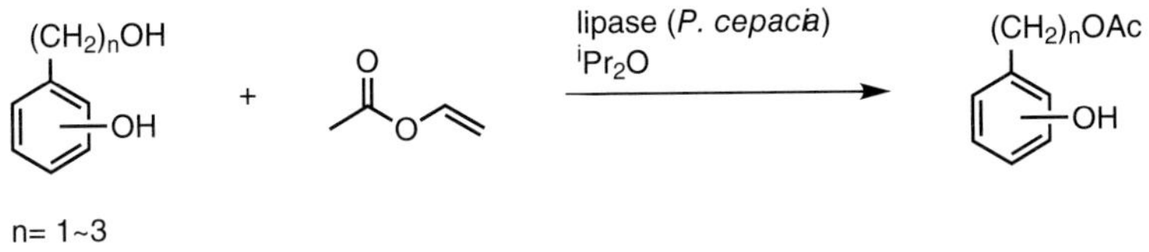

Scheme 7.19

In nonenzymatic treatments, differentiation between aliphatic and aromatic alcohols is dependent on the reaction conditions. Phenols are preferentially acylated over aliphatic alcohols under basic conditions. An example was described in the NaOH-promoted acetylation with 1-acetyl-v-triazolo/4,5-*b*/pyridine (see Section 1.4.3.1; Scheme 1.265) [668]. In addition, ω-hydroxyalkylphenol undergoes selective benzoylation of the phenol moiety on treatment with benzoyl chloride in the presence of K_2CO_3 in acetone (Scheme 7.20) [989, 990]. On the other hand, the reverse is true under acidic conditions.

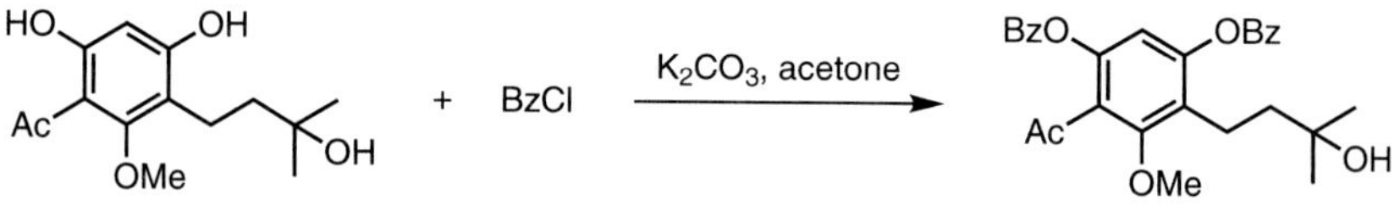

HO, OH, Ac, OMe, OH + BzCl → (K_2CO_3, acetone) BzO, OBz, Ac, OMe, OH

Scheme 7.20

Primary alcohols are typically acylated with acetic anhydride with catalysis by $Sc(OTf)_3$ (see Section 1.3.2.2) [503] or TMSOTf catalyst (see Section 1.3.2.2) [507]. Distannoxane-catalyzed acylation with vinyl esters takes place completely on the aliphatic alcohol, with no acylation on the phenol (Scheme 7.21) [350]. An acid-based technique is utilized in the synthesis of benzofuran adenosine antagonist XH-14 (Scheme 7.22) [991], while solid acids such as alumina (see Section 1.2.2.3; Scheme 1.98) and $NaHSO_4SiO_2$ (see Section 1.1.2.3; Scheme 1.22) [161] produce aliphatic carboxylic esters in competition with aromatic carboxylic acids in trans-esterification or esterification.

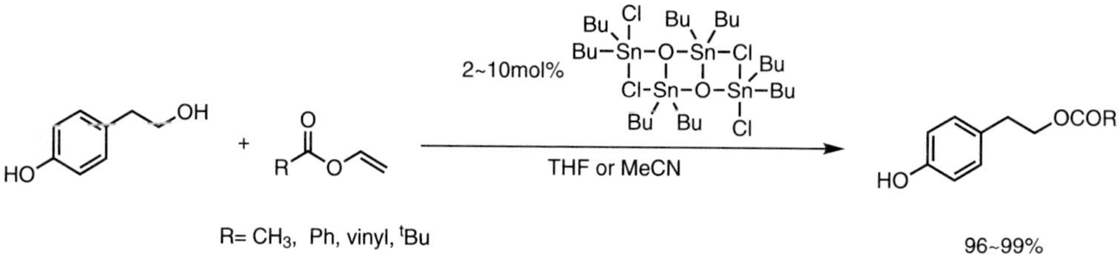

2~10mol% [ClBu₂Sn–O–SnBu₂Cl]₂ distannoxane; THF or MeCN

R= CH_3, Ph, vinyl, ^{t}Bu

96~99%

Scheme 7.21

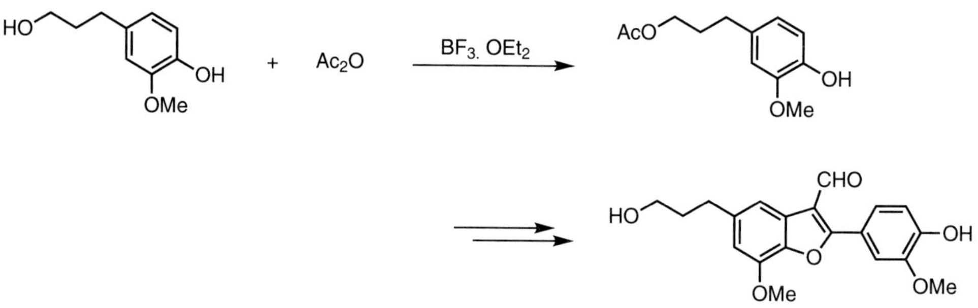

$BF_3.OEt_2$; Ac_2O

Scheme 7.22

7.1.2
Differentiation between Identical or Similar Functions

The selective transformation of one of two or more identical hydroxy groups is not easy to achieve. Subjection of an α,ω-diol to acylation, for instance, usually leads to a mixture of monoester, diester, and unreacted diol even with the use of one equivalent of acylating reagent. This notwithstanding, treatment of the diols adsorbed on silica gel with acetyl chloride in refluxing cyclohexane affords >99% yields of the corresponding monoacetate (Scheme 7.23) [992]. Monoacylation of *meso* and C_2-symmetric 1,2- and 1,3-diols can be achieved by the use of acid anhydrides and catalytic amounts of $CeCl_3$ (Scheme 7.24) [993]. Other lanthanide chlorides such as $DyCl_3$, $YbCl_3$, etc. are also effective for monoacylation of symmetrical diols [994].

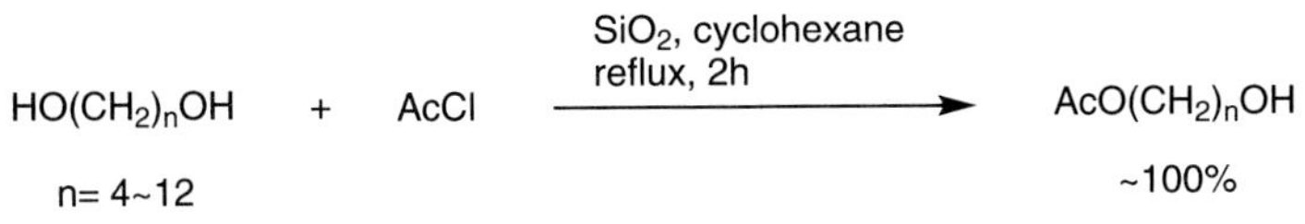

$$HO(CH_2)_nOH + AcCl \xrightarrow[\text{reflux, 2h}]{SiO_2\text{, cyclohexane}} AcO(CH_2)_nOH$$

n= 4~12 ~100%

Scheme 7.23

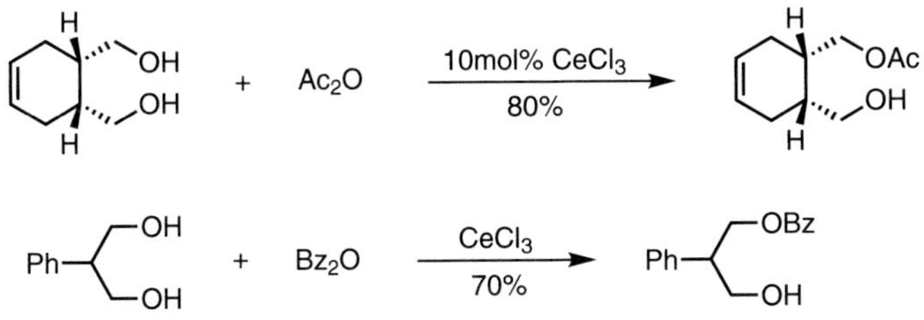

10mol% $CeCl_3$, Ac_2O, 80%; $CeCl_3$, Bz_2O, 70%

Scheme 7.24

Under distannoxane catalysis conditions, α,ω-diesters with fewer than four carbons are transformed into the corresponding monoesters on treatment with alcohol (Scheme 7.25) [995]. 2-Substituted propylene glycols also exhibit a considerable level of selectivity.

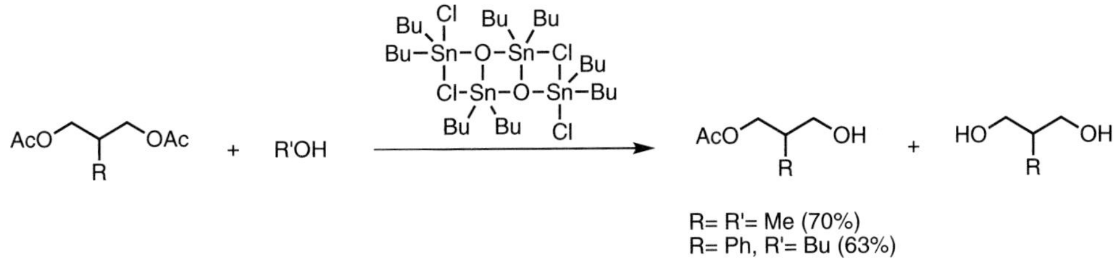

R= R'= Me (70%)
R= Ph, R'= Bu (63%)

Scheme 7.25

Although very similar difficulties arise with unsymmetrical substrates bearing different primary hydroxy groups at nonequivalent positions, because of the close reactivities involved, lipases act to overcome such problems. *Candida cylindracea* lipase selectively induces transesterification with isopropenyl acetate on the less sterically hindered sites in 2-alkylthio- and 2-dialkylamino-substituted 1,4- or 1,5-diols (see Section 1.2.5; Scheme 1.153) [458, 996]. Both PPL and CCL mediate acylation at the less hindered site with 2,2,2-trifluoroethyl butyrate (Scheme 7.26) [997]. Analogously, one of two hydroxymethyl groups on a pyridine ring may undergo acylation, as shown in Scheme 7.27 [998].

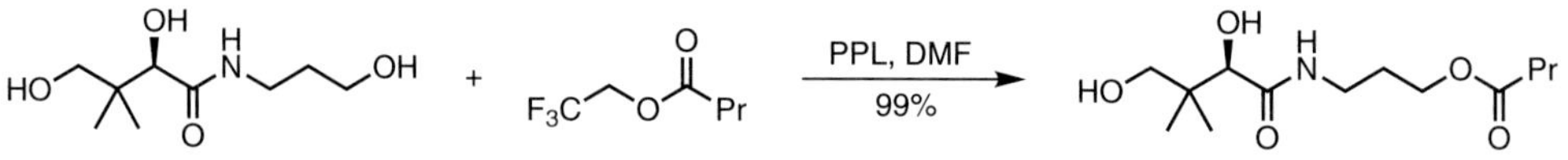

PPL, DMF
99%

Scheme 7.26

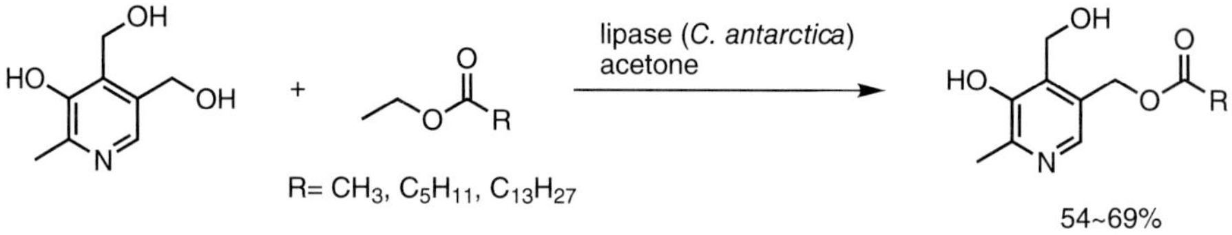

lipase (*C. antarctica*)
acetone

R= CH_3, C_5H_{11}, $C_{13}H_{27}$

54~69%

Scheme 7.27

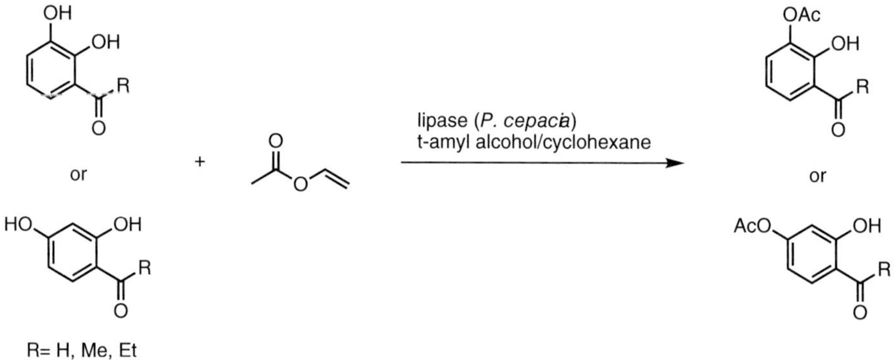

lipase (*P. cepacia*)
t-amyl alcohol/cyclohexane

R= H, Me, Et

Scheme 7.28

Regioselective acetylation takes place even with dihydroxy derivatives of phenyl ketones and benzaldehyde (Scheme 7.28) [999]. Treatment of these substrates with vinyl acetate in the presence of PCL results in selective acetylation of the *meta*- or *para*-hydroxy group, the *ortho*-hydroxy moiety being untouched. Treatment of (+)-catechin with vinyl acetate in the presence of immobilized PSL results in acetylation of the hydroxy groups on the A ring only, with neither diacetylation nor acetylation on the B ring occurring (Scheme 7.29) [1000].

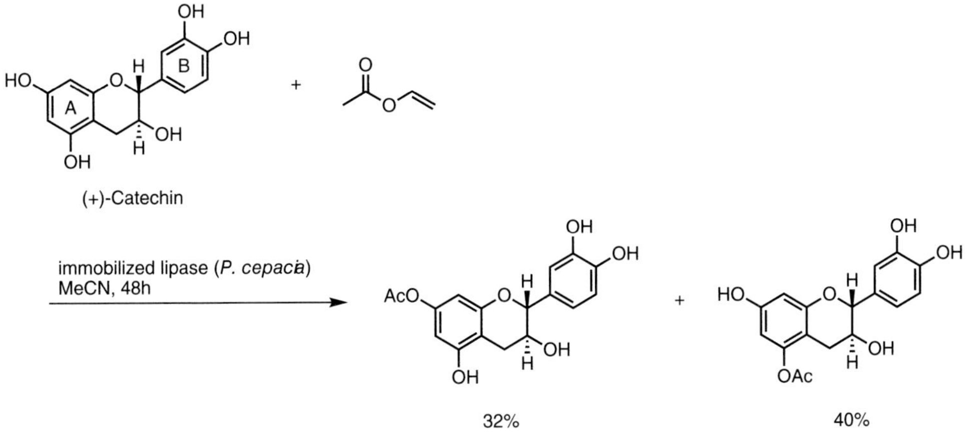

(+)-Catechin

immobilized lipase (*P. cepacia*)
MeCN, 48h

32%

40%

Scheme 7.29

Efficient differentiation between secondary hydroxy groups situated on different sites is required in a broad spectrum of organic syntheses. In particular, selective protection and deprotection of the hydroxy functions on the cyclohexane ring are among the most important in sugar, nucleoside, and steroid chemistry. As described in Chapter 2, the organotin procedure serves quite well to this end. The stannylene intermediate derived from 1,2-diol and dialkyltin oxide or dialkoxide undergoes regioselective acylation on one of the alkoxide sites upon treatment with acid halide or anhydride.

Benzoyl chloride in pyridine mediates selective benzoylation. Thus, *myo*-inositol is converted into the symmetrical 1,3,5-tri-*O*-benzoyl derivative (see Section 1.4.3.2; Scheme 1.273) [677]. Similarly, methyl 6-*O*-(*tert*-butyldiphenyl)silyl-α-D-mannopyranoside undergoes selective benzoylation at the 3-position (see Section 1.4.3.2; Scheme 1.274) [678]. One of two hydroxy groups on a pyrrolidinone ring is benzoylated under similar conditions (see Section 1.4.3.2; Scheme 1.270) [674].

Quinic and shikimic acid esters demand similar selectivity, because three hydroxy groups are present on the cyclohexane ring. *Chromobacterium viscosum* lipase allows selective acetylation of quinate esters at the 4-position by use of vinyl acetate (Scheme 7.30) [1001]. Methyl shikimate is also selectively acetylated at the

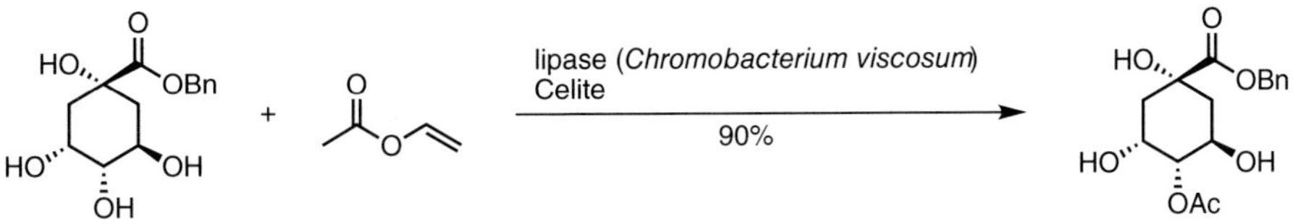

lipase (*Chromobacterium viscosum*)
Celite

90%

Scheme 7.30

4-position by use of *Candida antarctica* lipase A (Scheme 7.31) [1002]. Vinyl esters with longer chains give better results.

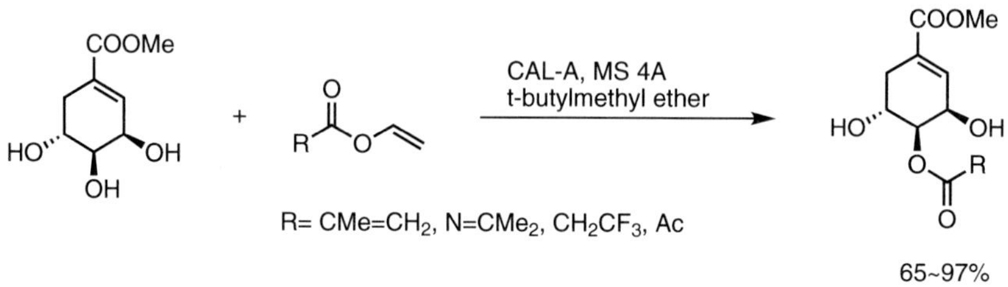

CAL-A, MS 4A
t-butylmethyl ether

R= CMe=CH$_2$, N=CMe$_2$, CH$_2$CF$_3$, Ac

65~97%

Scheme 7.31

Steroids are another class of compounds that frequently require selective acylation. *Candida antarctica* lipase B and 2,2,2-trifluoroethyl esters, for example, showed a marked preference for the alcoholic moiety on the A ring of the steroid skeleton (Scheme 7.32) [1003]. Immobilized enzyme (Novozym 435) catalyzes the regioselective acylation of (20*R*)-hydroxyecdysone at the C-2 OH (Scheme 7.33) [1004].

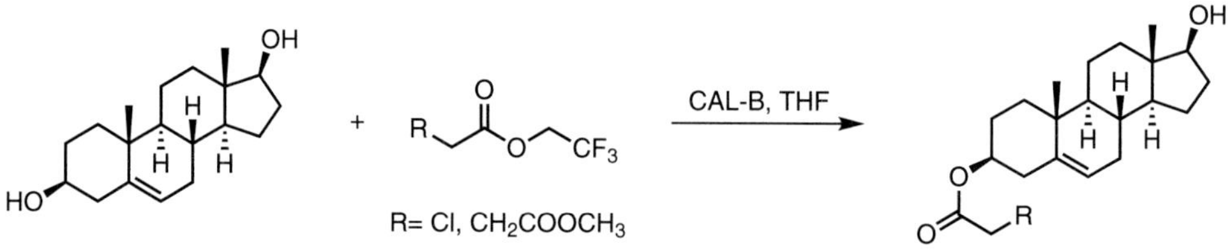

CAL-B, THF

R= Cl, CH$_2$COOCH$_3$

Scheme 7.32

Novozyme 435
t-amyl alcohol / Py

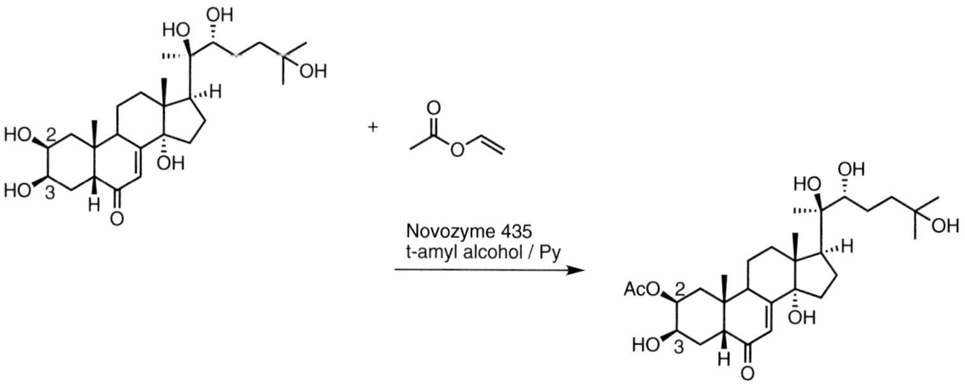

Scheme 7.33

Polycyclic compounds fairly easily undergo selective acylation at the less sterically demanding position, because of their steric congestion, through both enzymatic and nonenzymatic means. Some examples are given in Schemes 7.34–7.37 [1005–1008].

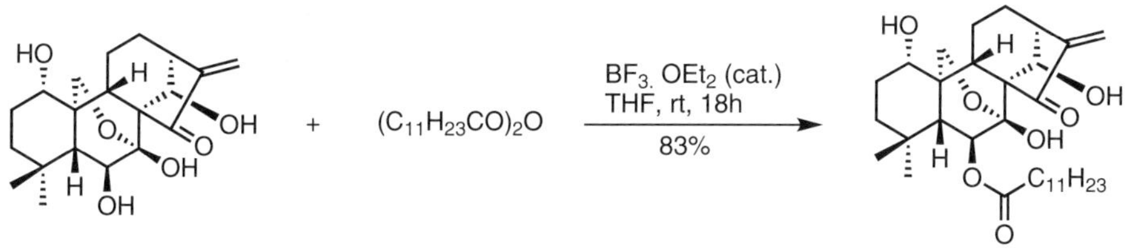

+ $(C_{11}H_{23}CO)_2O$

$BF_3 \cdot OEt_2$ (cat.)
THF, rt, 18h
83%

Scheme 7.34

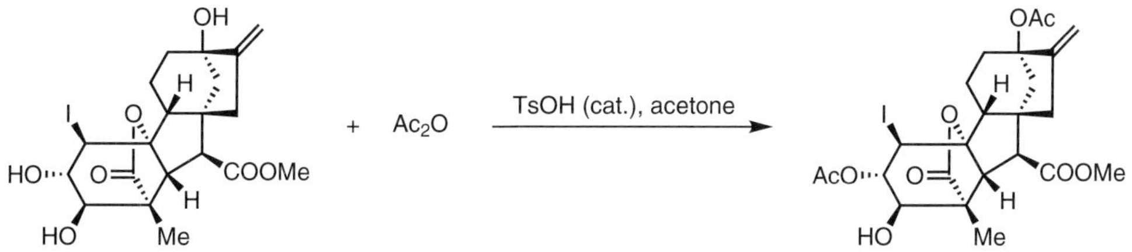

+ Ac_2O

TsOH (cat.), acetone

Scheme 7.35

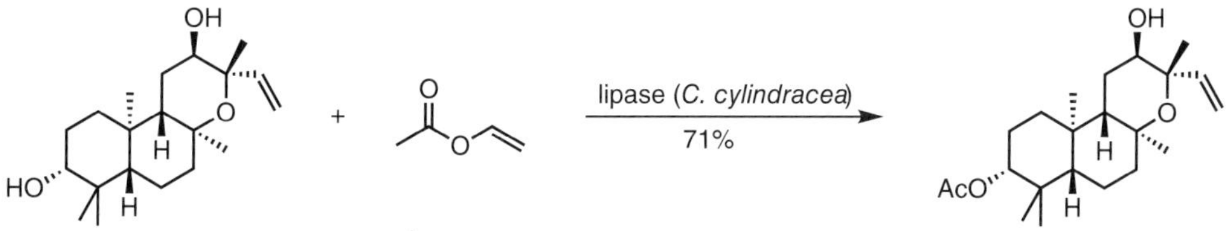

lipase (*C. cylindracea*)
71%

Scheme 7.36

+ Ac_2O

Et_3N or $^iPr_2EtN$
CH_2Cl_2

79% or 90%

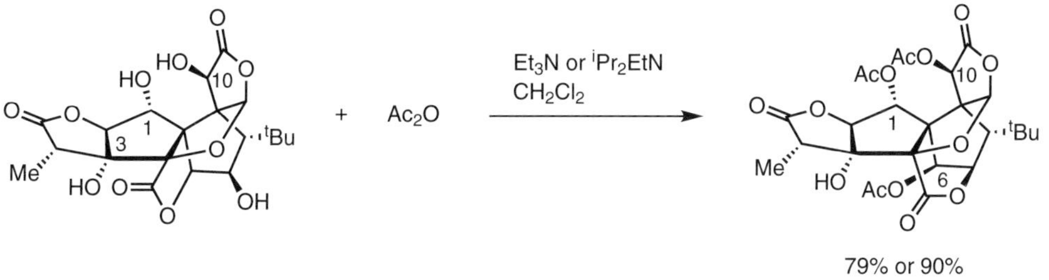

Scheme 7.37

7.2 Use of Theoretical Amounts of Reactants

The esterification reaction between carboxylic acid and alcohol and the transesterification reaction between ester and alcohol are both equilibrium processes, so it is common to use one of the reactants in excess and/or to remove the resulting alcohol or water constantly during the reaction to shift the equilibrium in favor of the product side. From the viewpoint of green chemistry, it is highly desirable to improve on such an inconvenient situation in esterification technology in order to save resources and energy. The use of the reactants in a strict 1:1 ratio is a first step toward this goal [1009]. The reaction between carboxylic acid and alcohol is promoted by graphite bisulfate (see Section 1.1.2.3) [123], an equimolar mixture of carboxylic acid and alcohol being stirred in the presence of graphite bisulfate in dry cyclohexane. The yield of the ester is more than 94% with simple reactants, but lower with secondary alcohols or cyclopropanecarboxylic acid (50–70%). Tertiary alcohols and phenol are not employable. $NaHSO_4 \cdot H_2O$ mediates reaction between acetic and propanoic acids with primary and secondary alcohols if the neat mixture is heated at reflux with an automatic water separator [1010]. The yields of esters are 85–96%. Microwave irradiation increases the yield of esters from *p*-toluenesulfonic acid-catalyzed esterification between equimolar amounts of carboxylic acid and alcohol, ranging from 82–97% [7]. $HfCl_4 \cdot 2THF$ is a versatile catalyst with which to effect esterification of an equimolar mixture of a carboxylic acid and an alcohol, although the use of a Soxhlet extractor with molecular sieves 4A is needed (see Section 1.1.2.2) [73]. Aliphatic, aromatic, α,β-unsaturated, and sterically demanding carboxylic acids are employable, as are primary, secondary, allylic, and propargylic alcohols. The yields of esters are generally greater than 90%, and are greater than 99% in the reactions between 4-phenylbutyric acid and benzyl and 1-phenethyl alcohols. This reaction is applicable to polycondensation of ω-hydroxycarboxylic acids to afford polyesters with $Mn = \sim 10^4$. $ZrCl_4 \cdot 2THF$ and $Zr(OEt)_4$ act similarly [74, 75]. The moisture sensitivity of the hafnium and zirconium complexes is improved by the use of $MOCl_2 \cdot 8H_2O$ (M = Zr, Hf) [76, 77]. $TiO(acac)_2$ is more active than the above hafnium and zirconium complexes [56]. $FeCl_3 \cdot 6H_2O$ is effective for condensation between long-chain carboxylic acids and alcohols [64]. $K_5CoW_{12}O_{40} \cdot 3H_2O$ (0.1 mol%) catalyzes esterification of an equimolar mixture of carboxylic acids and alcohols without solvent, though yields are less than 91% [393]. Phenyl esters are obtained by stirring an equimolar mixture of a carboxylic acid and phenol at 0 °C to room temperature in the presence of $AlCl_3$ (4 mol%)/$ZnCl_2$ (100 mol%) [42].

Diphenylammonium triflate catalyzes esterification with equimolar amounts of reactants (see Section 1.1.2.1) [18]. Heating of the reactants in toluene at 80–110 °C in the presence of 1 mol% of the catalyst furnishes the corresponding esters in 78–96% yield. Pentafluorophenylammonium triflate is also effective for esterification and transesterification with equimolar reactants [20]. Dimesitylammonium pentafluorobenzenesulfonate (1 mol%) catalyzes condensation of an equimolar mixture of carboxylic acids and alcohols without use of azeotropic dehydration [21].

High yields of esters are obtained from reaction between carboxylic acids and alcohols in a 1:1 mole ratio when the reaction is carried out in ionic liquids (see Section 7.3) [1011–1013].

It should be noted that the above procedure successfully achieves the use of reactants in 1:1 ratios, but the yield of the ester is not 100% except in the two cases of $HfCl_4 \cdot 2THF$-catalyzed reactions. Importantly, the 1:1 stoichiometry is truly effective only if 100% conversion is reached, since the two starting materials (recovered carboxylic acid and alcohol) otherwise have to be separated from the product mixture, which is a less favorable situation than that in which one of the reactants has been employed in excess. In this case, the recovery of only one reactant is necessary, on account of the more facile feasibility of the 100% conversion. As such, a 100% yield with use of equimolar amounts of the reactants is ideal because of the lack of a need for a purification process. Such conditions come close to fulfilment through the use of fluorous biphase technology. Heating a mixture of a methyl or ethyl ester and an alcohol in a 1:1 molar ratio in the presence of $[(ClRf_2SnOSnRf_2Cl)_2$: $(Rf = C_6F_{13}C_2H_4)]$ catalyst in FC-72 solvent at 150 °C provides the corresponding esters in 100% yields (see Section 1.1.2.2) [54]. Some representative results are given in Table 7.1. The perfect conversion is apparent from GLC analysis, which shows no reactants remaining at all. In addition, the tolerance of various functional groups is of great synthetic promise. The perfect conversion is attributable to the poor solubility of the liberated water in FC-72, so the equilibrium is shifted to the ester side. Sterically bulky carboxylic acids are reluctant to react under these conditions. This can result in complete discrimination in the competition between carboxylic acids of different steric bulk (Table 7.2). When a 1:1 mixture of two carboxylic acids of different size is treated with one equivalent of alcohol, only the smaller carboxylic acid reacts, to give a 100% yield of the corresponding ester, while no reaction at all takes place with the larger carboxylic acid. Similar perfect differentiation is also observed between aliphatic and aromatic carboxylic acids (Scheme 7.38). The unique features of this technology are described in further detail in the next section.

Table 7.1 Fluorous biphasic esterification.

Entry	RCOOH	R'OH	Yield of RCOOR' [%]	
			GLC	Isolated
1	$Ph(CH_2)_2COOH$	$PhCH_2OH$	>99.9	100
2	$Ph(CH_2)_2COOH$	$C_8H_{17}OH$	>99	100
3	$Ph(CH_2)_2COOH$	$TBSO(CH_2)_8OH$	>99	98
4	$Ph(CH_2)_2COOH$	$THPO(CH_2)_8OH$	>99	98
5	$Ph(CH_2)_2COOH$	geraniol	>99	100
6	$Ph(CH_2)_2COOH$	$PhCH=CHCH_2OH$	>99	99
7	$Ph(CH_2)_2COOH$	$PhC{\equiv}CCH_2OH$	>99	99
8	p-$NO_2C_6H_4COOH$	$PhCH_2OH$	>99.9	100
9	C_6F_5COOH	$PhCH_2OH$	>99.9	99
10	p-$CF_3C_6H_4COOH$	$PhCH_2OH$	>99.9	99
11	$CH_2=CH(CH_2)_8COOH$	$PhCH_2OH$	>99.9	98
12	2-(4-ClC_6H_4O)$OC(CH_3)_2COOH$	$PhCH_2OH$	>99.9	98

Table 7.2 Competition between carboxylic acids in fluorous biphasic esterification.

$$R^1COOH + R^2COOH + R^3OH \xrightarrow[Rf=C_6H_{13}C_2H_4]{[(ClRf_2SnOSnRf_2Cl)_2]} R^1COOR^3 + R^2COOR^3$$

Entry	R^1COOH	R^2COOH	R^3OH	Yield [%]		
				R^1COOR^2		R^1COOR^3
				GLC	Isolated	GLC
1	$Ph(CH_2)_2COOH$	$PhC(CH_3)COOH$	$C_8H_{17}OH$	>99.9	98	<0.1
2	$Ph(CH_2)_2COOH$	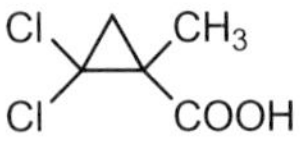 Cl, Cl, CH_3, COOH	$C_8H_{17}OH$	>99.9	100	<0.1
3	$Ph(CH_2)_2COOH$	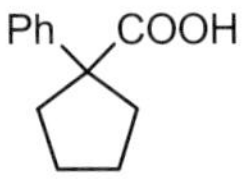 Ph, COOH	$C_8H_{17}OH$	>99.9	100	<0.1
4	$C_7H_{15}COOH$	$PhC(CH_3)COOH$	$C_8H_{17}OH$	>99.9	97	<0.1
5	p-$CF_3C_6H_4COOH$	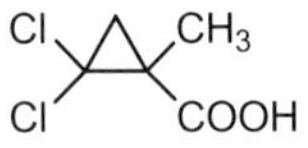 Cl, Cl, CH_3, COOH	$C_8H_{17}OH$	>99.9	99	<0.1
6	p-$CF_3C_6H_4COOH$	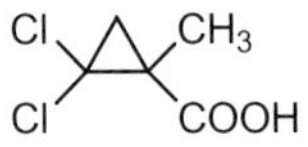 Cl, Cl, CH_3, COOH	$PhCH_2OH$	>99.9	98	<0.1
7	p-$CF_3C_6H_4COOH$	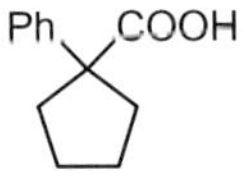 Ph, COOH	$C_8H_{17}OH$	>99.9	100	<0.1
8	p-$CF_3C_6H_4COOH$	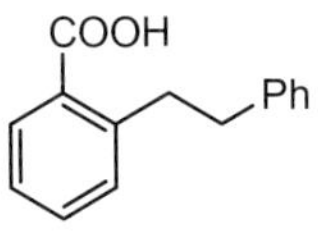 COOH, Ph	$C_8H_{17}OH$	>99.9	99	<0.1

COOH

$O(CH_2)_nCOOH$

+ ROH

$[(ClRf_2SnOSnRf_2Cl)_2]$
Rf= $C_6H_{13}C_2H_4$
FC-72

COOH

$O(CH_2)_nCOOR$

100%

R= $PhCH_2CH_2$, c-$C_6H_{11}CH_2CH_2$, C_8H_{17}

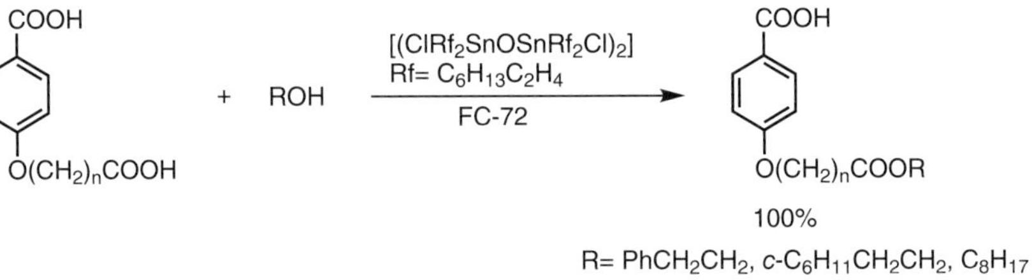

Scheme 7.38

As well as the above catalytic processes, activation of carboxylic acids by stoichiometric amounts of promoters is also an option (see Section 1.1.6). Treatment of carboxylic acid with dimethylstannocene affords a tin(II) dicarboxylate intermediate, which is converted into the corresponding ester by treatment with one equivalent of alcohol (Scheme 1.57) [256]. The yield of the esters ranges from 74–86%. Addition of $(Me_2SiO)_4$ effects esterification with equimolar amounts of carboxylic acid and alcohol [253]. The yield of the ester is 77–99% and silyl carboxylates are believed to be intermediates. Dimethylsulfamoyl chloride (2 equiv.) mediates the reaction between one equivalent each of carboxylic acid and alcohol in the presence of Et_3N (3 equiv.)/DMAP (0.1 equiv.) [236]. The yield is generally high (~95%). Esterification of a 1 : 1 mixture of carboxylic acid and alcohol is feasible in the presence of tosyl chloride (1.2 equiv.) and *N*-methylimidazole (3 equiv.) in 87–95% yields [232].

High levels of conversion are more difficult to achieve in transesterification than in esterification, because the alcohol formed is more difficult to remove from the reaction mixture than the water, because of its better compatibility with organic solvents. Nevertheless, the fluorous biphase approach with the fluoroalkyldistannoxane catalyst (see Section 1.2.2.2) allows perfect transesterification [352, 353]. Upon treatment of methyl or ethyl esters with one equivalent of alcohol heavier than ethanol in the presence of the fluoroalkyldistannoxane, the new esters are obtained in 100% yields (Table 7.3). The equilibrium is forced completely to one

Table 7.3 Transesterification in single fluorous phase system.

$$RCOOR' + R''OH \xrightarrow[Rf = C_6F_{13}C_2H_4,\ FC\text{-}72]{[(ClRf_2SnOSnRf_2Cl)_2]} RCOOR'' + R'OH$$

Entry	RCOOR′	R″OH	Yield (%)	
			GLC	Isolated
1	$Ph(CH_2)_2COOEt$	$C_8H_{17}OH$	>99	100
2		$PhCH{=}CHCH_2OH$	>99	100
3	$Ph(CH_2)_2COOMe$	$PhCH{=}CHCH_2OH$	>99	100
4		Geraniol	>99	98
5		$PhC{\equiv}CCH_2OH$	>99	100
6		$THPO(CH_2)_8OH$	>99	99
7		$TBSO(CH_2)_8OH$	>99	100
8		2-octanol	>99	100
9		Cyclohexanol	>99	99
10		Menthol	>99	
11		Borneol	>99	
12	PhCH CHCOOEt	$PhCH{=}CHCH_2OH$	>99	99
13	PhCOOEt	$PhCH{=}CHCH_2OH$	>99	100
14	PhCOOMe	$PhCH{=}CHCH_2OH$	>99	100

side because the methanol or ethanol produced is much less soluble in FC-72 than the heavier alcohols, so it is crucial to use esters with low-alcohol components. This technology is described in more detail in the next section.

Acylation by use of one equivalent of acid anhydirde is catalyzed by $InCl_3$ [529] and $BiOClO_4 \cdot xH_2O$ [523].

Finally, it should be noted that β-keto esters are unique substrates capable of undergoing facile transesterification with one equivalent of an alcohol (see Sections 1.2.2.3 and 1.2.4) [385, 389, 448].

7.3 New Reaction Media

(Trans)esterification reactions are usually conducted in organic solvents, but use of other reaction media occasionally gives better outcomes. As described in Section 1.4.3.3, acylation with acid halides can readily be conducted under phase-transfer conditions [689–693]. Other examples include acylation of steroids (Scheme 7.39) [1014], benzoylation of thiosugars (Scheme 7.40) [1015], formation of azacrown ethers (Scheme 7.41) [1016], benzoylation followed by condensation (Schemes 7.42 and 7.43) [1017, 1018], and benzylation of sodium benzoate [1019, 1020].

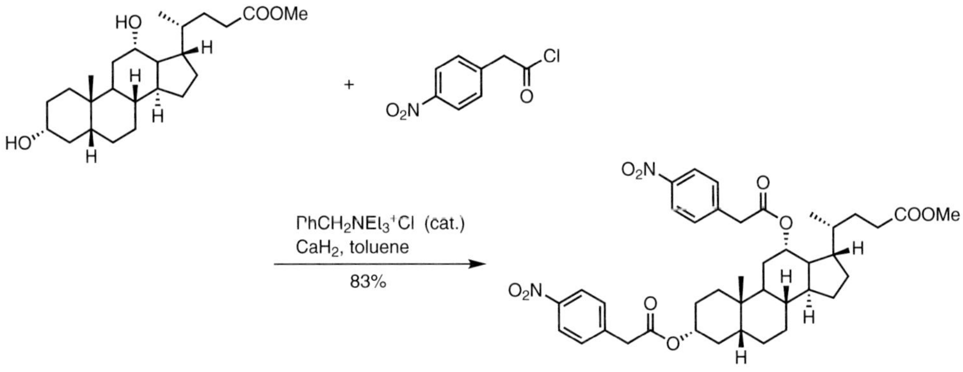

HO COOMe H H H HO H + O2N Cl O

PhCH2NEt3+Cl (cat.) CaH2, toluene 83%

O2N O O COOMe H H H O2N O O H

Scheme 7.39

OH O HO SEt BnO OBn + BzCl

NaOH, Bu4NHSO4 (cat.) CH2Cl2, H2O 87%

OBz O HO SEt BnO OBn

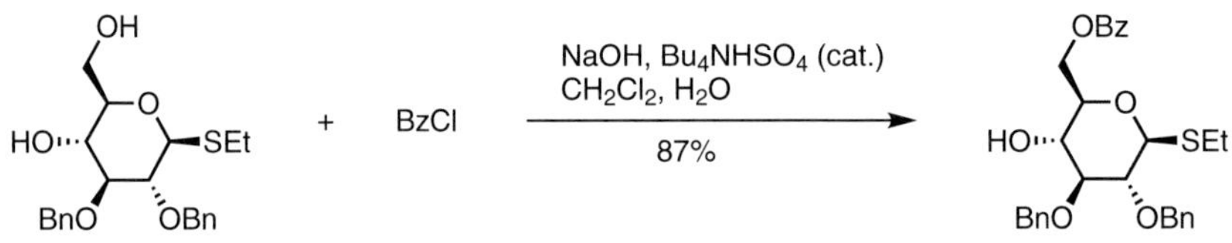

Scheme 7.40

This technology is also applicable to transesterification. Introduction of a gaseous mixture of the reactant ester and an alcohol into a column packed with solid potassium carbonate coated with a phase-transfer catalyst (Carbowax 6000 or 18-

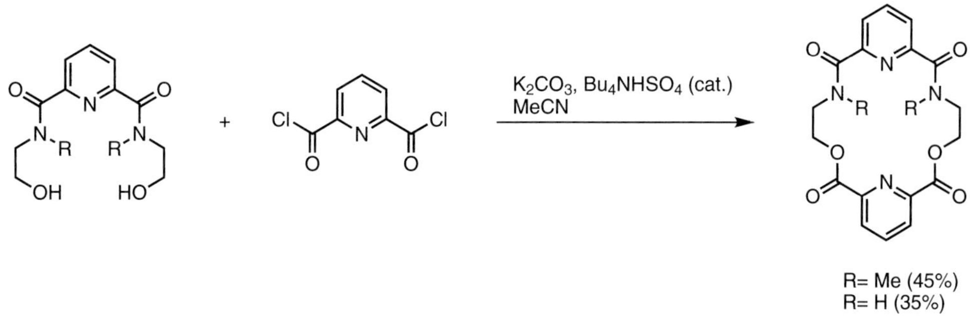

K_2CO_3, Bu_4NHSO_4 (cat.)
MeCN

R= Me (45%)
R= H (35%)

Scheme 7.41

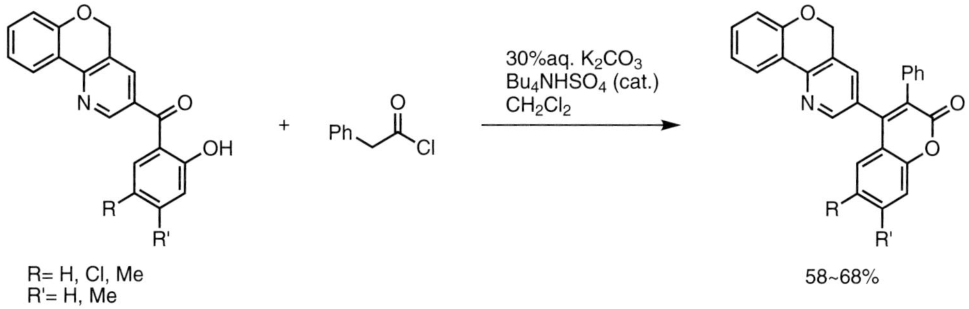

30%aq. K_2CO_3
Bu_4NHSO_4 (cat.)
CH_2Cl_2

R= H, Cl, Me
R'= H, Me

58~68%

Scheme 7.42

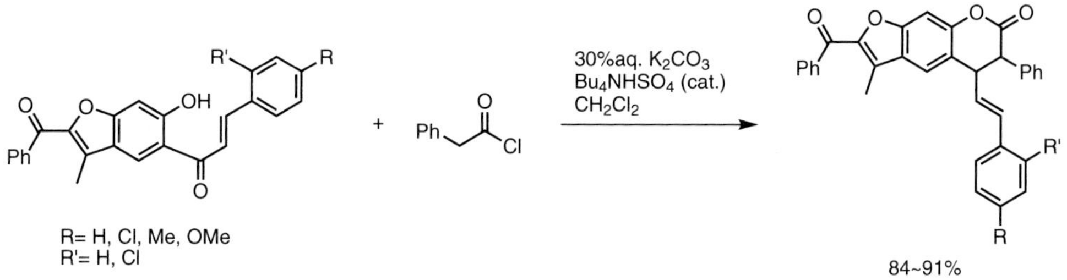

30%aq. K_2CO_3
Bu_4NHSO_4 (cat.)
CH_2Cl_2

R= H, Cl, Me, OMe
R'= H, Cl

84~91%

Scheme 7.43

crown-6) induces transesterification (Scheme 7.44) [398]. The yield of the esters, however, is modest (up to 65%). Transesterification of sugars under phase-transfer conditions is assisted by microwave irradiation (Scheme 7.45) [1021]. The primary and secondary hydroxy groups undergo acylation with methyl benzoate or laurate.

$$R^1COOR^2{}_{gas} + R^3OH_{gas} \xrightarrow[\text{(Carbowax 600 or 18-crown-6)}]{\text{PT catalyst}} R^1COOR^3{}_{gas} + R^2OH_{gas}$$

Scheme 7.44

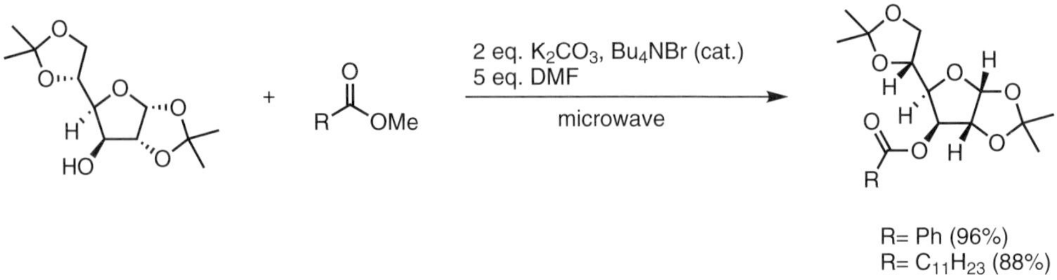

2 eq. K_2CO_3, Bu_4NBr (cat.)
5 eq. DMF
microwave

R= Ph (96%)
R= $C_{11}H_{23}$ (88%)

Scheme 7.45

p-Dodecylbenzenesulfonic acid acts as a surfactant-type Brønsted acid catalyst to permit direct esterification in water (Scheme 7.46) [1022, 1023]. In water, the surfactant-type catalyst and organic substrates form droplets with hydrophobic interior. The surfactants concentrate the catalytic species (such as a proton) onto the droplets' surfaces, where the reaction takes place, and then enhance the rate to reach equilibrium. For lipophilic substrates, the equilibrium position between the substrates and the products (esters) lies at the ester side, because water molecules are expelled out of the droplets because of the hydrophobic nature of their interior. As a result, quantitative yields of esters are obtained. This technique is less effective for transesterification [1024].

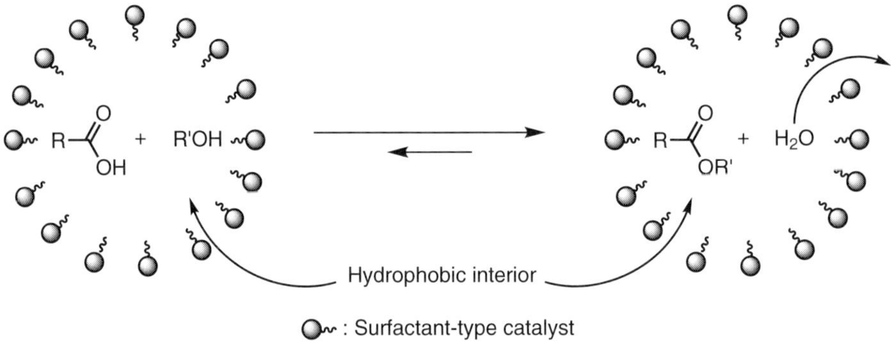

Hydrophobic interior
: Surfactant-type catalyst

Scheme 7.46

Fluorous biphase technology is useful for (trans)esterification. It has already been mentioned that fluoroalkyldistannoxanes are effective catalysts for both esterification and transesterification in various aspects, yet some further characteristic features in terms of fluorous chemistry deserve to be noted here. Transesterification is most conveniently performed simply by heating the ester and alcohol reactants in the presence of the catalyst in FC-72 (Scheme 7.47) [352, 353]. Reaction at 150 °C is crucial to make the reaction mixture homogenous. Usually the products are extracted into toluene after the reaction, but it is also possible instead, when the reaction is conducted on a large scale, to separate the products directly from the surface of the FC-72 layer. The yield is always quantitative. Although the catalyst can be recovered from the FC-72 solution, this solution itself can be used straightforwardly for the next reaction without isolation of the catalyst, no deterioration in the catalytic activity being observed after 20 recycles. Of course, the FC-72/toluene or alcohol binary solvent systems are employable. In the former case, use of a slight excess (~1.3 equiv.) of alcohol is required for >99% yield, while the reaction proceeds rapidly in the latter case. Reaction in toluene is also feasible, since the catalyst is soluble in hot toluene. The catalyst can be recovered by extraction with FC-72.

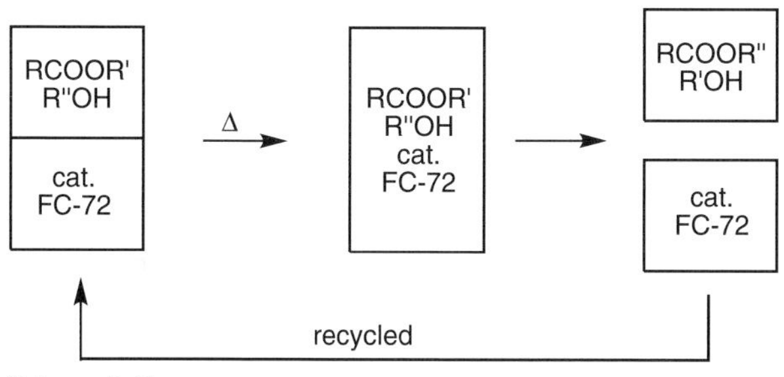

Scheme 7.47

Esterification is carried out similarly (Scheme 7.48) [54]. After the reaction, easily separable organic, aqueous, and fluorous phases result. The recovered catalyst is not the same as the original one, but modified in some way (probably carboxylated). However, this new form is also sufficiently active and so the catalyst solution in FC-72 can be reused repeatedly.

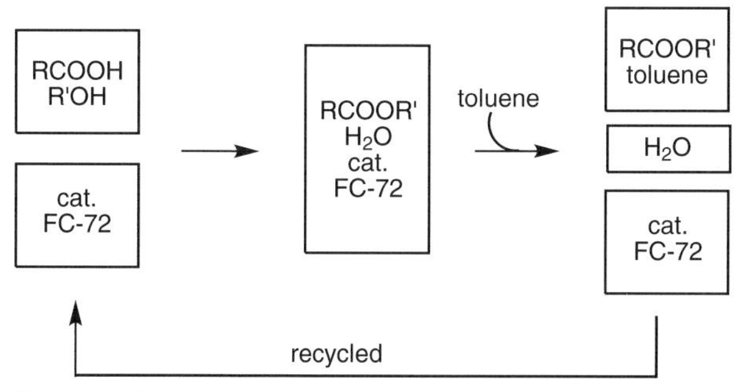

Scheme 7.48

Scandium and ytterbium tris(perfluorooctanesulfonyl)methide complexes $M[C(SO_2C_8F_{17})_3]_3$ (M = Sc, Yb) can be used for acetylation of cyclohexanol in both homogeneous and heterogeneous phases [517, 518]. A 1 : 1 mixture of cyclohexanol and acetic anhydride, together with the catalyst (1 mol%) in perfluoromethylcyclohexane (5 parts), toluene (5 parts) and perfluorobenzene (3 parts) constitutes a homogeneous phase at 40 °C. The reaction is complete in 15 min, and the reaction mixture then turns into two layers upon standing at 15 °C. Cyclohexyl acetate is obtained in quantitative yield from the upper layer (19.9% $CF_3C_6H_{11}$, 52.5%

toluene, 27.6% C_6F_6), while the catalyst is completely recovered from the lower layer (69.9% $CF_3C_6H_{11}$, 13.8% toluene, 16.3% C_6F_6). The same reaction is feasible in the heterogenous two-phase system made up of perfluoromethylcyclohexane and toluene in a 1 : 1 ratio. $M[N(SO_2C_8F_{17})_2]_4$ (M = Hf, Sn) are also efficient catalysts for fluorous biphasic esterification and transesterification [1025, 1026]. The hafnium catalyst can be employed for synthesis of methyl methacrylate directly from methacrylic acid and methanol [1027]. The fluorous biphasic esterification is also effected by fluorous ammonium triflate [1028].

The temperature-dependent solubility of a fluorous super Brønsted acid is utilized in acylation and esterification without fluorous solvents (Scheme 7.49) [1029]. 4-(1*H*,1*H*-Perfluorotetradecanoxy)-2,3,5,6-tetrafluorophenylbis(trifluoromethanesulfonyl)-methane is soluble in toluene at 70 °C but not soluble at room temperature. Thus, when a mixture of the reactants and catalyst in toluene is heated, a homogeneous solution is produced, in which the reaction takes place. After completion of the reaction, cooling of the reaction mixture to room temperature causes precipitation of the catalyst, which is separated by filtration. With the aid of this technology it is possible to conduct the benzoylation of methanol with benzoic anhydride and esterification of 3-phenylpropanoic acid in methanol are conducted to produce the esters quantitatively.

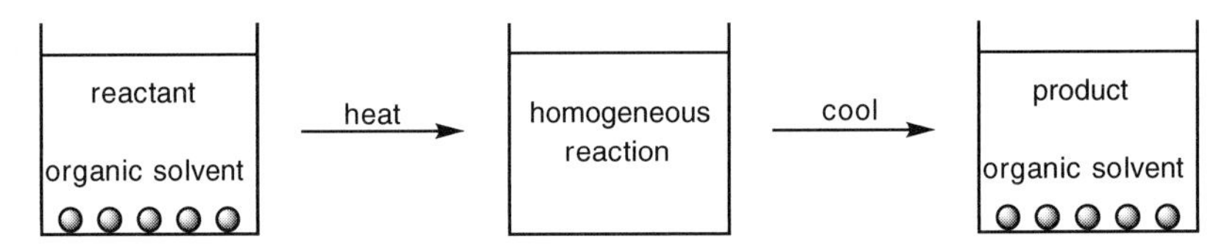

○ : perfluoroalkyl catalyst
recycle of catalyst by decantation

Scheme 7.49

The Mitsunobu reaction can be carried out under fluorous conditions. Bisfluoroalkyl azodicarboxylate effects esterification of benzoic acid (Scheme 7.50) [210]. After the reaction, solvent is evaporated and the residue is partitioned between dichloromethane and FC-72. The fluorous hydrazine co-product is not detected in the organic layer, thus leaving only triphenylphosphine oxide to be separated from the desired product. The combined use of fluorous phosphines $(C_6F_{13}C_2H_4C_6H_4)_2PPh$ or $C_8F_{17}C_2H_4C_6H_4PPh_2$ and fluorous azodicarboxylate allows easier separation of the product ester (Scheme 7.51) [211]. After the reaction, the solvent is evaporated, and the residue is taken up in MeOH and loaded onto fluorous silica. Elution with 80% MeOH gives the Mitsunobu adduct in 75–95% yield, free of any impurities. Further elution with ether gives a mixture of the fluorous phosphine oxide and fluorous hydrazine. This fluorous mixture can readily be separated on normal silica gel. These recovered fluorous reagents can be converted into the original fluorous phosphine and fluorous azocarboxylate pure enough to be reused. Also importantly, this technique can be applied in parallel reactions.

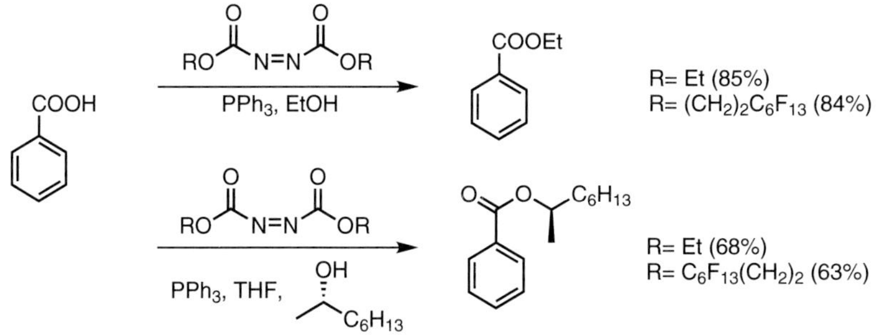

Scheme 7.50

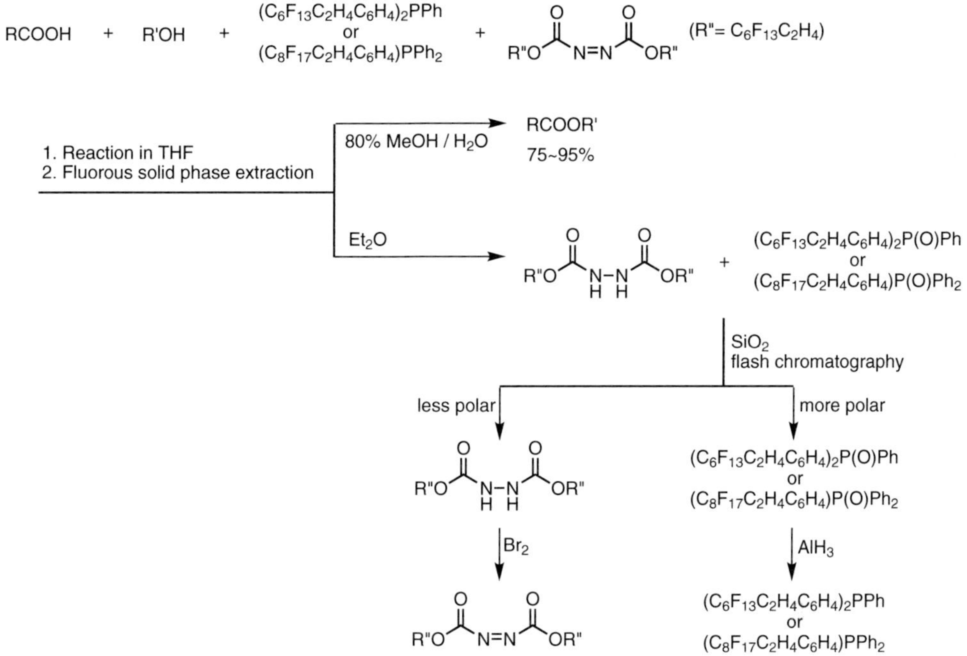

Scheme 7.51

The light fluorous technology is employable for esterification (Scheme 7.52) [1030]. Various carboxylic acids smoothly react with alcohols with the aid of a fluorous reagent, *N*-fluoroalkyl-2-choropyridinium triflate. The resulting fluorous pyridone can be readily separated by column chromatography on fluorous silica gel.

Kinetic resolution of racemic esters or carboxylic acids is conveniently achieved with the aid of fluorous biphase technology (Scheme 7.53) [1031]. A mixture of a hexane solution containing racemic ester or carboxylic acid and a perfluorohexane solution containing highly fluorinated decanol is heated at 40 °C with lipase from

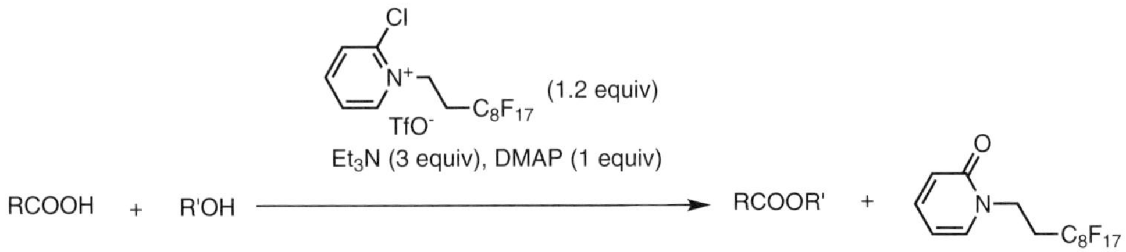

$$RCOOH + R'OH \xrightarrow[\text{Et}_3\text{N (3 equiv), DMAP (1 equiv)}]{\text{2-chloro-1-}(C_8F_{17}CH_2CH_2)\text{pyridinium TfO}^- \text{ (1.2 equiv)}} RCOOR' + \text{1-}(C_8F_{17}CH_2CH_2)\text{-2-pyridone}$$

Scheme 7.52

Candida rugosa. The reaction mixture is homogeneous at this stage. After removal of the lipase by filtration, the liquid phases are cooled to 0 °C, resulting in separation of the two solutions. From the fluorous solution the *S* fluoroalkyl esters are recovered in ca. 50% conversion with 72–95% *ee*. The unreacted esters or carboxylic acids recovered from the organic layer show 44–95% *ee*. In contrast to the above procedure, in which carboxylic acids are resolved, secondary alcohols can also be resolved with fluorous acyl donor (Scheme 7.54) [1032]. Reaction between vinyl cinnamate and benzyl alcohol with catalysis by PEG-lipase PL complex is accelerated when performed in perfluorooctane/isooctane [1033].

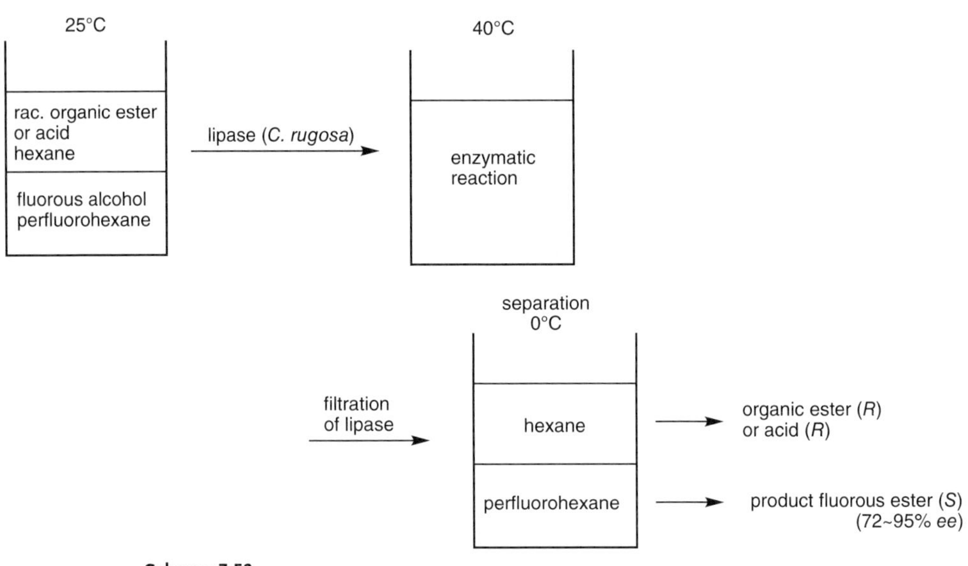

Scheme 7.53

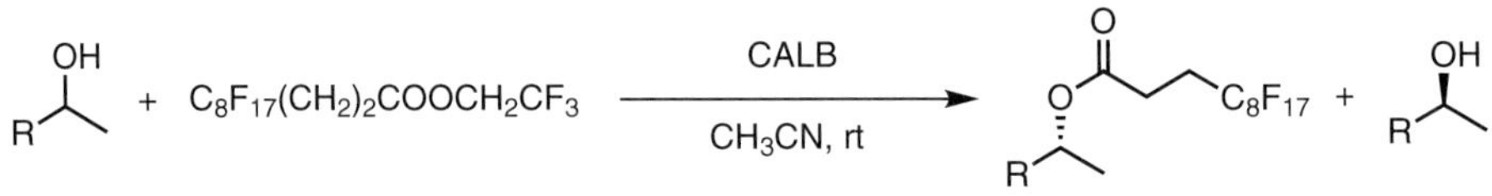

$$RCH(OH)CH_3 + C_8F_{17}(CH_2)_2COOCH_2CF_3 \xrightarrow[\text{CH}_3\text{CN, rt}]{\text{CALB}} RCH(CH_3)OCO(CH_2)_2C_8F_{17} + RCH(OH)CH_3$$

Scheme 7.54

Ionic liquids are attracting growing interest as alternative reaction media for a wide variety of synthetic processes because they are nonvolatile, thermally stable, and environmentally benign. A Brønsted acidic ionic liquid, 1-methylimidazolium tetrafluoroborate, $[HMIm]^+ BF_4^-$ catalyzes the condensation reaction between carboxylic acids and alcohols, affording the desired esters in high yields (Scheme 7.55) [1011].

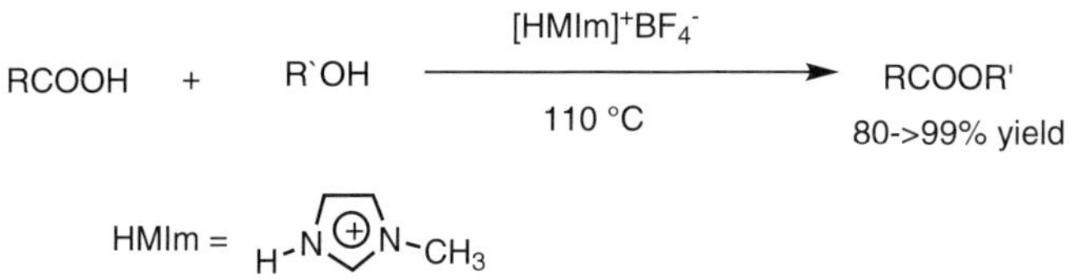

RCOOH + R'OH —[HMIm]$^+$BF$_4^-$, 110 °C→ RCOOR' (80->99% yield)

HMIm = H-N⊕N-CH$_3$ (imidazolium ring)

Scheme 7.55

The *O*-acetylation of alcohols and carbohydrates with acetic anhydride takes place smoothly in the ethylmethyl- and butylmethylimidazolium dicyanamides shown below [1034]. All hydroxy groups in saccharides such as glucoses, sucrose, and so on are acetylated. Notably, the dicyanamides function both as solvent and as catalyst.

[1-ethyl-3-methylimidazolium][NC-N$^-$-CN]   [1-butyl-3-methylimidazolium][NC-N$^-$-CN]

The DCC/DMAP technique for esterification between ferrocenemonocarboxylic acid and various phenols can be conducted in 1-butyl-3-methyl derivative, $[BMIm]^+ BF_4^-$, [1035]. Condensation between carboxylic acids and alcohols in a 1 : 1 mole ratio catalyzed by *p*-toluene sulfonic acid takes place smoothly in imidazolium ionic liquids [1012, 1013].

Acetylation of tertiary alcohols with acetic anhydride occurs without activator in $[BMIm]^+ BF_4^-$ [1036], and hemicelluloses are acetylated with catalysis by I_2 in $[BMIm]^+Cl^-$ [1037]. Reaction between sodium salts of carboxylic acids and alkyl halides occurs in high yield in $[MMIm]^+ CH_3SO_3^-$ or $[BMIm]^+ CH_3SO_3^-$ [1038]. Carboxylic acids react with alkyl halides in the presence of KF in $[BMIm]^+ PF_6^-$ [1039]. Triethyl orthoacetate is employable for alkylation of carboxylic acids (Scheme 7.56) [1040, 1041]. Of particular use is successful alkylation of sterically hindered carboxylic acids such as 1,3,5-triisopropylbenzoic acid.

2,4,6-triisopropylbenzoic acid (CO$_2$H) —CH$_3$C(OEt)$_3$, [BMIm]$^+$PF$_6^-$, 80 °C→ ethyl ester (CO$_2$Et)

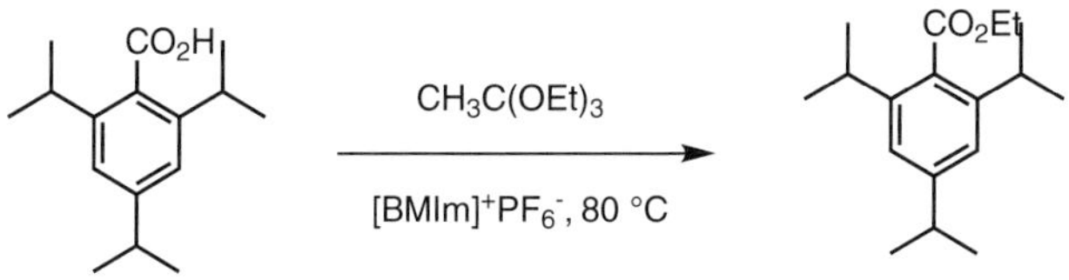

Scheme 7.56

Besides imidazolium salts, other ionic liquids are also employable. Carboxylic acids are esterified with alkyl halides in $C_{14}H_{29}(C_6H_{13})_3P^+(CF_3SO_2)_2N^-$ with the aid of Hunig's base [1042]. With a choline chloride/zinc chloride complex ionic liquid $[HOCH_2CH_2N(CH_3)_3Cl][ZnCl_2]$, monosaccharides and cellulose are acetylated with acetic anhydride [1043], and condensation between carboxylic acids and alcohols both having a long alkyl chain furnishes wax esters [1044].

As briefly described in Section 1.1.2.1, ionic liquids bearing an alkanesulfonic acid group act as dual solvent/catalysts. Brønsted acidic ionic liquids as shown below exhibit higher catalytic activity than conventional ones for reaction between carboxylic acids and alcohols [24, 25]. HSO_4^- is employable as a counter anion [1045], and pyridinium salts are also available [1046].

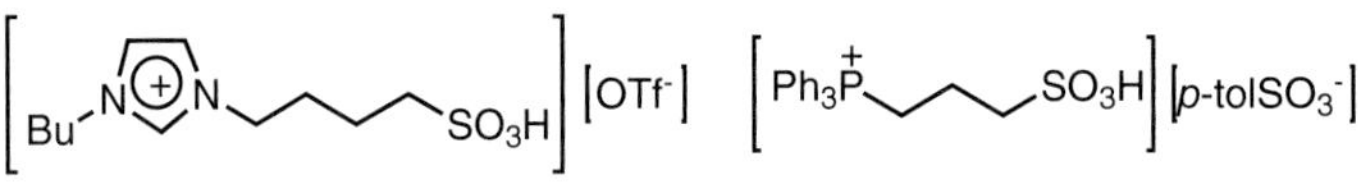

[OTf⁻] [p-tolSO₃⁻]

Enzymatic reactions in ionic liquids are also feasible. Lipase-catalyzed kinetic resolution of secondary alcohols in ethylmethyl- or butylmethylimidazolium ionic liquids can be up to 25 times more enantioselective than in organic solvents [475]. Lipase in ionic liquids exhibits catalytic activities comparable with those in organic solvents for acetylation of 2-hydroxymethyl-1,4-benzodioxane with vinyl acetate [1047]. In these reactions, the reaction mixture is usually worked up with organic solvents for purification of the products, although this defeats the original purpose of the use of ionic liquids. To bypass this problem, the extraction technique with supercritical carbon dioxide is available, both in batchwise and continuous flow processes [476]. PEG-modification of lipase enhances activity for acylation of benzyl alcohol with vinyl cinnamate [1048]. L-Ascorbic acid, which is not easily esterified on account of its poor solubility in organic solvents, undergoes CALB-catalyzed esterification with fatty acids in 1-*sec*-butyl-3-methylimidazolium tetrafluoroborate to give soluble esters [1049]. Glucose esters are obtained with the aid of CALB or PEG-modified CALB in $[BMIm][BF_4]$ containing 40% *tert*-butyl alcohol (Scheme 7.57) [1050]. Lipase-catalyzed regioselective esterification of 1-β-D-arabinofuranosylcytosine is feasible in ionic liquids with various alkyl chains (Scheme 7.58) [1051]. The highest yield is obtained in mixed solvents, 10% (v/v) $[BMIm][PF_6]$/THF.

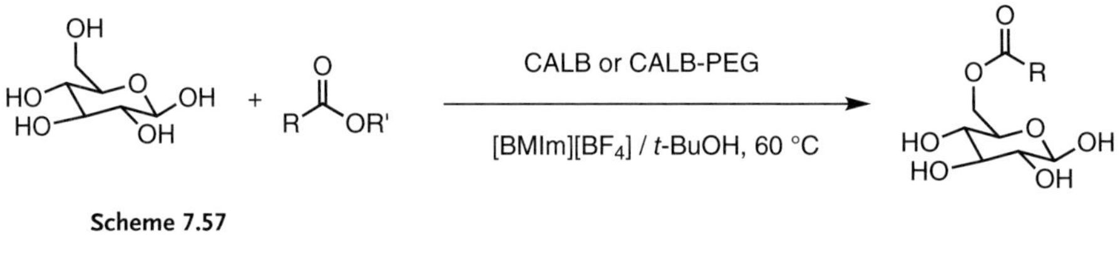

CALB or CALB-PEG

$[BMIm][BF_4]$ / *t*-BuOH, 60 °C

Scheme 7.57

α-Chymotrysin-catalyzed transesterification of *N*-acetyl-L-tyrosine with 1-propanol [1052] and protease-catalyzed esterification of amino acids also take place smoothly in ionic liquids [1053].

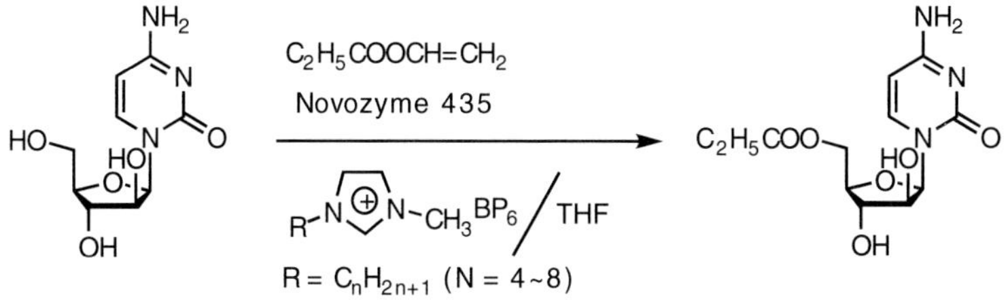

Scheme 7.58

The rate of lipase-catalyzed kinetic resolution of secondary alcohols is enhanced in [BMIm][PF_6] or [OMIm][PF_6], for example, the reaction of 2-substituted propanoic acid with 1-butanol by *Candida rugosa* [1054] and the reaction of benzyl alcohols with succinic anhydride by *Pseudomonas cepacia* [1055]. Kinetic resolution of ibuprofen is feasible in imidazolium ionic liquid/isooctane [1056, 1057].

Supercritical carbon dioxide ($scCO_2$) receives increasing attention as a green solvent [1058–1060]. Enantioselectivity of lipase-catalyzed esterification of 1-(*p*-chlorophenyl)-2,2,2-trifluoroethanol with vinyl acetate in $scCO_2$ changes the *E* value from 10 to 50 by decreasing the CO_2 pressure (Scheme 7.59) [1061, 1062]. Water activity has a very strong effect on Novozym-catalyzed esterification of geraniol with acetic acid in $scCO_2$ [1063]. Novozym 435 immobilized on mesoporous silica is more active than Amberlist 15 for methanolysis of oleic acid in $scCO_2$ [1064].

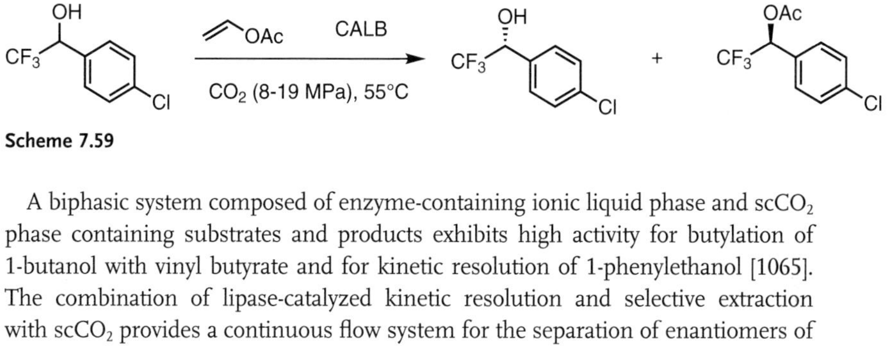

Scheme 7.59

A biphasic system composed of enzyme-containing ionic liquid phase and $scCO_2$ phase containing substrates and products exhibits high activity for butylation of 1-butanol with vinyl butyrate and for kinetic resolution of 1-phenylethanol [1065]. The combination of lipase-catalyzed kinetic resolution and selective extraction with $scCO_2$ provides a continuous flow system for the separation of enantiomers of esters and alcohols [1066]. The ionic liquid in this process can be replaced by poly(ethyleneglycol) [1067]. The use of a continuous-flow process for kinetic resolution of racemic alcohols with vinyl acetate with immobilized lipase improves the productivity of the optically active compounds over 400-fold compared to the corresponding batch reaction under $scCO_2$ [1068]. Scrutiny of lipase-catalyzed esterification of isoamyl alcohol with various acyl donors showed that initial reaction rate is faster in $scCO_2$ than in hexane, but final conversion is similar in both media [1069].

$scCO_2$ can be employed for esterification of long-chain fatty acids with alcohols [1070–1074]. Monoterpene lavandulol is acetylated with acetic acid with catalysis by CALB in $scCO_2$ [1075].

Lipase-catalyzed transesterification of methyl methacrylate and *N*-acetyl-L-phenylalanine with various alcohols takes place smoothly in supercritical fluoroform

[1076, 1077]. This medium can also be employed for kinetic resolution of phenethyl alcohol with lauric acid with the aid of lipid-coated lipase [1078] and transglycosylation with the aid of lipase-coated β-D-galactosidase (Scheme 7.60) [1079].

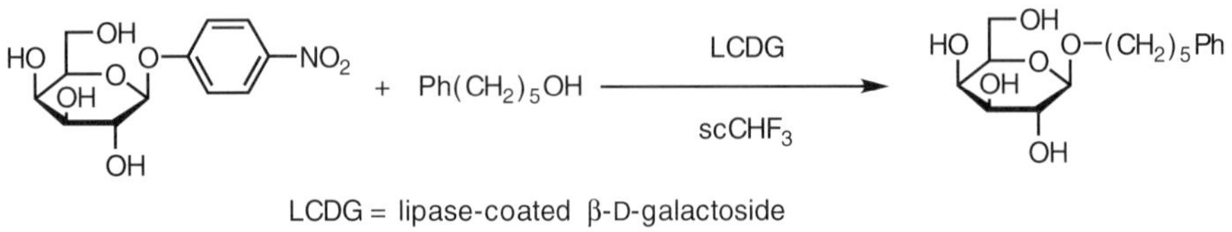

LCDG = lipase-coated β-D-galactoside

Scheme 7.60

Sub- or supercritical propane is also a medium of choice for lipase-catalyzed reaction between butyric acid and alcohols [1080, 1081]. For Novozym 435-catalyzed reaction between palmitic acid and octanol, supercritical methane is better than $scCO_2$ [1082]. Supercritical methanol is also used for transformation of rapeseed oil (Scheme 1.67) [323, 324]. Methyl esters are obtained without catalyst.

$scCO_2$ is also employable for nonenzymatic reactions. Since fluorous compounds are usually soluble in CO_2, acylation of cyclohexanol with acetic anhydride in the presence of $M[C(SO_2C_8F_{17})_3]_4$ and $M[N(SO_2C_8F_{17})_2]_4$ catalysts (M = Sc, Yb) takes place smoothly in $scCO_2$ [1083]. Continuous esterification of ethylhexanoic acid with 2-ethyl-1-hexanol is feasible with catalysis by Amberlist 15 in $scCO_2$ [1084]. An L-lactic acid oligomer is converted into high polymers on treatment with DCC/DMAP in $scCO_2$ [1085].

Finally, immobilization of catalyst, though not categorizable as a modification of reaction media in a strict sense, should be briefly mentioned. This technology can offset many drawbacks encountered in homogeneous reaction and is discussed in Part One, but in a number of different places. The studies relevant to immobilization of catalysts are therefore compiled here.

Brønsted acid: [17] (see Section 1.1.2.1); [501] (see Section 1.3.2.1).
Carbodiimide: [192] (see Section 1.1.4); [193] (see Section 1.1.4).
Sulfonyl chloride: [244] (see Section 1.1.6).
Lipase: [460] (see Section 1.2.5); [473] (see Section 1.2.5); [474] (see Section 1.2.5); [477] (see Section 1.2.5)

It should be noted that asymmetric ring opening of *meso* anhydrides occurs without solvent. That is, ball-milling of a mixture of anhydride and alcohol provides a chiral alcohol in modest to fair enantiomeric excesses (13–63% *ee*) (Scheme 7.61) [1086].

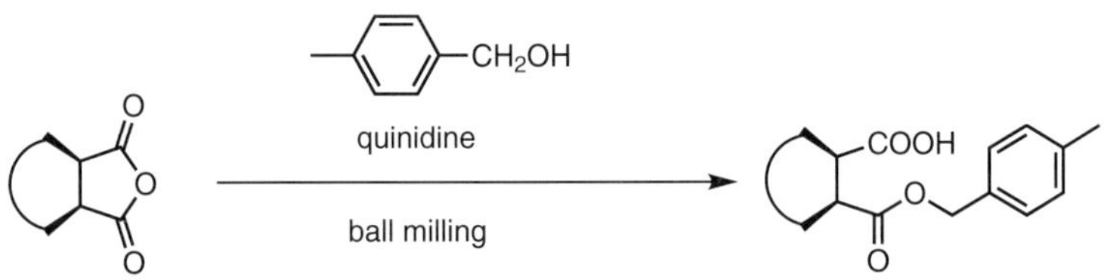

Scheme 7.61

7.4 New Technologies

Refinement of chemical processes is primarily performed in terms of chemistry itself such as design of synthetic routes, optimization of reaction conditions, etc., but engineering also can contribute to a great extent. Use of microwave irradiation is now one of the most popular means to increase reaction rate and chemical yields. A considerable number of applications of this technology to esterification are known. Brønsted acid-catalyzed reaction between carboxylic acids and alcohols occurs smoothly under microwave irradiation [1087, 1088]. Esterification catalyzed by Lewis acids such as $Zn(OTf)_2$ and $NiCl_2$ undergoes rate acceleration [45, 1089]. The microwave irradiation is effective for Lewis acids supported on silica or alumina [1090], and also for solid catalysts such as montmorillonite [1091], zeolite [1092], and ion-exchange resin [1093].

The microwave acceleration is experienced in various types of esterifications with base catalysts: DBU [1094], diisopropylcarbodiimide [1095], DMAP [1096], imidazole [1097], pyridine [1098], alkyl carbochloridate [1099], supported 2-chloro-1-methylpyridinium iodide [1100], and CsF/Celite [1101]. Combination of the microwave irradiation and Brønsted ionic liquid is useful for esterification of propanoic acid with *neo*-pentanol [1102]. Reactions between carboxylic acids and $RC(OR')_3$ proceed without catalyst under microwave irradiation [1103]. Enzymatic esterification of carboxylic acids with alcohols is also accelerated [1104, 1105]. Large-scale esterification under microwave irradiation is feasible in batch systems (500 mL or 3 mol) [1106, 1107], or in a continuous-flow reactor [796].

Sonication is another useful method for accelerating chemical reactions. Acetylation of phenols with acetyl chloride and $InCl_3$-catalyzed methanolysis of carboxylic acids are accelerated under ultrasound irradiation [47, 1108]. Acetylation of alcohols with acetic anhydride in imidazolium ionic liquids also experiences increase in reaction rate and chemical yield by sonication [1109]. Enzymatic reactions go similarly. The sonication enhances reaction rate and enantioselectivity significantly in lipase-catalyzed reactions such as kinetic resolution of vicinal azido alcohols [1110] and acylation of glucose with vinyl esters [1111].

Reactive distillation, by which reaction and distillative separation of products are simultaneously executed in one unit operation, serves for high conversion of equilibrium reactions. Despite being known since 1920s, this technology has found many applications to esterification only recently. Although this technology is of practical importance, its detailed description is beyond the scope of this book because it is totally a matter of engineering. The followings are representative literatures: [1112–1121].

Pervaporation is another technology effective for shifting equilibrium [1122, 1123]. Two or more components are separated by a thin polymeric membrane through which the diffusion rates are different, and this is followed by removal of the permeated component by evaporation, usually *in vacuo*. The water formed in an esterification reaction between carboxylic acid and alcohol can be removed selectively in high conversions. For example, a tubular water-permeable membrane is set in a distillation column connected to a reaction flask [1124, 1436]. A solution of oleic acid and an excess amount of ethanol in the presence of a catalytic amount of *p*-TsOH is heated under reflux. From the vaporizing components, only water permeates the membrane, so the reaction is driven to high conversion. Water-containing lactic acid is esterified with ethanol similarly [1126].

Unification of reaction and separation procedures in the same unit is more convenient. The membrane is set in the bottom of the reactor [1127, 1128]. As the reaction proceeds, the water is separated in a liquid phase in this case. A tubular membrane can be inserted to the reactor [1129]. Lipase-catalyzed esterification is feasible in the same manner [1130, 1131]. Ionic liquids are employable as solvents in this system [1132]. The ultimate goal is accessed by hybridization of the catalyst and membrane. Various catalysts can be coated on the membrane surface: zeolite [1125, 1133–1135]; sulfonated polystyrene [1136]; Amberlist 15 [1137, 1138].

The reaction/separation hybrid is also achievable with chromatographic reactors, in which a reactive column acts as both catalyst and separator. A column packed with Amberlist 15 is effective for reaction between glycerine and acetic acid [1139] and between lactic acid and ethanol [1119].

Microreactors can be used for esterification of carboxylic acids with alcohols [1140–1142].

7.5
Application to Natural Products Synthesis

There are numerous reports of natural products synthesis involving esterification. These are listed in Table 7.4 in the order of the sections in this book. Since it is not possible to describe all of them here, only the representative examples are given.

The synthesis of taxol and its analogs exemplifies how profoundly esterification contributes to this field. Although several total syntheses of taxol have been reported, the semi-synthesis starting from 10-deacetyl baccatin III is really practical.

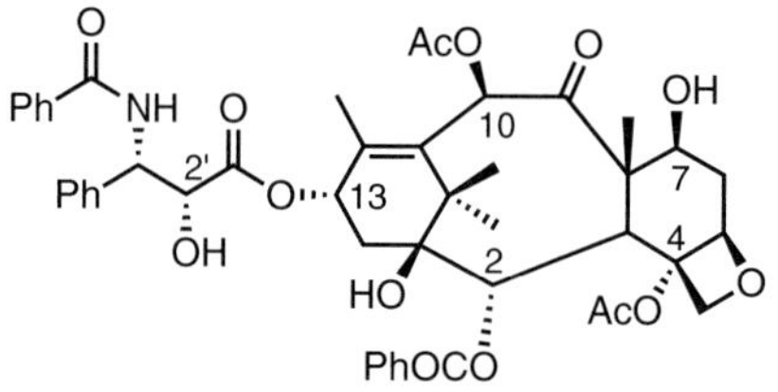

Taxol

Table 7.4 Natural products synthesis through esterification.

Method	References	Conditions	Target or related products
Reaction of Alcohols with Carboxylic Acids			
1.1.1	[1143]	At 190 °C	Quadrone
	[1144]	Reflux	Whisky lactone, Eldanolide
1.1.2.1	[1145]	TsOH	Parasorbic acid, Tridecanol acetate
	[1146]	HCl	(–)-Horsfiline
	[1147]	H_2SO_4	Pachylactone
	[1148]	TsOH	Anthridic acid
	[1149]	TsOH	(–)-Reserpine
1.1.4[a]	[1150]		Dihydromevinolin
	[1151]		Taxol
	[1152]		(–)-Valilactone
	[1153]		Antrimycin
	[1154]		Byssochlamic acid
	[1155]		Pyrrolizidine Alkaloid
	[1156]		Taxol
	[1157]		Pyrrolizidine Alkaloid
	[1158]		Taxol
	[1159]		Pyrrolizidine Alkaloid
	[1160]		Pyrrolizidine Alkaloid
	[1161]		Taxol
	[1162]		(±)-Myrocin
	[1163]		Taxol
	[1164]		Taxol
	[1165]		Taxol

Table 7.4 *Continued.*

Method	References	Conditions	Target or related products
	[1166]		Taxol
	[1167]		Taxol
	[1168]		Taxol
	[1169]		Taxol
	[1170]		Taxol
	[1171]		Taxol
	[1172]		Taxol
	[1173]		Taxol
	[1174]		Enterochelin
	[1175]		Dolabellin
	[1176]		(–)-Sypringolides
	[1177]		Taxol
	[1178]		Taxol
	[1179]		(–)-Virginiamycin
	[1180]		Maitotoxin
	[1181]		Cryptophycin
	[1182]		Tarchonanthuslactone
	[1183]		Taxol
	[1184]	EDC	Octalactin
	[1185]		Atractyligenin
	[1186]	EDC	Epothilone
	[1187]		Aerothionin
	[1188]		Taxol
	[1189]		Taxol

Table 7.4 *Continued.*

Method	References	Conditions	Target or related products
	[1190]	DIC	Taurospongin
	[1191]	EDC	Diazonamide A
	[1192]	EDC	Taxol
	[1193]	EDC	2-Ara-Gl (2-arachidonylglycerol)
1.1.5[b]	[1194]		Samanine
	[1195]		Combretastatin D
	[1196]	PBu_3, DEAD	Echinosporin
	[1197]		Latrunclin A and B
	[1144]		Whisky lactone, Eldanolide
	[1198]		Lipstatin
	[1199]		Pyrenophorin
	[1200]		Suspensolide
	[1201]		Combretastatin D
	[1202]		Geodiamolide A
	[1203]		Combretastatin D
	[1204]		(−)-Vermiculine
	[1205]		Nosiheptide
	[1206]		(−)-Lipstatin
	[1207]		Deoxymannojirimycin
	[1208]		Paniculide A
	[1209]	PPh_3, DIAD	Panclicin D
	[1210]	PPh_3, DIAD	(+)-Acetylphomalactone
	[1211]		Styllactone
	[1212]		Hastanecine

Table 7.4 *Continued.*

Method	References	Conditions	Target or related products
	[1213]		10-Oxo-11(*E*)-octadecen-13-diole
	[1214]		Combretastatin D
	[1215]		Macrolactins
	[1216]		Actinomycin Z_1
	[1217]		Lipoxin A
	[1218]		Dideemnin M
	[1219]		Solandelactones
	[1220]		Pyrrolizidine alkaloid
	[1221]	PPh_3, DIAD	UK-2A
	[1222]		Patulolide C
	[1223]	PPh_3, DIAD	Lobatamide C
	[1224]	PPh_3, DEAD	(+)-Leucascandrolide A
1.1.6	[1225]	2-Cl-1-Me-pyridinium iodide	(±)-Recifeiolide
	[1226]	di(2-Py)carbonate	Taxol
	[1227]	*N*-Me-2-Cl-pyridinium iodide	(−)-α-Kainic acid
	[1228]	*N*-Me-2-Cl-pyridinium iodide	Murrayaquinone A
	[1229]	$PhSO_2Cl$	Bourgeanic acid
	[1230]	*p*-Br-benzenesulfonyl-chloride	Obafluorin
	[1231]	*N*-Me-2-Cl-pyridinium iodide	Balanol
	[1232]	*N*-Me-2-Cl-pyridinium iodide	9-Propy-10-azacyclododecan-12-olide
	[1233]	di(2-Py)carbonate	Taxol

Table 7.4 *Continued.*

Method	References	Conditions	Target or related products
	[1234]	*N*-Me-2-Cl pyridinium iodide	Balanol
	[1235]	*N*-Me-2-Cl-pyridinium iodide	Balanol
	[1236]	di(2-pyridyl) thiocarbonate	Taxol
	[1237]	*N*-Me-2-Cl-pyridinium iodide	Balanol
1.1.7	[1238]	lipase from *C. rugosa*	(4*E*, 7*S*)-7-Methoxytetradec-4-enoic acid
Reaction with Esters			
1.2.1	[1239]	At 0 °C	Chorismic acid
	[1240]	Heat	no name natural product
	[1241]	At 210 °C	(−)-Oblongolide
	[1242]	Light	(*R*)-(+)-Umbelactone
	[1243]	At 130 °C	(+)-Heptelidic acid
	[321]	At 90 °C	Taxol
1.2.2.1	[1244]	MesOH	(+)-Eldanolide
	[1245]	HCl	(6*R*)-(−)-massoialactone, (4*R*, 6*R*)-(+)-4-hydroxy-6-pentylvalerolactone
	[1246]	TsOH	(±)-Mintlactone
	[1247]	TsOH	(±)-(1β, 6α, 9β, 10α)-9-chloro-10-hydroxy-8-(methyloxycarbonyl)-4-methylene-2,5-dioxabicyclo-[4.4.0]dec-7-en-3-one
	[1248]	Camphorsulfonic acid	Cryptocaryalactone
	[1249]	HF	(±)-Muricatacin
	[1250]	H_2SO_4	(±)-Cuanzine
	[1251]	TsOH	Taxol

Table 7.4 *Continued.*

Method	References	Conditions	Target or related products
	[333]	HF	(+)-Compactin
	[1252]	HCl	Tetrodotoxin
	[1253]	H_2SO_4	Calanolide A
	[1254]	TsOH	9-Acetoxyfukinanolide
	[1255]	TFA	(±)-Calanolide A
	[1256]	TsOH	*cis*-Hydrindane
	[1257]	TFA	(+)-Dihydrokawain
	[1258]	HCl	(3αS, 6α*S*)-Ethisolide, Whisky lactone, (−)-Avenacioide
1.2.2.2	[1259]	$(SCNBu_2SnOSn\text{-}Bu_2OH)_2$	Brefeldin
	[1260]	$(SCNBu_2SnOSn\text{-}Bu_2OH)_2$	Kadsurenone-ginkgolide hybrid
	[1261]	$(SCNBu_2Sn)_2O$	Library of polycyclic small molecules for use in chemical genetic assays
	[1262]	$BF_3.\ OEt_2$	Mycalamide
	[1263]	$(Bu_2SnCl)_2O$	Taxol
	[1432]	$(SCNBu_2Sn)_2O$	Library of polyketide-like macrolides
	[1433]	$(SCNBu_2Sn)_2O$	Amphidinolide P
	[1434]	$(SCNBu_2Sn)_2O$	Spirotryprostatin B
1.2.2.3	[1264]	Al_2O_3	Clavepictines A and B
	[1265]	Dowex 50Wx4	Naphthylisoquinoline alkaloid
1.2.3.1	[1266]	K_2CO_3	Acetogenins (Solamin, Reficulatacin)
	[402]	MeLi/CuI	Bafilomycin A
	[401]	LiOH	Seiridin
	[1267]	K_2CO_3	Harringtonolide
	[1268]	KI	Methyl picrotoxate

Table 7.4 *Continued.*

Method	References	Conditions	Target or related products
1.2.3.2	[1269]	DBU	Goniotriol, 8-Acetyl goniotriol
	[1270]	DMAP	(–)-Ptilomycalin A
	[1271]	DMAP	(±)-Heptelidic acid
1.2.5	[1272]	PPL	(–)-Massoialactone, 3-hydroxy-5-icosanolide, 3-hydroxy-5-decanolide
	[1273]	PPL	(3*Z*, 6*Z*)-Dodecanolide
	[1274]	Lipase from C. cylindracea	(*Z*, *Z*)-2-Hydroxy-4-oxohenicosa-12,15-dien-1-yl acetate
	[1275]	Lipase from *C. antarctica*	(2*S*, 3*R*)-3-Hydroxyproline
	[1276]	PPL	(3*Z*)-Dodecen-12-olide, (2*E*)-9-Hydroxydecenoic acid
	[1277]	Lipase from *C. rugosa*	(*R*)-Patulolide A
	[838]	Lipase PS	(+)-Strigol
	[1278]	Lipase PS	(+)-Crooksidine
	[1279]	Lipase PS	Nitraria alkaloid
	[960]	PPL	CC-1065, Duocarmycin
	[1280]	Lipase from *C. antarctica*	Taxol
	[961]	Lipase from *P. fluorescens*	(–)-Curcumanolide A
	[971]	Lipase from *P. fluorescens*	Oosponol
	[1234]	Lipase from *P. sp.*	Balanol
	[1004]	Lipase from *C. antarctica*	Ecdysteriods
	[1281]	Lipase from *C. antarctica*	Triptoquinone B and C
	[1237]	Lipase from *P. sp.*	Balanol

Table 7.4 *Continued.*

Method	References	Conditions	Target or related products
	[1282]	Lipase PS	(±)-(3aα, 8aα)-ethyl 8β-hydroxy-6β-methyl-2-oxooctahydro-2*H*-cyclohepta[*b*]furan-3α-carboxylate
	[1283]	Lipase LIP	Rengyoxide
	[846]	Lipase PS	Taxol
	[1284]	Lipase LIP	(−)-Neplanocin A
	[1285]	Lipase PS	Sceletium alkaloid
	[866]	Lipase from *C. rugosa*	Lactaranes, Marasmanes
	[1286]	Lipase from *C. rugosa*	Macrolide antibiotic A26771B
	[1287]	Novozym 435	(±)-Azamacrolide
	[868]	Lipase OF-360	(+)-Albocanol, etc.
Reaction with acid anhydrides			
1.3.2.1	[1006]	TsOH	Gibberellin, A81
	[1288]	H_2SO_4	Longifolene
	[493]	AcOH	Secosterol
1.3.2.2	[991]	BF_3. OEt_2	XH-14 (5-(3-hydroxypropyl)-7-methoxy-(3′-methoxy-4′-hydroxyphenyl)benzo[*b*]furan-3-carbaldehyde)
	[1289]	$ZnCl_2$	Taxol
	[1290]	$Ln(OTf)_3$	Taxol
1.3.3.1	[1291]	BuLi	Lasubine I, Subcosine I
	[562]	NaOAc	BMY40662 (benzothiazinone)
	[568]	KOAc	Coriandrin
1.3.3.2	[1292]	DMAP, NEt_3	(+)-Compactin
	[1293]	DMAP	(−)-Anisatin
	[1294]	Py	Bretonin

Table 7.4 *Continued.*

Method	References	Conditions	Target or related products
	[1295]	Py	3β, 6α-Dihydroxy-9-oxo-9,11-seco-5α-cholest-7-en-11al
	[1296]	Py	Ventilagone
	[1269]	Py	Goniotriol, 8-Acetyl goniotriol
	[1297]	Py, DMAP	Myrtine, Lasubinem, Subcosine
	[985]	DME	Pyrrolizidine alkaloid
	[1298]	Py	Salvinolone
	[1299]	Py, DMAP	Gephyrotoxin
	[1300]	DMAP	Epoxypolyynes
	[1301]	DMAP	Taxol
	[1302]	DMAP, NEt_3	Taxol
	[1303]	NEt_3	Specionin
	[1304]	DMAP, NEt_3	(*Z*)-6,8-Nonadien-2-ol
	[1305]	Py	(±)-Epibatidine
	[1306]	Py, DMAP	Drimane sesquiterpenes
	[1307]	Py, DMAP	Kjellmanianone
	[1308]	Py	Azamacrolide
	[1309]	Py	6,7-Diepicastanospermine
	[1310]	DMAP	Tautomycin
	[1311]	Py	Goniodiol-8-monoacetate
	[1312]	DMAP	Batzelladin A
	[1313]	DMAP	Carabrone
	[1314]	DMAP, NEt_3	GERI-BP001
	[577]	NEt_3	Tropane alkaloid
	[1315]	DMAP	(−)-(5*R*, 6*S*)-6-Acetoxy-5-hexadecanolide

Table 7.4 *Continued.*

Method	References	Conditions	Target or related products
	[1316]	Py	Paraconic acid
	[1317]	Py	Naturally occrring flavones
	[1318]	Py, DMAP	Taxol
	[1319]	Py	Acetoxycrenulide
	[1320]	Py, DMAP	(–)-Galbonolide B
	[1321]	Py	Clavepictine A and B
	[1322]	Py	Biopterin
	[1323]	Py	(+)-Goniodiol
	[1324]	Py	Melodorinol, Acetylmelodorinol
	[1325]	Py, DMAP	Tonghaosu
	[1326]	DMAP, NEt_3	Conagenin
	[1327]	Py	Nagastatin
	[1328]	Py, DMAP	(+)-Asperlin, (+)-Acetylphomalactone, (+)-(5*S*, 6*S*, 7*R*, 8*S*)-Asperlin
	[961]	Py, DMAP	Curcumanolide A
	[1329]	Py	Microcolin B
	[1330]	DMAP, NEt_3	Secosyrin, Syributin
	[1331]	Py	Salsolene oxide
	[1332]	DMAP, NEt_3	Monensin
	[1333]	Py	(5*R*, 6*S*)-6-Acetoxy-5-hexadecanolide
	[981]	Py	Melodorinol
	[1334]	Py	Duocarmycin SA
	[1335]	DMAP, NEt_3	Halichondrin B
	[1336]	Py, DMAP	(±)-Swainsonine
	[1337]	Py, DMAP	Coniochaetone A and B

Table 7.4 *Continued.*

Method	References	Conditions	Target or related products
	[1338]	DMAP, NEt_3	(+)-Phomalactone, (+)-Acetylphomalactone, Asperlin
	[1339]	Py, DMAP	Prevercynamin
	[1263]	Py, DMAP	Taxol
1.3.5[c]	[732]		Pyrrolizidine alkaloid
	[1340]		Rutamycin B
	[1308]		Epliachnene
	[1341]	ClCOOEt	Camptothecin
	[626]	$ClCOOCMe{=}CH_2$	Leualacin
	[1342]		Luffariolide *E*
	[1343]		Luffariolide *E*
	[1276]		(3*Z*)-Dodecen-12-olide, (2*E*)-9-Hydroxydecenoic acid
	[1344]		(–)-Colletol
	[1345]		Dolabellatrienone
	[1346]		(+)-Ikarugamycin
	[1347]		(+)-Laurencin
	[1348]		Epothilone A and B
	[1349]		Bafilomycin A_1
	[1350]		6-Deoxyerythromolide B
	[1351]		Halicholactone, Neohalicholactone
	[1352]		Epothilone
	[1353]		(–)-Disorazole
	[1354]		Macrolide dilactone isolated from *Verbascum undulatum*
	[1355]		Macroviracin D

Table 7.4 *Continued.*

Method	References	Conditions	Target or related products
	[1356]		(–)-Clavosolide A
	[1357]		Sporiolide A
	[1358]		(+)-Cladospolide C
	[1359]		2-*epi*-Amphidinolide *E*
Reaction with acid halides and related compounds			
1.4.1	[1360]	Reflux	(*R*)-(–)-Muscone
	[1361]	Reflux	Cyclo-2,3-diphospho-D-glycerate
	[1362]	At room temperature	Myriceric acid A
1.4.2.1	[1363]	AcOH	Cerpegin
	[1158]	HCl	Taxol
	[1364]	TsOH	(–)-Isoiridomyrmecin
	[1365]	MesOH	Lankacidin C
	[638]	HCl	Fusarentin methyl ethers
	[1366]	HCl	Taspine
1.4.3.1	[1367]	NaH	Taxol
	[1368]	BuLi	Taxol
	[1369]	BuLi	Taxol
	[1370]	$NNa(TMS)_2$	Taxol
	[1371]	$NLi(TMS)_2$	Taxol
	[1372]	NaH	Taxol
	[1373]	NaH	Taxol
	[1374]	$NNa(TMS)_2$	Taxol
	[1302]	$NNa(TMS)_2$	Taxol
	[1375]	$NNa(TMS)_2$	Taxol

Table 7.4 *Continued.*

Method	References	Conditions	Target or related products
	[1376]	KOtBu	Balanol
	[1377]	$NLi(TMS)_2$	Taxol
	[1378]	$NLi(TMS)_2$	Taxol
	[1379]	BuLi, $NLi(TMS)_2$	Taxol
	[1380]	$AgOCOCF_3$	(+)-Bengamide *E*
	[1381]	$NNa(TMS)_2$	Taxol
	[1382]	$NNa(TMS)_2$	Taxol
	[1383]	$NNa(TMS)_2$	Taxol
	[1384]	$AgOCOCF_3$	(+)-Bengamide *E*
	[984]	NaH and $AgOCOCF_3$	Pyrrolizidine alkaloid
	[1385]	NaH	Taxol
	[1386]	$NLi(TMS)_2$	Taxol
	[1387]	$NNa(TMS)_2$	Taxol
	[1388]	$NLi(TMS)_2$	Taxol
	[990]	K_2CO_3	Prenylisoflavones
	[1389]	$NLi(TMS)_2$	Taxol
	[1390]	$NLi(TMS)_2$	Taxol
	[1391]	$NLi(TMS)_2$	Taxol
	[1192]	$NLi(TMS)_2$	Taxol
1.4.3.2	[1226]	Py	Taxol
	[1294]	Py	Bretonin
	[1392]	Py	PFA (platelet-activating factor)
	[1393]	Py, DMAP	Taxol
	[1394]	Py, DMAP	Oleanolic acid, Erythrodiol, β-Amyrin

Table 7.4 *Continued.*

Method	References	Conditions	Target or related products
	[1395]	Py	Taxol
	[1396]	DMAP, NEt_3	Pancratistatin
	[1373]	DMAP	Taxol
	[1376]	NEt_3	Balanol
	[983]	Py, DMAP	Melodienone, 7-Hydroxy-6-hydromelodienone
	[673]	Py, DMAP	Taxol
	[1397]	NEt_3	Lycoperdic acid
	[1398]	NEt_3	Galanthamine
	[1399]	NEt_3	*m*-Hydroxycocaine, *m*-Hydroxybenzoylecgonine
	[1400]	Py	Bryostatin
	[1401]	Py	Mitotoxin
	[1254]	DMAP, NEt_3	9-Acetoxyfukinanolide
	[1402]	Py, DMAP	Alkanonic acid
	[1403]	Py, DMAP	Cladantholide, Estafiatin
	[1404]	NEt_3	Cocaine metabolite
	[1405]	Py	Anistatin
	[981]	NEt_3	Melodorinol
	[1285]	DMAP, NEt_3	Scletium alkaloid ((−)-Mesembrine, (+)-Sceleium A-4, (+)-Tortuosamine, (+)-*N*-Formyltortusamine)
Use of tin alkoxides			
2	[732]	Bu_2SnO	Pyrrolizidine alkaloid
	[733]	Bu_2SnO	Pyrrolizidine alkaloid
	[1159]	Bu_3SnOMe	Pyrrolizidine alkaloid

Table 7.4 *Continued.*

Method	References	Conditions	Target or related products
Reaction with diazomethane			
3.1	[1406]		Zoanthamine alkaloid
Reaction with alkyl halides			
3.2	[1407]		Kainic acid
	[1408]		Differanisole A
	[1409]		Phaseolinic acid

a DCC is employed in case no conditions are noted.
b The standard Mitsunobu reaction in case no conditions are noted.
c 2,4,6-Trichlorobenzoyl chloride is employed in case no conditions are noted.

A key step in this strategy is the incorporation of the (2*R*, 3*S*)-*N*-benzoyl-3-phenylisoserine side chain onto the highly sterically hindered 13-position, and the following methods have been reported.

1. Direct coupling between 10-deacetyl baccatin III and (2*R*, 3*S*)-*N*-benzoyl-*O*-(1-ethoxyethyl)-3-phenylisoserine [1156, 1161, 1164, 1226]. The condensation of 7-triethylsilyl baccatin III is effected by use of di-2-pyridyl carbonate and DMAP, in 80% yield at 50% conversion or 60% yield at 85% conversion [1226]. In addition to the unsatisfactory yields, this approach suffers from epimerization at carbon 2′.

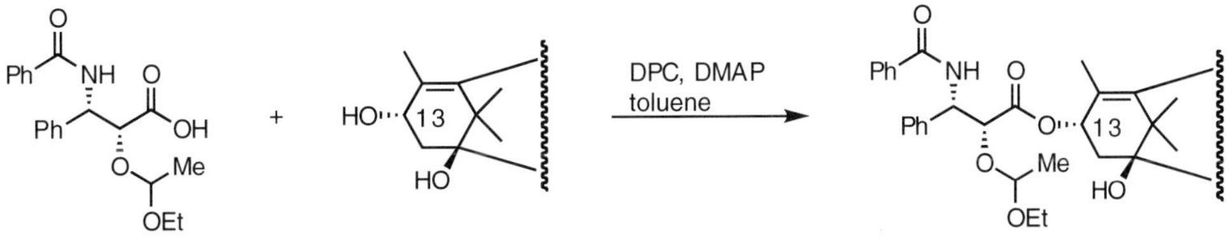

2. Use of β-lactam [1192, 1302, 1367–1374, 1377–1379, 1381–1383, 1385–1387, 1390, 1391, 1393, 1410–1412]. The drawbacks encountered in the above direct coupling technique are tackled by use of (3*R*, 4*S*)-3-(1-ethoxy)ethoxy-4-phenyl-2-azetidinone [1367, 1410]. The ring opening is promoted by DMAP/pyridine or more efficiently by use of sodium salt of the baccatin III.

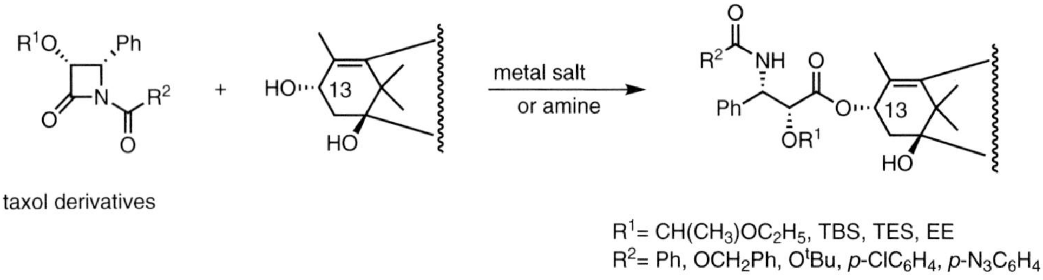

metal salt or amine

taxol derivatives

R^1= $CH(CH_3)OC_2H_5$, TBS, TES, EE
R^2= Ph, OCH_2Ph, O^tBu, *p*-ClC_6H_4, *p*-$N_3C_6H_4$

3. Use of 1,3-oxazolidine carboxylic acids [1156, 1158, 1161, 1163, 1165, 1166, 1168–1173, 1178, 1183, 1188, 1189, 1236, 1413]. Treatment of 2-substituted-1,3-oxazoline carboxylic acid with the C-7 triethylsilyl derivative of baccatin III in the presence of DCC/DMAP affords a 94% yield of the condensation product, which is transformed into taxol upon hydrolysis [1161].

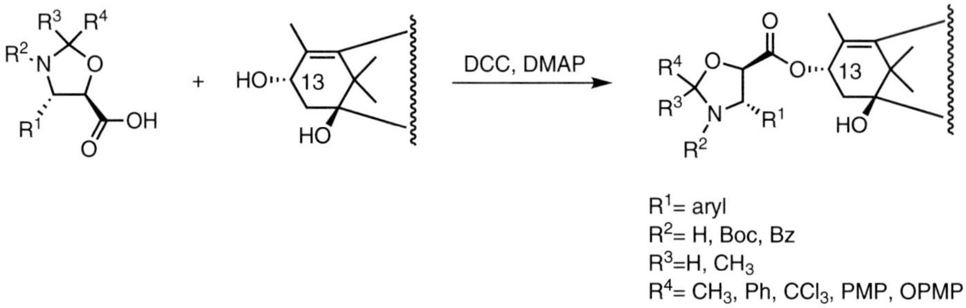

DCC, DMAP

R^1= aryl
R^2= H, Boc, Bz
R^3=H, CH_3
R^4= CH_3, Ph, CCl_3, PMP, OPMP

4. Use of oxazoline [1167, 1177, 1388]. 2-Substituted oxazolines are also useful protected side chains. Coupling between (4*S*,5*R*)-2,4-diphenyloxazoline-5-carboxylic acid and 7-(triethylsilyl)baccatin III in the presence of DCC/4-pyrrolidinopyridine provides the desired condensation product in 95% yield, and subsequent hydrolysis furnishes taxol in 75% yield [1167]. The thioesters of the oxazolidine and oxazoline are attached to the baccatin nucleus through promotion with lithium bis(trimethylsilyl)amide [1388].

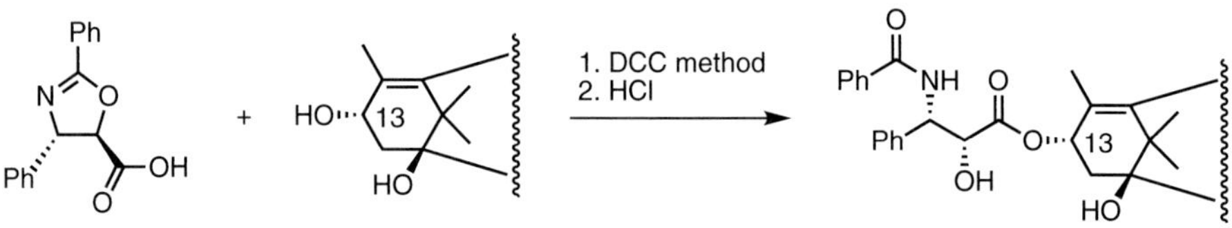

1. DCC method
2. HCl

5. Use of other side chain equivalents. Transesterification of ethyl benzoylacetate with protected baccatin III is carried out simply by heating at 90 °C (see Section 1.2.1; Scheme 1.2) [321] and the resulting ester is successfully transformed into taxol [1263]. 2′,2′-Difluoro derivatives of docetaxel are obtained by use of 3-amino-2,2-difluoropropanoic acids in the presence of dipyridyl carbonate/DMAP [1233].

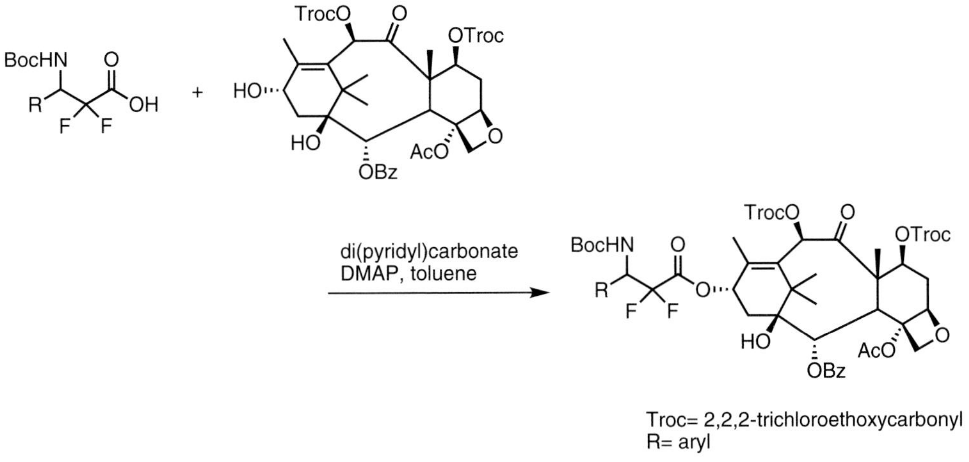

di(pyridyl)carbonate
DMAP, toluene

Troc= 2,2,2-trichloroethoxycarbonyl
R= aryl

The less sterically demanding cinnamic acid is readily incorporated at the 13-position of a taxol analog [1151].

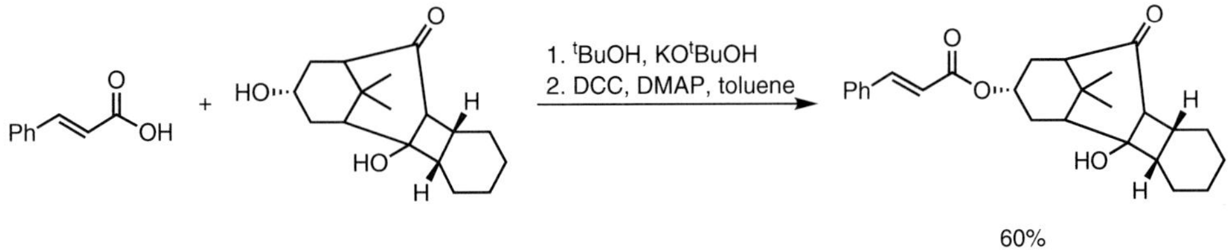

1. ^tBuOH, KO^tBuOH
2. DCC, DMAP, toluene

60%

When 1-deacetyl baccatin III is utilized, it is necessary to discriminate the C-13 OH from the three other hydroxy groups. The selective acylation of C-10 OH is achieved by the following method: with acid chloride [1379, 1395], or with acid anhydride [1263, 1289, 1290, 1301].

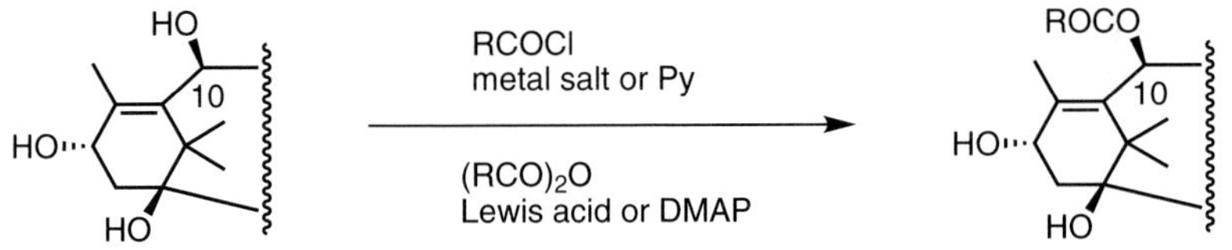

RCOCl
metal salt or Py

$(RCO)_2O$
Lewis acid or DMAP

The C-2 hydroxy group is acylated through the action of acid halides in the presence of bis(trimethylsilyl)amide [1389].

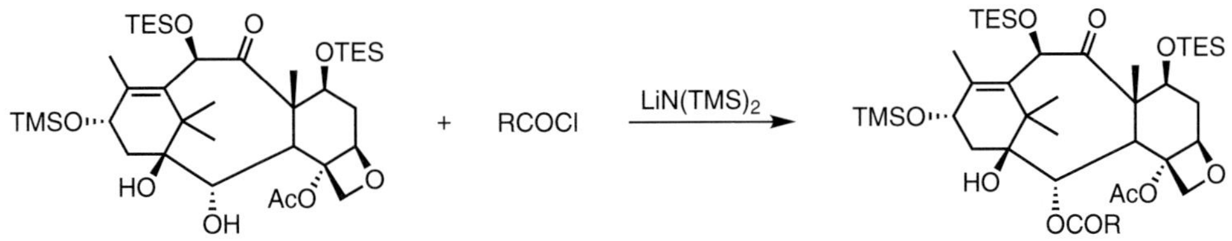

$LiN(TMS)_2$

The tertiary alcohol at the C-4 position is acetylated with acetic anhydride in pyridine [1318].

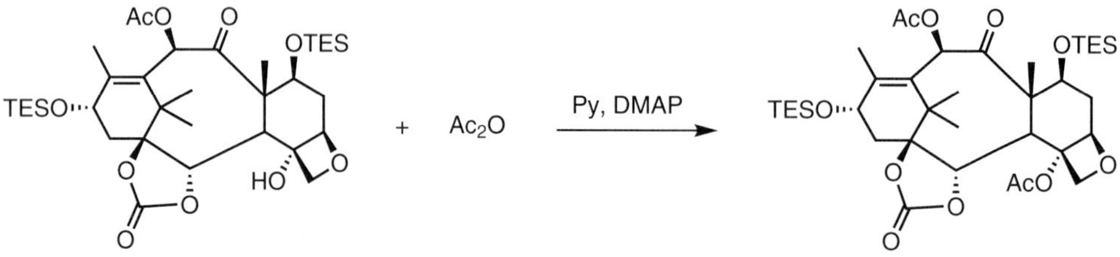

Py, DMAP

The next important application is seen in the synthesis of macrolides, because this is heavily dependent on esterification technology. The Yamaguchi technique (Section 1.3.5) serves quite well to this end. Two strategies are available for the final ring-closure in the synthesis of (–)-colletol. The first makes subtle use of various types of esterification technology (Scheme 7.62) [1414]: (i) distannoxane-catalyzed transesterification, (ii) DCC/DMAP condensation, and (iii) the Yamaguchi technique with 2,6-dichlorobenzoyl chloride/Et_3N/DMAP. The final lactonization proceeds in 84% yield. The second strategy employs the Yamaguchi reaction twice for the ester formation (Scheme 7.63) [1344]. The yield is 70% in the first esterification and 60% in the second. Total syntheses of bafilomycin A_1 (Scheme 7.64) [1349] and of epothilones A and B (Scheme 7.65) [1348] succeed through taking advantage of this approach.

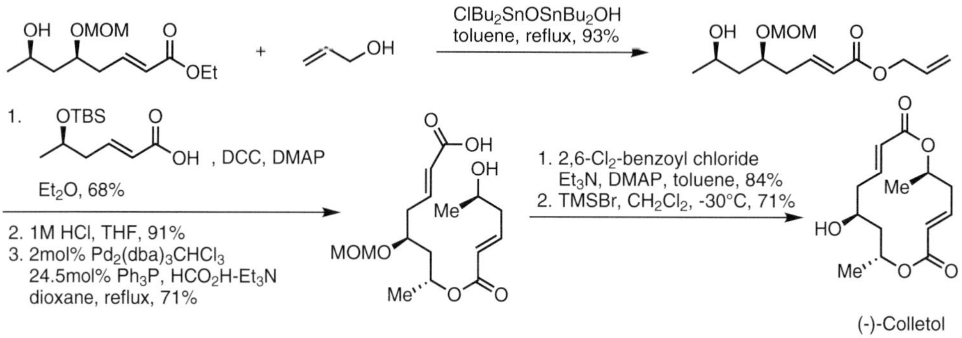

$ClBu_2SnOSnBu_2OH$, toluene, reflux, 93%

1. OTBS ... OH, DCC, DMAP, Et_2O, 68%
2. 1M HCl, THF, 91%
3. 2mol% $Pd_2(dba)_3CHCl_3$, 24.5mol% Ph_3P, HCO_2H-Et_3N, dioxane, reflux, 71%

1. 2,6-Cl_2-benzoyl chloride, Et_3N, DMAP, toluene, 84%
2. TMSBr, CH_2Cl_2, -30°C, 71%

(-)-Colletol

Scheme 7.62

The Yamaguchi reaction is made easier when the substrates have a suitable conformation (Yonemitsu modification). Thus, protection of 1,3-diol units as sterically demanding benzylidene acetals, coupled with DMAP activation, allows lactonization at room temperature in quantitative yield in the synthesis of

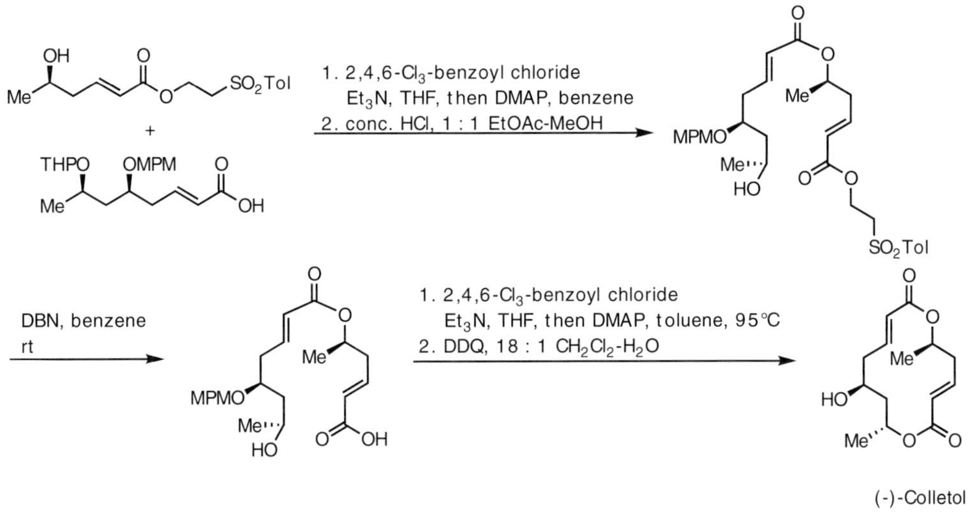

1. 2,4,6-Cl_3-benzoyl chloride
Et_3N, THF, then DMAP, benzene
2. conc. HCl, 1 : 1 EtOAc-MeOH

DBN, benzene
rt

1. 2,4,6-Cl_3-benzoyl chloride
Et_3N, THF, then DMAP, toluene, 95 °C
2. DDQ, 18 : 1 CH_2Cl_2-H_2O

(-)-Colletol

Scheme 7.63

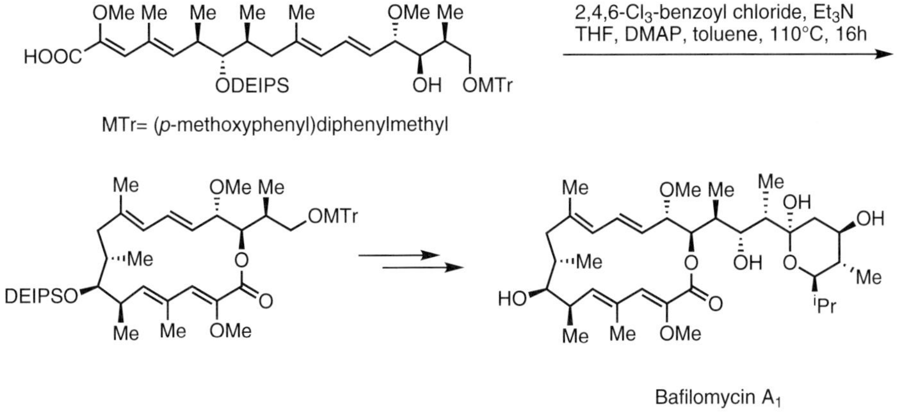

2,4,6-Cl_3-benzoyl chloride, Et_3N
THF, DMAP, toluene, 110 °C, 16h

MTr= (*p*-methoxyphenyl)diphenylmethyl

Bafilomycin A_1

Scheme 7.64

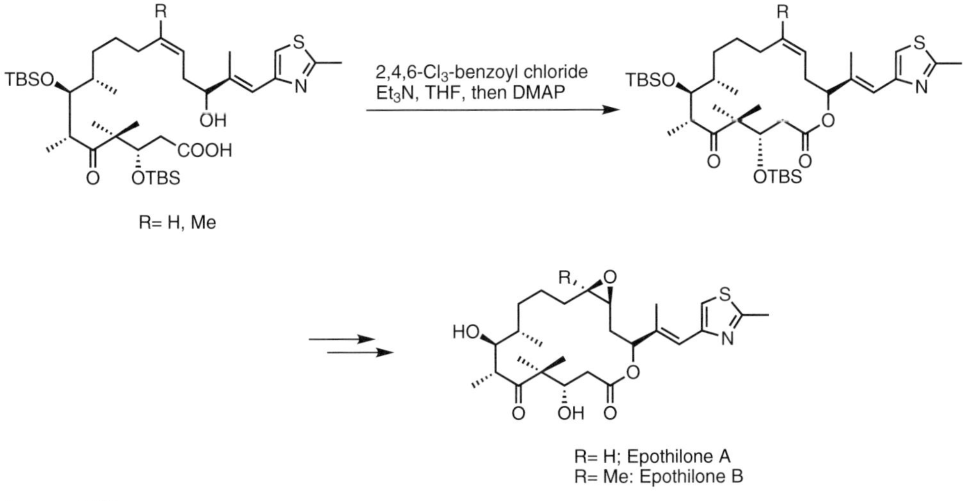

2,4,6-Cl_3-benzoyl chloride
Et_3N, THF, then DMAP

R= H, Me

R= H; Epothilone A
R= Me: Epothilone B

Scheme 7.65

9-dihydroerythronolide (Scheme 7.66) [1415]. This new version finds successful applications in the synthesis of macrolide antibiotics. The rutamycin B skeleton is constructed in 86% yield (Scheme 7.67) [1340]. Notably, the deconjugated macrolide inevitably forms in other methods. *p*-Methoxybenzylidene acetalization

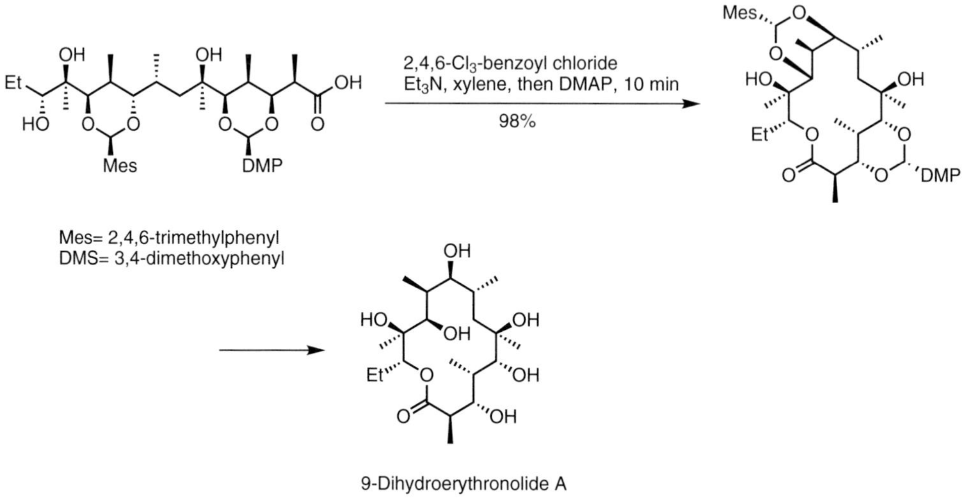

2,4,6-Cl_3-benzoyl chloride
Et_3N, xylene, then DMAP, 10 min
98%

Mes= 2,4,6-trimethylphenyl
DMS= 3,4-dimethoxyphenyl

9-Dihydroerythronolide A

Scheme 7.66

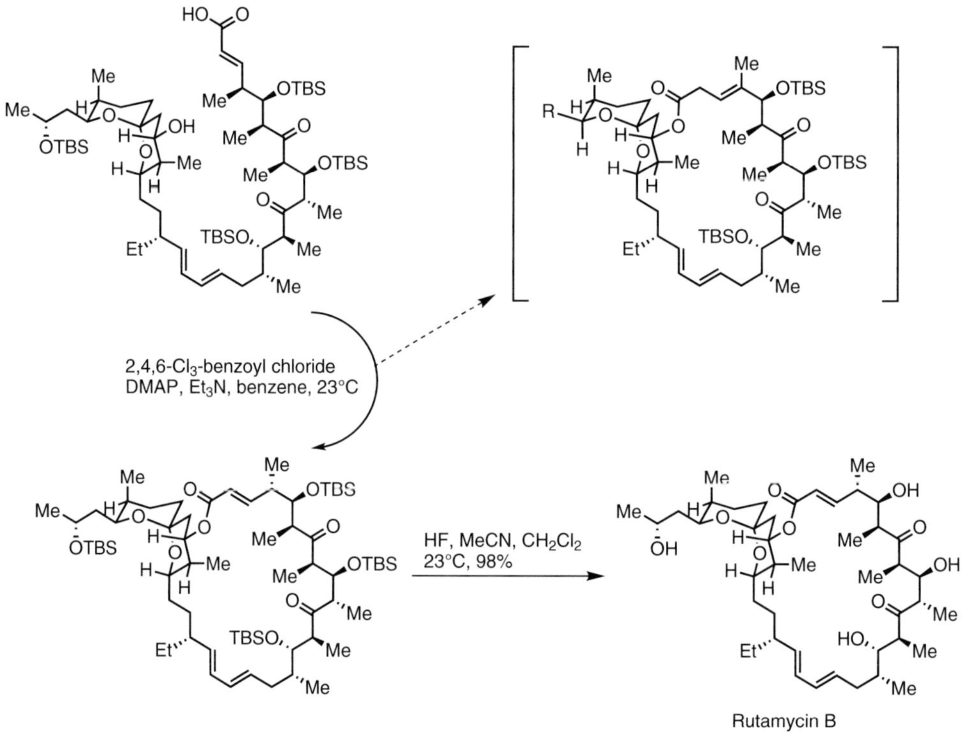

2,4,6-Cl_3-benzoyl chloride
DMAP, Et_3N, benzene, 23°C

HF, MeCN, CH_2Cl_2
23°C, 98%

Rutamycin B

Scheme 7.67

results in smooth lactonization in the synthesis of 6-deoxyerythronolide B (Scheme 7.68) [1350].

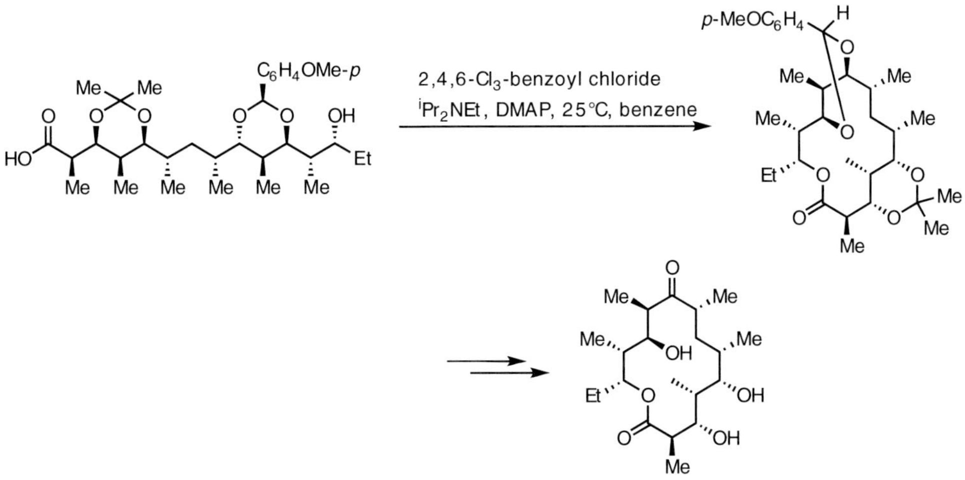

2,4,6-Cl_3-benzoyl chloride
$^iPr_2NEt$, DMAP, 25 °C, benzene

6-Deoxyerythronolide B

Scheme 7.68

The Mitsunobu reaction (Section 1.1.5) is useful for macrolide synthesis as well. The key intermediate for combretastatins D is constructed under the standard Mitsunobu conditions (Scheme 7.69) [1203, 1214]. The synthesis of (–)-echinosporin requires modified conditions (Scheme 7.70) [1196]. Of various phosphines, tributylphosphine is the best when coupled with diethyl azodicarboxylate at –15 °C. Dimerization of hydroxy acid is utilized for the synthesis of

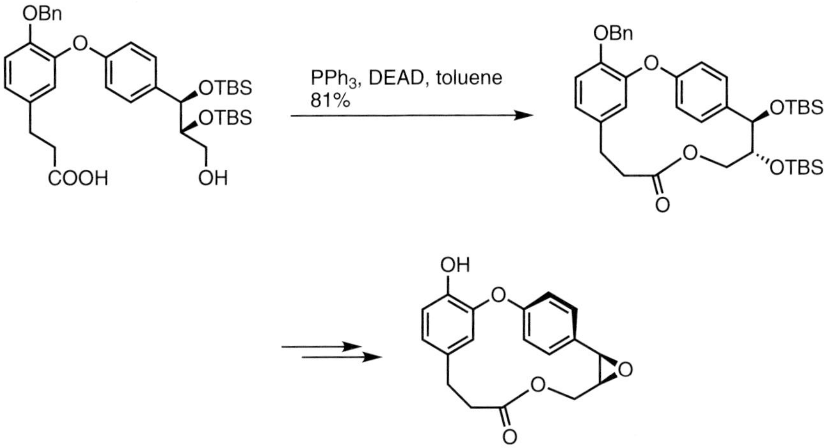

PPh_3, DEAD, toluene
81%

Combretastatin D-1

Scheme 7.69

(−)-pyrenophorin (Scheme 7.71) [1199]. In this route, the Mitsunobu reaction is employed twice, for the synthesis of secondary acetate and the final lactonization. In contrast, the analogous direct dimerization strategy furnishes only an 11% yield of (−)-vermiculine, owing to the concomitant formation of the trimer and tetramer. Use of the stepwise method instead gives rise to a better outcome (Scheme 7.72) [1204]. Under the Mitsunobu conditions, the first coupling is between hydroxy

PBu$_3$, DEAD, THF
MS 4A, -15°C→rt

(-)-Echinosporin (XK-213)

Scheme 7.70

AcOH, PPh$_3$, DEAD
Et$_2$O, rt, 40h, 80%

PPh$_3$, DEAD
toluene
-20~18°C, 90h
44%

(-)-Pyrenophorin

Scheme 7.71

PPh$_3$, DEAD
toluene
-40°C to 0°C
24h, 80%

PBu$_3$, DEAD
toluene
-40°C to 0°C
then rt, 62%

(-)-Vermiculine

Scheme 7.72

ester and silyloxy carboxylic acid, while the second lactonization, after protection of the masked functions, provides the diolide in 62% yield.

Needless to say, the Mitsunobu procedure is also applicable to intermolecular esterification. One example is provided in the synthesis of (−)-panclicin D (Scheme 7.73) [1209].

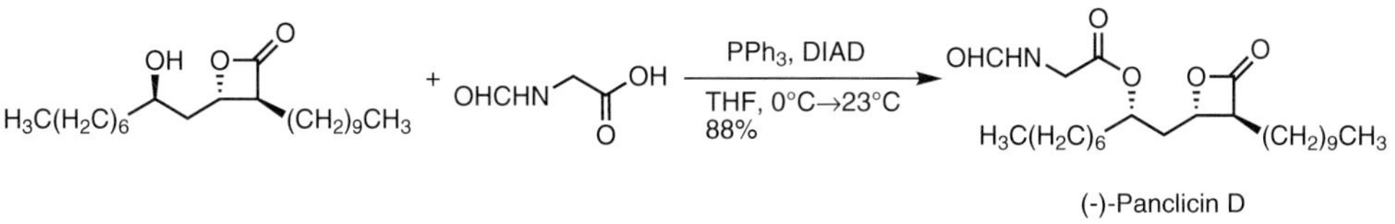

$H_3C(H_2C)_6$ (OH, β-lactone, $(CH_2)_9CH_3$) + OHCHN–CH(CO)–OH; PPh_3, DIAD; THF, 0°C→23°C; 88% → (-)-Panclicin D

Scheme 7.73

Lactonization is effected by the Mukaiyama technique with 2-chloro-1-methylpyridinium iodide (Section 1.1.6). (±)-Recifeiolide is obtained in high yield when 6-phenyl-2-pyridone and Et_3N are used as activators (Scheme 7.74) [1225]. 9-Propyl-10-azacyclododecan-12-olide, a minor component of the defensive secretion of the Mexican bean, is synthesized from the corresponding hydroxy acid (Scheme 7.75) [1232].

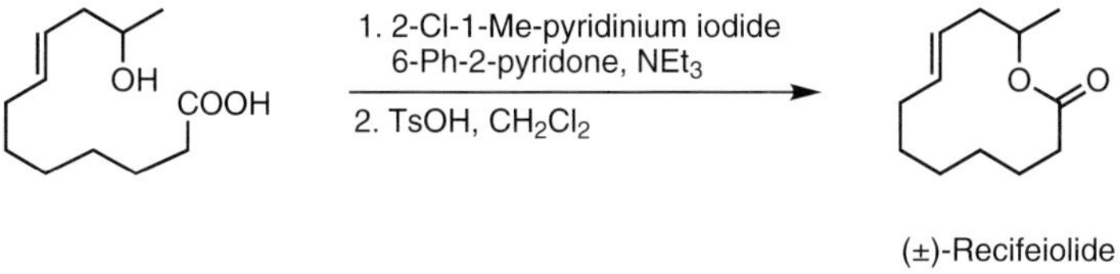

OH, COOH; 1. 2-Cl-1-Me-pyridinium iodide, 6-Ph-2-pyridone, NEt_3; 2. TsOH, CH_2Cl_2 → (±)-Recifeiolide

Scheme 7.74

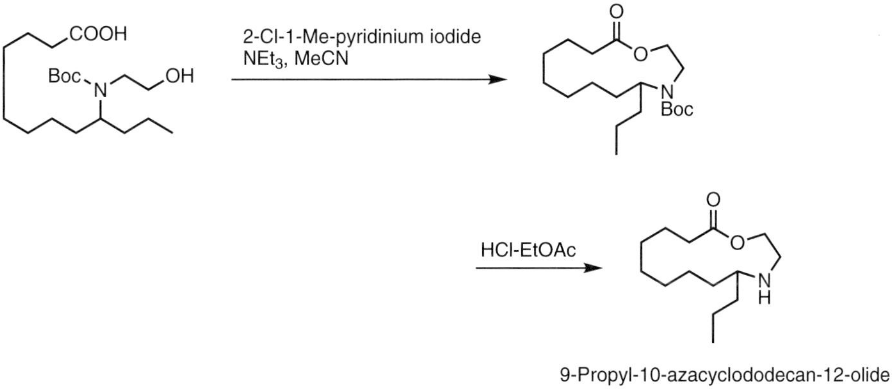

COOH, Boc–N–OH; 2-Cl-1-Me-pyridinium iodide, NEt_3, MeCN → N-Boc lactone; HCl-EtOAc → NH lactone

9-Propyl-10-azacyclododecan-12-olide

Scheme 7.75

The intermolecular version is utilized for the synthesis of (−)-balanol (Scheme 7.76) [1237].

In the above process, enzymatic kinetic resolution is invoked for preparation of the enantiopure hydroazepine. Another example of enzymatic kinetic resolution in natural product synthesis is illustrated in Scheme 7.77 [868]. Racemic albicanol

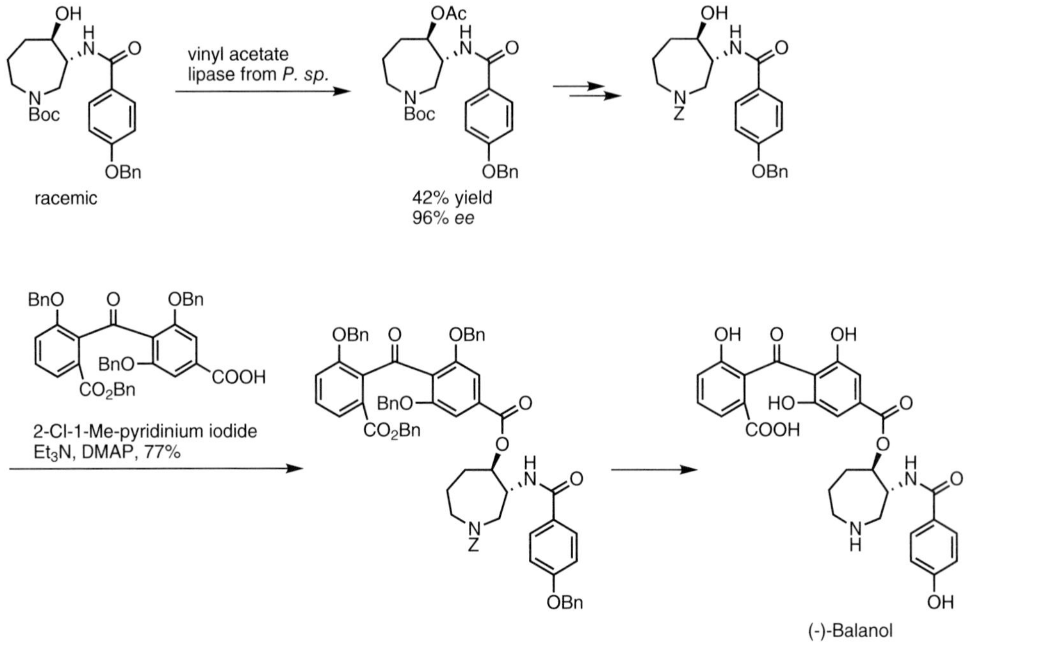

vinyl acetate
lipase from *P. sp.*

racemic

42% yield
96% *ee*

2-Cl-1-Me-pyridinium iodide
Et_3N, DMAP, 77%

(-)-Balanol

Scheme 7.76

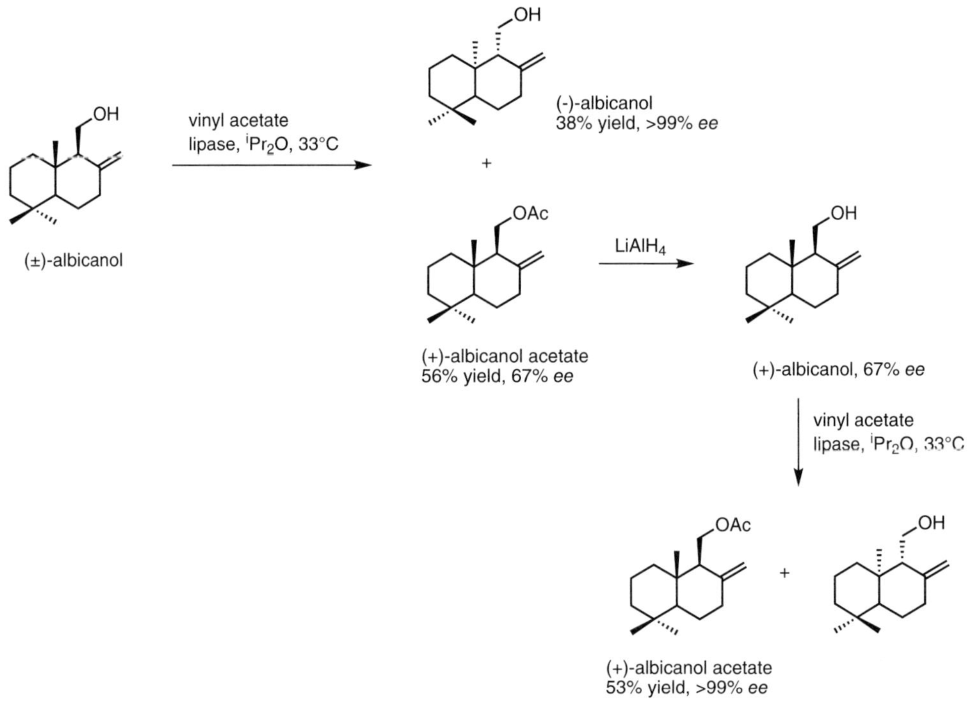

vinyl acetate
lipase, $^iPr_2O$, 33°C

(±)-albicanol

(-)-albicanol
38% yield, >99% *ee*

(+)-albicanol acetate
56% yield, 67% *ee*

$LiAlH_4$

(+)-albicanol, 67% *ee*

vinyl acetate
lipase, $^iPr_2O$, 33°C

(+)-albicanol acetate
53% yield, >99% *ee*

Scheme 7.77

is transformed into the enantiopure form. In addition, (−)-drimenol, (−)-drimenin, and (−)-ambrox are obtained by similar procedures.

DCC-promoted condensation (Section 1.1.4) is used in the synthesis of (−)-valilactone (Scheme 7.78) [1152] and (±)-myrocin C (Scheme 7.79) [1162]. EDC is the reagent of choice for preparation of a precursor for an olefin methathesis approach to epothilone A and its analogs (Scheme 7.80) [1186].

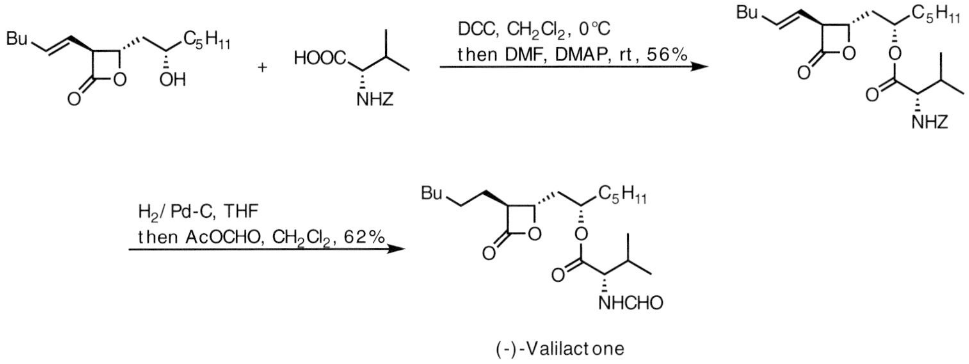
Bu
C5H11
O
OH
HOOC
NHZ
DCC, CH2Cl2, 0°C
then DMF, DMAP, rt, 56%
H2/ Pd-C, THF
then AcOCHO, CH2Cl2, 62%
NHCHO
(-)-Valilactone

Scheme 7.78

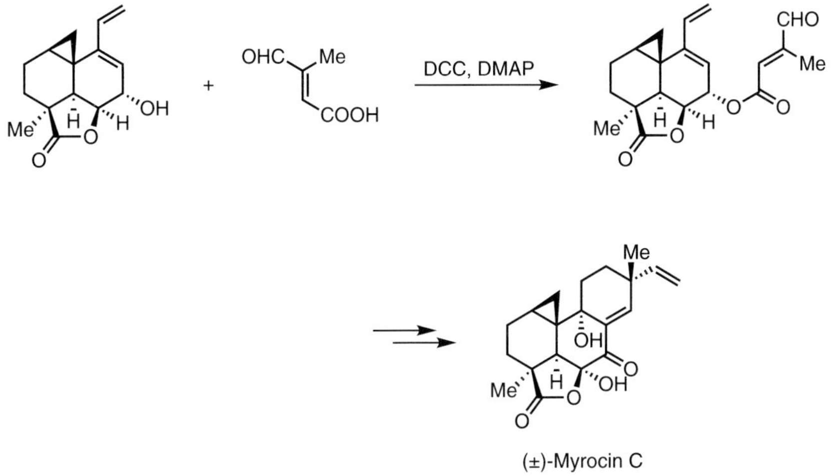
Me
H
OH
O
OHC
COOH
DCC, DMAP
CHO
(±)-Myrocin C

Scheme 7.79

Masamune's thioester procedure (Section 1.4.2.2) effects β-lactone formation, which constitutes a key step in total synthesis of (+)-bengamide *E* (Scheme 7.81) [1380].

3,4-Dimethoxycinnamic anhydride is a useful acylating agent in the synthesis of (±)- and (+)-subcosine I (Scheme 7.82) [1291, 1297].

Distannoxanes are extremely mild catalysts, with the aid of which transesterification can proceed under neutral conditions to enable the use of labile substrates.

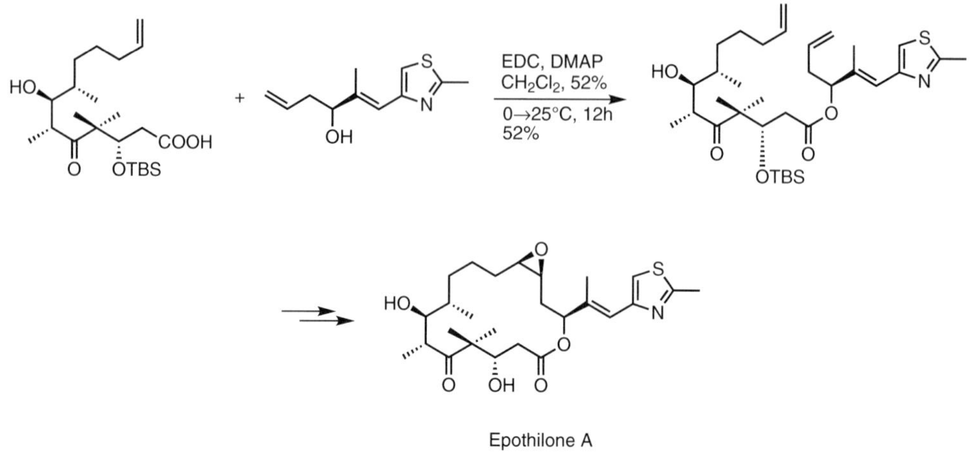

HO, COOH, O, OTBS + OH, S, N; EDC, DMAP, CH_2Cl_2, 52%, 0→25°C, 12h, 52%; HO, O, OTBS, O, O, S, N

HO, O, OH, O, O, S, N

Epothilone A

Scheme 7.80

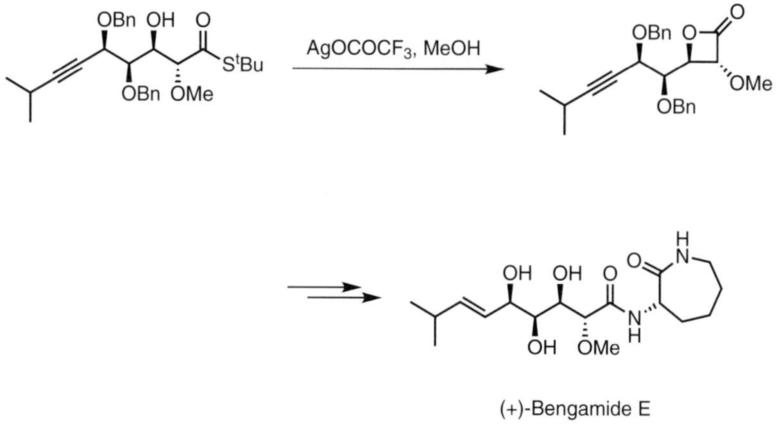

OBn OH O, S^tBu, OBn OMe; $AgOCOCF_3$, MeOH; OBn O O, OMe, OBn

OH OH O, O, H N, N H, OH OMe

(+)-Bengamide E

Scheme 7.81

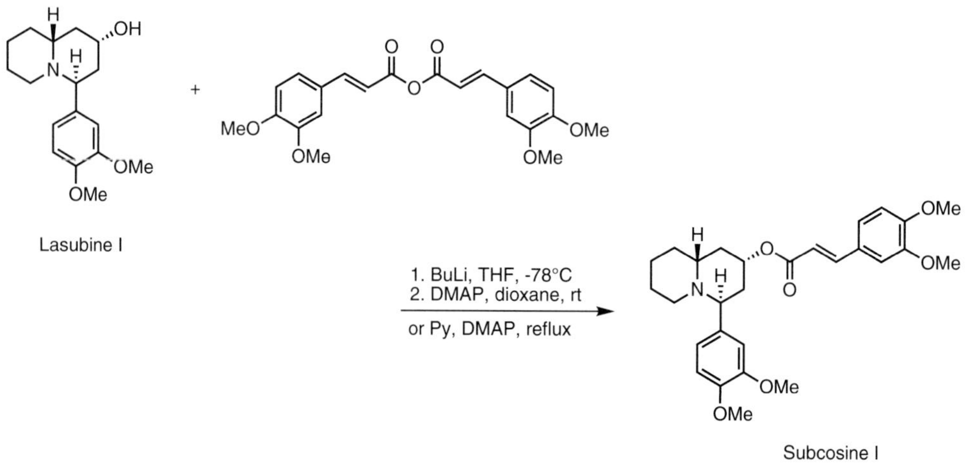

H, OH, N, H, OMe, OMe + O O, O, MeO, OMe, OMe, OMe

Lasubine I

1. BuLi, THF, -78°C
2. DMAP, dioxane, rt
or Py, DMAP, reflux

H, O, N, H, O, OMe, OMe, OMe, OMe

Subcosine I

Scheme 7.82

Highly sensitive methyl β-iodoacrylate successfully underwent distannoxane-catalyzed transesterification in the synthetic studies of the brefeldin series (Scheme 7.83) [1259], while no satisfactory results were obtained with other coupling agents. In the final step of the synthesis of kadsurenone, coupling between the lactol moiety and the butyl ester was achieved only with a distannoxane catalyst, to give a 1 : 1 mixture of readily separable bicyclic compounds isomeric at the benzylic carbon (Scheme 7.84) [1260]. A tetracylic template for a library of small polycyclic molecules for use in chemical genetic assays is obtained by a tandem transesterification-cycloaddition reaction of nitrone and expoxycyclohexanol (Scheme 7.85) [1261]. The transesterification is catalyzed by distannoxane in solution, but the solid-phase reaction proceeds sluggishly. In this case, the use of *N,N*-diisopropylamine/DMAP affords the satisfactory result.

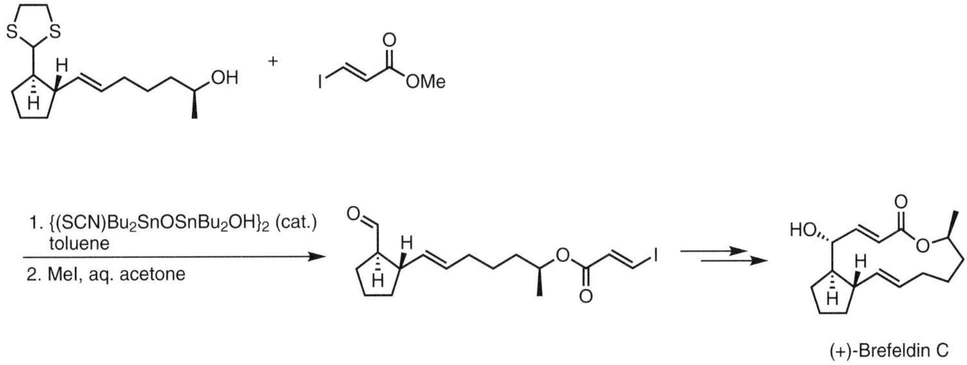

1. $\{(SCN)Bu_2SnOSnBu_2OH\}_2$ (cat.) toluene
2. MeI, aq. acetone

(+)-Brefeldin C

Scheme 7.83

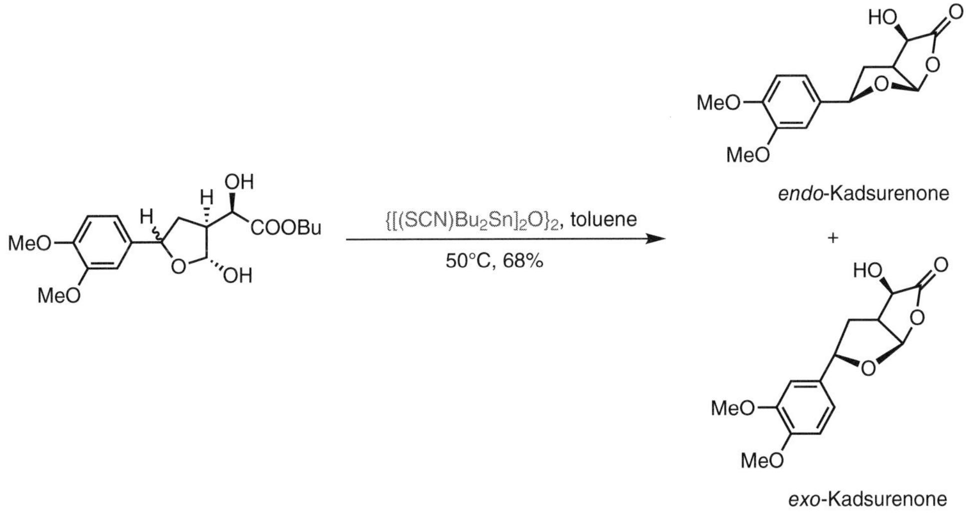

$\{[(SCN)Bu_2Sn]_2O\}_2$, toluene
50°C, 68%

endo-Kadsurenone

+

exo-Kadsurenone

Scheme 7.84

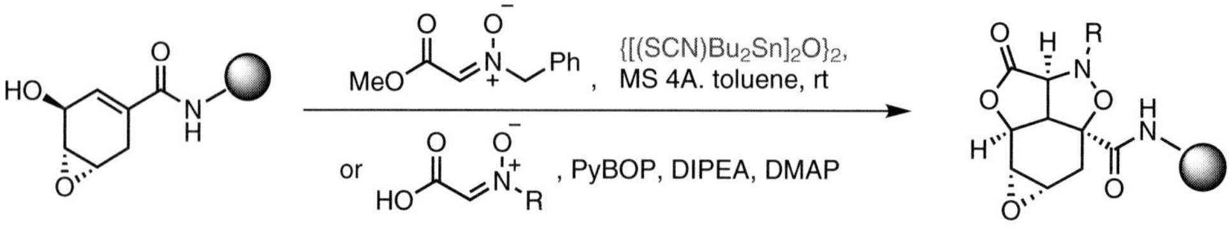

PyBOP= benzotriazol-1-yloxytripyrrolidinophosphonium hexafluorophosphate
DIPEA= *N,N*-diisopropylethylamine

Scheme 7.85

Stereochemically diversity-oriented synthesis of macrolides is feasible by cyclodimerization of hydroxy esters with distannoxane catalysts [1432]. Trans-esterification proceeds smoothyly to provide diolides in preference to triolides (Scheme 7.86). Heterodimeric macrolides are produced by use of two different monomers in the same procedure, together with the corresponding homodimers which are separable from the hetrodimers (Scheme 7.87).

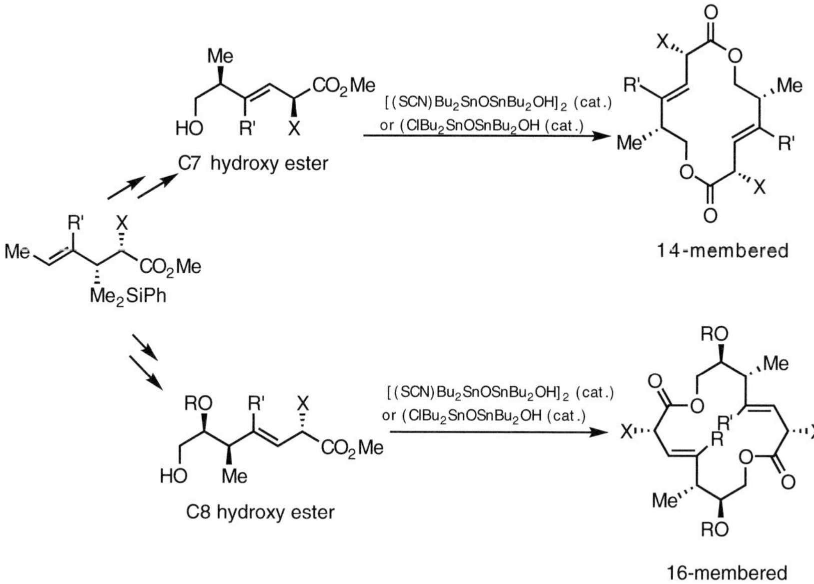

Scheme 7.86

In the synthesis of amphidinolide P, subjection of a hydroxy β-lactone to distannoxane-catalyzed transesterification results in exclusive formation of eight-membered lactone without formation of 15-membered macrolide (Scheme 7.88)

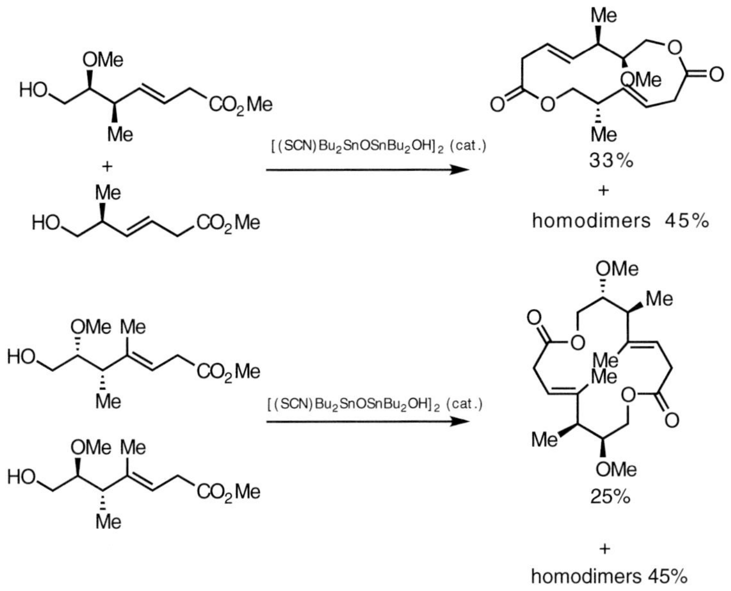

$[(SCN)Bu_2SnOSnBu_2OH]_2$ (cat.)

33% + homodimers 45%

$[(SCN)Bu_2SnOSnBu_2OH]_2$ (cat.)

25% + homodimers 45%

Scheme 7.87

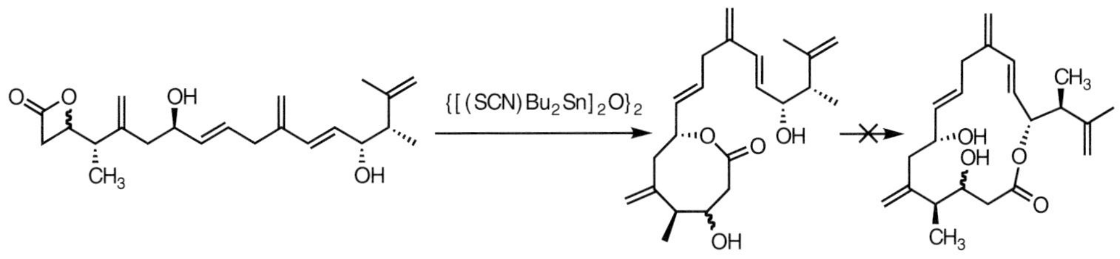

$\{[(SCN)Bu_2Sn]_2O\}_2$

Scheme 7.88

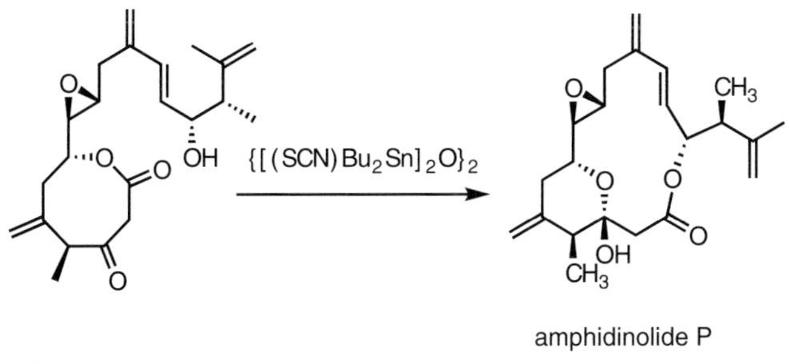

$\{[(SCN)Bu_2Sn]_2O\}_2$

amphidinolide P

Scheme 7.89

[1433]. The desired 15-membered macrolide, amphidinolide P, can be obtained by use of an epoxy substrate under similar conditions (Scheme 7-89).

Distannoxane-catalyzed transesterification is effective for synthesis of a prenyl ester intermediate for spirotryprostatin B (Scheme 7-90) [1434].

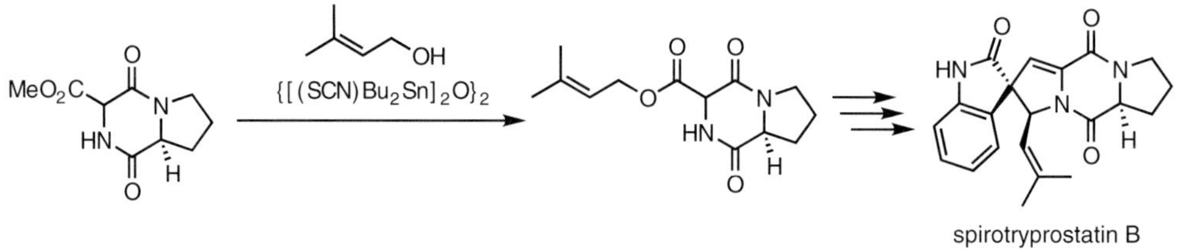

$\{[(SCN)Bu_2Sn]_2O\}_2$

spirotryprostatin B

Scheme 7.90

8
Industrial Uses

Esterification technology is utilized in a wide range of chemical industries, and it is not possible to cover all of them here. Moreover, many industrial processes are not fully disclosed, and thus it is not possible to give full details precisely and correctly. Consequently, only representative examples among them are described in this chapter, so that the readers may obtain a brief overview of the practical utilization of esterification.

8.1
Ethyl Acetate

Ethyl acetate is one of the most popular esters. It is used as a solvent in various industries for paints, inks, adhesives, and fragrances, with annual production amounting to 1 200 000 tons. It is freely miscible with most organic solvents such as ethanol, ether, benzene, hexane, etc., and, importantly, is expected to find broader applications in the replacement of toxic aromatic solvents.

It is commonly produced by acid-catalyzed esterification of acetic acid with ethanol (Scheme 8.1). This process is simple, but H_2SO_4, the most commonly employed acid catalyst, is strongly corrosive, and leaves sulfate residues. Much energy is also consumed for distillation of the water coproduced. Another popular process comprises Wacker oxidation of ethylene and subsequent Tishchenko reaction of the resulting acetaldehyde (Scheme 8.2). In this process, a large amount of aluminum residues, which are not easily separated, must be converted into aluminum hydroxide for disposal.

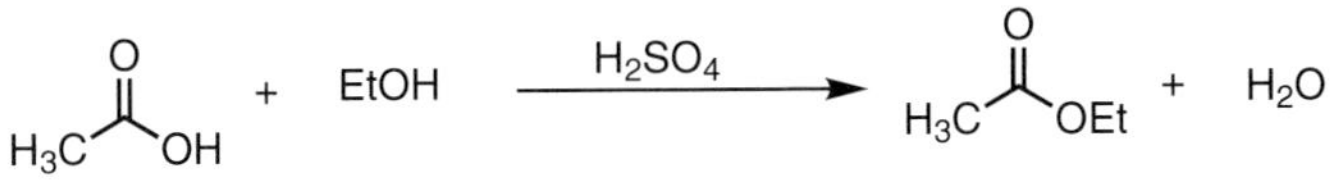

$$CH_3COOH + EtOH \xrightarrow{H_2SO_4} CH_3COOEt + H_2O$$

Scheme 8.1

It has long been known that low fatty acid esters are obtained by treatment of olefin with fatty acid in the presence of acid catalysts such as acidic ion-exchange

Esterification. Methods, Reactions, and Applications. 2nd Ed. J. Otera and J. Nishikido

ISBN: 978-3-527-32289-3

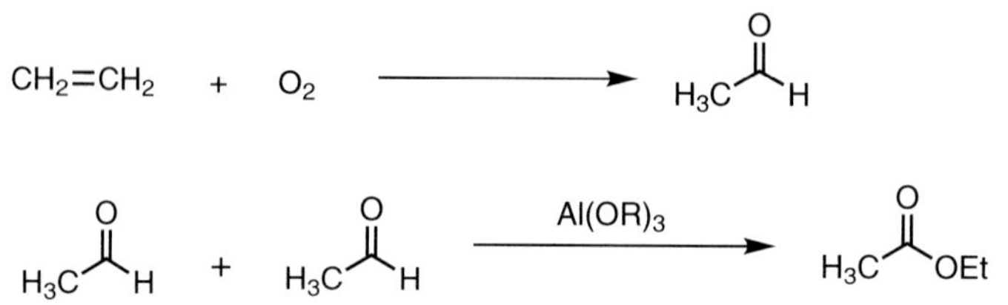

$CH_2{=}CH_2 + O_2 \longrightarrow H_3C\text{-}C(=O)\text{-}H$

$H_3C\text{-}C(=O)\text{-}H + H_3C\text{-}C(=O)\text{-}H \xrightarrow{Al(OR)_3} H_3C\text{-}C(=O)\text{-}OEt$

Scheme 8.2

resin, phosphoric acid, sulfuric acid, or heteropolyacids (Scheme 8.3). Now, a similar process for ethyl acetate from ethylene and acetic acid is available by use of improved heteropolyacid catalysts (Scheme 8.4). The heteropolyacids are supported on thin-layered silica to increase the specific surface [1416]. The catalytic activity is enhanced by immobilization. This process is superior to conventional ones in terms of atom economy, lower level of wastes, and safety.

$RCH{=}CH_2 + R'\text{-}C(=O)\text{-}OH \longrightarrow R'\text{-}C(=O)\text{-}OCH_2CH_2R$

Scheme 8.3

$CH_2{=}CH_2 + H_3C\text{-}C(=O)\text{-}OH \longrightarrow H_3C\text{-}C(=O)\text{-}OEt$

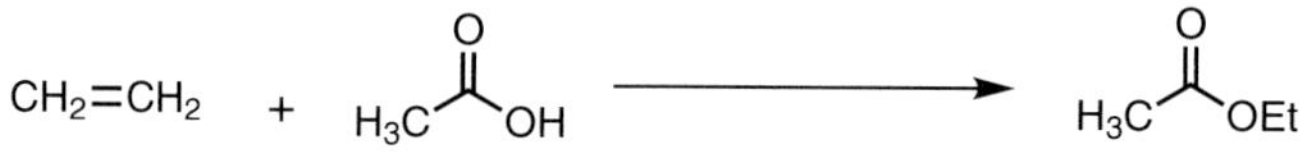

Scheme 8.4

8.2 Acrylic Esters

8.2.1 Methyl Methacrylate (MMA)

MMA is produced as a monomer (annual production 2 200 000 tons) for conversion to poly(methyl methacrylate). Thanks to its high transparency, good themoplasticity, and facile coloration, this polymer is used in place of glass and optical materials. The MMA monomer is also used in MBS, a ternary MMA/butadiene/styrene copolymer, which is used to modify poly(vinyl chloride) resin.

There are several industrial processes now available. Use of acetone cyanohydrin as an intermediate (ACH process) (Scheme 8.5), and ammoxidation of *tert*-butyl alcohol followed by methanolysis (Scheme 8.6) are most popular. However, an enormous quantity of ammonium sulfate is co-produced in these processes. The necessity to employ sulfuric acid in the ACH process can be circumvented by conversion of acetone cyanohydrin into methacrylamide followed by esterification with methyl formate.

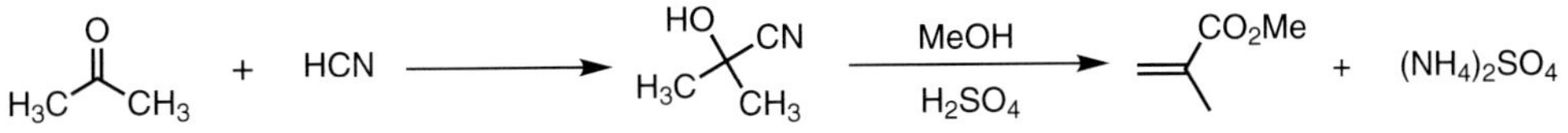

$$H_3C\text{-}CO\text{-}CH_3 + HCN \longrightarrow (H_3C)(CH_3)C(OH)CN \xrightarrow[H_2SO_4]{MeOH} CH_2{=}C(CH_3)CO_2Me + (NH_4)_2SO_4$$

Scheme 8.5

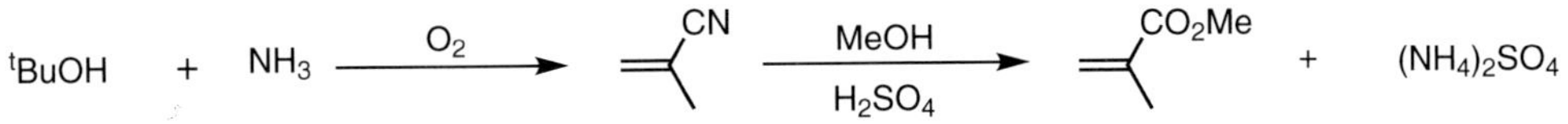

$$^tBuOH + NH_3 \xrightarrow{O_2} CH_2{=}C(CH_3)CN \xrightarrow[H_2SO_4]{MeOH} CH_2{=}C(CH_3)CO_2Me + (NH_4)_2SO_4$$

Scheme 8.6

A dual oxidation process, in which isobutene is oxidized and the resulting methacrolein is subsequently oxidized to methacrylic acid, is in operation (Scheme 8.7). MMA is also available by condensation between formaldehyde and methyl propanoate, the latter being obtained from ethylene, methanol, and carbon monoxide (Scheme 8.8).

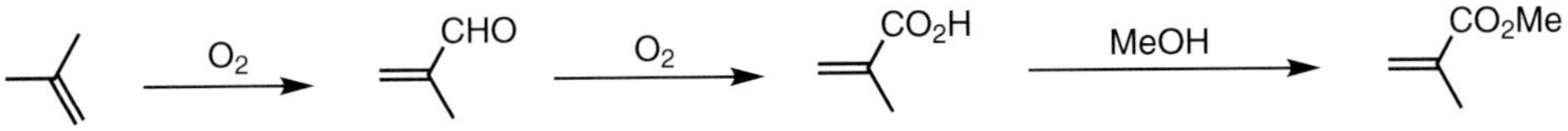

$$CH_2{=}C(CH_3)_2 \xrightarrow{O_2} CH_2{=}C(CH_3)CHO \xrightarrow{O_2} CH_2{=}C(CH_3)CO_2H \xrightarrow{MeOH} CH_2{=}C(CH_3)CO_2Me$$

Scheme 8.7

$$CH_2{=}CH_2 + MeOH + CO \longrightarrow CH_3CH_2CO_2Me$$

$$CH_3CH_2CO_2Me + HCHO \longrightarrow CH_2{=}C(CH_3)CO_2Me + H_2O$$

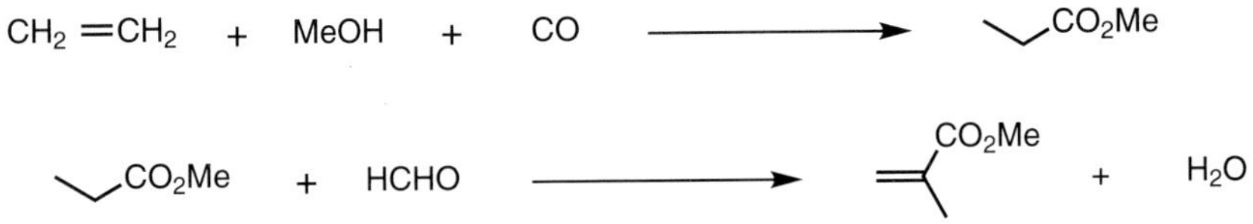

Scheme 8.8

A recent development in this field is a simple and green process comprising oxidation of isobutene and subsequent oxidative esterification (Scheme 8.9). The latter reaction is catalyzed by palladium metal, but accompanied by oxidation of the methacrolein (Scheme 8.10), resulting in only 30% selectivity for MMA. A bimetallic Pd/Pb catalyst, however, achieves smooth gas-liquid-solid triphasic oxidative decarbonylation, attaining >95% selectivity [1417].

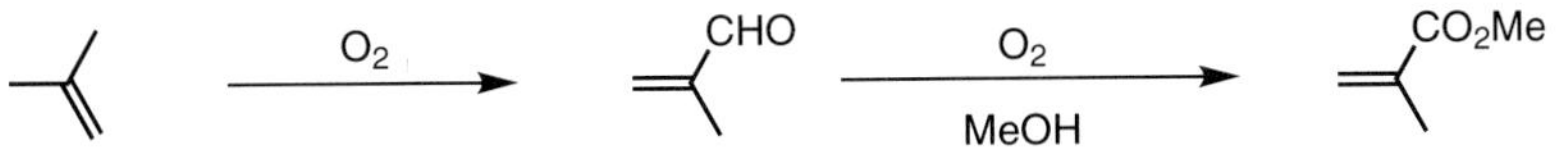

$$CH_2{=}C(CH_3)_2 \xrightarrow{O_2} CH_2{=}C(CH_3)CHO \xrightarrow[MeOH]{O_2} CH_2{=}C(CH_3)CO_2Me$$

Scheme 8.9

$$CH_2{=}C(CH_3)CHO \xrightarrow{O_2} CH_2{=}CHCH_3 + CO_2$$

Scheme 8.10

8.2.2 Alkyl Acrylates

Esters of acrylic acid with various alcohol components such as methanol, ethanol, butanol, 2-ethylhexanol, etc. are commercially important. Polymers of these esters are used for paints, adhesives, fibers, rubbers, electronics materials, and optical materials. These esters are mostly produced by condensation between acrylic acid and the corresponding alcohols (Scheme 8.11). In these reactions, sulfuric acid or *p*-toluenesulfonic acid is employed as a catalyst, which must be removed through neutralization followed by washing with water. Acrylic esters of high purity are required when used for electronics and optical purposes, and transesterfication by use of neutral catalysts is invoked in those cases, demanding neither neutralization nor washing operations (Scheme 8.12).

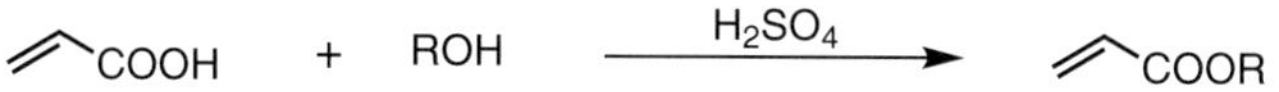

COOH + ROH —(H_2SO_4)→ COOR

Scheme 8.11

$COOR^1$ + R^2OH → $COOR^2$ + R^1OH

Scheme 8.12

8.3 Polyesters

Polyesters have a long history since the nineteenth century, and have experienced extensive development in the 1920s. As the fundamental production technology is fully established and well known, just a brief sketch is given here. Polyesters can be prepared through an exchange reaction between ester and hydroxy groups, usually called alcoholysis (Scheme 8.13). Alcoholysis proceeds very slowly in the absence of catalysts, even at high temperature. Of the many catalysts, the most effective are acetates of Pb(II), Pb(IV), Zn, Mg, Ca, Co and Cd, oxides such as

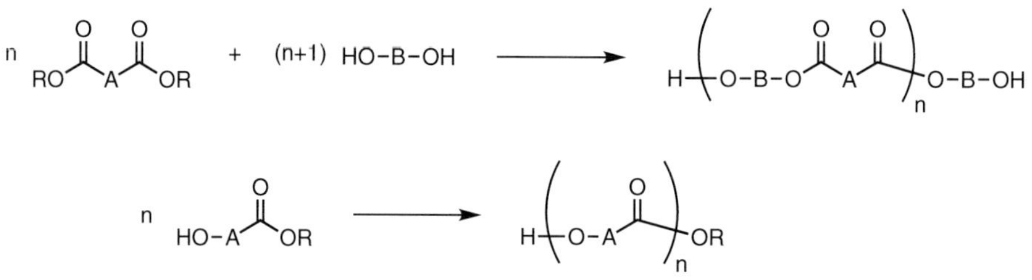

n RO(C=O)-A-(C=O)OR + (n+1) HO-B-OH → $H-(O-B-O-(C=O)-A-(C=O))_n-O-B-OH$

n HO-A-(C=O)OR → $H-(O-A-(C=O))_n-OR$

Scheme 8.13

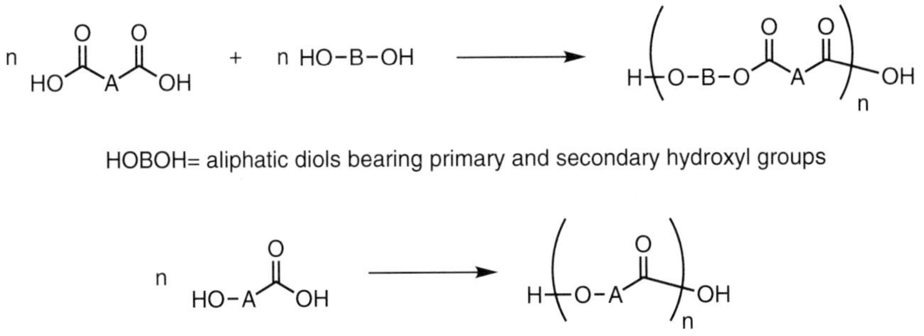

n HO–C(=O)–A–C(=O)–OH + n HO–B–OH ⟶ H–[O–B–O–C(=O)–A–C(=O)]$_n$–OH

HOBOH= aliphatic diols bearing primary and secondary hydroxyl groups

n HO–A–C(=O)–OH ⟶ H–[O–A–C(=O)]$_n$–OH

Scheme 8.14

Sb_2O_3 and GeO_2, and Ti alkoxides. This process used to be of great commercial importance, but – on account of higher costs resulting from more expensive raw materials, higher energy consumption and more expensive plants – the process actually employed at present is direct esterification. Most high-molecular-weight polyesters can be obtained by this process, from dicarboxylic acids and diols or from hydroxy acids at high temperature (Scheme 8.14). Reactions can take place at high temperature (180 ~ 230 °C) even in the absence of added catalysts. Small amounts (0.1 ~ 0.5 wt%) of catalyst are necessary in order to increase the reaction rate significantly. Many catalysts are reported; strong Brønsted acids (H_2SO_4, TsOH are most popular), oxides or salts of heavy metal ions (acetates are often preferred for their higher solubility), and organometallic compounds of Ti, Sn, Zr and Pb.

Polyesters are used in a variety of fields: textile fibers, films, bottles, resins, plastics, and so on. Notably, polyester fibers have since 1971 been the major material in synthetic fibers. The main component of the polyester fibers is the condensation polymer between ethylene glycol and terephthalic acid (PET). This is synthesized in two steps: esterification or transesterification with terephthalic acid or dimethyl terephthalate followed by polycondensation conducted under high temperature and vacuum conditions (Scheme 8.15).

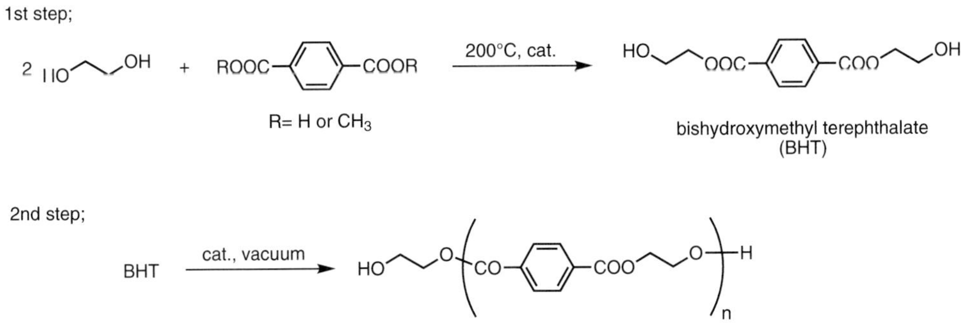

1st step;

2 HO–CH$_2$CH$_2$–OH + ROOC–C$_6$H$_4$–COOR —200°C, cat.⟶ HO–CH$_2$CH$_2$–OOC–C$_6$H$_4$–COO–CH$_2$CH$_2$–OH

R= H or CH_3

bishydroxymethyl terephthalate (BHT)

2nd step;

BHT —cat., vacuum⟶ HO–CH$_2$CH$_2$–O–[CO–C$_6$H$_4$–COO–CH$_2$CH$_2$–O]$_n$–H

Scheme 8.15

Some other polyesters are also used as fibers; these include polytrimethylene terephthalate (PTT) and polybutylene terephthalate (PBT), in which the ethylene glycol in PET is replaced by 1,3-propanediol and 1,4-butanediol, respectively. PTT fibers are very attractive, because the raw material (1,3-propanediol) is cheap and the fibers have shape stability, softness, and elongation recovery. PBT fibers are easy to dye and have good elasticity.

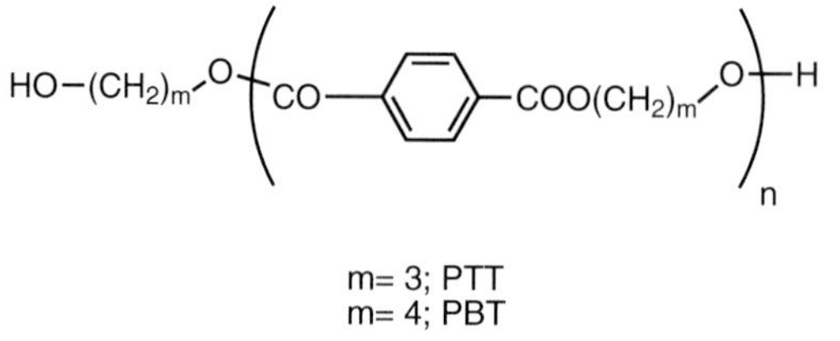

$HO-(CH_2)_m-O-[CO-C_6H_4-COO(CH_2)_m-O]_n-H$

m= 3; PTT
m= 4; PBT

Polyesters are used as materials in plastics as well. Unsaturated polyesters are produced from diacids such as maleic acid or fumaric acid and diols such as ethylene glycol. Styrene as a vinyl monomer and benzoyl peroxide as catalyst are added. Cross-linking and hardening occur at room temperature or above (Scheme 8.16). Unsaturated polyesters reinforced by glass fibers are used for pipes, helmets, car bodies, chairs, etc. Saturated polyesters such as PET and PBT are also used as plastics in electronics devices, bumpers, etc.

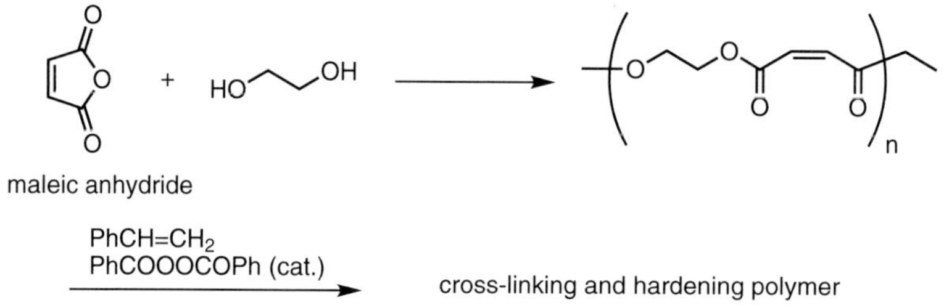

maleic anhydride + HO~~OH →

$PhCH{=}CH_2$
PhCOOOCOPh (cat.)
→ cross-linking and hardening polymer

Scheme 8.16

Polycondensates between polyhydric alcohols and polybasic carboxylic acids are known as alkyd resins. A variety of alcohols and acids, as shown below, are combined, and the resulting polycondensates are further treated with other oils or fatty acids. The properties of the resins are modified by this post-treatment. Drying oil resins, which are obtained by treatment with soybean oil at temperatures higher than 200 °C, are used for baking paint and air-drying coatings. Non-drying oil resins are obtained by treatment with castor oil or other non-drying fatty acids. These resins do not dry at room temperature and are used as components of thermosetting paints.

[Alcohols]

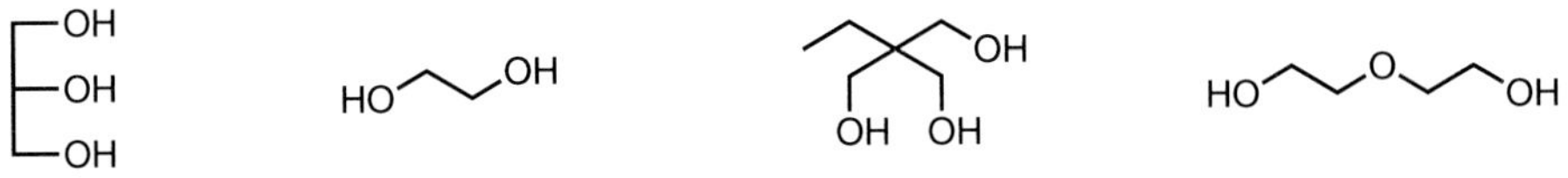

glycerol   ethylene glycol   trimethylol propane   diethylene glycol

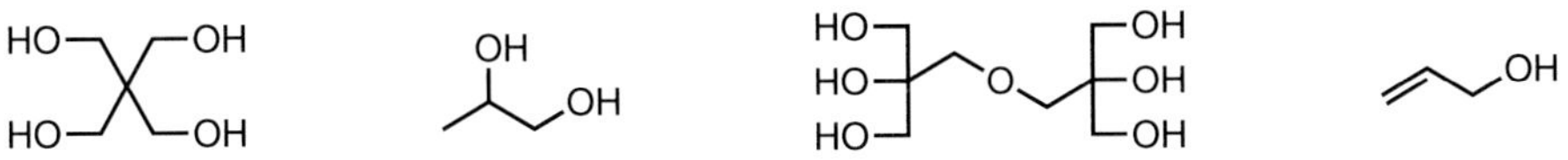

pentaerythritol   propylene glycol   dipentaerythritol   allyl alcohol

[Acids]

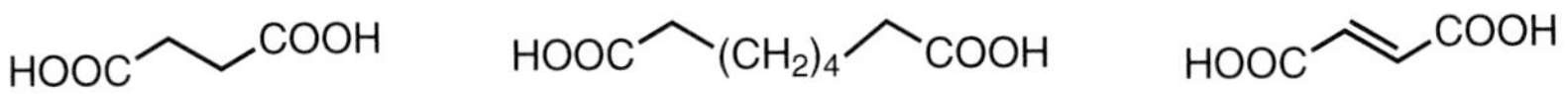

succinic acid   adipic acid   fumaric acid

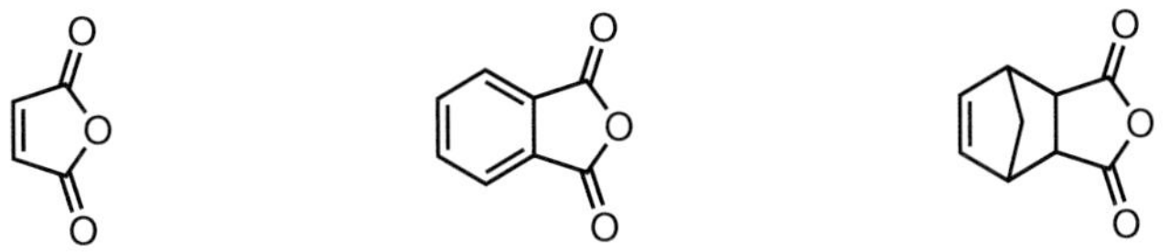

maleic anhydride   phthalic anhydride   carbic anhydride

Another important facet of polyesters is their potential for biodegradable polymers, and some of these have already been brought onto the market. These polymers find a wide spectrum of applications. Medical uses are most popular, for example, prosthetic devices such as bone plates and orthopedic pins and screws, artificial blood vessels, intravascular stents, nerve guides, surgical suture, and drug delivery system for controlled release. In addition to such specialist applications, the biodegradable polymers are of great promise for disposable plastics in general use, and their market should grow rapidly with the increasing demand for ecomaterials. However, the physical or mechanical properties of these polymers need to be improved as soon as possible to satisfy the requirements for commodity thermoplastics used in packaging and consumer goods. The first biopolymer to be commercialized was poly(3-hydroxybutyrate), while a copolymer with 3-hydroxyvalerate is also produced (Zeneca). Polycaprolactone, obtained by ring-opening of ε-caprolactone, is available from both UCC and from Daicel (Scheme 8.17). This polymer is used in the form of film or fiber, and its melting point is

ε-caprolactone —catalyst→ polycaprolactone

Scheme 8.17

fairly low (60 °C). This is a fault in terms of physical properties but an advantage in terms of processability, since it is easily moulded with hot water.

The condensation polymer of succinic acid and 1,4-butenediol cured with diisocyanate (Showa High Polymer) is also biodegradable (Scheme 8.18). This polymer possesses a high melting point (110 °C) as well as impact strength.

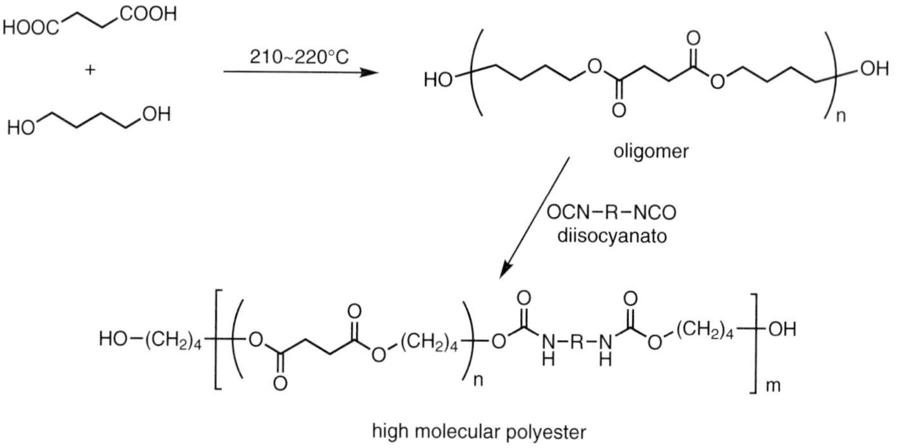

HOOC–COOH + HO–OH —210~220°C→ oligomer —OCN–R–NCO diisocyanato→ high molecular polyester

Scheme 8.18

Despite these precedents, the material that has recently received the most attention is poly(lactic acid). Several companies (Cargill Dow Polymers, Mitsui Chemical, Shimatdzu, etc) are setting out to commercialize this polymer. Poly(L-lactic acid) exhibits high biodegradability and the L-lactic acid monomer is produced in large quantity by fermentation of starch from corn or sugar beet. Polyesters do not exhibit satisfactory physical properties until the molecular weight is at least 25 000, but it is not easy to exceed this requirement by direct condensation of the monomer as the removal of the water formed in the last stage of the condensation is difficult due to the increased viscosity, preventing further increase in the molecular weight. The polymer is thus usually produced by ring-opening of the lactide, which is crystalline and easy to free from water (Scheme 8.19). This method is not straightforward, however, and the costs are rather high. More economical direct condensation has been achieved by the Mitsui Chemical group. In order to improve the physical properties, various copolymers with glycolide, β-propanolactone, γ-butyrolactone, δ-valerolactone, and ε-caprolactone are available.

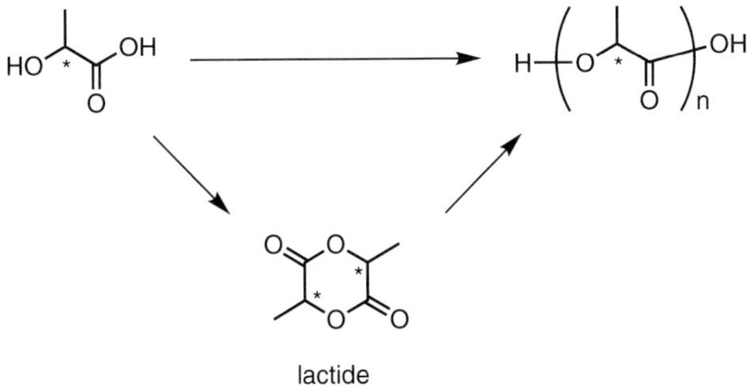

HO–CH(CH$_3$)*–C(=O)–OH → H–(O–CH(CH$_3$)*–C(=O))$_n$–OH

lactide

Scheme 8.19

8.4 Oils and Fats

Oils and fats are triglycerides, that is, triesters of glycerol with fatty acids. They are obtained from plants and animal products and are used in vast fields such as shortening, margarine, lard, plasticizer, materials of toilet and laundry soap, lubricating oil, and so on. Although natural oils and fats are very useful in their original forms, chemical modifications such as ester exchange and hydrogenation increase the utility further. Fatty acid esters, obtained by transesterification of oils and fats with alcohols, are also particularly important as surfactants for soaps, as food emulsifiers including foaming agents and antifoamers, and as dispersing agents.

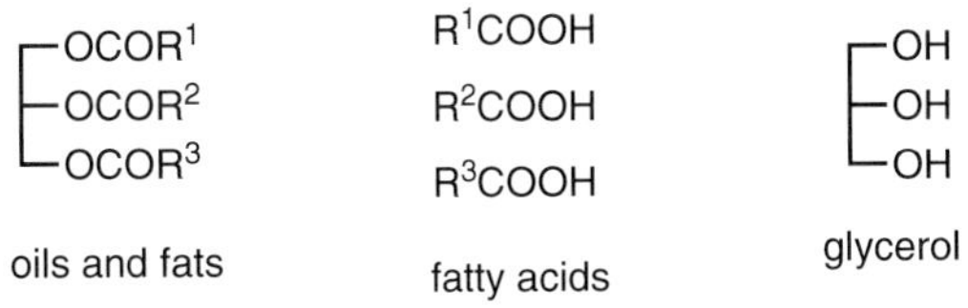

oils and fats	fatty acids	glycerol
$OCOR^1$	R^1COOH	OH
$OCOR^2$	R^2COOH	OH
$OCOR^3$	R^3COOH	OH

8.4.1 Food Emulsifiers

Acylglycerols, a mixture of mono- and diacylglycerols, are produced by transesterification of glycerol and fatty acids or oils. They are used as oil-in-water or water-in-oil emulsions, and antifoamers of bread and tofu. Fatty acid esters of polyglycerols are produced by transesterification of polyglycerol of a degree of polymerization of less than 10 with fatty acids, and give rise to emulsification at high salt levels or low pH values. Organic acid esters of monoacylglycerols are produced by esterification of the remaining hydroxyls in monoacylglycerols with organic acid such as acetic, lactic, citric, tartaric, and succinic acids. They are used in chewing gums and bread. Sucrose esters of fatty acids (sugar esters) are used in milk products and beverages. Sorbitan fatty acid esters, which are nonionic surfactants, are produced by treatment of sorbitol with fatty acids such as lauric, palmitic, stearic, or oleic acids. Propylene glycol esters serve for improvement of properties of oils and fats. Lecithin (1,2-diacyl-*sn*-glycerol 3-phosphocholine) is a phospholipid present in plants, animals, and yeast. Lecithin from soybeans is mainly used for emulsification of milk products and reduction of viscosity in chocolates.

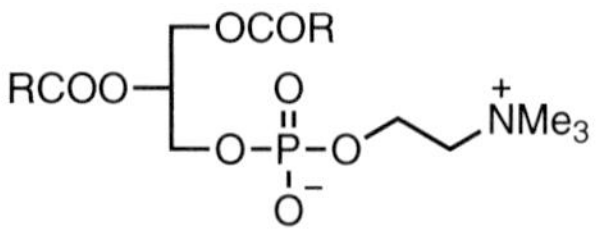

RCOO–[OCOR; O–P(=O)(O⁻)–O–CH₂CH₂–NMe₃⁺]

Lecithin

8.4.2
Soaps

Soaps are alkali salts of fatty acids, which are known and used as anionic surfactants. They are produced by several methods, one of which includes transesterification. Oils and fats are first transesterified to provide methyl esters of fatty acids, and these are then subjected to saponification to afford alkali salts of fatty acids and methanol (Scheme 8.20). Base catalysts such as NaOMe and NaOH (0.2 ~ 2.0%) are used at relatively low temperatures (50 °C ~ 90 °C); methanol can be recycled. This method has the following advantages; (i) purification is possible through transesterification, (ii) the oils and fats deteriorate less, because the reaction is performed at low temperature, and (iii) glycerol can be recovered in high concentrations.

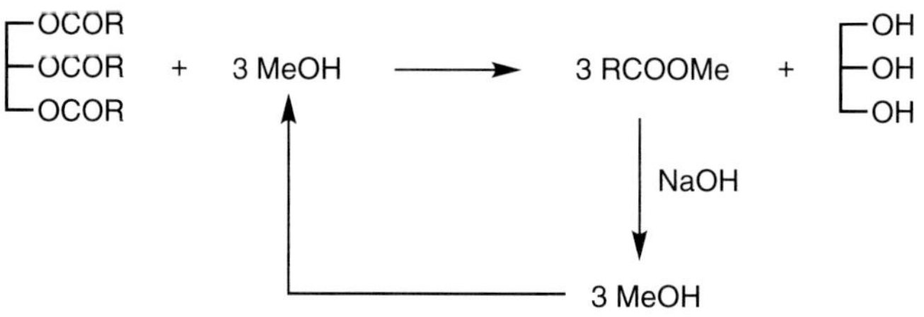

Triglyceride (OCOR ×3) + 3 MeOH → 3 RCOOMe + glycerol (OH ×3)

3 RCOOMe —NaOH→ 3 MeOH (recycled)

Scheme 8.20

8.5
Biodiesel Fuel

Biodiesel fuel (BDF), which is derived from renewable biomass resources, has been receiving increasing attention in the context of replacing fossil fuel. The replacement is believed to be effective for reducing carbon dioxide emission. BDF is a mixture of methyl esters of fatty acids such as oleic and linoleic acids, and produced by transesterification of triglycerides with methanol (Scheme 8.21). The

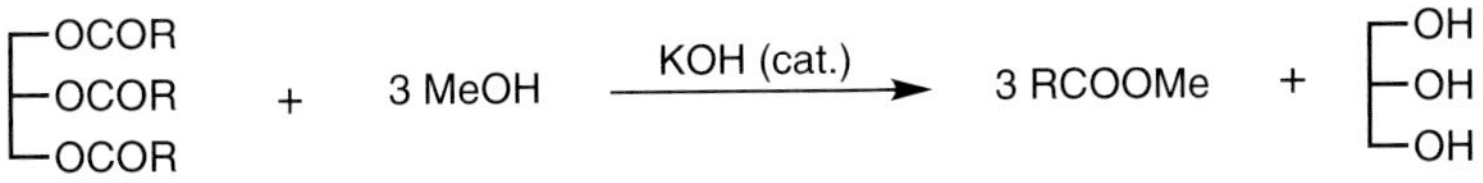

$$\begin{array}{l} \text{-OCOR} \\ \text{-OCOR} \\ \text{-OCOR} \end{array} + 3\,\text{MeOH} \xrightarrow{\text{KOH (cat.)}} 3\,\text{RCOOMe} + \begin{array}{l} \text{-OH} \\ \text{-OH} \\ \text{-OH} \end{array}$$

Scheme 8.21

conversion is most commonly carried out at 50–70 °C with an alkali catalyst such as KOH or NaOH, to give more than 98% yield. Since this exchange reaction takes place in a stepwise manner, di- and monoglycerides are co-produced if the reaction is not complete. Separation of glycerol (also known as glycerin or glycerine) is also an important technical issue. There are some other problems. If triglycerides containing free fatty acids (more than 2%) or waste triglycerides are employed, saponification of the fatty acids (Scheme 8.22) competes, decreasing the yield of the desired methyl esters. In such cases, removal of the fatty acids from the raw materials by suitable chemical or physical treatments or prior methylation of the fatty acids before transesterification is necessary. Another problem is dissolution of the alkali catalysts in the co-produced glycerol, resulting in difficulty in recovery of pure glycerol. Moreover, production of high-quality methyl esters requires purification by repeated washing with water, which is expensive.

$$\text{RCOOH} + \text{KOH} \longrightarrow \text{RCOOK} + H_2O$$

Scheme 8.22

Many other processes have been proposed. Increase in both yield of methyl esters and recovered glycerol is claimed by use of K_2CO_3 on alumina [1418], $CaCO_3$ [1419], and anion-exchange resin [1420]. Supercritical methanol brings about methyl ester formation without catalysts, though high pressure and high temperature are necessary [323, 324]. A milder process is available through hydrolysis of triglycerides in near-critical water followed by esterification in supercritical methanol (Scheme 8.23) [1421].

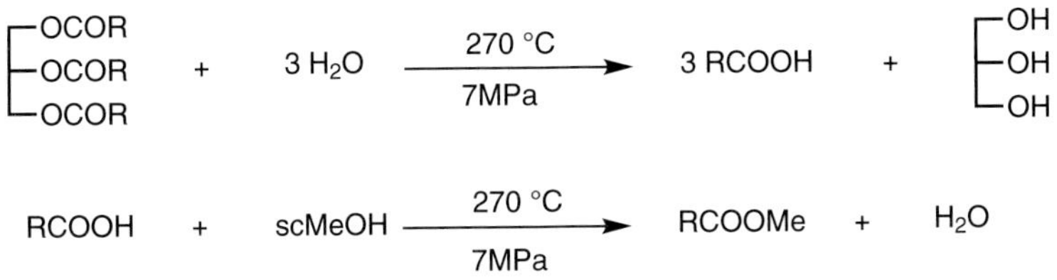

$$\begin{array}{l} \text{-OCOR} \\ \text{-OCOR} \\ \text{-OCOR} \end{array} + 3\,H_2O \xrightarrow[\text{7MPa}]{270\,^\circ\text{C}} 3\,\text{RCOOH} + \begin{array}{l} \text{-OH} \\ \text{-OH} \\ \text{-OH} \end{array}$$

$$\text{RCOOH} + \text{scMeOH} \xrightarrow[\text{7MPa}]{270\,^\circ\text{C}} \text{RCOOMe} + H_2O$$

Scheme 8.23

Immobilized lipases are also employable [1422, 1423]. This method is particularly useful for waste glyceride materials, but the preparation of immobilized lipases is expensive.

Sulfonated amorphous carbon catalyst derived from sugars is more active than conventional solid catalysts to give high-grade biodiesel [1424].

8.6 Amino Acid Esters

Amino acid esters are utilized in various fields such as sweeteners, drugs, cosmetics, toiletries, flavor substances, surfactants, fungicides, and so on. Generally, methyl or ethyl esters are produced by treating a suspension of amino acids in methanol or ethanol with HCl or $SOCl_2$ (Scheme 8.24). The latter approach is advantageous in that the reaction proceeds at lower temperatures and does not require the use of HCl gas. The product esters are isolated as HCl salts, the free esters being unstable. In cases in which the HCl salts are hygroscopic and crystallization is difficult, *p*-toluenesulfonic acid is frequently employed instead.

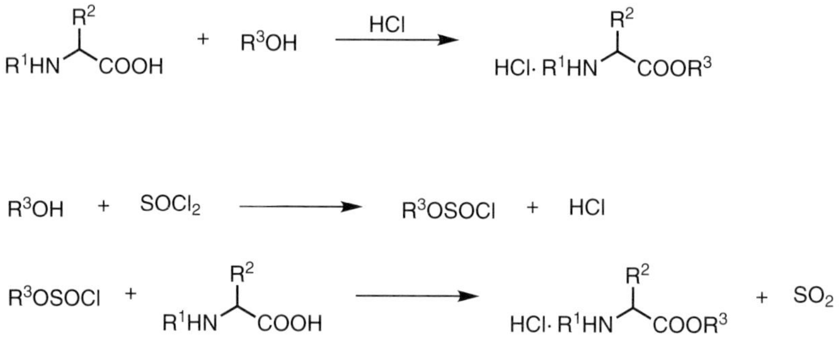

$$R^1HN\text{-}CH(R^2)\text{-}COOH + R^3OH \xrightarrow{HCl} HCl\cdot R^1HN\text{-}CH(R^2)\text{-}COOR^3$$

$$R^3OH + SOCl_2 \longrightarrow R^3OSOCl + HCl$$

$$R^3OSOCl + R^1HN\text{-}CH(R^2)\text{-}COOH \longrightarrow HCl\cdot R^1HN\text{-}CH(R^2)\text{-}COOR^3 + SO_2$$

Scheme 8.24

The corresponding benzyl and *p*-nitrobenzyl esters are obtained as *p*-toluenesulfonic acid salts by heating mixtures of amino acid, the benzyl alcohol, and *p*-toluenesulfonic acid in benzene (Scheme 8.25), while DCC-assisted coupling between *Z*-protected amino acids and *p*-methoxybenzyl alcohol is used for *p*-methoxybenzyl esters.

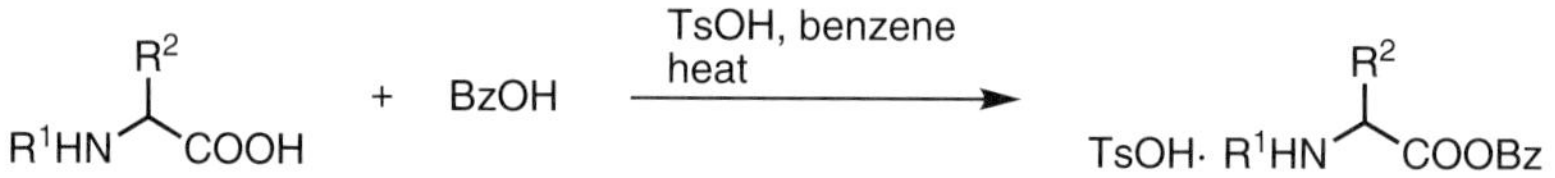

$$R^1HN\text{-}CH(R^2)\text{-}COOH + BzOH \xrightarrow[\text{heat}]{\text{TsOH, benzene}} TsOH\cdot R^1HN\text{-}CH(R^2)\text{-}COOBz$$

Scheme 8.25

The general way to produce *tert*-butyl esters is by treatment of *N*-protected amino acids with isobutene in the presence of H_2SO_4. Alternatively, the ester-interchange reaction between amino acid and *tert*-butyl acetate with H_2SO_4 or $HClO_4$ catalysis is also employed (Scheme 8.26).

$$ZHN\text{-}CH(R)\text{-}COOH + H_3C\text{-}C(=O)\text{-}O^tBu \xrightarrow{H^+} ZHN\text{-}CH(R)\text{-}COO^tBu$$

Scheme 8.26

Other methods, as shown in Scheme 8.27, are also feasible for practical production of amino acid esters.

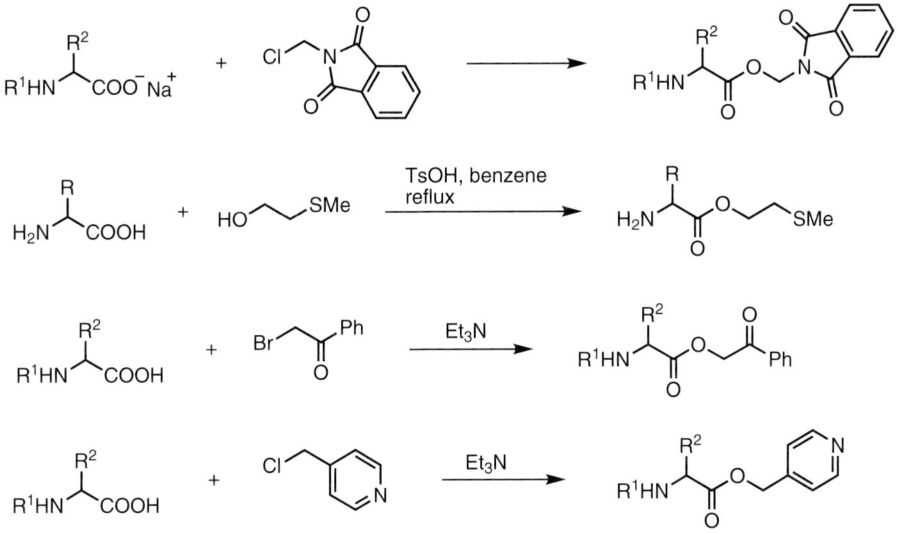

Scheme 8.27

Of commercially available amino acid esters, the most important one is α-L-aspartyl-L-phenylalanine methyl esters known as Aspartame. This amino acid ester was found to be 200 times sweeter than sugar by G. D. Searle & Co. in 1965 [1425], and was later marketed as a low-calorie sweetener in collaboration with Ajinomoto Co., Inc. A number of processes for this compound have been patented, one of them being shown in Scheme 8.28 [1426–1428]. L-Aspartic acid is transformed

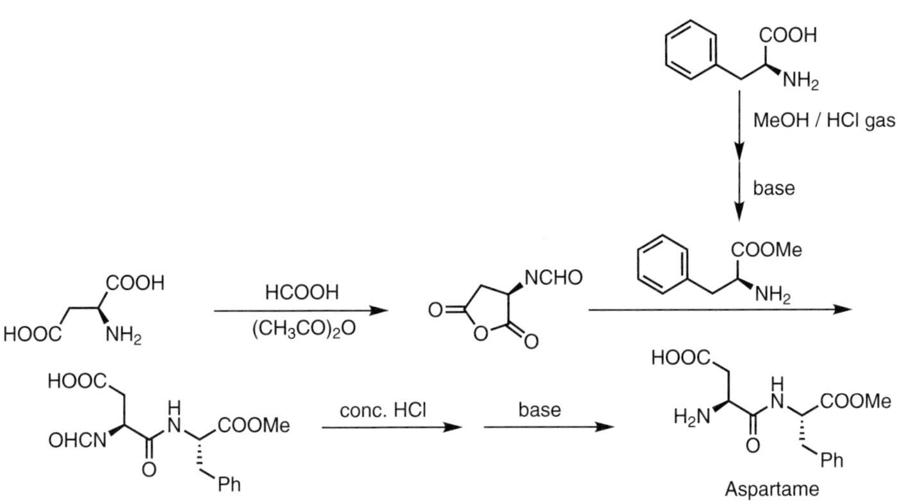

Scheme 8.28

into the corresponding *N*-formyl anhydride, which is then treated with methyl L-phenylalanate. Methyl *N*-formyl-α-L-aspartyl-L-phenylanate thus obtained is subjected to deformylation, followed by neutralization to give Aspartame.

It may be noted here that some amino acids function as blood pressure depressants. Enalapril from Merck and Captoril from Squibb are known as angiotensin converting enzyme inhibitors, although the latter is not an amino acid ester.

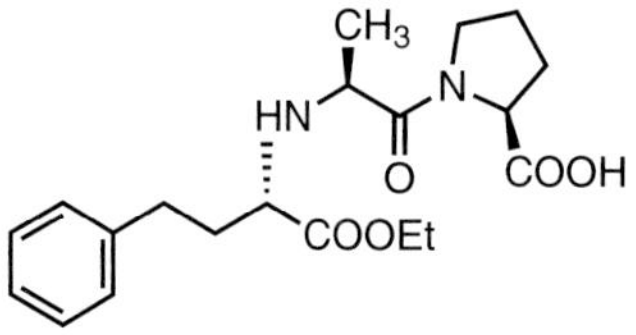

Enalapril

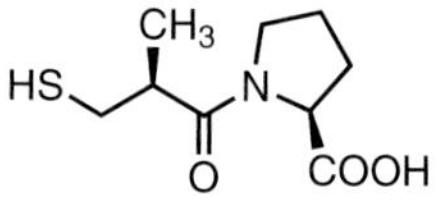

Captopril

Amino acid esters with long aliphatic alcohols, poly(ethylene glycol)s, and steroid alcohols, as shown below, are employed as surfactants for shampoo, cosmetic, and emollient agents.

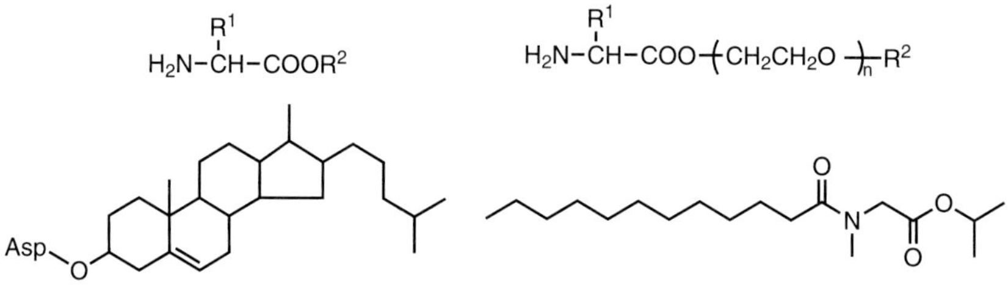

$H_2N\text{-}CH(R^1)\text{-}COOR^2$

$H_2N\text{-}CH(R^1)\text{-}COO\text{-}(CH_2CH_2O)_n\text{-}R^2$

8.7 Macrolides

Macrolide antibiotics function on bacteria ribosomes to inhibit protein synthesis, but not on human ribosomes, showing selective action only on bacteria. They involve a 14-, 15-, or 16-membered lactone ring as a core, to which at least one deoxy-sugar is attached. Erythromycin was the first macrolide antibiotic to be commercialized by Eli Lilly and Co. in 1952. As described in Section 7.5, there are many examples of macrolide synthesis, but the macrolide skeletons have so far been produced in practice by fermentation in industry. It should be noted, however, that some synthetic processes for macrolactonization are now in the last stage of development, and so will be commercialized in not too distant future. Remarkably, synthetic esterification technology frequently plays an important role in the incorporation of side chains, as mentioned in the case of taxol (see Section 7.5). A few more examples are given here, for which various esterification methods are invoked. A synthetic route for rokitamycin (3″-O-propanoylleucomycin A_5) is shown in Scheme 8.29 [1429]. Leucomycin A_5 undergoes acetylation at the 2′-hydroxy group, followed by silylation at the 3- and 9-hydroxy groups. Acylation

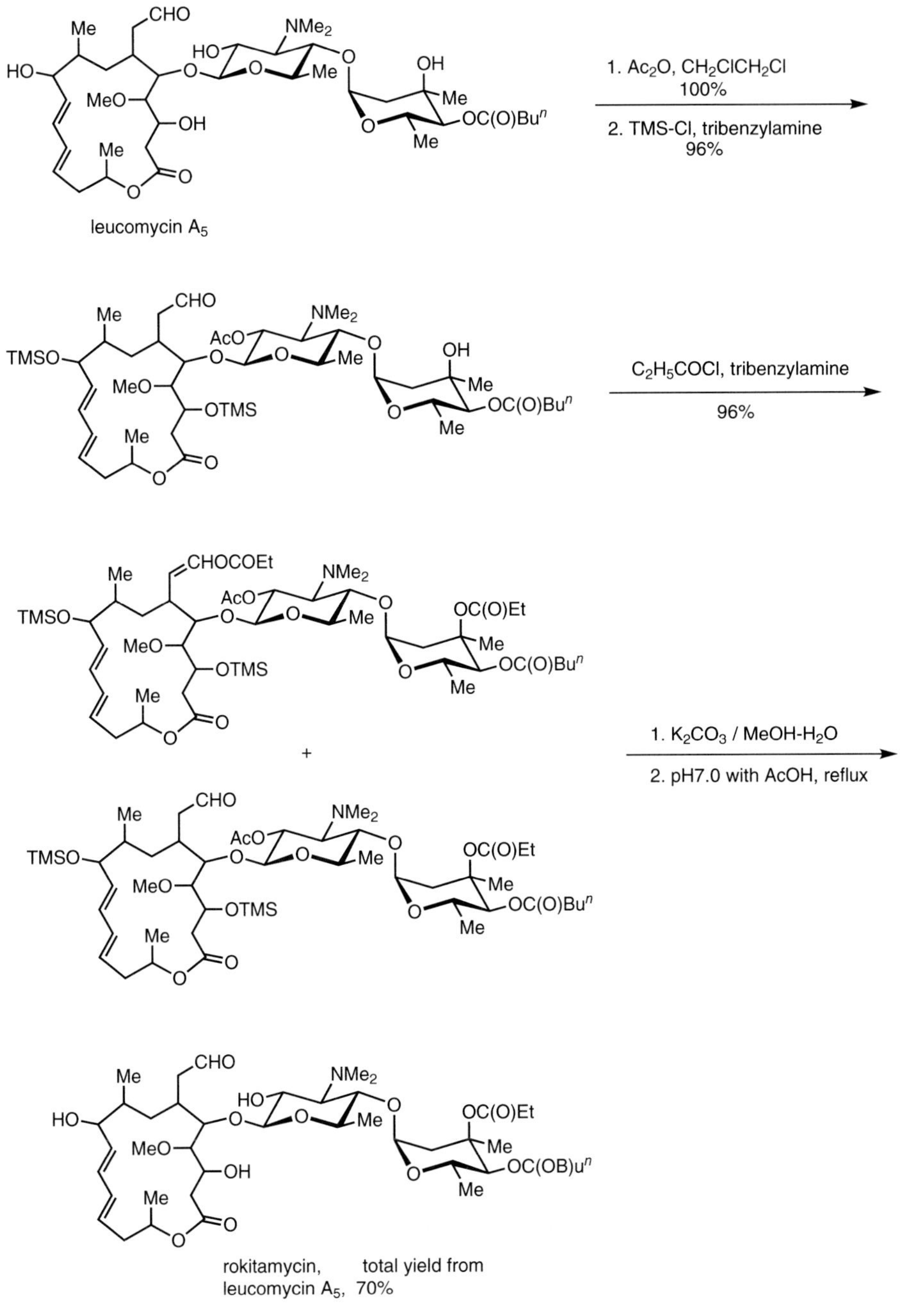

leucomycin A_5

1. Ac_2O, CH_2ClCH_2Cl 100%
2. TMS-Cl, tribenzylamine 96%

C_2H_5COCl, tribenzylamine 96%

+

1. K_2CO_3 / MeOH-H_2O
2. pH7.0 with AcOH, reflux

rokitamycin, total yield from leucomycin A_5, 70%

Scheme 8.29

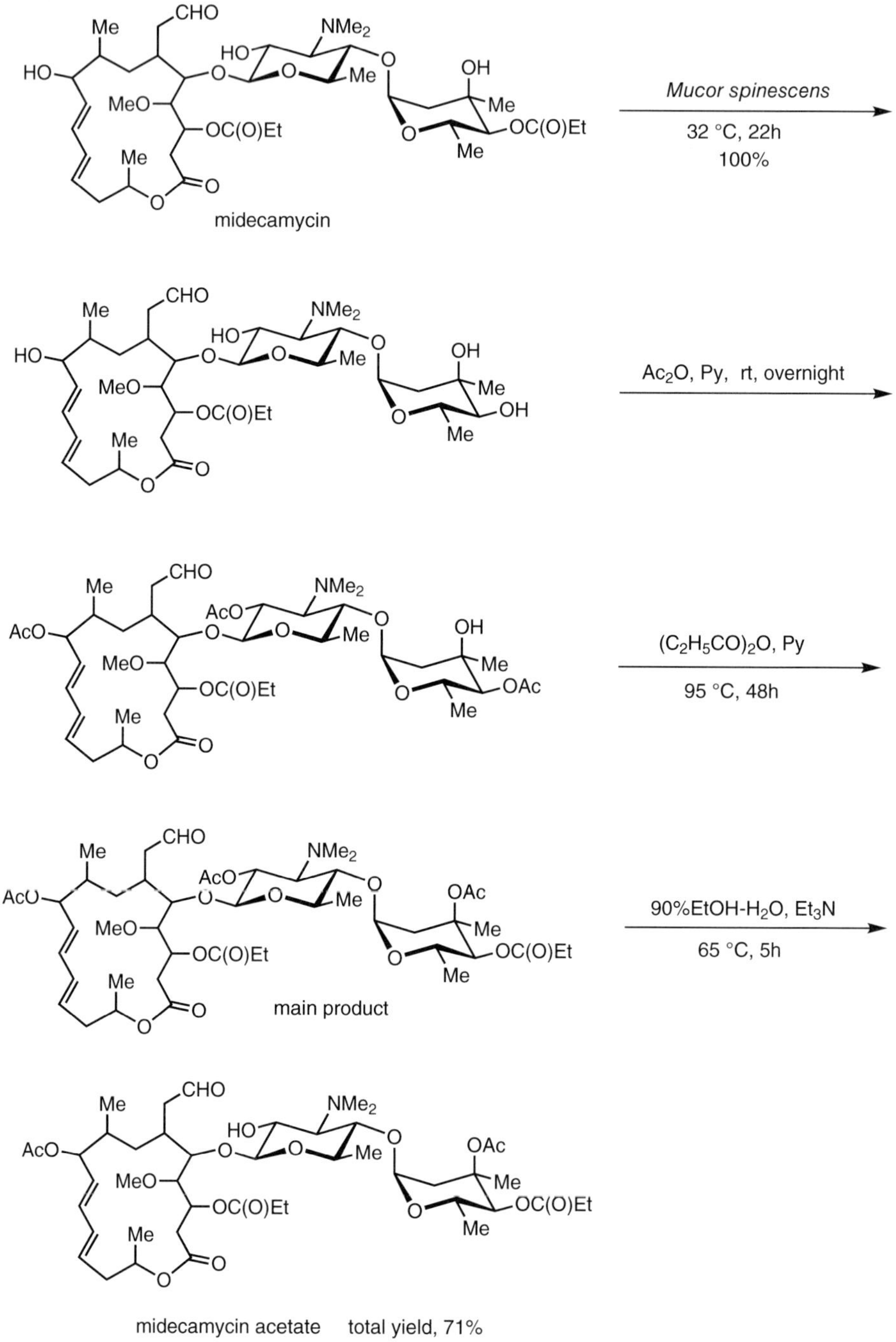

midecamycin

Mucor spinescens

32 °C, 22h

100%

Ac_2O, Py, rt, overnight

$(C_2H_5CO)_2O$, Py

95 °C, 48h

main product

90%EtOH-H_2O, Et_3N

65 °C, 5h

midecamycin acetate total yield, 71%

Scheme 8.30

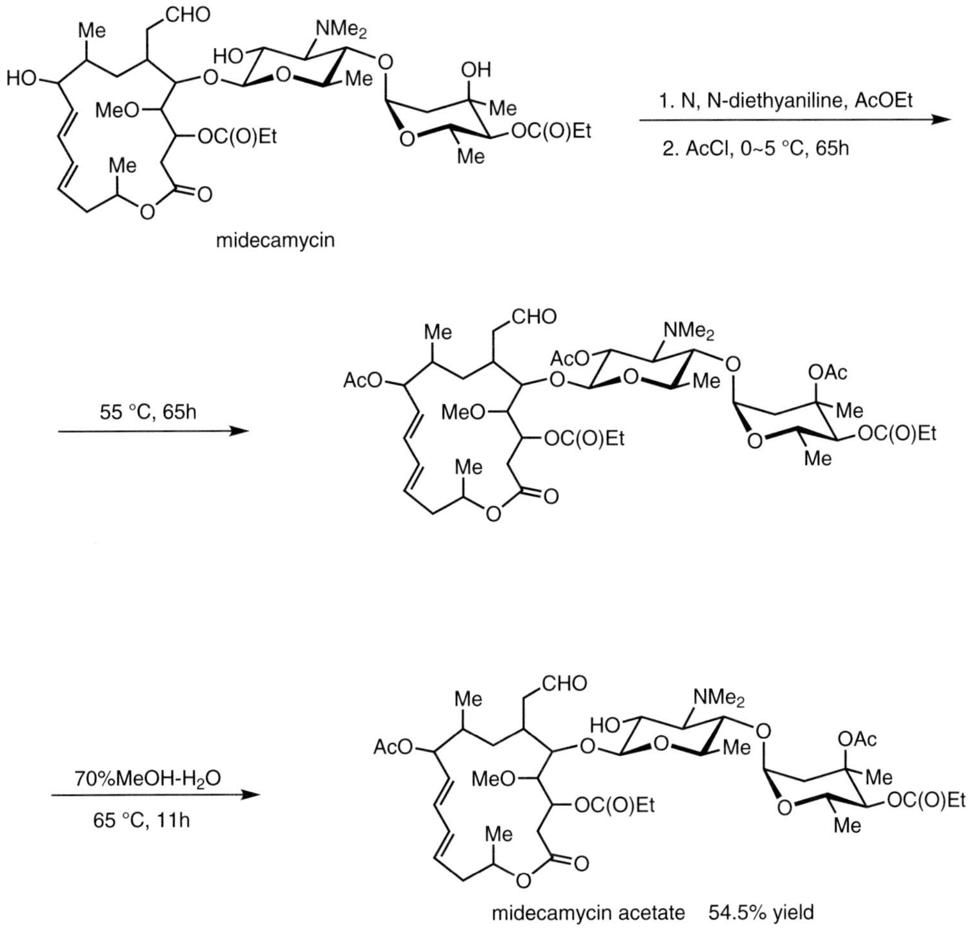

midecamycin

1. N, N-diethyaniline, AcOEt

2. AcCl, 0~5 °C, 65h

55 °C, 65h

70%MeOH-H_2O

65 °C, 11h

midecamycin acetate 54.5% yield

Scheme 8.31

with propanoyl chloride then gives 2′-*O*-acetyl-17,18-enol-18,3″-di-*O*-propanoyl-3,9-di-*O*-trimethylsilylleucomycin A_5 with a small amount of 2′-*O*-acetyl-3″-*O*-propanoyl-3,9-di-*O*-trimethylsilylleucomycin A_5. This mixture is subjected to manipulations for removal of trimethylsilyl, enol propanoyl, and acetyl groups, to give rokitamycin.

Scheme 8.30 shows a 4-step route for conversion of midecamycin into midecamycin acetate, involving a 4″-to-3″ acyl migration step [1430]. A simpler 2-step conversion is also available (Scheme 8.31) [1431].

8.8
Flavoring Agents and Fragrances

Flavoring agents are used in confectionery, chewing gums, beverages, alcoholic drinks, general foods, milk products, meat products, seasoning, and so on, while fragrances are used in perfumes, cosmetics, soaps, toothpaste, mouthwash, medical products, bath products, air care, aerosol products, insecticides, deodorants, paints, adhesive agents, rubber, plastics, leather, printing ink, textile industry, and so on. Most are synthetic flavors, and a great number of compounds are available.

Fragrance agents are classified into acyclic and cyclic compounds, depending on their structures, and also into another group, terpenes. Terpenes have a common $(C_5H_8)_n$ unit and are very important for flavoring. Monoterpenes $(C_5H_{10})_2$ and sesquiterpenes $(C_5H_{10})_3$ especially are present as odoriferous components in several plants, and are useful materials for various synthetic fragrance agents. Geraniol, with a rose-like odor, is a useful material for the synthesis of citral or other esters. Linalool is also important for the synthesis of ester fragrances, which have different odors depending on the absolute structure of the linalool.

OH

(*trans*)(*E*)-Geraniol

OH

(-)(3*S*)- or (+)(3*R*)-Linalool

These esters are prepared by the method demonstrated in previous sections (Sections 1.1, 1.2, 1.3, 1.4, and 3.1). The most popular processes are direct esterification between carboxylic acid and alcohol, or acylation of the alcohol with acid anhydride in the presence of a small amount of H_2SO_4 or TsOH at high temperature, toluene usually being used as solvent. If the reactants or the product esters are acid-sensitive, the esterification is conducted at higher temperatures without use of catalysts. Representative esters produced commercially in substantial amount are listed below.

Ethyl formate (ethyl methanoate)

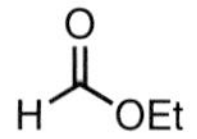

O

H OEt

(1) **Uses:** fruity flavor (peach, pineapple) for foods and butter, brandy and whisky (9.4 ~ 11 ppm); insecticide.

(2) **Synthesis:** usually prepared by esterification of ethyl alcohol and formic acid or by distillation of ethyl acetate and formic acid in the presence of concentrated H_2SO_4.

(3) **Organic/physical characteristics:** colorless, mobile, flammable liquid; ether-like sweet odor.

Butyl formate (butyl methoanoate)

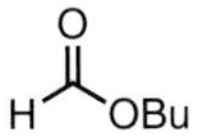

(1) **Uses:** top note of flowery flavors; fruity flavors (2.9 ~ 11 ppm); leathers.
(2) **Synthesis:** esterification of 1-butanol with formic acid in the presence of concentrated H_2SO_4.
(3) **Organic/physical characteristics:** colorless liquid; present in several natural products; ether- and rum-like odor, miscible in organic solvents.

Citronellyl formate (3,7-dimethyl-6-octen-1-yl formate)

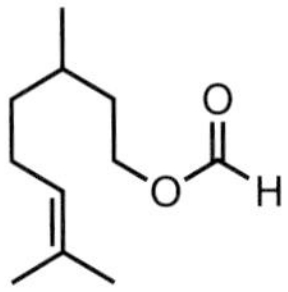

(1) **Uses:** geranium, bergamot, citrus, lavender flavors for soap; fruity, honey flavor for foods (14 ~ 32 ppm).

(2) **Synthesis:** prepared from citronellol and formic acid.

(3) **Organic/physical characteristics:** colorless oily liquid; present in geranium essential oil; strong fruity, rose-like odor with a sweet, fruity taste; soluble in alcohol, diethyl phthalate, oils; insoluble in water, glycerol.

Geranyl formate

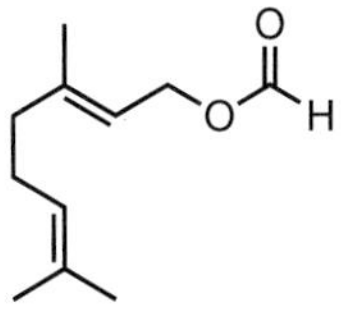

(1) **Uses:** top note of rose, tuberose, neroli, citrus, lavender flavors; apple, apricot, peach flavor for foods (0.8 ~ 7.5 ppm).
(2) **Synthesis:** esterification of geraniol with formic acid.
(3) **Organic/physical characteristics:** colorless to pale yellow liquid with rose-like odor; present in oil of geranium; insoluble in water.

Benzyl formate (benzyl methanoate)

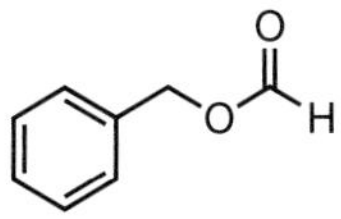

(1) **Uses:** flowery flavor (jasmine); cherry, apple, nut, strawberry flavor for foods (2.4 ~ 12 ppm).

(2) **Synthesis:** by heating a mixture of formic acetic anhydride and benzyl alcohol to 50 °C; by passing a mixture of formic acid and excess benzyl alcohol over a catalyst at high temperature.

(3) **Organic/physical characteristics:** colorless liquid; intense, pleasant, floral-fruity odor.

Ethyl acetate

O
Me OEt

(1) **Uses:** berry, fruits flavor for foods and butter, mint (~ 1500 ppm); industrial use (paints, etc.)
(2) **Synthesis:** esterification of 50% ethanol with 10% acetic acid; or catalytic oxidation of acetaldehyde.
(3) **Organic/physical characteristics:** colorless, mobile liquid; soluble in most organic solvents.

Isopropyl acetate

O
Me O^{i}Pr

(1) **Uses:** top note of citrus flavor.
(2) **Synthesis:** by direct acetylation of isopropyl alcohol in the presence of various catalysts (concentrated H_2SO_4, diethyl sulfate, chlorosulfonic acid, BF_3).
(3) **Organic/physical characteristics:** colorless liquid.

Isoamyl acetate (isopentyl acetate, 3-methyl-1-butyl acetate)

O
Me O

(1) **Uses:** fruits (banana, pear, apple, berries) flavor for foods, candy, butter, coconut, cola, rum (28 ~ 2700 ppm); odor for paint and insecticide.
(2) **Synthesis:** by esterification of commercial isoamyl alcohol with acetic acid in the presence of H_2SO_4 at high temperature.
(3) **Organic/physical characteristics:** colorless liquid with banana-like odor; almost insoluble in water, soluble in ether and most common organic solvents.

Linalyl acetate (3,7-dimethyl-1,6-octadien-3-yl acetate)

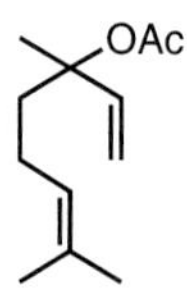

(1) **Uses:** bergamot, lavender, clay sage, jasmine flavor.
(2) **Synthesis:** almost direct acetylation of linalool with acetic anhydride.
(3) **Organic/physical characteristics:** colorless liquid with bergamot-lavender odor; soluble in essential oils and most common organic solvents.

Geranyl acetate (*trans*-3,7-dimethyl-2,6-octadien-1-yl acetate)

(1) **Uses:** flavor of soap, apple, apricot, banana, lemon, peach flavor for foods.

(2) **Synthesis:** esterification of geraniol with acetic acid.

(3) **Organic/physical characteristics:** colorless to pale yellow liquid; present in essential oils of lemongrass, neroli, Ceylon; pleasant, flowery odor; soluble in organic solvents.

Isobornyl acetate (*exo*-1,7,7-trimethylbicyclo[2.2.1]heptane-2-yl acetate)

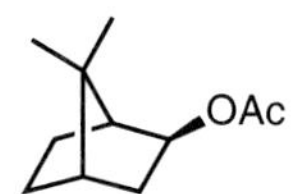

(1) **Uses:** soap, colon, shampoo, bath product, air care, woody and herbal flavor for aromatherapy.
(2) **Synthesis:** treatment of camphene with acetic acid, usually in the presence of catalyst; also by acetylation of isoborneol.
(3) **Organic/physical characteristics:** clear, colorless liquid, tends to yellow slightly on aging; readily soluble in organic solvents.

***p-tert*-Butylcyclohexyl acetate**

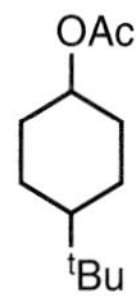

(1) **Uses:** soap, shampoo, toiletries because of cheapness and stability.
(2) **Synthesis:** Scheme 8.32
(3) **Organic/physical characteristics:** colorless liquid with orris and woody odor.

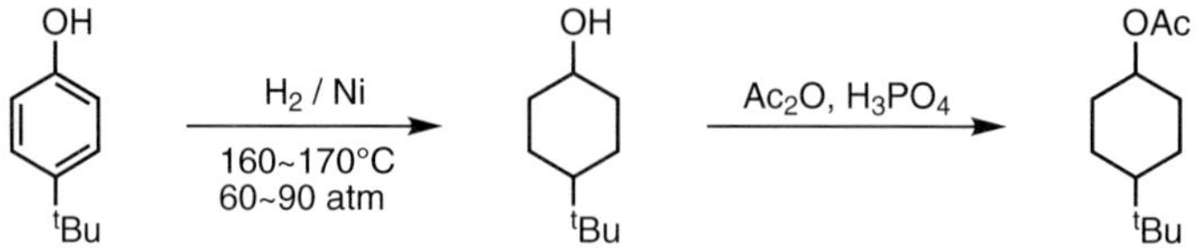

Scheme 8.32

Benzyl acetate (phenylmethyl acetate)

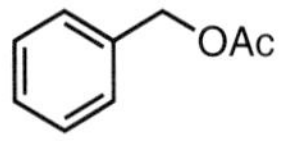

(1) **Uses:** industrial uses such as anesthetics, printing inks, lacquers; 5600 t produced annually worldwide.

(2) **Synthesis:** prepared from benzyl chloride and sodium acetate in the presence of a small amount of Ac_2O and pyridine at 160 ~ 170 °C; treatment of benzyl chloride with excess acetic acid in the presence of NaOH at 85 ~ 145 °C.

(3) **Organic/physical characteristics:** colorless liquid with flowery odor; present in jasmine, gardenia, hyacinth, apple, strawberry.

Ethyl acetoacetate

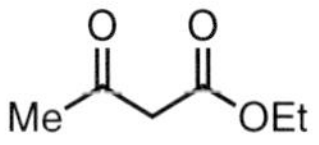

(1) **Uses:** pesticide, medicine, chelate crosslinking agent, etc.

(2) **Synthesis:** condensation of two molecules of ethyl acetate in the presence of sodium alkoxide; treatment of diketene with ethanol in the presence of acid or base catalyst.

(3) **Organic/physical characteristics:** colorless liquid with ether-like odor.

Geranyl propanoate (*trans*-3,7-dimethyl-2,6-octadien-1-yl propanoate)

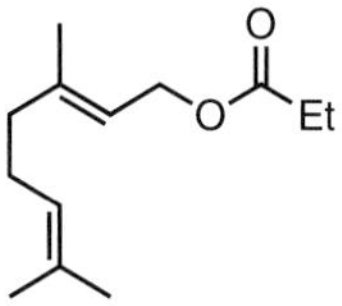

(1) **Uses:** essential for geranium and rose flavor, also used for gardenia, bergamot flavor.

(2) **Synthesis:** esterification of geraniol with propanoic acid in the presence of catalyst.
(3) **Organic/physical characteristics:** colorless liquid with fruity, rose-like odor.

Isoamyl isovalerate (isopentyl isovalerate, 3-methylbutyl 3-methylbutyrate)

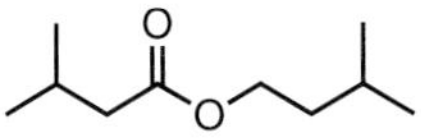

(1) **Uses:** essential for apple flavor; flavor of wine, honey, walnut.
(2) **Synthesis:** esterification of isoamyl alcohol with isovaleric acid.
(3) **Organic/physical characteristics:** colorless liquid; sweet, apple flavor.

Methyl benzoate

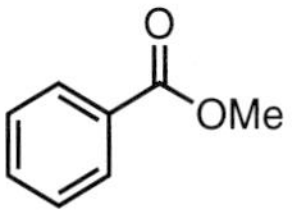

(1) **Uses:** base flavor of floral fragrance such as ylang-ylang; soap, industrial flavoring agent.
(2) **Synthesis:** direct esterification of methanol with benzoic acid.
(3) **Organic/physical characteristics:** colorless liquid; present in essential oil of ylang-ylang; insoluble in water, miscible with alcohol and ether.

Benzyl benzoate

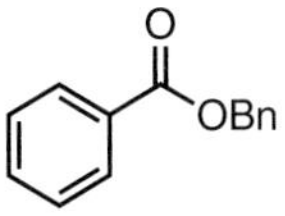

(1) **Uses:** floral fragrance such as tuberose, ylang-ylang; solvent, medicinal agents, platicizer.
(2) **Synthesis:** treatment of benzyl chloride with sodium acetate in the presence of NEt_3; esterification of benzyl alcohol with benzoic acid; Cannizaro reaction of benzaldehyde in the presence of alkali.
(3) **Organic/physical characteristics:** colorless liquid with light balsamic odor; present in oil of Tolu balsam and Peru balsam; insoluble in water or glycerol; soluble in alcohol, $CHCl_3$, and ether.

Methyl cinnamate (methyl 3-phenylpropanoate)

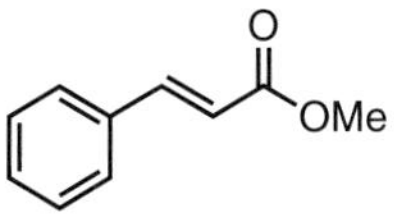

(1) **Uses:** applied to carnation fragrance; soap, mouthwash.

(2) **Synthesis:** direct esterification of ethanol with cinnamic acid in the presence of HCl; carboxymethylation from styrene.
(3) **Organic/physical characteristics:** white to slightly yellow, crystalline solid; soluble in propylene glycol, mineral oils; insoluble in water.

Methyl salicylate (methyl 2-hydroxybenzoate)

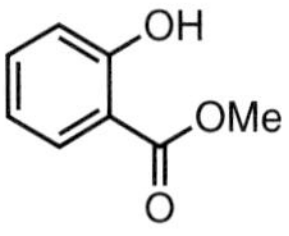

OH
OMe
O

(1) **Uses:** toothpaste, mousewash, medicinal agents.
(2) **Synthesis:** direct esterification of methanol with salicylic acid.
(3) **Organic/physical characteristics:** colorless liquid; found in oil of wintergreen.

Isoamyl salicylate (isopentyl salicylate)

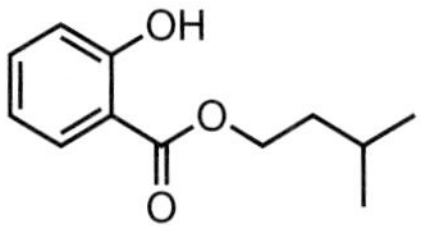

OH
O
O

(1) **Uses:** clover, carnation fragrance; cosmetics, perfumes, soaps.
(2) **Synthesis:** direct esterification of isoamyl alcohol with salicylic acid in the presence of catalyst.
(3) **Organic/physical characteristics:** colorless liquid; strong herbaceous, persistent odor.

Benzyl salicylate (benzyl *o*-hydroxybenzoate)

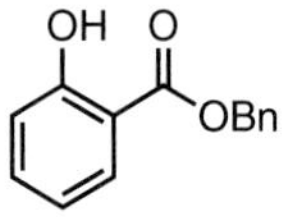

OH O
OBn

(1) **Uses:** flowery fragrance (jasmine, carnation); solvent of musk ambrette flavor.
(2) **Synthesis:** prepared from benzyl chloride and sodium salicylate; transesterification between methyl salicylate and benzyl alcohol (better yield).
(3) **Organic/physical characteristics:** colorless oily liquid; faint, sweet odor.

Allyl amyl glycolate (Allyl isoamyloxy acetate)
(1) **Uses:** perfume, shampoo.
(2) **Synthesis:** Scheme 8.33.
(3) **Organic/physical characteristics:** colorless liquid, fruity odor.

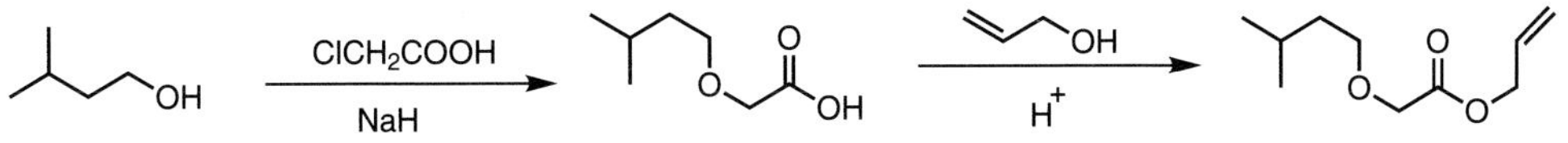

$ClCH_2COOH$ / NaH; allyl alcohol / H^+

Scheme 8.33

Rosamusk (α,3,3-trimethylcyclohexane-1-methyl acetate)

(1) **Uses:** cosmetic perfume.
(2) **Synthesis:** Scheme 8.34.
(3) **Organic physical characteristics:** colorless liquid, woody and floral odor.

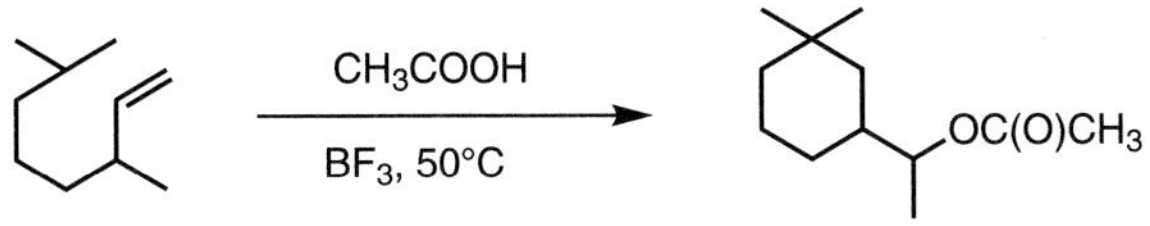

CH_3COOH / BF_3, 50°C → $OC(O)CH_3$

Scheme 8.34

Rosephenone (α-(trichloromethyl)benzyl acetate)

(1) **Uses:** Flavoring agent.
(2) **Synthesis:** Scheme 8.35.
(3) **Organic/physical characteristics:** white crystals, benzophenone-like odor.

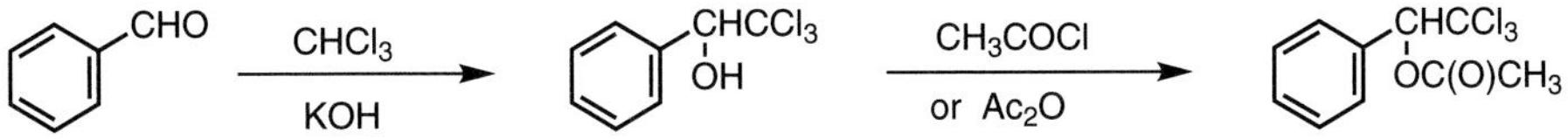

CHO → $CHCl_3$ / KOH → $CHCCl_3$, OH → CH_3COCl / or Ac_2O → $CHCCl_3$, $OC(O)CH_3$

Scheme 8.35

Jasmal (3-pentyltetrahydropyran-4-yl acetate)

(1) **Uses:** soap, detergent.
(2) **Synthesis:** Scheme 8.36.
(3) **Organic/physical characteristics:** colorless liquid, jasmine-like odor.

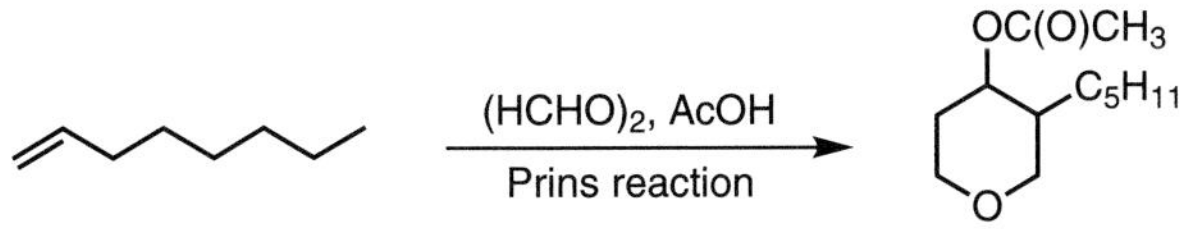

$(HCHO)_2$, AcOH / Prins reaction → $OC(O)CH_3$, C_5H_{11}, O

Scheme 8.36

Methyl cyclopentylidene acetate

(1) **Uses:** floral fragrances.
(2) **Synthesis:** Scheme 8.37.
(3) **Organic/physical characteristics:** colorless liquid, floral odor.

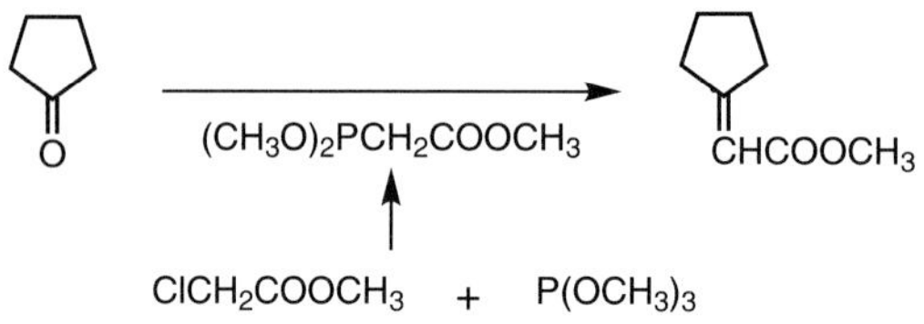

$(CH_3O)_2PCH_2COOCH_3$; $CHCOOCH_3$; $ClCH_2COOCH_3$ + $P(OCH_3)_3$

Scheme 8.37

Peranat (2-methyl-2-methylpentyl valerate)

(1) **Uses:** fragrances, lotion, soap.
(2) **Synthesis:** Scheme 8.38.
(3) **Organic/physical characteristics:** colorless liquid, floral odor.

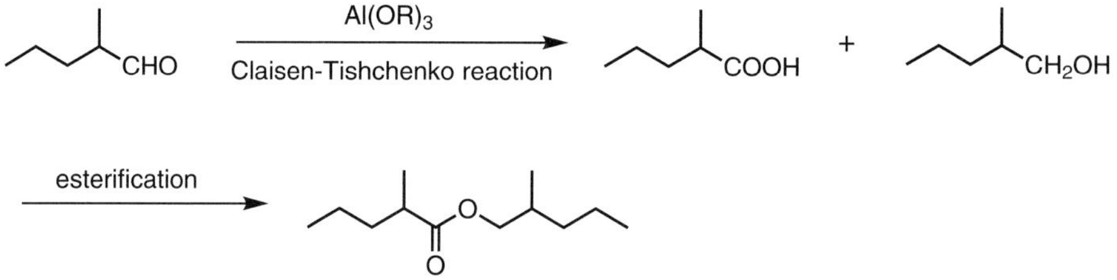

CHO ; $Al(OR)_3$; Claisen-Tishchenko reaction ; COOH + CH_2OH ; esterification

Scheme 8.38

Methyl dihydrojasmonate

(1) **Uses:** perfumes, toiletries, flavoring agents.
(2) **Synthesis:** Scheme 8.39.
(3) **Organic/physical characteristics:** colorless or pale yellow liquid, jasmine flavor.

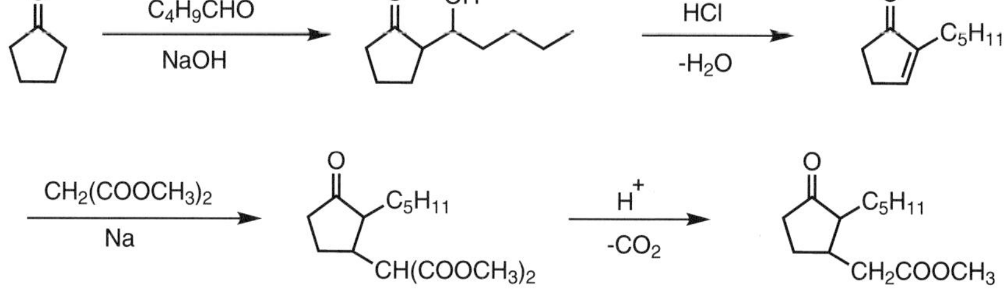

C_4H_9CHO ; NaOH ; OH ; HCl ; $-H_2O$; C_5H_{11} ; $CH_2(COOCH_3)_2$; Na ; C_5H_{11} ; $CH(COOCH_3)_2$; H^+ ; $-CO_2$; C_5H_{11} ; CH_2COOCH_3

Scheme 8.39

Methyl heptyn carbonate (methyl 2-octynoate)
(1) **Uses:** perfumes, flavoring agents.
(2) **Synthesis:** Scheme 8.40.
(3) **Organic/physical characteristics:** colorless or yellowish brown liquid, musk-like unpleasant odor.

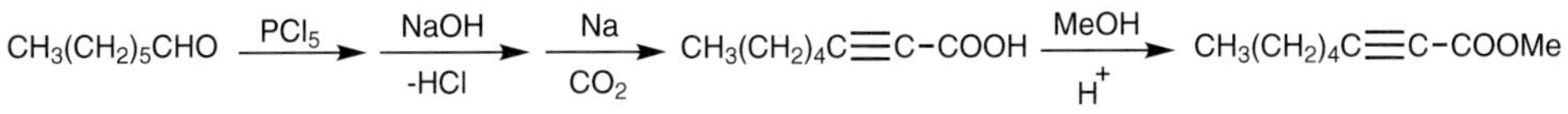

$$CH_3(CH_2)_5CHO \xrightarrow{PCl_5} \xrightarrow[-HCl]{NaOH} \xrightarrow[CO_2]{Na} CH_3(CH_2)_4C{\equiv}C{-}COOH \xrightarrow[H^+]{MeOH} CH_3(CH_2)_4C{\equiv}C{-}COOMe$$

Scheme 8.40

α-Angelicalactone (4-hydroxy-3-pentenoic acid γ-lactone)
(1) **Uses:** perfumes, flavoring agents.
(2) **Synthesis:** Scheme 8.41.
(3) **Organic/physical characteristics:** mp 18 °C.

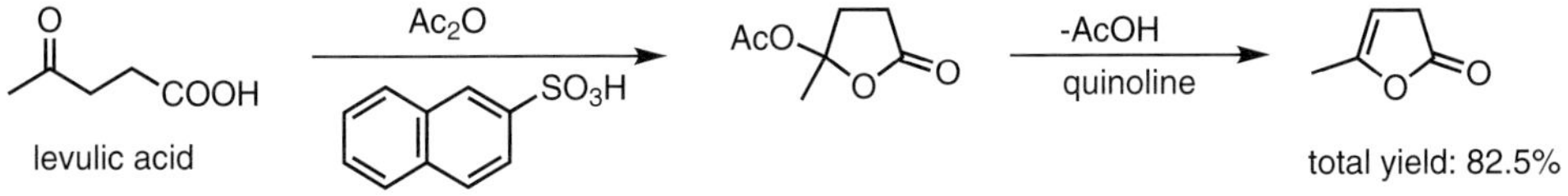

levulic acid — Ac_2O, naphthalene-SO_3H → AcO-lactone — -AcOH, quinoline → α-angelicalactone

total yield: 82.5%

Scheme 8.41

Whisky lactone (3-methyl-4-octanolide)
(1) **Uses:** perfumes, flavoring agents for nuts and liquor.
(2) **Synthesis:** Scheme 8.42.
(3) **Organic/physical characteristics:** colorless or pale yellow liquid.

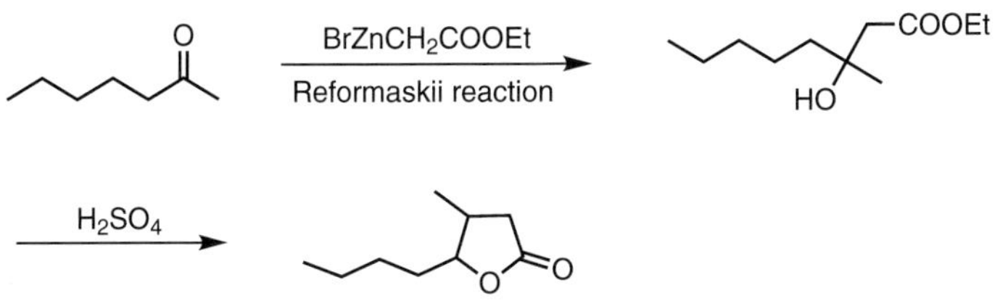

$BrZnCH_2COOEt$ / Reformaskii reaction → (COOEt, HO) → H_2SO_4 →

Scheme 8.42

Jasmine lactone (5-hydroxy-7-decenoic acid δ-lactone)

(1) **Uses:** perfumes, flavoring agents.
(2) **Synthesis:** Scheme 8.43.
(3) **Organic/physical characteristics:** major component of jasmine oil; colorless or pale yellow oil.

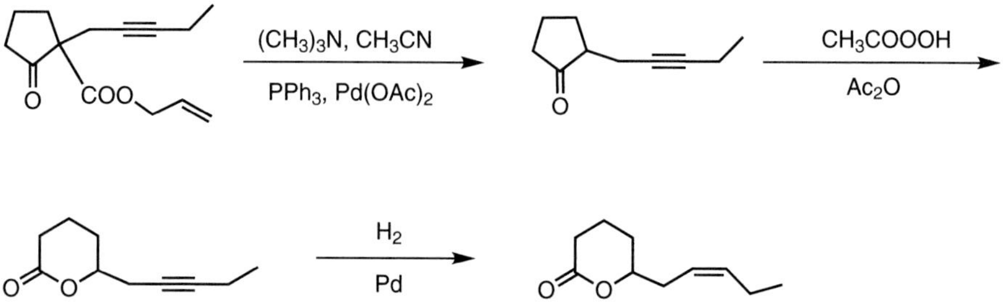

$(CH_3)_3N$, CH_3CN / PPh_3, $Pd(OAc)_2$; CH_3COOOH / Ac_2O ; H_2 / Pd

Scheme 8.43

δ-Undecalactone (5-hydroxyundecanoic acid lactone)

(1) **Uses:** fragrances for cosmetics, flavoring agents for butter, milk, and cream.
(2) **Synthesis:** Scheme 8.44.
(3) **Organic/physical characteristics:** colorless or pale yellow liquid.

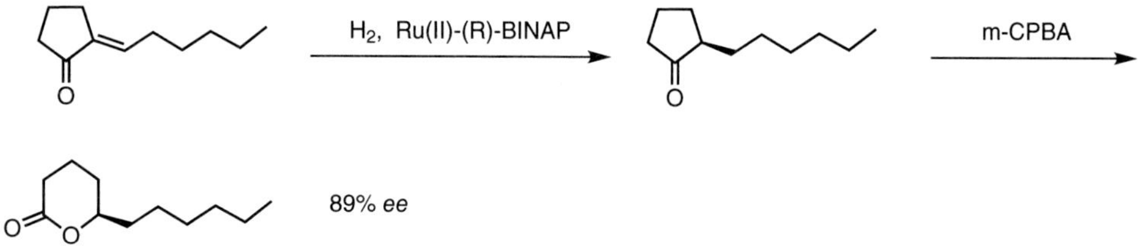

H_2, Ru(II)-(R)-BINAP ; m-CPBA

89% *ee*

Scheme 8.44

8.9 Pyrethroids

Pyrethroid esters, extracted from *Chrysanthemum cinerariaefolium*, are natural insecticides having six congeners (pyrethrin I and II, cinerin I and II, jasmolin I and II). It is expensive to extract these esters from their natural sources, so many synthetic pyrethroid esters have been developed, as shown below. NRDC-156, NRDC-149, and fenvalerate, which have strong insecticidal properties and are used as pesticide, are also produced.

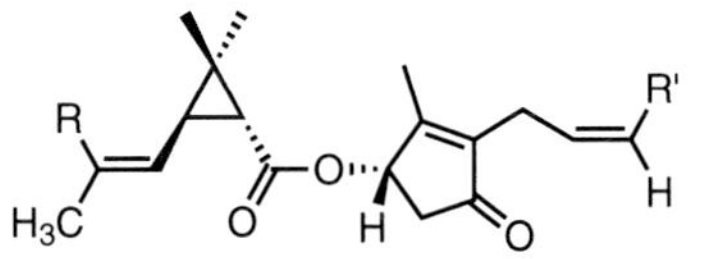

Pyrethrin I	R= CH_3, R'= $CH{=}CH_2$
Pyrethrin II	R= $COOCH_3$, R'= $CH{=}CH_2$
Cinerin I	R= CH_3, R'= CH_3
Cinerin II	R= $COOCH_3$, R'= CH_3
Jasmolin I	R= CH_3, R'= CH_2CH_3
Jasmolin II	R= $COOCH_3$, R'= CH_2CH_3

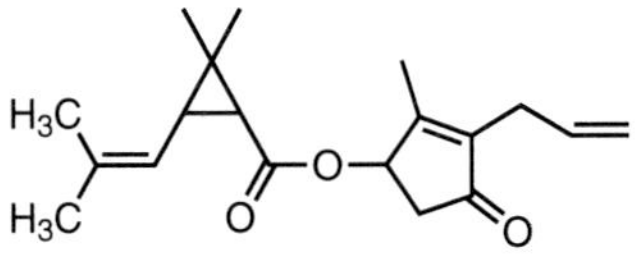

Allethrin

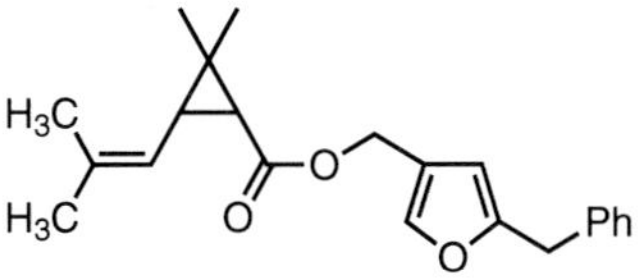

Resmethrin

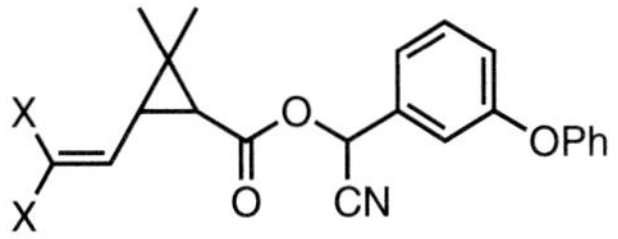

NRDC-156 X= Br
NRDC-149 X= Cl

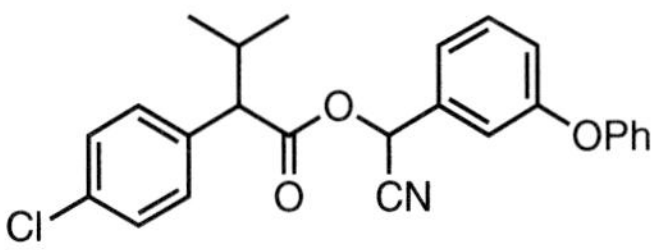

Fenvalerate

Pyrethroid ester components are produced in various ways. Some representative examples are described here. Classical methods are the azeotropic esterification of free acids and the reaction between acid chlorides and simple alcohols, but transesterification or ester interchange is also employable. Examples are: (i) treatment of methyl chrysanthemates with furfuryl acetate in the presence of NaOMe, and (ii) treatment of permethric acid ethyl ester with *meta*-phenoxybenzyl alcohol in the presence of NaOMe at elevated temperatures or the ester-interchange of the acetate of the latter alcohol in the presence of titanium tetraalkoxide to give permethrin (Scheme 8.45).

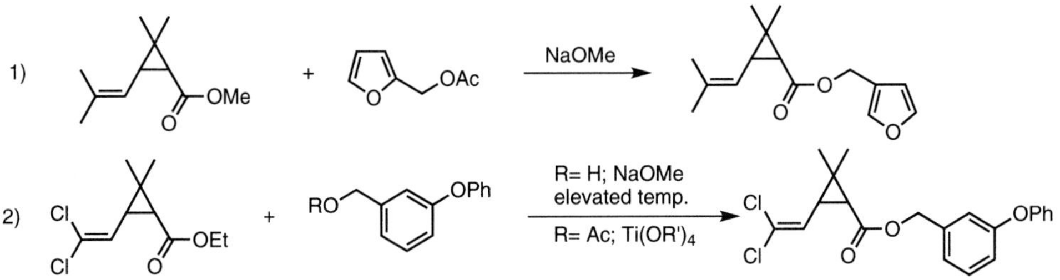

1) NaOMe

2) R= H; NaOMe elevated temp.
R= Ac; $Ti(OR')_4$

Permethrin

Scheme 8.45

Alkylation of acid salts is also usable (Scheme 8.46).

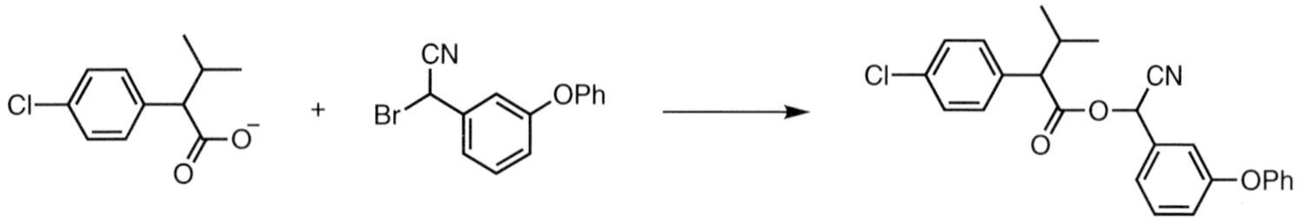

Scheme 8.46

NRDC-149 can be prepared by use of an anhydride (Scheme 8.47), and active fenvalerate can be prepared by making use of the silver salt of the carboxylic acid (Scheme 8.48).

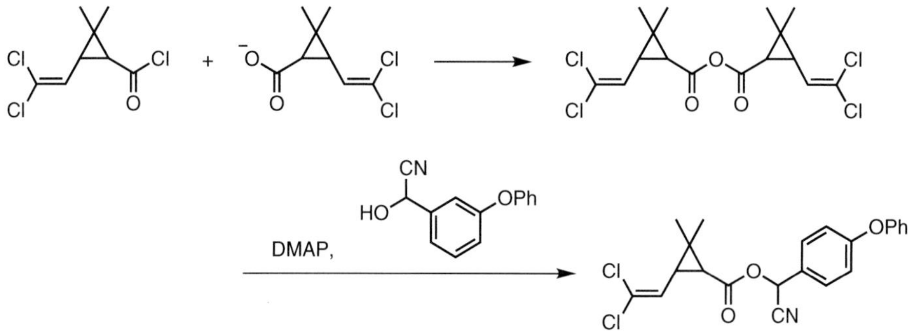

Scheme 8.47

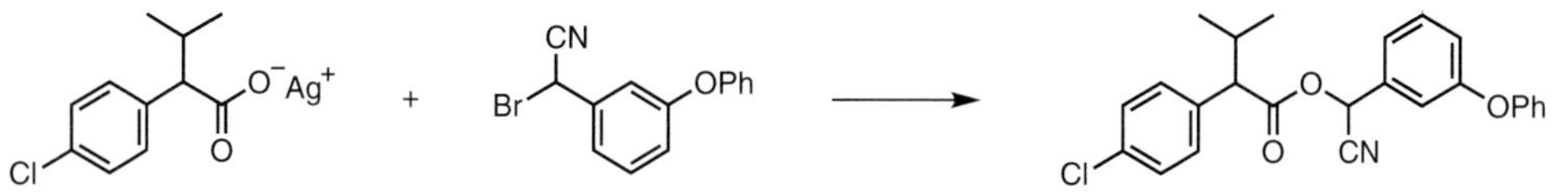

Scheme 8.48

Epilogue

The first edition concluded by pointing out several aspects of the esterification process that still need to be further improved or overcome. Some have gone well since then, and some have not. What happened is briefly summarized here.

We have seen great progress in the area of green chemistry. Access to greener catalysts has been the focus of attention, and a number of methods based on new solid acid catalysts (Section 1.1.2.3) and enzymes (Sections 1.1.7 and 1.2.5) have appeared. In the context of efficient utilization of resources, the employment of only the theoretical amounts of reactants has now become a popular topic (Section 7.2), although the ultimate goal of attaining 100% yield with this strict stoichiometry has unfortunately not always been achieved. In order to minimize or replace the use of organic solvents, various new reaction media have been applied to (trans)esterification (Section 7.3). Phase-transfer reaction as well as the use of surfactant-type catalysts enabled water to be used as a solvent. Fluorous catalysts and ionic liquids paved the way for new routes, even including enzymatic kinetic resolution. Also, supercritical carbon dioxide and hydrocarbons were successfully employed for enzyme-catalyzed (trans)esterification.

Selectivity was also an important issue. Many successful examples of enzymatic kinetic resolution appeared, together with some nonenzymatic processes (Chapter 5), showing considerable advancement in enantioselectivity. On other hand, little significant progress has been made in chemoselectivity, particularly in differentiating between identical or similar alcohols, an task which was recognized as a challenge in the first edition.

There has recently been a big leap forward in the biodiesel fuel industry. This is a good indication that even a classical chemical reaction such as (trans)esterification can still serve well for a modern industry. Nevertheless, the current processes, based on the use of alkali catalysts, are far from satisfactory in various respects and need to be updated by taking advantage of modern (trans)esterification techniques.

Esterification has continued to be of central importance in both organic synthesis and industry over the last decades, and it will never lose its significance in the future. However, it must again be emphasized that the demand for green esterification will continue to intensify.

Esterification. Methods, Reactions, and Applications. 2nd Ed. J. Otera and J. Nishikido

ISBN: 978-3-527-32289-3

References

1 Magerramov, M.N. (1995) *Zh. Prikl. Khim. (St Petersburg)*, **68**, 335–337.
2 Bromann, R., König, B. and Fischer, L. (1999) *Synth. Commun.*, **29**, 951–957.
3 Blanchard, L.A. and Brennecke, J.F. (2001) *Green Chem.*, **3**, 17–19.
4 Shimizu, M., Ishii, K. and Fujisawa, T. (1997) *Chem. Lett.*, 765–766.
5 Lowrance, W.W. Jr. (1971) *Tetrahedron Lett.*, **12**, 3453–3454.
6 Khurana, J.M., Sahoo, P.K. and Maikap, G.C. (1990) *Synth. Commun.*, **20**, 2267–2271.
7 Loupy, A., Petit, A., Ramdani, M. and Yvanaeff, C. (1993) *Can. J. Chem.*, **71**, 90–95.
8 Nudelman, A., Bechor, Y., Falb, E., Fischer, B., Wexler, B.A. and Nudelman, A. (1998) *Synth. Commun.*, **28**, 471–474.
9 Nakao, R., Oka, K. and Fukumoto, T. (1981) *Bull. Chem. Soc. Jpn.*, **54**, 1267–1268.
10 Brook, M.A. and Chan, T.H. (1983) *Synthesis*, 201–203.
11 Goswami, P., Hazarika, S., Borah, P. and Chowdhury, P. (2003) *Indian J. Chem., Sect. B*, **42**, 678–682.
12 Lee, A.S.-Y., Yang, H.-C. and Su, F.-Y. (2001) *Tetrahedron Lett.*, **42**, 301–303.
13 Hwu, J.R., Hsu, C.-Y. and Jain, M.L. (2004) *Tetrahedron Lett.*, **45**, 5151–5154.
14 Babler, J.H. and Coghlan, M.J. (1979) *Tetrahedron Lett.*, **20**, 1971–1974.
15 Ogiku, T., Yoshida, S., Ohmizu, H. and Iwasaki, T. (1995) *J. Org. Chem.*, **60**, 1148–1153.
16 Wright, S.W., Hageman, D.L., Wright, A.S. and McClure, L.D. (1997) *Tetrahedron Lett.*, **38**, 7345–7348.
17 Manabe, K. and Kobayashi, S. (2002) *Adv. Synth. Catal.*, **344**, 270–273.
18 Wakasugi, K., Misaki, T., Yamada, K. and Tanabe, Y. (2000) *Tetrahedron Lett.*, **41**, 5249–5252.
19 Gacem, B. and Jenner, G. (2003) *Tetrahedron Lett.*, **44**, 1391–1393.
20 Funatomi, T., Wakasugi, K., Misaki, T. and Tanabe, Y. (2006) *Green Chem.*, **8**, 1022–1027.
21 Ishihara, K., Nakagawa, S. and Sakakura, A. (2005) *J. Am. Chem. Soc.*, **127**, 4168–4169.
22 Palaniappan, S. and Ram, M.S. (2002) *Green Chem.*, **4**, 53–55.
23 Ram, M.S. and Palaniappan, S. (2003) *J. Mol. Catal. A*, **201**, 289–296.
24 Fraga-Dubreuil, J., Bourahla, K., Rahmouni, M., Bazureau, J.P. and Hamelin, J. (2002) *Catal. Commun.*, **3**, 185–190.
25 Cole, A.C., Jensen, J.L., Ntai, I., Tran, K.L.T., Weaver, K.J., Forbes, D.C. and Davis, J.H. Jr. (2002) *J. Am. Chem. Soc.*, **124**, 5962–5963.
26 Qiao, K., Hagiwara, H. and Yokoyama, C. (2006) *J. Mol. Catal. A*, **246**, 65–69.
27 Gu, Y., Ogawa, C. and Kobayashi, S. (2006) *Chem. Lett.*, **35**, 1176–1177.
28 Sowa, F.J. and Nieuwland, J.A. (1936) *J. Am. Chem. Soc.*, **58**, 271–272.
29 Kadaba, P.K., Carr, M., Tribo, M., Triplett, J. and Glasser, A.C. (1969) *J. Pharm. Sci.*, **58**, 1422–1423.
30 Marshall, J.L., Erickson, K.C. and Folsom, T.K. (1970) *Tetrahedron Lett.*, **11**, 4011–4012.
31 Kadaba, P.K. (1971) *Synthesis*, 316–317.

Esterification. Methods, Reactions, and Applications. 2nd Ed. J. Otera and J. Nishikido

ISBN: 978-3-527-32289-3

32 Kadaba, P.K. (1972) *Synthesis*, 628–630.
33 Shafiullah, Shamsuzzaman, Khan, B. and Ahmad, S. (1990) *Acta Chim. Hung.*, **127**, 705–710.
34 Iwagami, H. and Yasuda, N. (1991) *Chem. Express*, **6**, 277–280.
35 Bertho, J.-N., Ferrieres, V. and Plusquellec, D. (1995) *J. Chem. Soc., Chem. Commun.*, 1391–1393.
36 Kato, S. and Morie, T. (1996) *J. Heterocycl. Chem.*, **33**, 1171–1178.
37 Provent, C., Chautemps, P., Gellon, G. and Pierre, J.-L. (1996) *Tetrahedron Lett.*, **37**, 1393–1396.
38 Dyke, C.A. and Bryson, T.A. (2001) *Tetrahedron Lett.*, **42**, 3959–3961.
39 Ishihara, K., Ohara, S. and Yamamoto, H. (1996) *J. Org. Chem.*, **61**, 4196–4197.
40 Blossey, E.C., Turner, L.M. and Neckers, D.C. (1973) *Tetrahedron Lett.*, **14**, 1823–1826.
41 Karade, N.N., Shirodkar, S.G., Potrekar, R.A. and Karade, H.N. (2004) *Synth. Commun.*, **34**, 391–396.
42 Roy, H.N. and Al Mamun, A.H. (2006) *Synth. Commun.*, **36**, 2975–2981.
43 Chernova, I.K., Filimonova, E.I., Bychkov, B.N. and Solov'en, V.V. (1996) *Izv. Vyssh. U. Zaved, Khim. Khim. Tekhnol.*, **39**, 117–118.
44 Liao, X., Raghavan, G.S.V. and Yaylayan, V.A. (2002) *Tetrahedron Lett.*, **43**, 45–48.
45 Shekarriz, M., Taghipoor, S., Khalili, A.A. and Jamarani, M.S. (2003) *J. Chem. Res., Synop.*, 172–173.
46 Bartoli, G., Boeglin, J., Bosco, M., Locatelli, M., Massaccesi, M., Melchiorre, P. and Sambri, L. (2005) *Adv. Synth. Catal.*, **347**, 33–38.
47 Mineno, T. and Kansui, H. (2006) *Chem. Pharm. Bull.*, **54**, 918–919.
48 Cho, C.S., Kim, D.T., Choi, H.-J., Kim, T.-J. and Shim, S.C. (2002) *Bull. Korean Chem. Soc.*, **23**, 539–540.
49 Steliou, K., Szczygielska-Nowosielska, A., Favre, A., Poupart, M.A. and Hanessian, S. (1980) *J. Am. Chem. Soc.*, **102**, 7578–7579.
50 Steliou, K. and Poupart, M.-A. (1983) *J. Am. Chem. Soc.*, **105**, 7130–7138.
51 Otera, J., Yano, T., Himeno, Y. and Nozaki, H. (1986) *Tetrahedron Lett.*, **27**, 4501–4504.
52 Otera, J., Dan-oh, N. and Nozaki, H. (1991) *J. Org. Chem.*, **56**, 5307–5311.
53 Otera, J., Kawada, K. and Yano, T. (1996) *Chem. Lett.*, 225–226.
54 Xiang, J., Orita, A. and Otera, J. (2002) *Angew. Chem. Int. Ed.*, **41**, 4117–4119.
55 Kumar, A.K. and Chattopadhyay, T.K. (1987) *Tetrahedron Lett.*, **28**, 3713–3714.
56 Chen, C.-T. and Munot, Y.S. (2005) *J. Org. Chem.*, **70**, 8625–8627.
57 Gowda, S. and Rai, K.M.L. (2004) *J. Mol. Catal. A*, **217**, 27–29.
58 Kumar, B., Parmar, A. and Kumar, H. (1994) *Indian J. Chem., Sect. B*, **33**, 698–699.
59 Kumar, B., Kumar, H. and Parmar, A. (1992) *Synth. Commun.*, **22**, 1087–1094.
60 Heravi, M.M., Behbahani, F.K., Shoar, R.H. and Oskooie, H.A. (2006) *Catal. Commun.*, **7**, 136–139.
61 Zhang, G.-S. (1998) *Synth. Commun.*, **28**, 1159–1162.
62 Xu, Q.-H., Liu, W.-Y., Chen, B.-H. and Ma, Y.-X. (2001) *Synth. Commun.*, **31**, 2113–2117.
63 Sharma, G.V.M., Mahalingam, A.K., Nagarajan, M., Ilangovan, A. and Radhakrishna, P. (1999) *Synlett*, 1200–1202.
64 Mantri, K., Nakamura, R., Komura, K. and Sugi, Y. (2005) *Chem. Lett.*, **34**, 1502–1503.
65 Ram, R.N. and Charles, I. (1997) *Tetrahedron*, **53**, 7335–7340.
66 Ho, T.-L. (1989) *Synth. Commun.*, **19**, 2897–2898.
67 Iranpoor, N., Firouzabadi, H. and Zolfigol, M.A. (1998) *Synth. Commun.*, **28**, 1923–1934.
68 Saravanan, P. and Singh, V.K. (1999) *Tetrahedron Lett.*, **40**, 2611–2614.
69 Ma, J., Jiang, H. and Gong, H. (2005) *Org. Prep. Proced. Int.*, **37**, 87–92.
70 Nag, P., Bohra, R. and Mehrotra, R.C. (2002) *J. Chem. Res., Synop.*, 86–88.
71 Chandrasekhar, S., Sultana, S.S., Narsihmulu, C., Yadav, J.S., Gree, R. and Guillemin, J.C. (2002) *Tetrahedron Lett.*, **43**, 8335–8337.
72 Takasu, A., Oishi, Y., Iio, Y., Inai, Y. and Hirabayashi, T. (2003) *Macromolecules*, **36**, 1772–1774.
73 Ishihara, K., Ohara, S. and Yamamoto, H. (2000) *Science*, **290**, 1140–1142.

74 Ishihara, K., Nakayama, M., Ohara, S. and Yamamoto, H. (2001) *Synlett*, 1117–1120.

75 Ishihara, K., Nakayama, M., Ohara, S. and Yamamoto, H. (2002) *Tetrahedron*, **58**, 8179–8188.

76 Nakayama, M., Sato, A., Ishihara, K. and Yamamoto, H. (2004) *Adv. Synth. Catal.*, **346**, 1275–1279.

77 Mantri, K., Komura, K. and Sugi, Y. (2005) *Synthesis*, 1939–1944.

78 Sato, A., Nakamura, Y., Maki, T., Ishihara, K. and Yamamoto, H. (2005) *Adv. Synth. Catal.*, **347**, 1337–1340.

79 Nakamura, Y., Maki, T., Wang, X., Ishihara, K. and Yamamoto, H. (2006) *Adv. Synth. Catal.*, **348**, 1505–1510.

80 Ramalinga, K., Vijayalakshmi, P. and Kaimal, T.N.B. (2002) *Tetrahedron Lett.*, **43**, 879–882.

81 Olah, G.A., Keumi, T. and Meidar, D. (1978) *Synthesis*, 929–930.

82 Cho, B.R. and Yang, H.J. (1991) *Bull. Korean Chem. Soc.*, **12**, 9–10.

83 Cho, B.R. and Yang, H.J. (1992) *Bull. Korean Chem. Soc.*, **13**, 586–587.

84 Anand, R.C. and Selvapalam, N. (1994) *Synth. Commun.*, **24**, 2743–2747.

85 Anand, R.C. Vimal (1998) *Synth. Commun.*, **28**, 1963–1965.

86 Hoefnagel, A.J., Gunnewegh, E.A., Downing, R.S. and Van Bekkum, H. (1995) *J. Chem. Soc., Chem. Commun.*, 225–226.

87 Gangadwala, J., Mankar, S., Mahajani, S., Kienle, A. and Stein, E. (2003) *Ind. Eng. Chem. Res.*, **42**, 2146–2155.

88 Lichtenthaler, F.W., Klimesch, R., Mulet, V. and Kunz, M. (1993) *Liebigs Ann. Chem.*, 975–980.

89 Boyer, F.-D., Pancrazi, A. and Lallemand, J.-Y. (1995) *Synth. Commun.*, **25**, 1099–1108.

90 Delfourne, E., Despeyroux, P., Gorrichon, L. and Veronique, J. (1991) *J. Chem. Res., Synop.*, 56–57.

91 Effenberger, F., Hopf, M., Ziegler, T. and Hudelmayer, J. (1991) *Chem. Ber.*, **124**, 1651–1659.

92 Gumaste, V.K., Deshmukh, A.R.A.S. and Bhawal, B.M. (1996) *Indian J. Chem., Sect. B*, **35**, 1174–1179.

93 Ookoshi, T. and Onaka, M. (1998) *Tetrahedron Lett.*, **39**, 293–296.

94 Wegman, M.A., Elzinga, J.M., Neeleman, E., van Rantwijk, F. and Sheldon, R.A. (2001) *Green Chem.*, **3**, 61–64.

95 Ding, Y., Wu, R. and Lin, Q. (2002) *Synth. Commun.*, **32**, 2149–2153.

96 Kirumakki, S.R., Nagaraju, N. and Narayanan, S. (2004) *Appl. Catal. A*, **273**, 1–9.

97 Wu, K.-C. and Chen, Y.-W. (2004) *Appl. Catal. A*, **257**, 33–42.

98 Kirumakki, S.R., Nagaraju, N. and Chary, K.V.R. (2006) *Appl. Catal. A*, **299**, 185–192.

99 Palani, A. and Pandurangan, A. (2005) *J. Mol. Catal. A*, **226**, 129–134.

100 Jermy, B.R. and Pandurangan, A. (2005) *Appl. Catal. A*, **288**, 25–33.

101 Jermy, B.R. and Pandurangan, A. (2005) *J. Mol. Catal. A*, **237**, 146–154.

102 Palani, A. and Pandurangan, A. (2006) *J. Mol. Catal. A*, **245**, 101–105.

103 Palani, A., Palanichamy, M. and Pandurangan, A. (2007) *Catal. Lett.*, **115**, 40–45.

104 Waghmode, S.B., Thakur, V.V., Sudalai, A. and Sivasanker, S. (2001) *Tetrahedron Lett.*, **42**, 3145–3147.

105 Mbaraka, I.K., Radu, D.R., Lin, V.S.-Y. and Shanks, B.H. (2003) *J. Catal.*, **219**, 329–336.

106 Mbaraka, I.K. and Shanks, B.H. (2005) *J. Catal.*, **229**, 365–373.

107 Díaz, I., Mohino, F., Pérez-Pariente, J. and Sastre, E. (2003) *Appl. Catal. A*, **242**, 161–169.

108 Yang, Q., Kapoor, M.P., Inagaki, S., Shirokura, N., Kondo, J.N. and Domen, K. (2005) *J. Mol. Catal. A*, **230**, 85–89.

109 Nakajima, K., Tomita, I., Hara, M., Hayashi, S., Domen, K. and Kondo, J.N. (2006) *Catal. Today*, **116**, 151–156.

110 Alvaro, M., Corma, A., Das, D., Fornés, V. and García, H. (2004) *Chem. Commun.*, 956–957.

111 Alvaro, M., Corma, A., Das, D., Fornés, V. and García, H. (2005) *J. Catal.*, **231**, 48–55.

112 Kuriakose, G. and Nagaraju, N. (2004) *J. Mol. Catal. A*, **223**, 155–159.

113 Peters, T.A., Benes, N.E., Holmen, A. and Keurentjes, J.T.F. (2006) *Appl. Catal. A*, **297**, 182–188.

114 Suwannakarn, K., Lotero, E. and Goodwin, J.G. Jr. (2007) *Catal. Lett.*, **114**, 122–128.
115 Li, J.-T., Li, H.-Y. and Li, H.-Z. (2004) *J. Chem. Res.*, 416–417.
116 Sharghi, H., Sarvari, M.H. and Eskandari, R. (2005) *J. Chem. Res.*, 488–491.
117 Sathe, M. and Kaushik, M.P. (2006) *Catal. Commun.*, **7**, 644–646.
118 D'Souza, J. and Nagaraju, N. (2006) *Ind. J. Chem. Tech.*, **13**, 605–613.
119 Chen, Z.-H., Izuka, T. and Tanabe, K. (1984) *Chem. Lett.*, 1085–1088.
120 Hiyoshi, M., Lee, B., Lu, D., Hara, M., Kondo, J.N. and Domen, K. (2004) *Catal. Lett.*, **98**, 181–186.
121 Takahashi, K., Shibagaki, M. and Matsushita, H. (1989) *Bull. Chem. Soc. Jpn.*, **62**, 2353–2361.
122 Manohar, B., Reddy, V.R. and Reddy, B.M. (1998) *Synth. Commun.*, **28**, 3183–3187.
123 Olah, G.A., Liand, G. and Staral, J. (1974) *J. Am. Chem. Soc.*, **96**, 8113–8115.
124 Takagaki, A., Toda, M., Okamura, M., Kondo, J.N., Hayashi, S., Domen, K. and Hara, M. (2006) *Catal. Today*, **116**, 157–161.
125 Castanheiro, J.E., Ramos, A.M., Fonseca, I.M. and Vital, J. (2006) *Appl. Catal. A*, **311**, 17–23.
126 Choudary, B.M., Bhaskar, V., Kantam, M.L., Rao, K.K. and Raghavan, K.V. (2000) *Green Chem.*, **2**, 67–70.
127 Kantam, M.L., Bhaskar, V. and Choudary, B.M. (2002) *Catal. Lett.*, **78**, 185–188.
128 Kawabata, T., Mizugaki, T., Ebitani, K. and Kaneda, K. (2003) *Tetrahedron Lett.*, **44**, 9205–9208.
129 Srinivas, K.V.N.S. and Das, B. (2003) *J. Org. Chem.*, **68**, 1165–1167.
130 Reddy, C.R., Vijayakumar, B., Iyengar, P., Nagendrappa, G. and Prakash, B.S.J. (2004) *J. Mol. Catal. A*, **223**, 117–122.
131 Reddy, C.R., Iyengar, P., Nagendrappa, G. and Prakash, B.S.J. (2005) *Catal. Lett.*, **101**, 87–91.
132 Vijayakumar, B., Iyengar, P., Nagendrappa, G. and Prakash, B.S.J. (2005) *J. Indian Chem. Soc.*, **82**, 922–925.
133 Jiang, Y.-X., Chen, X.-M., Mo, Y.-F. and Tong, Z.-F. (2004) *J. Mol. Catal. A*, **213**, 231–234.
134 Banerjee, A., Sengupta, S., Adak, M.M. and Banerjee, G.C. (1983) *J. Org. Chem.*, **48**, 3106–3108.
135 Malakov, P.Y., De La Torre, M.C., Rodriguez, B. and Papanov, G.Y. (1991) *Tetrahedron*, **47**, 10129–10136.
136 Kollenz, G., Kappe, C.O. and Abd El Nabey, H. (1991) *Heterocycles*, **32**, 669–673.
137 Balasubramaniyan, V., Bhatia, V.G. and Wach, S.B. (1983) *Tetrahedron*, **39**, 1475–1485.
138 Banerjee, A., Mohan, M., Das, S., Banerjee, S. and Sengupta, S. (1987) *J. Indian Chem. Soc.*, **64**, 34–37.
139 Nomura, R., Miyazaki, S., Nakano, T. and Matsuda, H. (1991) *Appl. Organomet. Chem.*, **5**, 513–516.
140 Parikh, A. and Chudasama, U. (2003) *Ind. J. Chem. Tech.*, **10**, 44–47.
141 Hino, M., Takasaki, S., Furuta, S., Matsuhashi, H. and Arata, K. (2006) *Catal. Commun.*, **7**, 162–165.
142 Furuta, S., Matsuhashi, H. and Arata, K. (2004) *Appl. Catal. A*, **269**, 187–191.
143 Du, Y., Liu, S., Zhang, Y., Yin, C., Di, Y. and Xiao, F.-S. (2006) *Catal. Lett.*, **108**, 155–158.
144 Jin, T.-S., Ma, Y.-R., Li, Y., Sun, X. and Li, T.-S. (2001) *Synth. Commun.*, **31**, 2051–2054.
145 Zhang, Z., Pan, H., Hu, C., Fu, F., Sun, Y., Willem, R. and Gielen, M. (1991) *Appl. Organomet. Chem.*, **5**, 183–190.
146 Pizzio, L.R. and Blanco, M.N. (2003) *Appl. Catal. A*, **255**, 265–277.
147 Giri, B.Y., Rao, K.N., Devi, B.L.A.P., Lingaiah, N., Suryanarayana, I., Prasad, R.B.N. and Prasad, P.S.S. (2005) *Catal. Commun.*, **6**, 788–792.
148 Giri, B.Y., Devi, B.L.A.P., Gangadhar, K.N., Rao, K.N., Lingaiah, N., Prasad, P.S.S. and Prasad, R.B.N. (2006) *Synth. Commun.*, **36**, 7–11.
149 Bamoharram, F.F., Heravi, M.M., Roshani, M., Jahangir, M. and Gharib, A. (2006) *Appl. Catal. A*, **302**, 42–47.
150 Rafiee, E., Tangestaninejad, S., Habibi, M.H. and Mirkhani, V. (2004) *Bull. Korean Chem. Soc.*, **25**, 599–600.

151 Patel, S., Purohit, N. and Patel, A. (2003) *J. Mol. Catal. A*, **192**, 195–202.

152 Pizzio, L., Vázquez, P., Cáceres, C. and Blanco, M. (2001) *Catal. Lett.*, **77**, 233–239.

153 Bhatt, N. and Patel, A. (2005) *J. Mol. Catal. A*, **238**, 223–228.

154 Ramu, S., Lingaiah, N., Devi, B.L.A.P., Prasad, R.B.N., Suryanarayana, I. and Prasad, P.S.S. (2004) *Appl. Catal. A*, **276**, 163–168.

155 Rao, K.N., Sridhar, A., Lee, A.F., Tavener, S.J., Young, N.A. and Wilson, K. (2006) *Green Chem.*, **8**, 790–797.

156 Eshghi, H., Rafei, M. and Karimi, M.H. (2001) *Synth. Commun.*, **31**, 771–774.

157 El-Wahab, M.M.M.A. and Said, A.A. (2005) *J. Mol. Catal. A*, **240**, 109–118.

158 Sepúlveda, J.H., Yori, J.C. and Vera, C.R. (2005) *Appl. Catal. A*, **288**, 18–24.

159 Chin, S.Y., Ahmad, A.L., Mohamed, A.R. and Bhatia, S. (2006) *Appl. Cat. A*, **297**, 8–17.

160 Parmar, A., Kaur, J., Goyal, R., Kumar, B. and Kumar, H. (1998) *Synth. Commun.*, **28**, 2821–2826.

161 Das, B., Venkataiah, B. and Madhusudhan, P. (2000) *Synlett*, 59–60.

162 Yamazaki, O., Hao, X., Yoshida, A. and Nishikido, J. (2003) *Tetrahedron Lett.*, **44**, 8791–8795.

163 Mantri, K., Komura, K. and Sugi, Y. (2005) *Green Chem.*, **7**, 677–682.

164 Zhang, F.-M., Wang, J., Yuan, C.-S. and Ren, X.-Q. (2005) *Catal. Lett.*, **102**, 171–174.

165 Sharma, P., Vyas, S. and Patel, A. (2004) *J. Mol. Catal. A*, **214**, 281–286.

166 Izumi, Y. and Urabe, K. (1981) *Chem. Lett.*, 663–666.

167 Juan, J.C., Zhang, J., Jiang, Y., Cao, W. and Yarmo, M.A. (2007) *Catal. Lett.*, **117**, 153–158.

168 Engin, A., Haluk, H. and Gurkan, K. (2003) *Green Chem.*, **5**, 460–466.

169 Palaniappan, S. and Sairam, M. (2005) *J. Appl. Polym. Sci.*, **96**, 1584–1590.

170 Palaniappan, S. and John, A. (2005) *J. Mol. Catal. A*, **233**, 9–15.

171 Elson, K.E., Jenkins, I.D. and Loughlin, W.A. (2004) *Tetrahedron Lett.*, **45**, 2491–2493.

172 Zander, N., Gerhardt, J. and Frank, R. (2003) *Tetrahedron Lett.*, **44**, 6557–6560.

173 De La Hoz, A., Moreno, A. and Vazquez, E. (1999) *Synlett*, 608–610.

174 Kabza, K.G., Chapados, B.R., Gestwicki, J.E. and McGrath, J.L. (2000) *J. Org. Chem.*, **65**, 1210–1214.

175 Kraft, P. and Cadalbert, R. (1997) *Synlett*, 600–602.

176 Baldwin, B.W., Hirose, T., Wang, Z.-H., Uchimaru, T. and Yliniemela, A. (1997) *Bull. Chem. Soc. Jpn.*, **70**, 1895–1903.

177 Ogawa, S. and Morikawa, T. (1999) *Bioorg. Med. Chem. Lett.*, **9**, 1499–1504.

178 Buzas, A., Egnell, C. and Freon, P. (1962) *C. R. Acad. Sci.*, **255**, 945–947.

179 Holmberg, K. and Hansen, B. (1979) *Acta Chim. Scand. Ser.B*, **33**, 410–412.

180 Zengin, G. and Huffman, J.W. (2004) *Synthesis*, 1932–1934.

181 Neises, B. and Sterlich, W. (1978) *Angew. Chem. Int. Ed. Engl.*, **17**, 522–523.

182 Hassner, A. and Alexanian, V. (1978) *Tetrahedron Lett.*, **19**, 4475–4478.

183 Barker, D., Mcleod, M.D., Brimble, M.A. and Savage, G.P. (2001) *Tetrahedron Lett.*, **42**, 1785–1788.

184 Shimizu, T., Hiramoto, K. and Nakata, T. (2001) *Synthesis*, 1027–1034.

185 Sano, S., Harada, E., Azetsu, T., Ichikawa, T., Nakao, M. and Nagao, Y. (2006) *Chem. Lett.*, **35**, 1286–1287.

186 Nahmany, M. and Melman, A. (2001) *Org. Lett.*, **3**, 3733–3735.

187 Shelkov, R., Nahmany, M. and Melman, A. (2002) *J. Org. Chem.*, **67**, 8975–8982.

188 Shelkov, R., Nahmany, M. and Melman, A. (2004) *Org. Biomol. Chem.*, **2**, 397–401.

189 Zhao, H., Pendri, A. and Greenwald, R.B. (1998) *J. Org. Chem.*, **63**, 7559–7562.

190 Makita, Y., Kihara, N. and Takata, T. (2007) *Chem. Lett.*, **36**, 102–103.

191 Streinz, L., Koutek, B., Šaman, D. (2001) *Synlett*, 809–811.

192 Adamczyk, M., Fishpaugh, J.R. and Mattingly, P.G. (1995) *Tetrahedron Lett.*, **36**, 8345–8346.

193 Keck, G.E., Sanchez, C. and Wager, C.A. (2000) *Tetrahedron Lett.*, **41**, 8673–8676.

194 Mitsunobu, O. and Yamada, M. (1967) *Bull. Chem. Soc. Jpn.*, **40**, 2380–2382.

195 Mitsunobu, O. (1981) *Synthesis*, 1–28.

196 Hughes, D.L., Reamer, R.A., Bergan, J.J. and Grabowski, E.J.J. (1988) *J. Am. Chem. Soc.*, **110**, 6487–6491.

197 Swamy, K.C.K., Kumar, K.P. and Kumar, N.N.B. (2006) *J. Org. Chem.*, **71**, 1002–1008.

198 Hughes, D.L. and Reamer, R.A. (1996) *J. Org. Chem.*, **61**, 2967–2971.

199 Appendino, G., Minassi, A., Daddario, N., Bianchi, F. and Tron, G.C. (2002) *Org. Lett.*, **4**, 3839–3841.

200 Fitzjarrald, V.P. and Pongdee, R. (2007) *Tetrahedron Lett.*, **48**, 3553–3557.

201 Tapolcsányi, P., Wölfling, J., Mernyák, E. and Schneider, G. (2004) *Monatsh. Chem.*, **135**, 1129–1136.

202 Molinier, V., Fitremann, J., Bouchu, A. and Queneau, Y. (2004) *Tetrahedron: Asymmetry*, **15**, 1753–1762.

203 Ahn, C., Correia, R. and DeShong, P. (2002) *J. Org. Chem.*, **67**, 1751–1753.

204 Ahn, C. and DeShong, P. (2002) *J. Org. Chem.*, **67**, 1754–1758.

205 O'Neil, I.A., Thompson, S., Murray, C.L. and Kalindjian, S.B. (1998) *Tetrahedron Lett.*, **39**, 7787–7790.

206 Camp, D. and Jenkins, I.D. (1988) *Aust. J. Chem.*, **41**, 1835–1839.

207 Von Itzstein, M. and Mocerino, M. (1990) *Synth. Commun.*, **20**, 2049–2057.

208 Kiankarimi, M., Lowe, R., McCarthy, J.R. and Whitten, J.P. (1999) *Tetrahedron Lett.*, **40**, 4497–4500.

209 Arnold, L.D., Assil, H.I. and Vederas, J.C. (1989) *J. Am. Chem. Soc.*, **111**, 3973–3976.

210 Dobbs, A.P. and McGregor-Johnson, C. (2002) *Tetrahedron Lett.*, **43**, 2807–2810.

211 Dandapani, S. and Curran, D.P. (2002) *Tetrahedron*, **58**, 3855–3864.

212 Tsunoda, T., Yamamiya, Y. and Ito, S. (1993) *Tetrahedron Lett.*, **34**, 1639–1642.

213 Tsunoda, T., Yamamiya, Y., Kuwamura, Y. and Ito, S. (1995) *Tetrahedron Lett.*, **36**, 2529–2530.

214 Hulst, R., Van Basten, A., Fitzpatrick, K. and Kellogg, R.M. (1995) *J. Chem. Soc., Perkin Trans. 1*, 2961–2963.

215 Castro, J.L., Matassa, V.G. and Ball, R.G. (1994) *J. Org. Chem.*, **59**, 2289–2291.

216 Tsunoda, T., Ozaki, F. and Ito, S. (1994) *Tetrahedron Lett.*, **35**, 5081–5082.

217 McNulty, J., Capretta, A., Laritchev, V., Dyck, J. and Robertson, A.J. (2003) *J. Org. Chem.*, **68**, 1597–1600.

218 Torii, S., Okumoto, H., Fujikawa, M. and Rashid, M.A. (1992) *Chem. Express*, **7**, 933–936.

219 Mukaiyama, T., Usui, M. and Shimada, E. (1975) *Chem. Lett.*, 1045–1048.

220 Shoda, S. and Mukaiyama, T. (1980) *Chem. Lett.*, 391–392.

221 Mukaiyama, T., Usui, M. and Saigo, K. (1976) *Chem. Lett.*, 49–50.

222 Mukaiyama, T., Pai, F.-C., Onaka, M. and Narasaka, K. (1980) *Chem. Lett.*, 563–566.

223 Folmer, J.J. and Weinreb, S.M. (1993) *Tetrahedron Lett.*, **34**, 2737–2740.

224 Ueda, M., Oikawa, H. and Teshirogi, T. (1983) *Synthesis*, 908–909.

225 Kim, S., Lee, J.I. and Ko, Y.K. (1984) *Tetrahedron Lett.*, **25**, 4943–4946.

226 Mukaiyama, T., Oohashi, Y. and Fukumoto, K. (2004) *Chem. Lett.*, **33**, 552–553.

227 Oohashi, Y., Fukumoto, K. and Mukaiyama, T. (2005) *Chem. Lett.*, **34**, 190–191.

228 Ouihia, A., Rene, L., Guilhem, J., Pascard, C. and Badet, B. (1993) *J. Org. Chem.*, **58**, 1641–1642.

229 Kunishima, M., Morita, J., Kawachi, C., Iwasaki, F., Terao, K. and Tani, S. (1999) *Synlett*, 1255–1256.

230 Brewster, J.H. and Ciotti, C.J. Jr. (1955) *J. Am. Chem. Soc.*, **77**, 6214–6215.

231 Adam, W., Baeza, J. and Liu, J.-C. (1972) *J. Am. Chem. Soc.*, **94**, 2000–2006.

232 Wakasugi, K., Iida, A., Misaki, T., Nishii, Y. and Tanabe, Y. (2003) *Adv. Synth. Catal.*, **345**, 1209–1214.

233 Rad, M.N.S., Behrouz, S., Faghihi, M.A. and Khalafi-Nezhad, A. (2008) *Tetrahedron Lett.*, **49**, 1115–1120.

234 Chandrasekaran, S. and Turner, J.V. (1982) *Synth. Commun.*, **12**, 727–731.

235 Jacobi, P.A. and Sessions, E.H. Jr. (2003) *Synth. Commun.*, **33**, 2575–2579.

236 Wakasugi, K., Nakamura, A. and Tanabe, Y. (2001) *Tetrahedron Lett.*, **42**, 7427–7430.

237 Wakasugi, K., Nakamura, A., Iida, A., Nishii, Y., Nakatani, N., Fukushima, S. and Tanabe, Y. (2003) *Tetrahedron*, **59**, 5337–5345.

238 Olah, G.A., Narang, S.C. and Garcia-Luna, A. (1981) *Synthesis*, 790–791.

239 Tamayo, N., Echavarren, A.M. and Paredes, M.C. (1991) *J. Org. Chem.*, **56**, 6488–6491.

240 Kiely, J.S., Laborde, E., Lesheski, L.E. and Busch, R.A. (1991) *J. Heterocycl. Chem.*, **28**, 1581–1585.

241 D'Aniello, F., Mattii, D. and Taddei, M. (1993) *Synlett*, 119–121.

242 Kaul, S., Kumar, A., Sain, B. and Gupta, A.K. (2002) *Synth. Commun.*, **32**, 2885–2891.

243 Kim, J.-J., Park, Y.-D., Kweon, D.-H., Kang, Y.-J., Kim, H.-K., Lee, S.-G., Cho, S.-D., Lee, W.-S. and Yoon, Y.-J. (2004) *Bull. Korean Chem. Soc.*, **25**, 501–505.

244 Zander, N. and Frank, R. (2001) *Tetrahedron Lett.*, **42**, 7783–7785.

245 Hashimoto, S. and Furukawa, I. (1981) *Bull. Chem. Soc. Jpn.*, **54**, 2227–2228.

246 Jang, D.O., Cho, D.H. and Kim, J.-G. (2003) *Synth. Commun.*, **33**, 2885–2890.

247 Sucheta, K., Reddy, G.S.R., Ravi, D. and Rao, N.R. (1994) *Tetrahedron Lett.*, **35**, 4415–4416.

248 Ballester-Rodes, M. and Palomo-Coll, A.L. (1984) *Synth. Commun.*, **14**, 515–520.

249 Kim, M.H. and Patel, D.V. (1994) *Tetrahedron Lett.*, **35**, 5603–5606.

250 Niyogi, D.G., Singh, S. and Verma, R.D. (1994) *J. Fluorine Chem.*, **68**, 237–238.

251 Jaszay, Z.M., Petnehazy, I. and Toke, L. (1998) *Synth. Commun.*, **28**, 2761–2768.

252 Ohmori, H., Maeda, H., Kikuoka, M., Maki, T. and Masui, M. (1991) *Tetrahedron*, **47**, 767–776.

253 Izumi, J., Shiina, I. and Mukaiyama, T. (1995) *Chem. Lett.*, 141–142.

254 Mukaiyama, T., Izumi, J. and Shiina, I. (1997) *Chem. Lett.*, 187–188.

255 Kita, Y., Akai, S., Yamamoto, M., Taniguchi, M. and Tamura, Y. (1989) *Synthesis*, 334–337.

256 Mukaiyama, T., Ichikawa, J. and Asami, M. (1983) *Chem. Lett.*, 683–686.

257 Ogawa, T., Hikasa, T., Ikegami, T., Ono, N. and Suzuki, H. (1993) *Chem. Lett.*, 815–818.

258 Maki, T., Ishihara, K. and Yamamoto, H. (2007) *Tetrahedron*, **63**, 8645–8657.

259 Huang, Z., Reilly, J.E. and Buckle, R.N. (2007) *Synlett*, 1026–1030.

260 Keramane, E.-M., Boyer, B. and Roque, J.-P. (2001) *Tetrahedron Lett.*, **42**, 855–857.

261 Štefane, B., Ko evar, M. and Polanc, S. (2002) *Synth. Commun.*, **32**, 1703–1707.

262 Pan, W.-B., Chang, F.-R., Wei, L.-M., Wu, M.-J. and Wu, Y.-C. (2003) *Tetrahedron Lett.*, **44**, 331–334.

263 Kuttan, A., Nowshudin, S. and Rao, M.N.A. (2004) *Tetrahedron Lett.*, **45**, 2663–2665.

264 Kirchner, G., Scollar, M.P. and Klibanov, A.M. (1985) *J. Am. Chem. Soc.*, **107**, 7072–7076.

265 Makita, A., Nihira, T. and Yamada, Y. (1987) *Tetrahedron Lett.*, **28**, 805–808.

266 Guo, Z.-W. and Sih, C.J. (1988) *J. Am. Chem. Soc.*, **110**, 1999–2001.

267 Vija, H., Telling, A. and Tougu, V. (1997) *Bioorg. Med. Chem. Lett.*, **7**, 259–262.

268 Wang, Q., Li, Y. and Chen, Q. (2003) *Synth. Commun.*, **33**, 2125–2134.

269 Weber, N., Weitkamp, P. and Mukherjee, K.D. (2001) *J. Agric. Food Chem.*, **49**, 67–71.

270 Romano, A., Gandolfi, R., Molinari, F., Converti, A., Zilli, M. and Del Borghi, M. (2005) *Enzyme Microb. Technol.*, **36**, 432–438.

271 Wei, D., Gu, C., Song, Q. and Su, W. (2003) *Enzyme Microb. Technol.*, **33**, 508–512.

272 Sharma, A., Chattopadhyay, S. and Mamdapur, V.R. (1995) *Biotech. Lett.*, **17**, 939–942.

273 Liu, C.-F. and Tam, J.P. (2001) *Org. Lett.*, **3**, 4157–4159.

274 Gubicza, L., Kabiri-Badr, A., Keoves, E. and Belafi-Bako, K. (2000) *J. Biotech.*, **84**, 193–196.

275 Yan, Y., Bornscheuer, U.T. and Schmid, R.D. (2002) *Biotech. Bioeng.*, **78**, 31–34.

276 Giacometti, J., Giacometti, F., Milin, . and Vasi -Ra ki, Đ. (2001) *J. Mol. Catal. B*, **11**, 921–928.

277 Chamorro, S., Alcántara, A.R., de la Casa, R.M., Sinisterra, J.V. and Sánchez-Montero, J.M. (2001) *J. Mol. Catal. B*, **11**, 939–947.

278 Altreuter, D.H., Dordick, J.S. and Clark, D.S. (2002) *Enzyme Microb. Technol.*, **31**, 10–19.

279 Thakar, A. and Madamwar, D. (2005) *Process Biochem.*, **40**, 3263–3266.

280 Dandavate, V. and Madamwar, D. (2007) *Enzyme Microb. Technol.*, **41**, 265–270.

281 Moniruzzaman, M., Hayashi, Y., Talukder, M.R. and Kawanishi, T. (2007) *Biocatal. Biotransform.*, **25**, 51–58.

282 Weber, N., Weitkamp, P. and Mukherjee, K.D. (2001) *J. Agric. Food Chem.*, **49**, 5210–5216.

283 Langone, M.A.P., De Abreu, M.E., Rezende, M.J.C. and Sant'Anna, G.L. Jr. (2002) *Appl. Biochem. Biotechnol.*, **98–100**, 987–996.

284 Martínez, I., Markovits, A., Chamy, R. and Markovits, A. (2004) *Appl. Biochem. Biotechnol.*, **112**, 55–62.

285 Petersson, A.E.V., Gustafsson, L.M., Nordblad, M., Börjesson, P., Mattiasson, B. and Adlercreutz, P. (2005) *Green Chem.*, **7**, 837–843.

286 Vosmann, K., Weitkamp, P. and Weber, N. (2006) *J. Agric. Food Chem.*, **54**, 2969–2976.

287 Sun, S., Shan, L., Jin, Q., Liu, Y. and Wang, X. (2007) *Biotech. Lett.*, **29**, 945–949.

288 Pereira, E.B., De Castro, H.F., De Moraes, F.F. and Zanin, G.M. (2002) *Appl. Biochem. Biotechnol.*, **98–100**, 977–986.

289 Foresti, M.L. and Ferreira, M.L. (2007) *Enzyme Microb. Technol.*, **40**, 769–777.

290 Harun, A., Basri, M., Ahmad, M.B. and Salleh, A.B. (2004) *J. Appl. Polym. Sci.*, **92**, 3381–3386.

291 Trubiano, G., Borio, D. and Ferreira, M.L. (2004) *Biomacromolecules*, **5**, 1832–1840.

292 Foresti, M.L. and Ferreira, M.L. (2005) *Catal. Today*, **107–108**, 23–30.

293 Foresti, M.L., Alimenti, G.A. and Ferreira, M.L. (2005) *Enzyme Microb. Technol.*, **36**, 338–349.

294 Pierre, A. and Buisson, P. (2001) *J. Mol. Catal. B*, **11**, 639–647.

295 Maury, S., Buisson, P. and Pierre, A.C. (2002) *J. Mol. Catal. B*, **19–20**, 269–278.

296 Park, E.Y., Sato, M. and Kojima, S. (2006) *Enzyme Microb. Technol.*, **39**, 889–896.

297 Chen, J.-P. and Lin, W.-S. (2003) *Enzyme Microb. Technol.*, **32**, 801–811.

298 Pires, E.L., Miranda, E.A. and Valença, G.P. (2002) *Appl. Biochem. Biotechnol.*, **98–100**, 963–976.

299 Peres, C., Harper, N., da Silva, M.D.R.G. and Barreiros, S. (2005) *Enzyme Microb. Technol.*, **37**, 145–149.

300 Rahman, M.B.A., Basri, M., Hussein, M.Z., Rahman, R.N.Z.A., Zainol, D.H. and Salleh, A.B. (2004) *Appl. Biochem. Biotechnol.*, **118**, 313–320.

301 Rahman, M.B.A., Yunus, N.M.M., Hussein, M.Z., Rahman, R.N.Z.R.A., Salleh, A.B. and Basri, M. (2005) *Biocatal. Biotransform.*, **23**, 233–239.

302 Rahman, M.B.A., Basri, M., Hussein, M.Z., Idris, M.N.H., Rahman, R.N.Z.R.A. and Salleh, A.B. (2004) *Catal. Today*, **93–95**, 405–410.

303 Ghamgui, H., Karra-Chaâbouni, M. and Gargouri, Y. (2004) *Enzyme Microb. Technol.*, **35**, 355–363.

304 Ghamgui, H., Miled, N., Rebaï, A., Karra-Chaâbouni, M. and Gargouri, Y. (2006) *Enzyme Microb. Technol.*, **39**, 717–723.

305 Sánchez, A., del Río, J.L., Valero, F., Lafuente, J., Faus, I. and Solà, C. (2000) *J. Biotech.*, **84**, 1–12.

306 Piao, J. and Adachi, S. (2004) *Biocatal. Biotransform.*, **22**, 269–274.

307 Kiran, K.R. and Divakar, S. (2001) *J. Biotech.*, **87**, 109–121.

308 Castillo, E., Pezzotti, F., Navarro, A. and López-Munguía, A. (2003) *J. Biotech.*, **102**, 251–259.

309 Chang, C.-S., Su, C.-C., Zhuang, J.-R. and Tsai, S.-W. (2004) *J. Mol. Catal. B*, **30**, 151–157.

310 Passicos, E., Santarelli, X. and Coulon, D. (2004) *Biotech. Lett.*, **26**, 1073–1076.

311 Chang, S.-W., Shaw, J.-F., Yang, K.-H., Shih, I.-L., Hsieh, C.-H. and Shieh, C.-J. (2005) *Green Chem.*, **7**, 547–551.

312 Langrand, G., Secci, M., Buono, G., Baratti, J. and Triantaphylides, C. (1985) *Tetrahedron Lett.*, **26**, 1857–1860.

313 Sonnet, P.E. (1987) *J. Org. Chem.*, **52**, 3477–3479.

314 Ueji, S., Nishimura, M., Kudo, R., Matsumi, R., Watanabe, K. and Ebara, Y. (2001) *Chem. Lett.*, 912–913.

315 Morrone, R., Piattelli, M. and Nicolosi, G. (2001) *Eur. J. Org. Chem.*, 1441–1443.

316 Iranpoor, N. and Mottaghinejad, E. (1995) *Synth. Commun.*, **25**, 2253–2260.

317 Masaki, Y., Tanaka, N. and Miura, T. (1997) *Chem. Lett.*, 55–56.

318 Bader, A.R., Cummings, L.O. and Vogel, H.A. (1951) *J. Am. Chem. Soc.*, **73**, 4195–4197.

319 Witzeman, J.S. (1990) *Tetrahedron Lett.*, **31**, 1401–1404.

320 Witzeman, J.S. and Nottingham, W.D. (1991) *J. Org. Chem.*, **56**, 1713–1718.

321 Mandai, T., Kuroda, A., Okumoto, H., Nakanishi, K., Mikuni, K., Hara, K.-J. and Hara, K.-Z. (2000) *Tetrahedron Lett.*, **41**, 239–242.

322 Koval, L.I., Dzyuba, V.I., Ilnitska, O.L. and Pekhnyo, V.I. (2008) *Tetrahedron Lett.*, **49**, 1645–1647.

323 Saka, S. and Kusdina, D. (2001) *Fuel*, **80**, 225–231.

324 Kusdiana, D. and Saka, S. (2001) *Fuel*, **80**, 693–698.

325 Rao, C.C. and Lalitha, N. (1994) *Indian J. Chem., Sect. B*, **33**, 3.

326 Wernick, D.L., Savion, Z. and Levy, J. (1988) *J. Incl. Phenom.*, 483–490.

327 Hwang, S.W., Adiyaman, M., Khanapure, S., Schio, L. and Rokach, J. (1994) *J. Am. Chem. Soc.*, **116**, 10829–10830.

328 Harrowven, D.C. and Hannam, J.C. (1999) *Tetrahedron*, **55**, 9341–9346.

329 Eras, J., Llovera, M., Ferran, X. and Canela, R. (1999) *Synth. Commun.*, **29**, 1129–1133.

330 Hagiwara, H., Morohashi, K., Sakai, H., Suzuki, T. and Ando, M. (1998) *Tetrahedron*, **54**, 5845–5852.

331 Olah, G.A., Narang, S.C., Salem, G.F. and Gupta, B.G.B. (1981) *Synthesis*, 142–143.

332 Kolb, M. and Barth, J. (1981) *Synth. Commun.*, **11**, 763–767.

333 Hagiwara, H., Nakano, T., Konno, M. and Uda, H. (1995) *J. Chem. Soc., Perkin Trans. 1*, 777–783.

334 Jakiwczyk, O.M., Nielsen, K.E., De Carvalho, H.N. and Dmitrieko, G.I. (1997) *Tetrahedron Lett.*, **38**, 6541–6544.

335 Wulferding, A., Jankowski, J.H. and Hoffmann, H.M.R. (1994) *Chem. Ber.*, **127**, 1275–1281.

336 Kita, Y., Maeda, H., Takahashi, F. and Fukui, S. (1993) *J. Chem. Soc., Chem. Commun.*, 410–412.

337 Kita, Y., Maeda, H., Omori, K., Okuno, T. and Tamura, Y. (1993) *Synlett*, 273–274.

338 Trost, B.M. and Chisholm, J.D. (2002) *Org. Lett.*, **4**, 3743–3745.

339 Chenera, B., West, M.L., Finkelstein, J.A. and Dreyer, G.B. (1993) *J. Org. Chem.*, **58**, 5605–5606.

340 Seebach, D., Hungerbuhler, E., Naef, R., Schnurrenberger, P., Weidmann, B. and Züger, M. (1982) *Synthesis*, 138–141.

341 Schnurrenberger, P., Züger, M. and Seebach, D. (1982) *Helv. Chim. Acta*, **65**, 1197–1201.

342 Krasik, P. (1998) *Tetrahedron Lett.*, **39**, 4223–4226.

343 Valizadeh, H. and Shockravi, A. (2005) *Tetrahedron Lett.*, **46**, 3501–3503.

344 Tamura, O., Yamaguchi, T., Noe, K. and Sakamoto, M. (1993) *Tetrahedron Lett.*, **34**, 4009–4010.

345 Pereyre, M., Colin, G. and Delvigne, J.-P. (1969) *Bull. Soc. Chim. Fr.*, 262–267.

346 Poller, R.C. and Retout, S.P. (1979) *J. Organomet. Chem.*, **173**, C7–C8.

347 Pilati, F., Munari, A. and Manaresi, P. (1984) *Polymer Commun.*, **25**, 187–189.

348 Otera, J., Yano, T., Kawabata, A. and Nozaki, H. (1986) *Tetrahedron Lett.*, **27**, 2383–2386.

349 Orita, A., Mitsutome, A. and Otera, J. (1998) *J. Org. Chem.*, **63**, 2420–2421.

350 Orita, A., Sakamoto, K., Hamada, Y., Mitsutome, A. and Otera, J. (1999) *Tetrahedron*, **55**, 2899–2910.

351 Shirae, Y., Mino, T., Hasegawa, T., Sakamoto, M. and Fujita, T. (2005) *Tetrahedron Lett.*, **46**, 5877–5879.

352 Xiang, J., Toyoshima, S., Orita, A. and Otera, J. (2001) *Angew. Chem. Int. Ed.*, **40**, 3670–3672.

353 Xiang, J., Orita, A. and Otera, J. (2002) *Adv. Synth. Catal.*, **344**, 84–90.

354 Orita, A., Sakamoto, K., Hamada, Y. and Otera, J. (2000) *Synlett*, 140–142.

355 Orita, A., Hamada, Y., Nakano, T., Toyoshima, S. and Otera, J. (2001) *Chem. Eur. J.*, **7**, 3321–3327.

356 Furlan, R.L.E., Mata, E.G. and Mascaretti, O.A. (1998) *Tetrahedron Lett.*, **39**, 2257–2260.

357 Baumhof, P., Mazitschek, R. and Giannis, A. (2001) *Angew. Chem. Int. Ed.*, **40**, 3672–3674.

358 Trost, B.M. and McIntosh, M.C. (1995) *J. Am. Chem. Soc.*, **117**, 7255–7256.

359 Kumar, B., Kumar, H. and Parmar, A. (1993) *Indian J. Chem., Sect. B*, **32**, 292–293.

360 Kim, S. and Lee, J.I. (1984) *J. Org. Chem.*, **49**, 1712–1716.

361 Hanamoto, T., Sugimoto, Y., Yokoyama, Y. and Inanaga, J. (1996) *J. Org. Chem.*, **61**, 4491–4492.

362 Vedejs, E. and Cshen, X. (1996) *J. Am. Chem. Soc.*, **118**, 1809–1810.

363 Kaminska, J.E., Kaminski, Z.J. and Gora, J. (1999) *Synthesis*, 593–596.

364 Okano, T., Miyamoto, K. and Kiji, J. (1995) *Chem. Lett.*, 246.

365 Nelson, S.G., Wan, Z., Peelen, T.J. and Spencer, K.L. (1999) *Tetrahedron Lett.*, **40**, 6535–6539.

366 Lin, M.-H. and RajanBabu, T.V. (2000) *Org. Lett.*, **2**, 997–1000.

367 Lin, M.-H. and RajanBabu, T.V. (2002) *Org. Lett.*, **4**, 1607–1610.

368 Ishii, Y., Takeno, M., Kawasaki, Y., Muromachi, A., Nishiyama, Y. and Sakaguchi, S. (1996) *J. Org. Chem.*, **61**, 3088–3092.

369 Tashiro, D., Kawasaki, Y., Sakaguchi, S. and Ishii, Y. (1997) *J. Org. Chem.*, **62**, 8141–8144.

370 Fort, Y., Remy, M. and Caubere, P. (1993) *J. Chem. Res., Synop.*, 418–419.

371 Remme, N., Koschek, K. and Schneider, C. (2007) *Synlett*, 491–493.

372 Nava, R., Halachev, T., Rodríguez, R. and Castaño, V.M. (2002) *Appl. Catal. A*, **231**, 131–149.

373 Solhy, A., Clark, J.H., Tahir, R., Sebti, S. and Larzek, M. (2006) *Green Chem.*, **8**, 871–874.

374 Posner, G.H., Okada, S.S., Babiak, K.A., Miura, K. and Rose, R.K. (1981) *Synthesis*, 789–790.

375 Posner, G.H. and Oda, M. (1981) *Tetrahedron Lett.*, **22**, 5003–5006.

376 Rana, S.S., Barlow, J.J. and Matta, K.L. (1981) *Tetrahedron Lett.*, **22**, 5007–5010.

377 Nishiguchi, T. and Taya, H. (1989) *J. Am. Chem. Soc.*, **111**, 9102–9103.

378 Nishiguchi, T., Kawamine, K. and Ohtsuka, T. (1992) *J. Org. Chem.*, **57**, 312–316.

379 Breton, G.W. (1997) *J. Org. Chem.*, **62**, 8952–8954.

380 Waghoo, G., Jayaram, R.V. and Joshi, M.V. (1999) *Synth. Commun.*, **29**, 513–520.

381 Shah, P., Ramaswamy, A.V., Lazar, K. and Ramaswamy, V. (2004) *Appl. Catal. A*, **273**, 239–248.

382 Iranpoor, N., Firouzabadi, H. and Jamalian, A. (2005) *Tetrahedron Lett.*, **46**, 7963–7966.

383 Kuno, H., Shibagaki, M., Takahashi, K., Honda, I. and Matsushita, H. (1992) *Chem. Lett.*, 571–574.

384 Kuno, H., Shibagaki, M., Takahashi, K. and Matsushita, H. (1993) *Bull. Chem. Soc. Jpn.*, **66**, 1305–1307.

385 Balaji, B.S., Sasidharan, M., Kumar, R. and Chanda, B. (1996) *Chem. Commun.*, 707–708.

386 Chavan, S.P., Zubaidha, P.K., Dantale, S.W., Keshavaraja, A., Ramaswamy, A.V. and Ravindranathan, T. (1996) *Tetrahedron Lett.*, **37**, 233–236.

387 Ponde, D.E., Deshpande, V.H., Bulbule, V.J., Sudalai, A. and Gajare, A.S. (1998) *J. Org. Chem.*, **63**, 1058–1063.

388 Reddy, B.M., Reddy, V.R. and Manohar, B. (1999) *Synth. Commun.*, **29**, 1235–1239.

389 Kumar, P. and Pandey, R.K. (2000) *Synlett*, 251–253.

390 Srinivas, K.V.N.S., Mahender, I. and Das, B. (2003) *Synlett*, 2419–2421.

391 Parmar, A., Goyal, R., Kumar, B. and Kumar, H. (1999) *Synth. Commun.*, **29**, 139–143.

392 Tayebee, R. and Alizadeh, M.H. (2006) *Monatsh. Chem.*, **137**, 1063–1069.

393 Bose, D.S., Satyender, A., Das, A.P.R. and Mereyala, H.B. (2006) *Synthesis*, 2392–2396.

394 Otto, M.-C. (1986) *J. Chem. Soc., Chem. Commun.*, 695–697.

395 Katoh, A., Lu, T., Devadas, B., Adams, S.P., Gordon, J.I. and Gokel, G.W. (1991) *J. Org. Chem.*, **56**, 731–735.

396 Xu, Y.-C., Bizuneh, A. and Walker, C. (1996) *J. Org. Chem.*, **61**, 9086–9089.

397 Gonzalez, A.G., Jorge, Z.D. and Dorta, H.L. (1981) *Tetrahedron Lett.*, **22**, 335–336.

398 Angeletti, E., Tundo, P. and Venturello, P. (1983) *J. Org. Chem.*, **48**, 4106–4108.

399 Burke, S.D., McDermott, T.S. and O'Donnell, C.J. (1998) *J. Org. Chem.*, **63**, 2715–2718.

400 Frater, G. and Müller, U. (1993) *Tetrahedron Lett.*, **34**, 2753–2756.

401 Bonini, C., Chiummiento, L., Evidente, A. and Funicello, M. (1995) *Tetrahedron Lett.*, **36**, 7285–7286.

402 Hanessian, S., Meng, Q. and Olivier, E. (1994) *Tetrahedron Lett.*, **35**, 5393–5396.

403 Lawrence, R.M. and Perlmutter, P. (1992) *Chem. Lett.*, 305–308.

404 Brady, P.A., Bonar-Law, R.P., Rowan, S.J., Suckling, C.J. and Sanders, J.K.M. (1996) *Chem. Commun.*, 319–320.

405 Rowan, S.J., Brady, P.A. and Sanders, J.K.M. (1996) *Angew. Chem. Int. Ed. Engl.*, **35**, 2143–2145.

406 Rowan, S.J., Hamilton, D.G., Brady, P.A. and Sanders, J.K.M. (1997) *J. Am. Chem. Soc.*, **119**, 2578–2579.

407 Adam, W., Albert, R., Hasemann, L., Salgado, V.O.N., Nestler, B., Peters, E.-M., Peters, K., Prechtl, F. and Von Schnering, H.G. (1991) *J. Org. Chem.*, **56**, 5782–5785.

408 Chen, Y. and Vogel, P. (1994) *J. Org. Chem.*, **59**, 2487–2496.

409 Mukerjee, A.K., Joseph, K., Homami, S.-S. and Tikdari, A.M. (1991) *Heterocycles*, **32**, 1317–1325.

410 Iqbal, N. and Knaus, E.E. (1996) *J. Heterocycl. Chem.*, **33**, 157–160.

411 Yiu, S. and Knaus, E.E. (1996) *J. Med. Chem.*, **39**, 4576–4582.

412 West, F.G. and Naidu, B.N. (1994) *J. Org. Chem.*, **59**, 6051–6056.

413 Ghosh, A.K. and Liu, C. (1997) *Chem. Commun.*, 1743–1744.

414 Bhatt, R.K., Chauhan, K., Wheelan, P., Murphy, R.C. and Flack, J.R. (1994) *J. Am. Chem. Soc.*, **116**, 5050–5056.

415 Belzecki, C., Urbanski, R., Urbanczyk-Lipkowska, Z. and Chmielewski, M. (1997) *Tetrahedron*, **53**, 14153–14168.

416 Kim, S., Chang, H. and Kim, W.J. (1985) *J. Org. Chem.*, **50**, 1751–1752.

417 Battistelli, C.L., De Castro, C., Iadonisi, A., Lanzetta, R., Mangoni, L. and Parrilli, M. (1999) *J. Carbohydr. Chem.*, **18**, 69–86.

418 Nacro, K., Baltas, M., Escudier, J.-M. and Gorrichon, L. (1997) *Tetrahedron*, **53**, 659–672.

419 Hagiwara, H., Koseki, A., Isobe, K., Shimizu, K., Hoshi, T. and Suzuki, T. (2004) *Synlett*, 2188–2190.

420 Taber, D.F., Amedio, J.C. Jr. and Patel, Y.K. (1985) *J. Org. Chem.*, **50**, 3618–3619.

421 Gilbert, J.C. and Kelly, T.A. (1988) *J. Org. Chem.*, **53**, 449–450.

422 Hatakeyama, S., Satoh, K., Sakurai, K. and Takano, S. (1987) *Tetrahedron Lett.*, **28**, 2713–2716.

423 Hatakeyama, S., Satoh, K., Sakurai, K. and Takano, S. (1987) *Tetrahedron Lett.*, **28**, 2717–2720.

424 Chen, S.-T., Chen, S.-Y., Chen, S.-J. and Wang, K.-T. (1994) *Tetrahedron Lett.*, **35**, 3583–3584.

425 Gaudino, J.J. and Wilcox, C.S. (1990) *Carbohydr. Res.*, **206**, 233–250.

426 Shustov, G.V., Krutius, O.N., Voznesenskii, V.N., Chervin, I.I., Eremeev, A.V., Kostyanovskii, R.G. and Polyak, F.D. (1990) *Tetrahedron*, **46**, 6741–6752.

427 Avenoza, A., Busto, J.H., Cativiela, C. and Peregrina, J.M. (1994) *Tetrahedron*, **50**, 12989–12998.

428 Yang, Z. and Zhou, W. (1995) *Tetrahedron*, **51**, 1429–1436.

429 Dubois, K.J. and Hoornaert, G.J. (1996) *Tetrahedron*, **52**, 6997–7002.

430 Dubois, K.J., Fannes, C.C., Toppet, S.M. and Hoornaert, G.J. (1996) *Tetrahedron*, **52**, 12529–12540.

431 Bunuel, E., Cativiela, C. and Diaz-De-Villegas, M.D. (1996) *Tetrahedron: Asymmetry*, **7**, 1431–1436.

432 Yoshida, Y., Ukaji, Y., Fujinami, S. and Inomata, K. (1998) *Chem. Lett.*, 1023–1024.
433 Seebach, D., Thaler, A., Blaser, D. and Ko, S.Y. (1991) *Helv. Chim. Acta*, **74**, 1102–1118.
434 Kazmaier, U. (1997) *Chem. Commun.*, 2305–2306.
435 Kunesch, N., Miet, C. and Poisson, J. (1987) *Tetrahedron Lett.*, **28**, 3569–3572.
436 Babiak, K.A., Ng, J.S., Dygos, J.H. and Weyker, C.L. (1990) *J. Org. Chem.*, **55**, 3377–3381.
437 Ellervik, U. and Magnusson, G. (1997) *Tetrahedron Lett.*, **38**, 1627–1628.
438 Nowakowski, M. and Hoffmann, H.M.R. (1997) *Tetrahedron Lett.*, **38**, 1001–1004.
439 Neveux, M., Bruneau, C., Lecolier, S. and Dixneuf, P.H. (1993) *Tetrahedron*, **49**, 2629–2640.
440 Hashimoto, K., Kitaguchi, J., Mizuno, Y., Kobayashi, T. and Shirahama, H. (1996) *Tetrahedron Lett.*, **37**, 2275–2278.
441 Ilankumaran, P. and Verkade, J.G. (1999) *J. Org. Chem.*, **64**, 3086–3089.
442 Palanichamy, I. and Verkade, J.G. (1999) *J. Org. Chem.*, **64**, 9063–9066.
443 Yoshimoto, K., Kawabata, H., Nakamichi, N. and Hayashi, M. (2001) *Chem. Lett.*, 934–935.
444 Hans, J.J., Driver, R.W. and Burke, S.D. (1999) *J. Org. Chem.*, **64**, 1430–1431.
445 Kharchafi, G., Jérôme, F., Douliez, J.-P. and Barrault, J. (2006) *Green Chem.*, **8**, 710–716.
446 Gijsen, H.J.M. and Wong, C.-H. (1994) *J. Am. Chem. Soc.*, **116**, 8422–8423.
447 Bonjoch, J., Catena, J., Isabal, E., Lopez-Canet, M. and Valls, N. (1996) *Tetrahedron: Asymmetry*, **7**, 1899–1902.
448 Bandgar, B.P., Uppalla, L.S. and Sadavarte, V.S. (2001) *Synlett*, 1715–1718.
449 de Sairre, M.I., Bronze-Uhle, É.S. and Donate, P.M. (2005) *Tetrahedron Lett.*, **46**, 2705–2708.
450 Bosco, J.W.J. and Saikia, A.K. (2004) *Chem. Commun.*, 1116–1117.
451 Yanada, R., Bessho, K. and Yanada, K. (1995) *Synlett*, 443–444.
452 Huang, Y., Zhang, Y. and Wang, Y. (1997) *Tetrahedron Lett.*, **38**, 1065–1066.
453 Banik, B.K., Mukhopadhyay, C., Venkatraman, M.S. and Becker, F.F. (1998) *Tetrahedron Lett.*, **39**, 7243–7246.
454 Singh, R., Kissling, R.M., Letellier, M.-A. and Nolan, S.P. (2004) *J. Org. Chem.*, **69**, 209–212.
455 Therisod, M. and Klibanov, A.M. (1986) *J. Am. Chem. Soc.*, **108**, 5638–5640.
456 Gutman, A.L., Zuboi, K. and Boltansky, A. (1987) *Tetrahedron Lett.*, **28**, 3861–3864.
457 Yamada, H., Ohsawa, S., Sugai, T. and Ohta, H. (1989) *Chem. Lett.*, 1775–1776.
458 Fuganti, C., Pedrocchi-Fantoni, G. and Servi, S. (1990) *Chem. Lett.*, 1137–1140.
459 Hisano, T., Onodera, K., Toyabe, Y., Mase, N., Yoda, H. and Takabe, K. (2005) *Tetrahedron Lett.*, **46**, 6293–6295.
460 Cordova, A. and Janda, K.D. (2001) *J. Org. Chem.*, **66**, 1906–1909.
461 Ferreira-Dias, S., Correia, A.C., da Fonseca, M.M.R. (2003) *J. Mol. Catal. B*, **21**, 71–80.
462 Framis, V., Camps, F. and Clapés, P. (2004) *Tetrahedron Lett.*, **45**, 5031–5033.
463 Hsu, A.-F., Jones, K.C., Foglia, T.A. and Marmer, W.N. (2004) *Biotech. Lett.*, **26**, 917–921.
464 Hazarika, S. and Dutta, N.N. (2004) *Org. Process Res. Dev.*, **8**, 229–237.
465 García, J., Fernández, S., Ferrero, M., Sanghvi, Y.S. and Gotor, V. (2004) *Tetrahedron Lett.*, **45**, 1709–1712.
466 Negishi, S., Shirasawa, S., Arai, Y., Suzuki, J. and Mukataka, S. (2003) *Enzyme Microb. Technol.*, **32**, 66–70.
467 Ghanem, A. (2003) *Org. Biomol. Chem.*, **1**, 1282–1291.
468 Stokes, T.M. and Oehlschlager, A.C. (1987) *Tetrahedron Lett.*, **28**, 2091–2094.
469 Francalanci, F., Cesti, P., Cabri, W., Bianchi, D., Martinengo, T. and Foa, M. (1987) *J. Org. Chem.*, **52**, 5079–5082.
470 Hiratake, J., Inagaki, M., Nishioka, T. and Oda, J. (1988) *J. Org. Chem.*, **53**, 6130–6133.
471 Cantele, F., Restelli, A., Riva, S., Tentorio, D. and Villa, M. (2001) *Adv. Synth. Catal.*, **343**, 721–725.
472 Rotticci, D., Rotticci-Mulder, J.C., Denman, S., Norin, T. and Hult, K. (2001) *Chembiochem*, **2**, 766–770.
473 Kumar, A. and Gross, R.A. (2000) *J. Am. Chem. Soc.*, **122**, 11767–11770.

474 Cordova, A., Tremblay, M.R., Clapham, B. and Janda, K.D. (2001) *J. Org. Chem.*, **66**, 5645–5648.

475 Kim, K.-W., Song, B., Choi, M.-Y. and Kim, M.-J. (2001) *Org. Lett.*, **3**, 1507–1509.

476 Reetz, M.T., Weisenhöfer, W., Francio, G. and Leitner, W. (2002) *Chem. Commun.*, 992–993.

477 Vedejs, E. and Rozners, E. (2001) *J. Am. Chem. Soc.*, **123**, 2428–2429.

478 Pamies, O. and Bäckvall, J.-E. (2001) *Adv. Synth. Catal.*, **343**, 726–731.

479 Cambou, B. and Klibanov, A.M. (1984) *J. Am. Chem. Soc.*, **106**, 2687–2692.

480 Kremnický, L., Mastihuba, V. and Côté, G.L. (2004) *J. Mol. Catal. B*, **30**, 229–239.

481 Potier, P., Bouchu, A., Descotes, G. and Queneau, Y. (2001) *Synthesis*, 458–462.

482 Potier, P., Bouchu, A., Gagnaire, J. and Queneau, Y. (2001) *Tetrahedron: Asymmetry*, **12**, 2409–2419.

483 Senanayake, C.H., Bill, T.J., Larsen, R.D., Leazer, J. and Reider, P.J. (1992) *Tetrahedron Lett.*, **33**, 5901–5904.

484 Riva, S., Chopineau, J., Kieboom, A.P.G. and Klibanov, A.M. (1988) *J. Am. Chem. Soc.*, **110**, 584–589.

485 Wong, C.-H., Chen, S.-T., Hennen, W.J., Bibbs, J.A., Wang, Y.-F., Liu, J.L.-C., Pantoliano, M.W., Whitlow, M. and Bryan, P.N. (1990) *J. Am. Chem. Soc.*, **112**, 945–953.

486 Singh, H.K., Cote, G.L. and Hadfield, T.M. (1994) *Tetrahedron Lett.*, **35**, 1353–1356.

487 Wang, Y.-F., Yakovlevsky, K., Zhang, B. and Margolin, A.L. (1997) *J. Org. Chem.*, **62**, 3488–3495.

488 Herradon, B. and Valverde, S. (1995) *Synlett*, 599–602.

489 Faraldos, J., Arroyo, E. and Herradon, B. (1997) *Synlett*, 367–370.

490 Veum, L., Kuster, M., Telalovic, S., Hanefeld, U. and Maschmeyer, T. (2002) *Eur. J. Org. Chem.*, 1516–1522.

491 Kammoun, N., Bigot, Y.L., Delmas, M. and Boutevin, B. (1997) *Synth. Commun.*, **27**, 2777–2781.

492 Ku, Y.-Y., Riley, D., Patel, H., Yang, C.X. and Liu, J.-H. (1997) *Bioorg. Med. Chem. Lett.*, **7**, 1203–1206.

493 Kuhl, A. and Kreiser, W. (1998) *Tetrahedron Lett.*, **39**, 1145–1148.

494 Ghali, N., Johnston, B., Beauchamp, L., Naseree, T., Scott, T., Flanagan, R. and Rodriguez, M. (1995) *Nucleosides Nucleotides*, **14**, 1591–1600.

495 Cope, A.C. and Herrick, E.C. (1963) *Org. Synth.*, **4**, 304.

496 Baer, H.H. and Mateo, F.H. (1990) *Can. J. Chem.*, **68**, 2055–2059.

497 Uno, H., Shiraishi, Y. and Matsushima, Y. (1991) *Bull. Chem. Soc. Jpn.*, **64**, 842–850.

498 Jin, T.-S., Ma, Y.-R., Zhang, Z.-H. and Li, T.-S. (1998) *Synth. Commun.*, **28**, 3173–3177.

499 Fleming, I. and Ghosh, S.K. (1992) *J. Chem. Soc., Chem. Commun.*, 1775–1777.

500 Fedorov, B.S. and Arakcheeva, V.V. (1996) *Izv. Akad. Nauk. SSSR, Ser. Khim.*, 1321–1322.

501 Ishihara, K., Hasegawa, A. and Yamamoto, H. (2001) *Angew. Chem. Int. Ed.*, **40**, 4077–4079.

502 Ishihara, K., Kubota, M., Kurihara, H. and Yamamoto, H. (1995) *J. Am. Chem. Soc.*, **117**, 4413–4414.

503 Ishihara, K., Kubota, M., Kurihara, H. and Yamamoto, H. (1996) *J. Org. Chem.*, **61**, 4560–4567.

504 Norsikian, S., Holmes, I., Lagasse, F. and Kagan, H.B. (2002) *Tetrahedron Lett.*, **43**, 5715–5717.

505 Ishihara, K., Kubota, M. and Yamamoto, H. (1996) *Synlett*, 265–266.

506 Procopiou, P.A., Baugh, S.P.D., Flack, S.S. and Inglis, G.G.A. (1996) *Chem. Commun.*, 2625–2626.

507 Procopiou, P.A., Baugh, S.P.D., Flack, S.S. and Inglis, G.G.A. (1998) *J. Org. Chem.*, **63**, 2342–2347.

508 Kumareswaran, R., Gupta, A. and Vankar, Y.D. (1997) *Synth. Commun.*, **27**, 277–282.

509 Orita, A., Tanahashi, C., Kakuda, A. and Otera, J. (2000) *Angew. Chem. Int. Ed.*, **39**, 2877–2879.

510 Orita, A., Tanahashi, C., Kakuda, A. and Otera, J. (2001) *J. Org. Chem.*, **66**, 8926–8934.

511 Mohammadpoor, B.-I., Aliyan, H. and Khosropur, A.R. (2001) *Tetrahedron*, **57**, 5851–5854.

512 Ohshima, M. and Mukaiyama, T. (1987) *Chem. Lett.*, 377–380.

513 Chen, C.-T., Kuo, J.-H., Li, C.-H., Barhate, N.B., Hon, S.-W., Li, T.-W., Chao, S.-D., Liu, C.-C., Chang, I.-H., Lin, J.-S., Liu, C.-J. and Chou, Y.-C. (2001) *Org. Lett.*, **3**, 3729–3732.

514 Chauhan, K.K., Frost, C.G., Love, I. and Waite, D. (1999) *Synlett*, 1743–1744.

515 Mikami, K., Kotera, O., Motoyama, Y., Sakaguchi, H. and Maruta, M. (1996) *Synlett*, 171–172.

516 Nishikido, J., Nakajima, H., Saeki, T., Ishii, A. and Mikami, K. (1998) *Synlett*, 1347–1348.

517 Mikami, K., Mikami, Y., Matsumoto, Y., Nishikido, J., Yamamoto, F. and Nakajima, H. (2001) *Tetrahedron Lett.*, **42**, 289–292.

518 Mikami, K., Mikami, Y., Matsuzawa, H., Matsumoto, Y., Nishikido, J., Yamamoto, F. and Nakajima, H. (2002) *Tetrahedron*, **58**, 4015–4021.

519 Nie, J., Zhao, Z., Xu, J. and Liu, D. (1999) *J. Chem. Res., Synop.*, 160–161.

520 Nakae, Y., Kusaki, I. and Sato, T. (2001) *Synlett*, 1584–1586.

521 Bartoli, G., Bosco, M., Dalpozzo, R., Marcantoni, E., Massaccesi, M., Rinaldi, S. and Sambri, L. (2003) *Synthesis*, 39–42.

522 Bartoli, G., Bosco, M., Dalpozzo, R., Marcantoni, E., Massaccesi, M. and Sambri, L. (2003) *Eur. J. Org. Chem.*, 4611–4617.

523 Chakraborti, A.K. and Gulhane, R. (2003) *Shivani Synlett*, 1805–1808.

524 Ahmad, S. and Iqbal, J. (1987) *J. Chem. Soc., Chem. Commun.*, 114–115.

525 Sabitha, G., Reddy, B.V.S., Srividya, R. and Yadav, J.S. (1999) *Synth. Commun.*, **29**, 2311–2315.

526 Pansare, S., Malusare, M.G. and Rai, A.N. (2000) *Synth. Commun.*, **30**, 2587–2592.

527 Vedejs, E. and Daugulis, O. (1996) *J. Org. Chem.*, **61**, 5702–5703.

528 Chowdhury, P.K. (1993) *J. Chem. Res., Synop.*, 338–339.

529 Chakraborti, A.K. and Gulhane, R. (2003) *Tetrahedron Lett.*, **44**, 6749–6753.

530 Choudhary, V.R., Mantri, K. and Jana, S.K. (2001) *Catal. Commun.*, **2**, 57–61.

531 Salavati-Niasari, M., Khosousi, T. and Hydarzadeh, S. (2005) *J. Mol. Catal. A*, **235**, 150–153.

532 De, S.K. (2004) *Tetrahedron Lett.*, **45**, 2919–2922.

533 Dalpozzo, R., De Nino, A., Maiuolo, L., Oliverio, M., Procopio, A., Russo, B. and Tocci, A. (2007) *Aust. J. Chem.*, **60**, 75–79.

534 Ogawa, T., Yoshikawa, A., Wada, H., Ogawa, C., Ono, N. and Suzuki, H. (1995) *J. Chem. Soc., Chem. Commun.*, 1407–1408.

535 Orita, A., Ito, T., Yasui, Y. and Otera, J. (1999) *Synlett*, 1927–1929.

536 Sakamoto, K., Hamada, Y., Akashi, H., Orita, A. and Otera, J. (1999) *Organometallics*, **18**, 3555–3557.

537 Durand, S., Sakamoto, K., Fukuyama, T., Orita, A., Otera, J., Duthie, A., Daktermieks, D., Schulte, M. and Jurkschat, K. (2000) *Organometallics*, **19**, 3220–3223.

538 Tangestaninejad, S., Habibi, M.H., Mirkhani, V. and Moghadam, M. (2002) *Synth. Commun.*, **32**, 1337–1343.

539 Moghadam, M., Tangestaninejad, S., Mirkhani, V., Mohammadpour-Baltork, I. and Shaibani, R. (2004) *J. Mol. Catal. A*, **219**, 73–78.

540 Kartha, K.P.R. and Field, R.A. (1997) *Tetrahedron*, **53**, 11753–11766.

541 Ballini, R., Bosica, G., Carloni, S., Ciaralli, L., Maggi, R. and Sartori, G. (1998) *Tetrahedron Lett.*, **39**, 6049–6052.

542 Bhaskar, P.M. and Loganathan, D. (1999) *Synlett*, 129–131.

543 Bhagiyalakshmi, M., Shanmugapriya, K., Palanichamy, M., Arabindoo, B. and Murugesan, V. (2004) *Appl. Catal. A*, **267**, 77–86.

544 Udayakumar, S., Pandurangan, A. and Sinha, P.K. (2005) *J. Mol. Catal. A*, **240**, 139–154.

545 Zhao, Z.H. (2001) *J. Mol. Catal. A*, **168**, 147–152.

546 Rao, Y.V.S., Kulkarni, S.J., Subrahmanyam, M. and Rao, A.V.R. (1993) *J. Chem. Soc., Chem. Commun.*, 1456–1457.

547 Wei, R.-B., Yuan, Z.-Y. and Li, H.-X. (1997) *Gazz. Chim. Ital.*, **127**, 811–813.

548 Li, A.-X., Li, T.-S. and Ding, T.H. (1997) *Chem. Commun.*, 1389–1390.

549 Li, T.-S. and Li, A.-X. (1998) *J. Chem. Soc., Perkin Trans. 1*, 1913–1917.

550 Bhaskar, P.M. and Loganathan, D. (1998) *Tetrahedron Lett.*, **39**, 2215–2218.

551 Reddy, C.R., Iyengar, P., Nagendrappa, G. and Prakash, B.S.J. (2005) *J. Mol. Catal. A*, **229**, 31–37.

552 Kumareswaran, R., Pachamuthu, K. and Vankar, Y.D. (2000) *Synlett*, 1652–1654.

553 Srivastava, V., Tandon, A. and Ray, S. (1992) *Synth. Commun.*, **22**, 2703–2710.

554 Kumar, P., Pandey, R.K., Bodas, M.S. and Dongare, M.K. (2001) *Synlett*, 206–209.

555 Chandrasekhar, S., Ramachader, T. and Takhi, M. (1998) *Tetrahedron Lett.*, **39**, 3263–3266.

556 Curini, M., Epifano, F., Marcotullio, M.C., Rosati, O. and Rossi, M. (2000) *Synth. Commun.*, **30**, 1319–1329.

557 Habibi, M.H., Tangestaninejad, S., Mirkhani, V. and Yadollahi, B. (2002) *Synth. Commun.*, **32**, 863–867.

558 Mirkhani, V., Tangestaninejad, S., Moghadam, M., Yadollahi, B. and Alipanah, L. (2004) *Monatsh. Chem.*, **135**, 1257–1263.

559 Sejidov, F.T., Mansoori, Y. and Goodarzi, N. (2005) *J. Mol. Catal. A*, **240**, 186–190.

560 Kok, G.B., Groves, D. and Von Itzstein, M. (1999) *J. Chem. Soc., Perkin Trans. 1*, 2109–2115.

561 Chilin, A., Rodighiero, P., Pastorini, G. and Guiotto, A. (1991) *J. Org. Chem.*, **56**, 980–983.

562 Kelly, T.R., Kim, M.H. and Curtis, A.D.M. (1993) *J. Org. Chem.*, **58**, 5855–5857.

563 Caffieri, S., Moor, A.C.E., Beijersbergen Van Henegouwen, G.M.J., Dall'Acqua, F., Guiotto, A., Chilin, A. and Rodighiero, P. (1995) *Z. Naturforsch. B: Chem. Sci.*, **50**, 1257–1264.

564 Khan, M.S.Y. and Sharma, P. (1993) *Indian J. Chem., Sect. B*, **32**, 817–821.

565 Majumdar, K.C. and Choudhury, P.K. (1992) *Synth. Commun.*, **22**, 3013–3027.

566 Majumdar, K.C. and Kundu, A.K. (1996) *Synth. Commun.*, **26**, 4023–4037.

567 Majumdar, K.C. and Bhattacharyya, T. (1998) *Synth. Commun.*, **28**, 2907–2923.

568 Kraus, G.A. and Ridgeway, J. (1994) *J. Org. Chem.*, **59**, 4735–4737.

569 Sun, J.-H., Teleha, C.A., Yan, J.-S., Rodgers, J.D. and Nugiel, D.A. (1997) *J. Org. Chem.*, **62**, 5627–5629.

570 Cambie, R.C., Higgs, P.I., Rutledge, P.S. and Woodgate, P.D. (1994) *Aust. J. Chem.*, **47**, 1815–1831.

571 McAllister, G.D., Hartley, R.C., Dawson, M.J. and Knaggs, A.R. (1998) *J. Chem. Soc., Perkin Trans. 1*, 3453–3457.

572 Shin, J.-S., Kim, K.-S., Kim, M.-B., Jeong, J.-H. and Kim, B.-K. (1999) *Bioorg. Med. Chem. Lett.*, **9**, 869–874.

573 Sagara, T., Ozawa, S., Kushiyama, E., Koike, K., Takayanagi, I. and Kanematsu, K. (1995) *Bioorg. Med. Chem. Lett.*, **5**, 1505–1508.

574 Wack, H., Drury, W.J., Taggi, A.E., Ferraris, D. and Lectka, T. (1999) *Org. Lett.*, **1**, 1985–1988.

575 Ohtani, M., Matsuura, T., Watanabe, F. and Narisada, M. (1991) *J. Org. Chem.*, **56**, 2122–2127.

576 Suda, Y., Yago, S., Shiro, M. and Taguchi, T. (1992) *Chem. Lett.*, 389–392.

577 Majewski, M. and Lazny, R. (1995) *J. Org. Chem.*, **60**, 5825–5830.

578 Cheng, M.-H., Ho, Y.-H., Wang, S.-L., Cheng, C.-Y., Peng, S.-M., Lee, G.-H. and Liu, R.-S. (1992) *J. Chem. Soc., Chem. Commun.*, 45–46.

579 Tomassy, B. and Zwierzak, A. (1998) *Synth. Commun.*, **28**, 1201–1214.

580 Tacke, R., Wiesenberger, F., Becker, B., Rohr-Aehle, R., Schneider, P.B., Ulbrich, U., Sarge, S.M., Cammenga, H.K., Koslowski, T. and Von Niessen, W. (1992) *Chem. Ber.*, **125**, 591–605.

581 Hoye, T.R., Cermohous, J.J. and Pfister, K.P. (1998) *Tetrahedron Lett.*, **39**, 1857–1860.

582 Steglich, W. and Höfle, G. (1969) *Angew. Chem. Int. Ed. Engl.*, **8**, 981.

583 Bohlmann, F. and Franke, H. (1971) *Chem. Ber.*, **104**, 3229–3233.

584 Höfle, G. and Steglich, W. (1972) *Synthesis*, 619–621.

585 Höfle, G., Steglich, W. and Vorbrüggen, H. (1978) *Angew. Chem. Int. Ed. Engl.*, **17**, 569–583.

586 Fischer, C.B., Xu, S. and Zipse, H. (2006) *Chem. Eur. J.*, **12**, 5779–5784.

587 Shimizu, T., Kobayashi, R., Ohmori, H. and Nakata, T. (1995) *Synlett*, 650–652.

588 Ruble, J.C., Tweddell, J. and Fu, G.C. (1998) *J. Org. Chem.*, **63**, 2794–2795.

589 Tao, B., Ruble, J.C., Hoic, D.A. and Fu, G.C. (1999) *J. Am. Chem. Soc.*, **121**, 5091–5092.

590 Spivey, A.C., Fekner, T. and Spey, S.E. (2000) *J. Org. Chem.*, **65**, 3154–3159.

591 Kawabata, T., Nagato, M., Takasu, K. and Fuji, K. (1997) *J. Am. Chem. Soc.*, **119**, 3169–3170.

592 Miller, S.J., Copeland, G.T., Papaioammou, N., Horstmann, T. and Ruel, E.M. (1998) *J. Am. Chem. Soc.*, **120**, 1629–1630.

593 Copeland, G.T., Jarvo, E.R. and Miller, S.J. (1998) *J. Org. Chem.*, **63**, 6784–6785.

594 Hiratake, J., Yamamoto, Y. and Oda, J. (1985) *J. Chem. Soc., Chem. Commun.*, 1717–1719.

595 Hiratake, J., Inagaki, M., Yamamoto, Y. and Oda, J. (1987) *J. Chem. Soc., Perkin Trans. 1*, 1053–1058.

596 Aitken, R.A. and Gopal, J. (1990) *Tetrahedron: Asymmetry*, **1**, 517–520.

597 Aitken, R.A. and Hirst, J.A. (1988) *J. Chem. Soc., Chem. Commun.*, 632–634.

598 Bolm, C., Schiffers, I., Dinter, C.L. and Gerlach, A. (2000) *J. Org. Chem.*, **65**, 6984–6991.

599 Chen, Y., Tian, S.-K. and Deng, L. (2000) *J. Am. Chem. Soc.*, **122**, 9542–9543.

600 Chen, Y. and Deng, L. (2001) *J. Am. Chem. Soc.*, **123**, 11302–11303.

601 Shimizu, M., Matsukawa, K. and Fujisawa, T. (1993) *Bull. Chem. Soc. Jpn.*, **66**, 2128–2130.

602 Zolfigol, M.A., Khazaei, A., Choghamarani, A.G., Rostami, A. and Hajjami, M. (2006) *Catal. Commun.*, **7**, 399–402.

603 Vedejs, E. and Diver, S.T. (1993) *J. Am. Chem. Soc.*, **115**, 3358–3359.

604 Vedejs, E., Bennett, N.S., Conn, L.M., Diver, S.T., Gingras, M., Lin, S., Oliver, P.A. and Peterson, M.J. (1993) *J. Org. Chem.*, **58**, 7286–7288.

605 Vedejs, E., Daugulis, O. and Diver, S.T. (1996) *J. Org. Chem.*, **61**, 430–431.

606 Vedejs, E. and Daugulis, O. (1999) *J. Am. Chem. Soc.*, **121**, 5813–5814.

607 D'sa, B.A. and Verkade, J.G. (1996) *J. Org. Chem.*, **61**, 2963–2966.

608 Bianchi, D., Cesti, P. and Battistel, E. (1988) *J. Org. Chem.*, **53**, 5531–5534.

609 Berger, B., Rabiller, D.G., Königsberger, K., Faber, K. and Griengl, H. (1990) *Tetrahedron: Asymmetry*, **1**, 541–546.

610 Ramaswamy, S., Morgan, B. and Oehlschlager, A.C. (1990) *Tetrahedron Lett.*, **31**, 3405–3408.

611 Inagawa, J., Hirata, K., Saeki, H., Katsuki, T. and Yamaguchi, M. (1979) *Bull. Chem. Soc. Jpn.*, **52**, 1989–1993.

612 Seebach, D., Brändli, U. and Schnurrenberger, P. (1988) *Helv. Chim. Acta*, **71**, 155–167.

613 Mori, K. and Ebata, T. (1986) *Tetrahedron*, **42**, 4685–4689.

614 Gil, P., Razkin, J. and Gonzalez, A. (1998) *Synthesis*, 386–392.

615 Leemhuis, F.M.C., Thijs, L. and Zwanenburg, B. (1993) *J. Org. Chem.*, **58**, 7170–7179.

616 Shiina, I., Miyoshi, S., Miyashita, M. and Mukaiyama, T. (1994) *Chem. Lett.*, 515–518.

617 Shiina, I., Ibuka, R. and Kubota, M. (2002) *Chem. Lett.*, 286–287.

618 Shiina, I., Kubota, M., Oshiumi, H. and Hashizume, M. (2004) *J. Org. Chem.*, **69**, 1822–1830.

619 Shiina, I., Oshiumi, H., Hashizume, M., Yamai, Y.-S. and Ibuka, R. (2004) *Tetrahedron Lett.*, **45**, 543–547.

620 Shiina, I. (2004) *Tetrahedron*, **60**, 1587–1599.

621 Parish, C. and Stock, L.M. (1965) *J. Org. Chem.*, **30**, 927–929.

622 Mota, A.J., Robles, R., de Cienfuegos, L.Á. and Lamenca, A. (2004) *Tetrahedron Lett.*, **45**, 3349–3353.

623 Banerjee, S. and Trivedi, G.K. (1992) *Tetrahedron*, **48**, 9939–9950.

624 Kanth, J.V.B. and Periasamy, M. (1995) *Synth. Commun.*, **25**, 1523–1530.

625 Joulin, P., Castro, B., Zeggaf, C. and Pantaloni, A. (1987) *Tetrahedron Lett.*, **28**, 1661–1664.

626 McLaren, K.L. (1995) *J. Org. Chem.*, **60**, 6082–6084.

627 Gooßen, L.J. and Döhring, A. (2004) *Synlett*, 263–266.
628 Nagarajan, M., Kumar, V.S. and Rao, B.V. (1997) *Tetrahedron Lett.*, **38**, 5835–5838.
629 Zacharie, B., Connolly, T.P. and Penney, C.L. (1995) *J. Org. Chem.*, **60**, 7072–7074.
630 Duibe-Jampel, E. and Bassir, M. (1994) *Tetrahedron Lett.*, **35**, 421–422.
631 Ogawa, T., Nakazato, A., Tsuchida, K. and Hatakeyama, K. (1993) *Chem. Pharm. Bull.*, **41**, 1049–1054.
632 Saito, Y., Yamaki, T., Kohashi, F., Watanabe, T., Ouchi, H. and Takahata, H. (2005) *Tetrahedron Lett.*, **46**, 1277–1279.
633 Hussain, M.A., Liebert, T. and Heinze, T. (2004) *Macromol. Rapid Commun.*, **25**, 916–920.
634 Gololobov, Y.G. and Galkina, M.A. (1995) *Izv. Akad. Nauk. SSSR, Ser. Khim.*, 779–780.
635 Krasnov, V.P., Bukrina, I.M., Zhdanova, E.A., Kodess, M.I. and Korolyova, M.A. (1994) *Synthesis*, 961–964.
636 Yamada, S., Sugai, T. and Matsuzaki, K. (1996) *J. Org. Chem.*, **61**, 5932–5938.
637 Katritzky, A.R., Yang, B. and Semenzin, D. (1997) *J. Org. Chem.*, **62**, 726–728.
638 McNicholas, C., Simpson, T.J. and Willett, N.J. (1996) *Tetrahedron Lett.*, **37**, 8053–8056.
639 Palomo, C., Arrieta, A., Cossio, F.P., Aizpurua, J.M., Mielgo, A. and Aurrekoetxea, N. (1990) *Tetrahedron Lett.*, **31**, 6429–6432.
640 Dixit, A.N., Tandel, S.K. and Rajappa, S. (1994) *Tetrahedron Lett.*, **35**, 6133–1634.
641 Oriyama, T., Imai, K., Hosoya, T. and Sano, T. (1996) *Tetrahedron Lett.*, **37**, 8543–8546.
642 Habata, Y., Fujishiro, F. and Akabori, S. (1996) *J. Chem. Soc., Perkin Trans. 1*, 953–957.
643 Prabhakar, P., Suryakiran, N. and Venkateswarlu, Y. (2007) *Chem. Lett.*, **36**, 732–733.
644 Reddy, T.S., Narasimhulu, M., Suryakiran, N., Mahesh, K.C., Ashalatha, K. and Venkateswarlu, Y. (2006) *Tetrahedron Lett.*, **47**, 6825–6829.
645 Torregiani, E., Seu, G., Minassi, A. and Appendino, G. (2005) *Tetrahedron Lett.*, **46**, 2193–2196.
646 Fisher, L.E., Caroon, J.M., Stabler, S.R., Lundberg, S., Zaidi, S., Sorensen, C.M., Sparacino, M.L. and Muchowski, J.M. (1994) *Can. J. Chem.*, **72**, 142–145.
647 Fukuzawa, S. and Hongo, Y. (1998) *Tetrahedron Lett.*, **39**, 3521–3524.
648 Kocevar, M., Mihorko, P. and Polanc, S. (1995) *J. Org. Chem.*, **60**, 1466–1469.
649 Masamune, S., Kamata, S. and Schilling, W. (1975) *J. Am. Chem. Soc.*, **97**, 3515–3516.
650 Masamune, S., Hayase, Y., Schilling, W., Chan, W.K. and Bates, G.S. (1977) *J. Am. Chem. Soc.*, **99**, 6756–6758.
651 Aggarwal, V.K., Thomas, A. and Franklin, R.J. (1994) *J. Chem. Soc., Chem. Commun.*, 1653–1654.
652 Muzard, M. and Portella, C. (1993) *J. Org. Chem.*, **58**, 29–31.
653 Zadmard, R., Aghapoor, K., Bolourtchian, M. and Saidi, M.R. (1998) *Synth. Commun.*, **28**, 4495–4499.
654 Nagasawa, K., Yoshitake, S., Amiya, T. and Ito, K. (1990) *Synth. Commun.*, **20**, 2033–2040.
655 Breton, G.W., Kurtz, M.J. and Kurtz, S.L. (1997) *Tetrahedron Lett.*, **38**, 3825–3828.
656 Tamaddon, F., Amrollahi, M.A. and Sharafat, L. (2005) *Tetrahedron Lett.*, **46**, 7841–7844.
657 Greenlee, W.J. and Thoesett, E.D. (1981) *J. Org. Chem.*, **46**, 5351–5353.
658 Kaiser, E.M. and Woodruff, R.A. (1970) *J. Org. Chem.*, **35**, 1198–1199.
659 Van Loon, J.-D., Kraft, D., Ankone, M.J.K., Verboom, W., Harkema, S., Vogt, W., Böhmer, V. and Reinhoudt, D.N. (1990) *J. Org. Chem.*, **55**, 5176–5179.
660 Itoh, T. and Chika, J. (1995) *J. Org. Chem.*, **60**, 4968–4969.
661 Itoh, T., Chika, J., Shirakami, S., Ito, H., Yoshida, T., Kubo, Y. and Uenishi, J. (1996) *J. Org. Chem.*, **61**, 3700–3705.
662 Itoh, T., Shirakami, S., Nakao, Y. and Yoshida, T. (1998) *Chem. Lett.*, 979–980.
663 Moutel, S. and Prandi, J. (1998) *Tetrahedron Lett.*, **39**, 9667–9670.

664 Mayer, S.C., Pfizenmayer, A.J. and Joullie, M.M. (1996) *J. Org. Chem.*, **61**, 1655–1664.

665 Barua, A.B., Huselton, C.A. and Olson, J.A. (1996) *Synth. Commun.*, **26**, 1355–1361.

666 Flynn, D.L., Zelle, R.E. and Grieco, P.A. (1983) *J. Org. Chem.*, **48**, 2424–2426.

667 Alcaide, B., Miranda, M., Perez-Castells, J., Polanco, C. and Sierra, M.A. (1994) *J. Org. Chem.*, **59**, 8003–8010.

668 Paradisi, M.P., Zecchini, G.P. and Torrini, I. (1986) *Tetrahedron Lett.*, **27**, 5029–5032.

669 Takimoto, S., Inanaga, J., Katsuki, T. and Yamaguchi, M. (1976) *Bull. Chem. Soc. Jpn.*, **49**, 2335–2336.

670 Huang, Z.-T. and Wang, G.-Q. (1994) *Synth. Commun.*, **24**, 11–22.

671 Nieman, J.A. and Keay, B.A. (1996) *Tetrahedron: Asymmetry*, **7**, 3521–3526.

672 Barros, M.T., Januario-Charmier, M.O., Maycock, C.D. and Pires, M. (1996) *Tetrahedron*, **52**, 7861–7874.

673 Swindell, C.S., Fan, W. and Klimko, P.G. (1994) *Tetrahedron Lett.*, **35**, 4959–4962.

674 Almeida, J.F., Grande, M., Moran, J.R., Anaya, J., Mussons, L. and Caballero, M.C. (1993) *Tetrahedron: Asymmetry*, **4**, 2483–2494.

675 Allevi, P., Anastasia, M., Fiecchi, A., Sanvito, A.M. and Scala, A. (1991) *Synthesis*, 438–440.

676 Oriyama, T., Imai, K., Hosoya, T. and Sano, T. (1998) *Tetrahedron Lett.*, **39**, 397–400.

677 Watanabe, Y., Fujimoto, T., Shinohara, T. and Ozaki, S. (1991) *J. Chem. Soc., Chem. Commun.*, 428–429.

678 Chung, S.-K. and Yu, S.-H. (1996) *Bioorg. Med. Chem. Lett.*, **6**, 1461–1464.

679 Luo, J. and Sun, Y. (2006) *J. Appl. Polym. Sci.*, **100**, 3288–3296.

680 Hajipour, A.R. and Mazloumi, G. (2002) *Synth. Commun.*, **32**, 23–30.

681 Takahashi, M., Kuroda, T., Ogiku, T., Ohmizu, H., Kondo, K. and Iwasaki, T. (1991) *Tetrahedron Lett.*, **32**, 6919–6922.

682 Takahashi, M., Kuroda, T., Ogiku, T., Ohmizu, H., Kondo, K. and Iwasaki, T. (1992) *Heterocycles*, **34**, 2061–2064.

683 Ishihara, K., Kurihara, H. and Yamamoto, H. (1993) *J. Org. Chem.*, **58**, 3791–3793.

684 Santoyo-Gonzalez, F., Garcia-Calvo-Flores, F., Isac-Garcia, J., Robles-Diaz, R. and Vargas-Berenguel, A. (1994) *Synthesis*, 97–101.

685 Iida, T. and Itaya, T. (1993) *Tetrahedron*, **49**, 10511–10530.

686 Itaya, T. and Iida, T. (1994) *J. Chem. Soc., Perkin Trans. 1*, 1671–1672.

687 Sano, T., Ohashi, K. and Oriyama, T. (1999) *Synthesis*, 1141–1144.

688 Nakatsuji, H., Morita, J., Misaki, T. and Tanabe, Y. (2006) *Adv. Synth. Catal.*, **348**, 2057–2062.

689 Illi, V.O. (1979) *Tetrahedron Lett.*, **20**, 2431–2432.

690 Hashimoto, I. and Higashi, T. (1993) *Chem. Express*, **8**, 445–448.

691 Wei, T., Chen, J., Wang, X. and Zhang, Y. (1996) *Synth. Commun.*, **26**, 1447–1454.

692 Malanga, C., Pagliai, L. and Menicagli, R. (1990) *Synth. Commun.*, **20**, 2821–2826.

693 Branalt, J., Kvarnstrom, I., Niklasson, G. and Svensson, S.C.T. (1994) *J. Org. Chem.*, **59**, 1783–1788.

694 Jia, X.-S., Wang, H.-L., Huang, Q., Kong, L.-L. and Zhang, W.-H. (2006) *J. Chem. Res.*, 135–138.

695 Shah, S.T.A., Khan, K.M., Hussain, H., Anwar, M.U., Fecker, M. and Voelter, W. (2005) *Tetrahedron*, **61**, 6652–6656.

696 Corey, E.J. and Nicolaou, K.C. (1974) *J. Am. Chem. Soc.*, **96**, 5614–5616.

697 Evans, D.A., Anderson, J.C. and Taylor, M.K. (1993) *Tetrahedron Lett.*, **34**, 5563–5566.

698 Yamada, S. and Ohe, T. (1996) *Tetrahedron Lett.*, **37**, 6777–6780.

699 Yamada, S. and Katsumata, H. (1999) *J. Org. Chem.*, **64**, 9365–9373.

700 Orita, A., Nagano, Y., Hirano, J. and Otera, J. (2001) *Synlett*, 637–639.

701 Prakash, O., Sharma, V. and Sadana, A. (1996) *J. Chem. Res., Synop.*, 100–101.

702 Stefane, B., Kocevar, M. and Polanc, S. (1999) *Tetrahedron Lett.*, **40**, 4429–4432.

703 Anelli, P.L., Brocchetta, M., Palano, D. and Visigalli, M. (1997) *Tetrahedron Lett.*, **38**, 2367–2368.

704 Bachi, M.D., Korshin, E.E., Ploypradith, P., Cumming, J.N., Xie, S., Shapiro, T.A. and Posner, G.H. (1998) *Bioorg. Med. Chem. Lett.*, **8**, 903–908.

705 Yamaguchi, M., Tsukamoto, Y. and Minami, T. (1990) *Chem. Lett.*, 1223–1226.

706 Pittelkow, M., Kamounah, F.S., Boas, U., Pedersen, B. and Christensen, J.B. (2004) *Synthesis*, 2485–2492.

707 Mattson, A., Oehrner, N., Hult, K. and Norin, T. (1993) *Tetrahedron: Asymmetry*, **4**, 925–930.

708 Akita, H., Kato, K., Umezawa, I. and Matsukura, H. (1997) *Chem. Pharm. Bull.*, **45**, 2085–2088.

709 Ghogare, A. and Kumar, G.S. (1989) *J. Chem. Soc., Chem. Commun.*, 1533–1535.

710 Santoyo-Gonzalez, F., Uriel, C. and Calvo-Asin, J.A. (1998) *Synthesis*, 1787–1792.

711 Valade, J. and Pereyre, M. (1962) *C. R. Acad. Sci. Paris*, **254**, 3693–3695.

712 Wagner, D., Verheyden, J.P.H. and Moffatt, J.G. (1974) *J. Org. Chem.*, **39**, 24–30.

713 Munavu, R.M. and Szmant, H.H. (1976) *J. Org. Chem.*, **41**, 1832–1836.

714 Ogawa, T. and Matsui, M. (1981) *Tetrahedron*, **37**, 2363–2369.

715 Holzapfel, C.W., Koekemoer, J.M. and Matais, C.F. (1984) *S. Afr. J. Chem.*, **37**, 19–26.

716 Zapata, A., De La Pradilla, R.F., Martin-Lomas, M. and Penades, S. (1991) *J. Org. Chem.*, **56**, 444–447.

717 Crombez-Robert, C., Benazza, M., Frechou, C. and Demailly, G. (1998) *Carbohydr. Res.*, **307**, 355–359.

718 Corrie, J.E.T. (1993) *J. Chem. Soc., Perkin Trans. 1*, 2161–2166.

719 Mukaiyama, T., Tomioka, I. and Shimizu, M. (1984) *Chem. Lett.*, 49–52.

720 Mukaiyama, T. and Tanabe, Y. (1984) *Chem. Lett.*, 401–404.

721 Iwasaki, F., Maki, T., Nakashima, W., Onomura, O. and Matsumura, Y. (1999) *Org. Lett.*, **1**, 969–972.

722 Ricci, A., Roelens, S. and Vannucchi, A. (1985) *J. Chem. Soc., Chem. Commun.*, 1457–1458.

723 Reginato, G., Ricci, A., Roelens, S. and Scapecchi, S. (1990) *J. Org. Chem.*, **55**, 5132–5139.

724 Maki, T., Iwasaki, F. and Matsumura, Y. (1998) *Tetrahedron Lett.*, **39**, 5601–5604.

725 Shanzer, A. and Mayer-Shochet, N. (1980) *J. Chem. Soc., Chem. Commun.*, 176–177.

726 Shanzer, A., Mayer-Shochet, N., Frolow, F. and Rabinovich, D. (1981) *J. Org. Chem.*, **46**, 4662–4665.

727 Shanzer, A. and Berman, E. (1980) *J. Chem. Soc., Chem. Commun.*, 259–260.

728 Shanzer, A., Libman, J., Gottlieb, H. and Frolow, F. (1982) *J. Am. Chem. Soc.*, **104**, 4220–4225.

729 Shazner, A. and Libman, J. (1982) *J. Organomet. Chem.*, **239**, 301–306.

730 Mukaiyama, T., Ichikawa, J. and Asami, M. (1983) *Chem. Lett.*, 293–296.

731 Ishikawa, J., Asami, M. and Mukaiyama, T. (1984) *Chem. Lett.*, 949–952.

732 Niwa, H., Miyachi, Y., Okamoto, O., Uosaki, Y., Kuroda, A., Ishiwata, H. and Yamada, K. (1992) *Tetrahedron*, **48**, 393–412.

733 Niwa, H., Sakata, T. and Yamada, K. (1994) *Bull. Chem. Soc. Jpn.*, **67**, 1990–1993.

734 Anchisi, C., Corda, L., Maccioni, A. and Podda, G. (1982) *J. Heterocycl. Chem.*, **19**, 141–144.

735 Anchisi, C., Corda, L., Fedda, A.M. and Maccioni, A. (1984) *J. Heterocycl. Chem.*, **21**, 577–581.

736 Gridley, J.J., Osborn, H.M.I. and Suthers, W.G. (1999) *Tetrahedron Lett.*, **40**, 6991–6994.

737 Ansell, R.J., Barrett, S.A., Meegan, J.E. and Warriner, S.L. (2007) *Chem. Eur. J.*, **13**, 4654–4664.

738 Mansoori, Y., Tataroglu, F.S. and Sadaghian, M. (2005) *Green Chem.*, **7**, 870–873.

739 Ohshima, T., Iwasaki, T., Maegawa, Y., Yoshiyama, A. and Mashima, K. (2008) *J. Am. Chem. Soc.*, **130**, 2944–2945.

740 Von Pechmann, H. (1894) *Chem. Ber.*, **27**, 1888.

741 Von Pechmann, H. (1895) *Chem. Ber.*, **28**, 855.

742 Fieser, L.F. and Fieser, M. (1967) *Reagents Org. Synth.*, **1**, 191.
743 Hashimoto, N., Aoyama, T. and Shioiri, T. (1981) *Chem. Pharm. Bull.*, **29**, 1475–1478.
744 Presser, A. and Hüfner, A. (2004) *Monatsh. Chem.*, **135**, 1015–1022.
745 Shiozaki, M., Deguchi, N., Mochizuki, T. and Nishijima, M. (1996) *Tetrahedron Lett.*, **37**, 3875–3876.
746 Czernecki, S., Valery, J.-M. and Wilkens, R. (1996) *Bull. Chem. Soc. Jpn.*, **69**, 1347–1351.
747 Ogawa, H., Chihara, T. and Taya, K. (1985) *J. Am. Chem. Soc.*, **107**, 1365–1369.
748 Furrow, M.E. and Myers, A.G. (2004) *J. Am. Chem. Soc.*, **126**, 12222–12223.
749 Pfeffer, P.E. and Silbert, L.S. (1976) *J. Org. Chem.*, **41**, 1373–1379.
750 Pfeffer, P.E., Foglia, T.A., Barr, P.A., Schmeltz, I. and Silbert, L.S. (1972) *Tetrahedron Lett.*, **13**, 4063–4066.
751 Kunerth, D.C. and Sherry, J.J. (1973) *Tetrahedron Lett.*, **14**, 689–692.
752 Marko, I.E., Seres, P., Swarbrick, T.M., Staton, I. and Adams, H. (1992) *Tetrahedron Lett.*, **33**, 5649–5652.
753 Bocchi, V., Casnati, G., Dossena, A. and Marchelli, R. (1979) *Synthesis*, 961–962.
754 Sivakumar, S., Pangarkar, V.G. and Sawant, S.B. (2002) *Org. Process Res. Dev.*, **6**, 149–151.
755 Guo, W., Li, J., Fan, N., Wu, W., Zhou, P. and Xia, C. (2005) *Synth. Commun.*, **35**, 145–152.
756 Avila-Zárraga, J.G. and Martínez, R. (2001) *Synth. Commun.*, **31**, 2177–2183.
757 Clark, J.H. and Miller, J.M. (1977) *J. Am. Chem. Soc.*, **99**, 498–504.
758 Clark, J.H. and Emsley, J. (1975) *J. Chem. Soc., Dalton Trans.*, 2129–2134.
759 Clark, J.H., Emsley, J. and Hoyte, O.P.A. (1977) *J. Chem. Soc., Perkin Trans. 1*, 1091–1094.
760 Clark, J.H. and Miller, J.M. (1977) *Tetrahedron Lett.*, **18**, 599–602.
761 Tam, J.P., Kent, S.B.H., Wong, T.W. and Merrifield, R.B. (1979) *Synthesis*, 955–957.
762 Clark, J.H. (1978) *J. Chem. Soc., Chem. Commun.*, 789–791.
763 Gisin, B.F. (1973) *Helv. Chim. Acta*, **56**, 1476–1482.
764 Wang, S.-S., Gisin, B.F., Winter, D.P., Makofske, R., Kylesha, I.D., Tzougraki, C. and Meienhofer, J. (1977) *J. Org. Chem.*, **42**, 1286–1290.
765 Lerchen, H.-G. and Kunz, H. (1985) *Tetrahedron Lett.*, **26**, 5257–5260.
766 Kunz, H. and Lerchen, H.-G. (1987) *Tetrahedron Lett.*, **28**, 1873–1876.
767 Kruizinga, W.H. and Kellogg, R.M. (1981) *J. Am. Chem. Soc.*, **103**, 5183–5189.
768 Dijkstra, G., Kruizinga, W.H. and Kellogg, R.M. (1987) *J. Org. Chem.*, **52**, 4230–4234.
769 Galli, C. and Mandolini, L. (1991) *J. Org. Chem.*, **56**, 3045–3047.
770 Luthman, K., Orbe, M., Waglund, T. and Claesson, A. (1987) *J. Org. Chem.*, **52**, 3777–3784.
771 Sato, T., Otera, J. and Nozaki, H. (1992) *J. Org. Chem.*, **57**, 2166–2169.
772 Lee, J.C. and Choi, Y. (1998) *Synth. Commun.*, **28**, 2021–2026.
773 Ooi, T., Sugimoto, H., Doda, K. and Matsuoka, K. (2001) *Tetrahedron Lett.*, **42**, 9245–9248.
774 Ono, N., Yamada, T., Saito, T., Tanaka, K. and Kaji, A. (1978) *Bull. Chem. Soc. Jpn.*, **51**, 2401–2404.
775 Shono, T., Ishige, O., Uyama, H. and Kashimura, S. (1986) *J. Org. Chem.*, **51**, 546–549.
776 Bocchi, V., Casnati, G., Dossena, A. and Marchelli, R. (1979) *Synthesis*, 957–961.
777 Puntambekar, H.M., Naik, D.G. and Kapadi, A.H. (1993) *Indian J. Chem., Sect. B*, **32**, 793–794.
778 Chidambaram, M., Sonavane, S.U., da la Zerda, J. and Sasson, Y. (2007) *Tetrahedron*, **63**, 7696–7701.
779 Raber, D.J. and Gariano, P. (1971) *Tetrahedron Lett.*, **12**, 4741–4744.
780 Raber, D.J., Gariano, P., Brod, A.O. and Gariano, A. (1979) *J. Org. Chem.*, **44**, 1149–1154.
781 Taber, D.F., Gerstenhaber, D.A. and Zhao, X. (2006) *Tetrahedron Lett.*, **47**, 3065–3066.
782 Yamauchi, K., Tanabe, T. and Kinoshita, M. (1979) *J. Org. Chem.*, **44**, 638–639.

783 Badet, B., Julia, M., Ramirez-Munoz, M. and Sarrazin, C.A. (1983) *Tetrahedron*, **39**, 3111–3125.

784 Yamauchi, K., Nakamura, K. and Kinoshita, M. (1979) *Tetrahedron Lett.*, **20**, 1787–1790.

785 Denney, D.B., Melis, R. and Pendse, A.D. (1978) *J. Org. Chem.*, **43**, 4672–4673.

786 Mohacsi, E. (1982) *Synth. Commun.*, **12**, 453–456.

787 Grundy, B.G. and Pattenden, G. (1972) *Tetrahedron Lett.*, **13**, 757–758.

788 Ballini, R. and Carotti, A. (1983) *Synth. Commun.*, **13**, 1197–1201.

789 Chakraborti, A.K., Basak, A. and Grover, V. (1999) *J. Org. Chem.*, **64**, 8014–8017.

790 Trujillo, J.I. and Gopalan, A.S. (1993) *Tetrahedron Lett.*, **34**, 7355–7358.

791 Pasto, M., Castejon, P., Moyano, A., Pericas, M.A. and Riera, A. (1996) *J. Org. Chem.*, **61**, 6033–6037.

792 Ikejiri, M., Miyashita, K., Tsunemi, T. and Imanishi, T. (2004) *Tetrahedron Lett.*, **45**, 1243–1246.

793 Renga, J.M. (1984) *Synth. Commun.*, **14**, 77–82.

794 Lee, Y. and Shimizu, I. (1998) *Synlett*, 1063–1064.

795 Shieh, W.-C., Dell, S. and Repic, O. (2002) *J. Org. Chem.*, **67**, 2188–2191.

796 Shieh, W.-C., Dell, S. and Repic, O. (2002) *Tetrahedron Lett.*, **43**, 5607–5609.

797 Kirumakki, S.R., Nagaraju, N., Murthy, K.V.V.S.B.S.R. and Narayanan, S. (2002) *Appl. Catal. A*, **226**, 175–182.

798 D'Souza, J. and Nagaraju, N. (2007) *Ind. J. Chem. Tech.*, **14**, 292–300.

799 Selva, M. and Tundo, P. (2006) *J. Org. Chem.*, **71**, 1464–1470.

800 Rivero, I.A., Heredia, S. and Ochoa, A. (2001) *Synth. Commun.*, **31**, 2169–2175.

801 Von Vorbrüggen, H. (1963) *Angew. Chem. Int. Ed. Engl.*, **75**, 296–297.

802 Brechbuehler, H., Buechi, H., Hatz, E., Schreiber, J. and Eschemnoser, A. (1965) *Helv. Chim. Acta*, **48**, 1746–1771.

803 Büchi, H., Steen, K. and Eschenmoser, A. (1964) *Angew. Chem. Int. Ed. Engl.*, **3**, 62–63.

804 Widmer, U. (1983) *Synthesis*, 135–136.

805 Ludwig, J. and Lehr, M. (2004) *Synth. Commun.*, **34**, 3691–3695.

806 Baldwin, J.E., Farthing, C.N., Russell, A.T., Schofield, C.J. and Spivey, A.C. (1996) *Tetrahedron Lett.*, **37**, 3761–3764.

807 Mangia, A. and Scandroglio, A. (1978) *Tetrahedron Lett.*, **19**, 5219–5220.

808 Crosignani, S., White, P.D. and Linclau, B. (2002) *Org. Lett.*, **4**, 2961–2963.

809 Crosignani, S., White, P.D. and Linclau, B. (2004) *J. Org. Chem.*, **69**, 5897–5905.

810 Kirsch, S.F. and Overman, L.E. (2005) *J. Am. Chem. Soc.*, **127**, 2866–2867.

811 Fritsche, A., Deguara, H. and Lehr, M. (2006) *Synth. Commun.*, **36**, 3117–3123.

812 Mukaiyama, T., Kikuchi, W. and Shintou, T. (2003) *Chem. Lett.*, **32**, 300–301.

813 Mukaiyama, T., Shintou, T. and Fukumoto, K. (2003) *J. Am. Chem. Soc.*, **125**, 10538–10539.

814 Mukaiyama, T., Shintou, T. and Kikuchi, W. (2002) *Chem. Lett.*, 1126–1127.

815 Shintou, T. and Mukaiyama, T. (2003) *Chem. Lett.*, **32**, 1100–1101.

816 Hanessian, S., Mascitti, V., Lu., P.-P. and Ishida, H. (2002) *Synthesis*, 1959–1968.

817 Okano, T., Hayashizaki, Y. and Kiji, J. (1993) *Bull. Chem. Soc. Jpn.*, **66**, 1863–1865.

818 Stanton, M.G. and Gagne, M.R. (1997) *J. Am. Chem. Soc.*, **119**, 5075–5076.

819 Stanton, M.G., Allen, C.B., Kissling, R.M., Lincoln, A.L. and Gagne, M.R. (1998) *J. Am. Chem. Soc.*, **120**, 5981–5989.

820 Stanton, M.G. and Gagne, M.R. (1997) *J. Org. Chem.*, **62**, 8240–8242.

821 Kissling, R.M. and Gagne, M.R. (2000) *Org. Lett.*, **2**, 4209–4212.

822 Kissling, R.M. and Gagne, M.R. (2001) *J. Org. Chem.*, **66**, 9005–9010.

823 Kagan, H.B. and Fiaud, J.C. (1988) *Topics in Stereochemistry*, Vol. 18, John Wiley & Sons, Inc., p. 249.

824 Chen, C.-S., Fujimoto, Y., Girdaukas, G. and Sih, C.J. (1982) *J. Am. Chem. Soc.*, **104**, 7294–7299.

825 Okamoto, T., Yasuhito, E. and Ueji, S. (2006) *Org. Biomol. Chem.*, **4**, 1147–1153.

826 Saul, S., Corr, S. and Micklefield, J. (2004) *Angew. Chem. Int. Ed.*, **43**, 5519–5523.

827 Ueji, S., Tanaka, H., Hanaoka, T., Ueda, A., Watanabe, K., Kaihatsu, K. and Ebara, Y. (2001) *Chem. Lett.*, 1066–1067.

828 Liu, H.-L. and Anthonsen, T. (2002) *Chirality*, **14**, 25–27.

829 Bouzemi, N., Aribi-Zouioueche, L. and Fiaud, J.-C. (2006) *Tetrahedron: Asymmetry*, **17**, 797–800.

830 Wimmer, Z., Skouridou, V., Zarevúcka, M., Šaman, D. and Kolisis, F.N. (2004) *Tetrahedron: Asymmetry*, **15**, 3911–3917.

831 Sanfilippo, C., D'Antona, N. and Nicolosi, G. (2006) *Tetrahedron: Asymmetry*, **17**, 12–14.

832 Legros, J.-Y., Toffano, M., Drayton, S.K., Rivard, M. and Fiaud, J.-C. (1997) *Tetrahedron Lett.*, **38**, 1915–1918.

833 Sakai, T., Takayama, T., Ohkawa, T., Yoshio, O., Ema, T. and Utaka, M. (1997) *Tetrahedron Lett.*, **38**, 1987–1990.

834 Jeromin, G.E. and Welsch, V. (1995) *Tetrahedron Lett.*, **36**, 6663–6664.

835 Ling, L. and Ozaki, S. (1995) *Bull. Chem. Soc. Jpn.*, **68**, 1200–1205.

836 Izumi, T. and Fukaya, K. (1993) *Bull. Chem. Soc. Jpn.*, **66**, 1216–1221.

837 Jouglet, B. and Rousseau, G. (1993) *Tetrahedron Lett.*, **34**, 2307–2310.

838 Thuring, J.W.J.F., Nefkens, G.H.L., Wegman, M.A., Klunder, A.J.H. and Zwanenburg, B. (1996) *J. Org. Chem.*, **61**, 6931–6935.

839 Ema, T., Maeno, S., Takaya, Y., Sakai, T. and Utaka, M. (1996) *J. Org. Chem.*, **61**, 8610–8616.

840 Takagi, Y., Teramoto, J., Kihara, H., Itoh, T. and Tsukube, H. (1996) *Tetrahedron Lett.*, **37**, 4991–4992.

841 Takagi, Y., Ino, R., Kihara, H., Itoh, T. and Tsukube, H. (1997) *Chem. Lett.*, 1247–1248.

842 Kita, Y., Takebe, Y., Murata, K., Naka, T. and Akai, S. (1996) *Tetrahedron Lett.*, **37**, 7369–7372.

843 Schudok, M. and Kretzschmar, G. (1997) *Tetrahedron Lett.*, **38**, 387–388.

844 Sakai, T., Kawabata, I., Kishimoto, T., Ema, T. and Utaka, M. (1997) *J. Org. Chem.*, **62**, 4906–4907.

845 Kawanami, Y., Iizuna, N. and Okamo, K. (1998) *Chem. Lett.*, 1231–1232.

846 Lee, D. and Kim, M.-J. (1998) *Tetrahedron Lett.*, **39**, 2163–2166.

847 Ema, T., Jittani, M., Sakai, T. and Utaka, M. (1998) *Tetrahedron Lett.*, **39**, 6311–6314.

848 Allevi, P., Ciuffreda, P., Tarocco, G. and Anastasia, M. (1996) *J. Org. Chem.*, **61**, 4144–4147.

849 Hirose, K., Naka, H., Yano, M., Ohashi, S., Naemura, K. and Tobe, Y. (2000) *Tetrahedron: Asymmetry*, **11**, 1199–1210.

850 Berkowitz, D.B., Pumphrey, J.A. and Shen, Q. (1994) *Tetrahedron Lett.*, **35**, 8743–8746.

851 Herradon, B. (1993) *Synlett*, 108–110.

852 Sanfilippo, C. and Nicolosi, G. (2007) *Tetrahedron: Asymmetry*, **18**, 1828–1832.

853 Johnson, C.R. and Sakaguchi, H. (1992) *Synlett*, 813–816.

854 Hamada, H., Shiromoto, M., Funahashi, M., Itoh, T. and Nakamura, K. (1996) *J. Org. Chem.*, **61**, 2332–2336.

855 Garcia-Urdiales, E., Rebolledo, F. and Gotor, V. (2001) *Adv. Synth. Catal.*, **343**, 646–654.

856 Lambusta, D., Nicolosi, G., Patti, A. and Piattelli, M. (1996) *Tetrahedron Lett.*, **37**, 127–130.

857 Barnier, J.P., Rayssac, V., Morisson, V. and Blanco, L. (1997) *Tetrahedron Lett.*, **38**, 8503–8506.

858 Krishna, S.H., Persson, M. and Bornscheuer, U.T. (2002) *Tetrahedron: Asymmetry*, **13**, 2693–2696.

859 Oshitari, T. and Mandai, T. (2003) *Synlett*, 2374–2376.

860 de Gonzalo, G., Brieva, R., Sánchez, V.M., Bayod, M. and Gotor, V. (2003) *J. Org. Chem.*, **68**, 3333–3336.

861 Cruz Silva, M.M. and Riva, S. and Sáe Melo, M.L. (2004) *Tetrahedron: Asymmetry*, **15**, 1173–1179

862 Sorgedrager, M.J., Malpique, R., van Rantwijk, F. and Sheldon, R.A. (2004) *Tetrahedron: Asymmetry*, **15**, 1295–1299.

863 Lee, Y.S., Hong, J.H., Jeon, N.Y., Won, K. and Kim, B.T. (2004) *Org. Process Res. Dev.*, **8**, 948–951.

864 Pinot, E., Guy, A., Guyon, A.-L., Rossi, J.-C. and Durand, T. (2005) *Tetrahedron: Asymmetry*, **16**, 1893–1895.

865 Vieira, T.O., Ferraz, H.M.C., Andrade, L.H. and Porto, A.L.M. (2006) *Tetrahedron: Asymmetry*, **17**, 1990–1994.

866 Franssen, M.C.R., Jongejan, H., Kooijman, H., Spek, A.L., Bell, R.P.L., Wijnberg, J.B.P.A. and De Groot, A. (1999) *Tetrahedron: Asymmetry*, **10**, 2729–2738.

867 Isleyen, A., Tanyeli, C. and Dogan, Ö. (2006) *Tetrahedron: Asymmetry*, **17**, 1561–1567.

868 Akita, H., Nozawa, M., Mitsuda, A. and Ohsawa, H. (2000) *Tetrahedron: Asymmetry*, **11**, 1375–1388.

869 Sharma, A., Pawar, A.S. and Chattopadhyay, S. (1996) *Synth. Commun.*, **26**, 19–25.

870 Lin, G. and Lin, W.-Y. (1998) *Tetrahedron Lett.*, **39**, 4333–4336.

871 Ors, M., Morcuende, A., Jimenez-Vacas, M.I., Valverde, S. and Herradon, B. (1996) *Synlett*, 449–451.

872 Bakker, M., Spruijt, A.S., Van Rantwijk, F. and Sheldon, R.A. (2000) *Tetrahedron: Asymmetry*, **11**, 1801–1808.

873 Fukui, T., Kawamoto, T. and Tanaka, A. (1994) *Tetrahedron: Asymmetry*, **5**, 73–82.

874 Ema, T., Okada, R., Fukumoto, M., Jittani, M. and Ishida, M. (1999) *Tetrahedron Lett.*, **40**, 4367–4370.

875 Miyazawa, T., Onishi, K., Murashima, T., Yamada, T. and Tsai, S.-W. (2005) *Tetrahedron: Asymmetry*, **16**, 2569–2573.

876 Engel, K.-H. (1991) *Tetrahedron: Asymmetry*, **2**, 165–168.

877 Berglund, P., Holmquist, M., Hedenström, E., Hult, K. and Högberg, H.-E. (1993) *Tetrahedron: Asymmetry*, **4**, 1869–1878.

878 Morrone, R., Nicolosi, G., Patti, A. and Piattelli, M. (1995) *Tetrahedron: Asymmetry*, **6**, 1773–1778.

879 Fadnavis, N.W. and Koteshwar, K. (1997) *Tetrahedron: Asymmetry*, **8**, 337–339.

880 Nguyen, B.-V. and Hedenström, E. (1999) *Tetrahedron: Asymmetry*, **10**, 1821–1826.

881 Hedenström, E., Nguyen, B.-V. and Silks, L.A. III (2002) *Tetrahedron: Asymmetry*, **13**, 835–844.

882 Yang, H., Henke, E. and Bornscheuer, U.T. (1999) *J. Org. Chem.*, **64**, 1709–1712.

883 Adam, W., Groer, P. and Saha-Möller, C.R. (1997) *Tetrahedron: Asymmetry*, **8**, 833–836.

884 Kim, Y.H., Cheong, C.S., Lee, S.H. and Kim, K.S. (2001) *Tetrahedron: Asymmetry*, **12**, 1865–1869.

885 Guibe-Jampel, E., Chalecki, Z., Bassir, M. and Gelo-Pujic, M. (1996) *Tetrahedron*, **52**, 4397–4402.

886 Hirose, Y., Kariya, I., Sasaki, I., Kurono, Y. and Achiwa, K. (1993) *Tetrahedron Lett.*, **34**, 5915–5918.

887 Cheng, Y.-C. and Tsai, S.-W. (2004) *Tetrahedron: Asymmetry*, **15**, 2917–2920.

888 Ruppert, S. and Gias, H.-J. (1997) *Tetrahedron: Asymmetry*, **8**, 3657–3664.

889 Furukawa, S., Ono, T., Iijima, H. and Kawakami, K. (2002) *J. Mol. Catal. B*, **17**, 23–28.

890 Guo, Z. and Sun, Y. (2004) *Biotechnol. Prog.*, **20**, 500–506.

891 Sakai, T., Hayashi, K., Yano, F., Takami, M., Ino, M., Korenaga, T. and Ema, T. (2003) *Bull. Chem. Soc. Jpn.*, **76**, 1441–1446.

892 Sakai, T., Matsuda, A., Korenaga, T. and Ema, T. (2003) *Bull. Chem. Soc. Jpn.*, **76**, 1819–1821.

893 Ema, T., Kageyama, M., Korenaga, T. and Sakai, T. (2003) *Tetrahedron: Asymmetry*, **14**, 3943–3947.

894 Irimescu, R., Saito, T. and Kato, K. (2004) *J. Mol. Catal. B*, **27**, 69–73.

895 Kawanami, Y. and Itoh, K. (2005) *Chem. Lett.*, **34**, 682–683.

896 Li, X.-G., Lähitie, M., Päiviö, M. and Kanerva, L.T. (2007) *Tetrahedron: Asymmetry*, **18**, 1567–1573.

897 Ujang, Z., Husain, W.H., Seng, M.C. and Rashid, A.H.A. (2003) *Process Biochem.*, **38**, 1483–1488.

898 Sano, T., Imai, K., Ohashi, K. and Oriyama, T. (1999) *Chem. Lett.*, 265–266.

899 Priem, G., Pelotier, B., Macdonald, S.J.F., Anson, M.S. and Campbell, I.B. (2003) *J. Org. Chem.*, **68**, 3844–3848.

900 Dálaigh, C.Ó., Hynes, S.J., Maher, D.J. and Connon, S.J. (2005) *Org. Biomol. Chem.*, **3**, 981–984.

901 Vedejs, E. and MacKay, J.A. (2001) *Org. Lett.*, **3**, 535–536.
902 Vedejs, E. and Daugulis, O. (2003) *J. Am. Chem. Soc.*, **125**, 4166–4173.
903 Ruble, J.C. and Fu, G.C. (1996) *J. Org. Chem.*, **61**, 7230–7231.
904 Ruble, J.C., Lathan, H.A. and Fu, G.C. (1997) *J. Am. Chem. Soc.*, **119**, 1492–1493.
905 Garrett, C.E. and Fu, G.C. (1998) *J. Am. Chem. Soc.*, **120**, 7479–7483.
906 Fu, G.C. (2000) *Acc. Chem. Res.*, **33**, 412–420.
907 Spivey, A.C., Arseniyadis, S., Fekner, T., Maddaford, A. and Leese, D.P. (2006) *Tetrahedron*, **62**, 295–301.
908 Poisson, T., Penhoat, M., Papamicaël, C., Dupas, G., Dalla, V., Marsais, F. and Levacher, V. (2005) *Synlett*, 2285–2288.
909 Jarvo, E.R., Copeland, G.T., Papaioannou, N., Bonitatebus, P.J. Jr. and Miller, S.J. (1999) *J. Am. Chem. Soc.*, **121**, 11638–11643.
910 Vasbinder, M.N., Jarvo, E.R. and Miller, S.J. (2001) *Angew. Chem. Int. Ed.*, **40**, 2824–2827.
911 Ishihara, K., Kosugi, Y. and Akakura, M. (2004) *J. Am. Chem. Soc.*, **126**, 12212–12213.
912 Birman, V.B., Uffman, E.W., Hui, J., Li, X.-M. and Kilbane, C.J. (2004) *J. Am. Chem. Soc.*, **126**, 12226–12227.
913 Varkonyi-Schlovicsko, E., Takacs, K. and Hermecz, I. (1997) *J. Heterocycl. Chem.*, **34**, 1065–1066.
914 Birman, V.B. and Guo, L. (2006) *Org. Lett.*, **8**, 4859–4861.
915 Suzuki, Y., Muramatsu, K., Yamauchi, K., Morie, Y. and Sato, M. (2006) *Tetrahedron*, **62**, 302–310.
916 Berlin, W.K., Zhang, W.-S. and Shen, T.Y. (1991) *Tetrahedron*, **47**, 1–20.
917 Vacca, J.P., De Solms, S.J., Huff, J.R., Billington, D.C., Baker, R., Kulagowski, J.J. and Mawer, I.M. (1989) *Tetrahedron*, **45**, 5679–5702.
918 Jiang, B. and Zhao, X.-L. (2004) *Tetrahedron: Asymmetry*, **15**, 1141–1143.
919 Dro, C., Bellemin-Laponnaz, S., Welter, R. and Gade, L.H. (2004) *Angew. Chem. Int. Ed.*, **43**, 4479–4482.
920 Bevinakatti, H., Newadkar, R.V. and Banerji, A.A. (1990) *J. Chem. Soc., Chem. Commun.*, 1091–1092.
921 Crich, J.Z., Brieva, R., Marquart, P., Gu, R.-L., Flemming, S. and Sih, C.J. (1993) *J. Org. Chem.*, **58**, 3252–3258.
922 Parker, M.-C., Brown, S.A., Robertson, L. and Turner, N.J. (1998) *Chem. Commun.*, 2247–2248.
923 Thuring, J.W.J.F., Klunder, A.J.H., Nefkens, G.H.L., Wegman, M.A. and Zwanenburg, B. (1996) *Tetrahedron Lett.*, **37**, 4759–4760.
924 Van Der Deen, H., Cuiper, A.D., Hof, R.P., Van Oeveren, A., Feringa, B.L. and Kellogg, R.M. (1996) *J. Am. Chem. Soc.*, **118**, 3801–3803.
925 Um, P.-J. and Drueckhammer, D.G. (1998) *J. Am. Chem. Soc.*, **120**, 5605–5610.
926 Strauss, U.T. and Faber, K. (1999) *Tetrahedron: Asymmetry*, **10**, 4079–4081.
927 Stürmer, R. (1997) *Angew. Chem. Int. Ed. Engl.*, **36**, 1173–1174.
928 Dihn, P.M., Howard, J.A., Hudnott, A.R. and Williams, J.M.J. (1996) *Tetrahedron Lett.*, **37**, 7623–7626.
929 Persson, B.A., Larsson, A.L.E., Ray, M.L. and Bäckvall, J.-E. (1999) *J. Am. Chem. Soc.*, **121**, 1645–1650.
930 Persson, B.A., Huerta, F.F. and Bäckvall, J.-E. (1999) *J. Org. Chem.*, **64**, 5237–5240.
931 Huerta, F., Laxmi, Y.R.S. and Bäckvall, J.-E. (2000) *Org. Lett.*, **2**, 1037–1040.
932 Kim, M.-J., Choi, Y.K., Choi, M.Y., Kim, M.J. and Park, J. (2001) *J. Org. Chem.*, **66**, 4736–4738.
933 Pamies, O. and Bäckvall, J.-E. (2001) *J. Org. Chem.*, **66**, 4022–4025.
934 Runmo, A.L., Pamies, O., Faber, K. and Bäckvall, J.-E. (2002) *Tetrahedron Lett.*, **43**, 2983–2986.
935 Pàmies, O. and Bäckvall, J.-E. (2002) *J. Org. Chem.*, **67**, 1261–1265.
936 Martín-Matute, B. and Bäckvall, J.-E. (2004) *J. Org. Chem.*, **69**, 9191–9195.
937 Verzijl, G.K.M., de Vries, J.G. and Broxterman, Q.B. (2005) *Tetrahedron: Asymmetry*, **16**, 1603–1610.
938 Hoyos, P., Fernández, M., Sinisterra, J.V. and Alcántara, A.R. (2006) *J. Org. Chem.*, **71**, 7632–7637.

939 Huerta, F.F. and Bäckvall, J.-E. (2001) *Org. Lett.*, **3**, 1209–1212.

940 Martín-Matute, B., Edin, M., Bogár, K. and Bäckvall, J.-E. (2004) *Angew. Chem. Int. Ed.*, **43**, 6535–6539.

941 Martín-Matute, B., Edin, M., Bogár, K., Kaynak, F.B. and Bäckvall, J.-E. (2005) *J. Am. Chem. Soc.*, **127**, 8817–8825.

942 Kim, N., Ko, S.-B., Kwon, M.S., Kim, M.-J. and Park, J. (2005) *Org. Lett.*, **7**, 4523–4526.

943 Choi, J.H., Kim, Y.H., Nam, S.H., Shin, S.T., Kim, M.-J. and Park, J. (2002) *Angew. Chem. Int. Ed.*, **41**, 2373–2376.

944 Choi, J.H., Choi, Y.K., Kim, Y.H., Park, E.S., Kim, E.J., Kim, M.-J. and Park, J. (2004) *J. Org. Chem.*, **69**, 1972–1977.

945 Koh, J.H., Jung, H.M., Kim, M.-J. and Park, J. (1999) *Tetrahedron Lett.*, **40**, 6281–6284.

946 Riermeier, T.H., Gross, P., Monsees, A., Hoff, M. and Trauthwein, H. (2005) *Tetrahedron Lett.*, **46**, 3403–3406.

947 Akai, S., Tanimoto, K., Kanao, Y., Egi, M., Yamamoto, T. and Kita, Y. (2006) *Angew. Chem. Int. Ed.*, **45**, 2592–2595.

948 Veum, L. and Hanefeld, U. (2005) *Synlett*, 2382–2384.

949 Wuyts, S., De Temmerman, K., De Vos, D. and Jacobs, P. (2003) *Chem. Commun.*, 1928–1929.

950 Wuyts, S., De Temmerman, K., De Vos, D.E. and Jacobs, P.A. (2005) *Chem. Eur. J.*, **11**, 386–397.

951 Seebach, D., Jaeschke, G., Gottwald, K., Matsuda, K., Formisano, R., Chaplin, D.A., Breuning, M. and Bringmann, G. (1997) *Tetrahedron*, **53**, 7539–7556.

952 Liang, J., Ruble, J.C. and Fu, G.C. (1998) *J. Org. Chem.*, **63**, 3154–3155.

953 Ammazzalorso, A., Amoroso, R., Bettoni, G., De Filippis, B., Giampietro, L., Pierini, M. and Tricca, M.L. (2002) *Tetrahedron Lett.*, **43**, 4325–4328.

954 Vedejs, E. and Chen, X. (1997) *J. Am. Chem. Soc.*, **119**, 2584–2585.

955 Davies, S.G., Garner, A.C., Long, M.J.C., Smith, A.D., Sweet, M.J. and Withey, J.M. (2004) *Org. Biomol. Chem.*, **2**, 3355–3362.

956 García, J., Fernández, S., Ferrero, M., Sanghvi, Y.S. and Gotor, V. (2004) *Org. Lett.*, **6**, 3759–3762.

957 Guo, J., Wu, J., Siuzdak, G. and Finn, M.G. (1999) *Angew. Chem. Int. Ed.*, **38**, 1755–1758.

958 Tombo, G.M.R., Schar, H.-P., Busquets, F.I. and Ghisalba, O. (1986) *Tetrahedron Lett.*, **27**, 5707–5710.

959 Wang, Y.-F., Lalonde, J.J., Momongan, M., Bergbreiter, D.E. and Wang, C.-H. (1988) *J. Am. Chem. Soc.*, **110**, 7200–7205.

960 Chenevert, R. and Courchesne, G. (1997) *Chem. Lett.*, 11–12.

961 Fujita, T., Tanaka, M., Norimine, Y. and Suemune, H. (1997) *J. Org. Chem.*, **62**, 3824–3830.

962 Akai, S., Naka, T., Fujita, T., Tanabe, Y., Tsujino, T. and Kita, Y. (2002) *J. Org. Chem.*, **67**, 411–419.

963 Ishihara, K., Kubota, M. and Yamamoto, H. (1994) *Synlett*, 611–614.

964 Yamada, S. and Katsumata, H. (1998) *Chem. Lett.*, 995–996.

965 Oriyama, T., Taguchi, H., Terakado, D. and Sano, T. (2002) *Chem. Lett.*, 26–27.

966 Seebach, D., Jaeschke, G. and Wang, Y.M. (1995) *Angew. Chem. Int. Ed. Engl.*, **34**, 2395–2396.

967 Jaeschke, G. and Seebach, D. (1998) *J. Org. Chem.*, **63**, 1190–1197.

968 Imado, H., Ishizuka, T. and Kunieda, T. (1995) *Tetrahedron Lett.*, **36**, 931–934.

969 Bolm, C., Gerlach, A. and Dinter, C.L. (1999) *Synlett*, 195–196.

970 Ottolina, G., Carrea, G. and Riva, S. (1990) *J. Org. Chem.*, **55**, 2366–2369.

971 Kovacs, T. and Sonnenbichler, J. (1997) *Liebigs Ann. Chem.*, 211–212.

972 Takayama, S., Livingston, P.O. and Wong, C.-H. (1996) *Tetrahedron Lett.*, **37**, 9271–9274.

973 Sugai, T., Okazaki, H., Kuboki, A. and Ohta, H. (1997) *Bull. Chem. Soc. Jpn.*, **70**, 2535–2540.

974 Gotor, V. and Pulido, R. (1991) *J. Chem. Soc., Perkin Trans. 1*, 491–492.

975 Pulido, R. and Gotor, V. (1992) *J. Chem. Soc., Perkin Trans. 1*, 2891–2898.

976 Moris, F. and Gotor, V. (1993) *J. Org. Chem.*, **58**, 653–660.

977 Gotor, V. and Moris, F. (1992) *Synthesis*, 626–628.

978 Jeong, L.S. and Marquez, V.E. (1995) *J. Org. Chem.*, **60**, 4276–4279.

979 Shing, T.K.M., Zhu, X.Y. and Mak, T.C.W. (1996) *Tetrahedron: Asymmetry*, 7, 673–676.
980 Schow, S.R., Rossignol, D.P., Lund, A.E. and Schnee, M.E. (1997) *Bioorg. Med. Chem. Lett.*, **7**, 181–186.
981 Lu, X., Chen, G., Xia, L. and Guo, G. (1997) *Tetrahedron: Asymmetry*, **8**, 3067–3072.
982 Kim, H.-S., Kim, I.-C. and Lee, S.-O. (1997) *Tetrahedron*, **53**, 8129–8136.
983 Barco, A., Benetti, S., De Risi, G., Pollini, G.P., Romagnoli, R. and Zanirato, V. (1994) *Tetrahedron*, **50**, 10491–10496.
984 Chou, W.-C. and Fang, J.-M. (1996) *J. Org. Chem.*, **61**, 1473–1477.
985 Kurth, M.J., Milco, L.A. and Miller, R.B. (1992) *Tetrahedron*, **48**, 1407–1416.
986 Rizzo, C.J. and Smith, A.B. III (1991) *J. Chem. Soc., Perkin Trans. 1*, 969–979.
987 Nambu, M. and White, J.D. (1996) *Chem. Commun.*, 1619–1620.
988 Allevi, P., Ciuffreda, P., Longo, A. and Anastasia, M. (1998) *Tetrahedron: Asymmetry*, **9**, 2915–2924.
989 Tsukayama, M., Kikuchi, M. and Yoshioka, S. (1993) *Chem. Lett.*, 1895–1898.
990 Tsukayama, M., Li, H., Tsurumoto, K., Nishiuchi, M. and Kawamura, Y. (1998) *Bull. Chem. Soc. Jpn.*, **71**, 2673–2680.
991 Hutchinson, S.A., Luetjens, H. and Scammells, P.J. (1997) *Bioorg. Med. Chem. Lett.*, **7**, 3081–3084.
992 Ogawa, H., Amano, M. and Chihara, T. (1998) *Chem. Commun.*, 495–496.
993 Clarke, P.A. (2002) *Tetrahedron Lett.*, **43**, 4761–4763.
994 Clarke, P.A., Kayaleh, N.E., Smith, M.A., Baker, J.R., Bird, S.J. and Chan, C. (2002) *J. Org. Chem.*, **67**, 5226–5231.
995 Otera, J., Dan-oh, N. and Nozaki, H. (1991) *J. Chem. Soc., Chem. Commun.*, 1742–1743.
996 Pedrocchi-Fantoni, G. and Servi, S. (1992) *J. Chem. Soc., Perkin Trans. 1*, 1029–1033.
997 Parma, V.S., Sinha, R., Bisht, K.S., Gupta, S., Prasad, A.K. and Tanja, P. (1993) *Tetrahedron*, **49**, 4107–4116.
998 Baldessari, A., Mangone, C.P. and Gros, E.G. (1998) *Helv. Chim. Acta*, **81**, 2407–2413.
999 Nicolosi, G., Piattelli, M. and Sanfilippo, C. (1993) *Tetrahedron*, **49**, 3143–3148.
1000 Lambusta, D., Nicolosi, G., Patti, A. and Piattelli, M. (1993) *Synthesis*, 1155–1158.
1001 Danieli, B. and De Bellis, P. (1992) *Helv. Chim. Acta*, **75**, 1297–1304.
1002 Armesto, N., Ferrero, M., Fernandez, S. and Gotor, V. (2002) *J. Org. Chem.*, **67**, 4978–4981.
1003 Bertinotti, A., Carrea, G., Ottolina, G. and Riva, S. (1994) *Tetrahedron*, **50**, 13165–13172.
1004 Danieli, B., Lasma, G. and Luisetti, M. (1997) *Tetrahedron*, **53**, 5855–5862.
1005 Nagao, Y., Fujita, E., Kohno, T. and Yagi, M. (1981) *Chem. Pharm. Bull.*, **29**, 3202–3207.
1006 Beale, M.H., McMillan, J., Makinson, I.K. and Willis, C.L. (1991) *J. Chem. Soc., Perkin Trans. 1*, 1191–1195.
1007 Garcia-Granados, A., Martinez, A. and Quiros, R. (1999) *Tetrahedron*, **55**, 8567–8578.
1008 Jaracz, S., Stromgaard, K. and Nakanishi, K. (2002) *J. Org. Chem.*, **67**, 4623–4626.
1009 Otera, J. (2001) *Angew. Chem. Int. Ed.*, **40**, 2044–2045.
1010 Li, Y.-Q. (1999) *Synth. Commun.*, **29**, 3901–3903.
1011 Zhu, H.-P., Yang, F., Tang, J. and He, M.-Y. (2003) *Green Chem.*, **5**, 38–39.
1012 Jiang, T., Chang, Y., Zhao, G., Han, B. and Yang, G. (2004) *Synth. Commun.*, **34**, 225–230.
1013 Nguyen, H.-P., Znifeche, S. and Baboulène, M. (2004) *Synth. Commun.*, **34**, 2085–2093.
1014 Maitra, U. and Bag, B.G. (1992) *J. Org. Chem.*, **57**, 6979–6981.
1015 Garegg, P.J., Kvarnstroem, I., Niklasson, A., Niklasson, G. and Svensson, S.C.T. (1993) *J. Carbohydr. Chem.*, **12**, 933–953.
1016 Kumar, S., Singh, R. and Singh, H. (1993) *Bioorg. Med. Chem. Lett.*, **3**, 363–364.
1017 Reddy, K.V., Sabitha, G. and Rao, A.V.S. (1996) *Org. Prep. Proced. Int.*, **28**, 325–332.
1018 Rao, P.S., Reddy, K.V.V. and Ashok, D. (1997) *Synth. Commun.*, **27**, 3181–3189.

1019 Yang, H.-M. and Lin, C.-L. (2003) *J. Mol. Catal. A*, **206**, 67–76.

1020 Park, S.-W., Kim, T.-Y., Park, D.-W. and Lee, J.-W. (2003) *Catal. Today*, **87**, 179–186.

1021 Limousin, C., Cleophax, J., Loupy, A. and Petit, A. (1998) *Tetrahedron*, **54**, 13567–13578.

1022 Manabe, K., Sun, X.-M. and Kobayashi, S. (2001) *J. Am. Chem. Soc.*, **123**, 10101–10102.

1023 Manabe, K., Iimura, S., Sun, X.-M. and Kobayashi, S. (2002) *J. Am. Chem. Soc.*, **124**, 11971–11978.

1024 Han, Y. and Chu, Y. (2005) *J. Mol. Catal. A*, **237**, 232–237.

1025 Hao, X., Yoshida, A. and Nishikido, J. (2004) *Tetrahedron Lett.*, **45**, 781–785.

1026 Hao, X., Yoshida, A. and Nishikido, J. (2006) *J. Fluorine Chem.*, **127**, 193–199.

1027 Hao, X., Yoshida, A. and Nishikido, J. (2004) *Green Chem.*, **6**, 566–569.

1028 Mercs, L., Pozzi, G. and Quici, S. (2007) *Tetrahedron Lett.*, **48**, 3053–3056.

1029 Ishihara, K., Hasegawa, A. and Yamamoto, H. (2002) *Synlett*, 1299–1301.

1030 Matsugi, M., Hasegawa, M., Sadachika, D., Okamoto, S., Tomioka, M., Ikeya, Y., Masuyama, A. and Mori, Y. (2007) *Tetrahedron Lett.*, **48**, 4147–4150.

1031 Beier, P. and O'Hagan, D. (2002) *Chem. Commun.*, 1680–1681.

1032 Hungerhoff, B., Sonnenschein, H. and Theil, F. (2002) *J. Org. Chem.*, **67**, 1781–1785.

1033 Maruyama, T., Kotani, T., Yamamura, H., Kamiya, N. and Goto, M. (2004) *Org. Biomol. Chem.*, **2**, 524–527.

1034 Forsyth, S.A., MacFarlane, D.R., Thomson, R.J. and Von Itzstein, M. (2002) *Chem. Commun.*, 714–715.

1035 Imrie, C., Elago, E.R.T., McCleland, C.W. and Williams, N. (2002) *Green Chem.*, **4**, 159–160.

1036 Duan, Z., Gu, Y. and Deng, Y. (2006) *J. Mol. Catal. A*, **246**, 70–75.

1037 Ren, J.L., Sun, R.C., Liu, C.F., Cao, Z.N. and Luo, W. (2007) *Carbohydr. Polym.*, **70**, 406–414.

1038 Brinchi, L., Germani, R. and Savelli, G. (2003) *Tetrahedron Lett.*, **44**, 2027–2029.

1039 Brinchi, L., Germani, R. and Savelli, G. (2003) *Tetrahedron Lett.*, **44**, 6583–6585.

1040 Yoshino, T. and Togo, H. (2004) *Synlett*, 1604–1606.

1041 Yoshino, T., Imori, S. and Togo, H. (2006) *Tetrahedron*, **62**, 1309–1317.

1042 McNulty, J., Cheekoori, S., Nair, J.J., Larichev, V., Capretta, A. and Robertson, A.J. (2005) *Tetrahedron Lett.*, **46**, 3641–3644.

1043 Abbott, A.P., Bell, T.J., Handa, S. and Stoddart, B. (2005) *Green Chem.*, **7**, 705–707.

1044 Sunitha, S., Kanjilal, S., Reddy, P.S. and Prasad, R.B.N. (2007) *Tetrahedron Lett.*, **48**, 6962–6965.

1045 Gui, J., Cong, X., Liu, D., Zhang, X., Hu, Z. and Sun, Z. (2004) *Catal. Commun.*, **5**, 473–477.

1046 Xing, H., Wang, T., Zhou, Z. and Dai, Y. (2005) *Ind. Eng. Chem. Res.*, **44**, 4147–4150.

1047 Nara, S.J., Harjani, J.R. and Salunkhe, M.M. (2002) *Tetrahedron Lett.*, **43**, 2979–2982.

1048 Maruyama, T., Nagasawa, S. and Goto, M. (2002) *Biotech. Lett.*, **24**, 1341–1345.

1049 Park, S., Viklund, F., Hult, K. and Kazlauskas, R.J. (2003) *Green Chem.*, **5**, 715–719.

1050 Ganske, F. and Bornscheuer, U.T. (2005) *Org. Lett.*, **7**, 3097–3098.

1051 Li, X.-F., Lou, W.-Y., Smith, T.J., Zong, M.-H., Wu, H. and Wang, J.-F. (2006) *Green Chem.*, **8**, 538–544.

1052 Lozano, P., de Diego, T., Guegan, J.-P., Vaultier, M. and Iborra, J.L. (2001) *Biotech. Bioeng.*, **75**, 563–569.

1053 Noritomi, H., Nishida, S. and Kato, S. (2007) *Biotech. Lett.*, **29**, 1509–1512.

1054 Ulbert, O., Fráter, T., Bélafi-Bakó, K. and Gubicza, L. (2004) *J. Mol. Catal. B*, **31**, 39–45.

1055 Rasalkar, M.S., Potdar, M.K. and Salunkhe, M.M. (2004) *J. Mol. Catal. B*, **27**, 267–270.

1056 Yu, H., Wu, J. and Ching, C.B. (2005) *Chirality*, **17**, 16–21.

1057 Contesini, F.J. and Carvalho, P.O. (2006) *Tetrahedron: Asymmetry*, **17**, 2069–2073.

1058 Nakamura, K., Chi, Y.M., Yamada, Y. and Yano, T. (1986) *Chem. Eng. Commun.*, **45**, 207–212.

1059 Matsuda, T., Harada, T. and Nakamura, K. (2004) *Green Chem.*, **6**, 440–444.

1060 Matsuda, T., Watanabe, K., Harada, T. and Nakamura, K. (2004) *Catal. Today*, **96**, 103–111.

1061 Matsuda, T., Kanamaru, R., Watanabe, K., Harada, T. and Nakamura, K. (2001) *Tetrahedron Lett.*, **42**, 8319–8321.

1062 Matsuda, T., Kanamaru, R., Watanabe, K., Kamitanaka, T., Harada, T. and Nakamura, K. (2003) *Tetrahedron: Asymmetry*, **14**, 2087–2091.

1063 Peres, C., Da Silva, D.R.G. and Barreiros, S. (2003) *J. Agric. Food Chem.*, **51**, 1884–1888.

1064 Jackson, M.A., Mbaraka, I.K. and Shanks, B.H. (2006) *Appl. Catal. A*, **310**, 48–53.

1065 Lozano, P., de Diego, T., Carrié, D., Vaultier, M. and Iborra, J.L. (2002) *Chem. Commun.*, 692–693.

1066 Reetz, M.T., Wiesenhöfer, W., Franciò, G. and Leitner, W. (2003) *Adv. Synth. Catal.*, **345**, 1221–1228.

1067 Reetz, M.T. and Wiesenhöfer, W. (2004) *Chem. Commun.*, 2750–2751.

1068 Matsuda, T., Watanabe, K., Harada, T., Nakamura, K., Arita, Y., Misumi, Y., Ichikawa, S. and Ikariya, T. (2004) *Chem. Commun.*, 2286–2287.

1069 Romero, M.D., Calvo, L., Alba, C., Habulin, M., Primoži , M. and Knez, Ž. (2005) *J. Supercrit. Fluids*, **33**, 77–84.

1070 Oliveira, D. and Oliveira, J.V. (2001) *J. Supercrit. Fluids*, **19**, 141–148.

1071 Srivastava, S., Madras, G. and Modak, J. (2003) *J. Supercrit. Fluids*, **27**, 55–64.

1072 Nagesha, G.K., Manohar, B. and Sankar, K.U. (2004) *J. Supercrit. Fluids*, **32**, 137–145.

1073 Šabeder, S., Habulin, M. and Knez, Ž. (2005) *Ind. Eng. Chem. Res.*, **44**, 9631–9635.

1074 Knez, Ž., Laudani, C.G., Habulin, M. and Reverchon, E. (2007) *Biotech. Bioeng.*, **97**, 1366–1375.

1075 Olsen, T., Kcrton, F., Marriott, R. and Grogan, G. (2006) *Enzyme Microb. Technol.*, **39**, 621–625.

1076 Kamat, S.V., Beckman, E.J. and Russell, A.J. (1993) *J. Am. Chem. Soc.*, **115**, 8845–8846.

1077 Chaudhary, A.K., Kamat, S.V., Beckman, E.J., Nurok, D., Kleyle, R.M., Hajdu, P. and Russell, A.J. (1996) *J. Am. Chem. Soc.*, **118**, 12891–12901.

1078 Mori, T., Funasaki, M., Kobayashi, A. and Okahata, Y. (2001) *Chem. Commun.*, 1832–1833.

1079 Mori, T., Li, M., Kobayashi, A. and Okahata, Y. (2002) *J. Am. Chem. Soc.*, **124**, 1188–1189.

1080 Novak, Z., Habulin, M., Krmelj, V. and Knez, Ž. (2003) *J. Supercrit. Fluids*, **27**, 169–178.

1081 Habulin, M. and Knez, Ž. (2001) *J. Chem. Technol. Biotechnol.*, **76**, 1260–1266.

1082 Madras, G., Kumar, R. and Modak, J. (2004) *Ind. Eng. Chem. Res.*, **43**, 7697–7701.

1083 Nishikido, J., Kamishima, M., Matsuzawa, H. and Mikami, K. (2002) *Tetrahedron*, **58**, 8345–8349.

1084 Ghaziaskar, H.S., Daneshfar, A. and Calvo, L. (2006) *Green Chem.*, **8**, 576–581.

1085 Yoda, S., Bratton, D. and Howdle, S.M. (2004) *Polymer*, **45**, 7839–7843.

1086 Rantanen, T., Schiffers, I. and Bolm, C. (2007) *Org. Process Res. Dev.*, **11**, 592–597.

1087 Zhang, Z., Zhou, L., Zhang, M., Wu, H. and Chen, Z. (2001) *Synth. Commun.*, **31**, 2435–2439.

1088 Pérez, E.R., Carnevalli, N.C., Cordeiro, P.J., Rodrigues-Filho, U.P. and Franco, D.W. (2001) *Org. Prep. Proced. Int.*, **33**, 395–400.

1089 Constantinou-Kokotou, V. and Peristeraki, A. (2004) *Synth. Commun.*, **34**, 4227–4232.

1090 Barbosa, S.L., Dabdoub, M.J., Hurtado, G.R., Klein, S.I., Baroni, A.C.M. and Cunha, C. (2006) *Appl. Catal. A*, **313**, 146–150.

1091 Mallavadhani, U.V., Sahoo, L. and Roy, S. (2004) *Indian J. Chem., Sect. B*, **43**, 2175–2177.

1092 Mohan, K.V.V.K., Narender, N. and Kulkarni, S.J. (2006) *Green Chem.*, **8**, 368–372.

1093 Toukoniitty, B., Mikkola, J.-P., Eränen, K., Salmi, T. and Murzin, D.Y. (2005) *Catal. Today*, **100**, 431–435.

1094 Rajabi, F. and Saidi, M.R. (2004) *Synth. Commun.*, **34**, 4179–4188.

1095 Stadler, A. and Kappe, C.O. (2001) *Tetrahedron*, **57**, 3915–3920.

1096 Satgé, C., Verneuil, B., Branland, P., Granet, R., Krausz, P., Rozier, J. and Petit, C. (2002) *Carbohydr. Polym.*, **49**, 373–376.

1097 Hirose, T., Kopek, B.G., Wang, Z.-H., Yusa, R. and Baldwin, B.W. (2003) *Tetrahedron Lett.*, **44**, 1831–1833.

1098 Biswas, A., Shogren, R.L., Kim, S. and Willett, J.L. (2006) *Carbohydr. Polym.*, **64**, 484–487.

1099 Pathania, V., Sharma, A. and Shinha, A.K. (2005) *Helv. Chim. Acta*, **88**, 811–816.

1100 Donati, D., Morelli, C. and Taddei, M. (2005) *Tetrahedron Lett.*, **46**, 2817–2819.

1101 Polshettiwar, V. and Kaushik, M.P. (2005) *Catal. Commun.*, **6**, 191–194.

1102 Arfan, A. and Bazureau, J.P. (2005) *Org. Process Res. Dev.*, **9**, 743–748.

1103 Lee, J.C., Song, I.-G. and Park, J.Y. (2002) *Synth. Commun.*, **32**, 2209–2213.

1104 Huang, W., Xia, Y.-M., Gao, H., Fang, Y.-J., Wang, Y. and Fang, Y. (2005) *J. Mol. Catal. B*, **35**, 113–116.

1105 Yadav, G.D. and Lathi, P.S. (2005) *Synth. Commun.*, **35**, 1699–1705.

1106 Pipuš, G., Plazl, I. and Koloini, T. (2002) *Ind. Eng. Chem. Res.*, **41**, 1129–1134.

1107 Amore, K.M. and Leadbeater, N.E. (2007) *Macromol. Rapid Commun.*, **28**, 473–477.

1108 Rama, K. and Pasha, M.A. (2002) *Indian J. Chem., Sect. B*, **41**, 2604–2605.

1109 Gholap, A.R., Venkatesan, K., Daniel, T., Lahoti, R.J. and Srinivasan, K.V. (2003) *Green Chem.*, **5**, 693–696.

1110 Brenelli, E.C.S. and Fernandes, J.L.N. (2003) *Tetrahedron: Asymmetry*, **14**, 1255–1259.

1111 Xiao, Y.-M., Wu, Q., Cai, Y. and Lin, X.-F. (2005) *Carbohydr. Res.*, **340**, 2097–2103.

1112 Okur, H. and Bayramoglu, M. (2001) *Ind. Eng. Chem. Res.*, **40**, 3639–3646.

1113 Hanika, J., Smejkal, Q. and Kolena, J. (2001) *Catal. Today*, **66**, 219–223.

1114 Kenig, E.Y., Bäder, H., Górak, A., Beßling, B., Adrian, T. and Schoenmakers, H. (2001) *Chem. Eng. Sci.*, **56**, 6185–6193.

1115 Nijhuis, T.A., Beers, A.E.W., Kapteijn, F. and Moulijn, J.A. (2002) *Chem. Eng. Sci.*, **57**, 1627–1632.

1116 Omota, F., Dimian, A.C. and Bliek, A. (2003) *Chem. Eng. Sci.*, **58**, 3159–3174.

1117 Steinigeweg, S. and Gmehling, J. (2003) *Ind. Eng. Chem. Res.*, **42**, 3612–3619.

1118 Omota, F., Dimian, A.C. and Bliek, A. (2003) *Chem. Eng. Sci.*, **58**, 3175–3185.

1119 Asthana, N., Kolah, A., Vu, D.T., Lira, C.T. and Miller, D.J. (2005) *Org. Process Res. Dev.*, **9**, 599–607.

1120 Singh, A., Hiwale, R., Mahajani, S.M. and Gudi, R.D. (2005) *Ind. Eng. Chem. Res.*, **44**, 3042–3052.

1121 Hung, W.-J., Lai, I.-K., Chen, Y.-W., Hung, S.-B., Huang, H.-P., Lee, M.-J. and Yu, C.-C. (2006) *Ind. Eng. Chem. Res.*, **45**, 1722–1733.

1122 Jennings, J.F. and Binning, R.C. (1960) U.S. Patent 2956070.

1123 Pearse, G.K. (1987) European Patent 0210055.

1124 Kita, H., Tanaka, K., Okamoto, K. and Yamamoto, M. (1987) *Chem. Lett.*, 2053–2056.

1125 Bernal, M.P., Coronas, J., Menéndez, M. and Santamaría, J. (2002) *Chem. Eng. Sci.*, **57**, 1557–1562.

1126 Benedict, D.J., Parulekar, S.J. and Tsai, S.-P. (2003) *Ind. Eng. Chem. Res.*, **42**, 2282–2291.

1127 Kita, H., Sasaki, S., Tanaka, K., Okamoto, K. and Yamamoto, M. (1988) *Chem. Lett.*, 2025–2028.

1128 Okamoto, K., Yamamoto, M., Otoshi, Y., Semoto, T., Yano, M., Tanaka, K. and Kita, H. (1993) *J. Chem. Eng. Jpn.*, **26**, 475–481.

1129 Tanaka, K., Yoshikawa, R., Ying, C., Kita, H. and Okamoto, K. (2001) *Catal. Today*, **67**, 121–125.

1130 Bartling, K., Thompson, J.U.S., Pfromm, P.H., Czermak, P. and Rezac, M.E. (2001) *Biotech. Bioeng.*, **75**, 676–681.

1131 Hazarika, S., Dutta, A. and Dutta, N.N. (2003) *Biocatal. Biotransform.*, **21**, 101–113.

1132 Gubicza, L., Nemestóthy, N., Fráter, T. and Bélafi-Bakó, K. (2003) *Green Chem.*, **5**, 236–239.

1133 Peters, T.A., Benes, N.E. and Keurentjes, J.T.F. (2005) *Ind. Eng. Chem. Res.*, **44**, 9490–9496.

1134 Inoue, T., Nagase, T., Hasegawa, Y., Kiyozumi, Y., Sato, K., Kobayashi, K.,

Nishioka, M., Hamakawa, S. and Mizukami, F. (2006) *Chem. Lett.*, **35**, 76–77.

1135 Inoue, T., Nagase, T., Hasegawa, Y., Kiyozumi, Y., Sato, K., Nishioka, M., Hamakawa, S. and Mizukami, F. (2007) *Ind. Eng. Chem. Res.*, **46**, 3743–3750.

1136 Shah, T.N. and Ritchie, S.M.C. (2005) *Appl. Catal. A*, **296**, 12–20.

1137 Nemec, D. and van Gemert, R. (2005) *Ind. Eng. Chem. Res.*, **44**, 9718–9726.

1138 Peters, T.A., Benes, N.E. and Keurentjes, J.T.F. (2007) *Appl. Catal. A*, **317**, 113–119.

1139 Gelosa, D., Ramaioli, M. and Valente, G. (2003) *Ind. Eng. Chem. Res.*, **42**, 6536–6544.

1140 Watts, P., Wiles, C., Haswell, S.J., Pombo-Villar, E. and Styring, P. (2001) *Chem. Commun.*, 990–991.

1141 Watts, P., Wiles, C., Haswell, S.J. and Pombo-Villar, E. (2002) *Tetrahedron*, **58**, 5427–5439.

1142 Brivio, M., Oosterbroek, R.E., Verboom, W., Goedbloed, M.H., van den Berg, A. and Reinhoudt, D.N. (2003) *Chem. Commun.*, 1924–1925.

1143 Smith, A.B. III, Konopelski, J.P., Wexler, B.A. and Sprengeler, P.A. (1991) *J. Am. Chem. Soc.*, **113**, 3533–3542.

1144 Suzuki, Y., Mori, W., Ishizone, H., Naito, K. and Honda, T. (1992) *Tetrahedron Lett.*, **33**, 4931–4932.

1145 Gopalan, A.S. and Jacobs, H.K. (1990) *Tetrahedron Lett.*, **31**, 5575–5578.

1146 Pellegrini, C., Straessler, C., Weber, M. and Borschberg, H.-J. (1994) *Tetrahedron: Asymmetry*, **5**, 1979–1992.

1147 Booker-Milburn, K.I., Jimenez, F.D. and Sharpe, A. (1995) *Synlett*, 735–736.

1148 Shimano, M., Nagaoka, H. and Yamada, Y. (1996) *Tetrahedron Lett.*, **37**, 7553 7556.

1149 Hanessian, S., Pan, J., Carnell, A., Bouchard, H. and Lesage, L. (1997) *J. Org. Chem.*, **62**, 465–473.

1150 Hecker, S.J. and Heathcock, C.H. (1986) *J. Am. Chem. Soc.*, **108**, 4586–4594.

1151 Blechert, S. and Kleine-Klausing, A. (1991) *Angew. Chem. Int. Ed. Engl.*, **30**, 412–414.

1152 Bates, R.W., Fernandez-Moro, R. and Ley, S.V. (1991) *Tetrahedron*, **47**, 9929–9938.

1153 Nakamura, Y. and Shin, C.-G. (1992) *Chem. Lett.*, 49–52.

1154 White, J.D., Dillon, M.P. and Butlin, R.J. (1992) *J. Am. Chem. Soc.*, **114**, 9673–9674.

1155 Niwa, H., Ogawa, T., Okamoto, O. and Yamada, K. (1992) *Tetrahedron*, **48**, 10531–10548.

1156 Commerçon, A., Bezard, D., Bernard, F. and Bourzat, J.D. (1992) *Tetrahedron Lett.*, **33**, 5185–5188.

1157 Ogawa, T., Niwa, H. and Yamada, K. (1993) *Tetrahedron*, **49**, 1571–1578.

1158 Bourzat, J.C. and Commerçon, A. (1993) *Tetrahedron Lett.*, **34**, 6049–6052.

1159 Niwa, H., Sakata, T. and Yamada, K. (1994) *Bull. Chem. Soc. Jpn.*, **67**, 2345–2347.

1160 Niwa, H., Kunitani, K., Nagoya, T. and Yamada, K. (1994) *Bull. Chem. Soc. Jpn.*, **67**, 3094–3099.

1161 Kanazawa, A.M., Denis, J.-N. and Greene, A.E. (1994) *J. Chem. Soc., Chem. Commun.*, 2591–2592.

1162 Chu-Moyer, M.Y., Danishefsky, S.J. and Schulte, G.K. (1994) *J. Am. Chem. Soc.*, **116**, 11213–11228.

1163 Kanazawa, A.M., Denis, J.-N. and Greene, A.E. (1994) *J. Org. Chem.*, **59**, 1238–1240.

1164 Denis, J.-N., Kanazawa, A.M. and Greene, A.E. (1994) *Tetrahedron Lett.*, **35**, 105–108.

1165 Didier, E., Fouque, E., Taillepied, I. and Commerçon, A. (1994) *Tetrahedron Lett.*, **35**, 2349–2352.

1166 Didier, E., Fouque, E. and Commerçon, A. (1994) *Tetrahedron Lett.*, **35**, 3063–3064.

1167 Kingston, D.G.I., Chaudhary, A.G., Gunatilaka, A.A.L. and Middleton, M.L. (1994) *Tetrahedron Lett.*, **35**, 4483–4484.

1168 Klein, L.L., Yeung, C.M., Li, L. and Plattner, J.J. (1994) *Tetrahedron Lett.*, **35**, 4707–4710.

1169 Neidigh, K.A., Gharpure, M.M., Rimoldi, J.M. and Kingston, D.G.I. (1994) *Tetrahedron Lett.*, **35**, 6839–6842.

1170 Johnson, R.A., Nidy, E.G., Dobrowolski, P.J., Gebhard, I., Qualls, S.J.,

Wicnienski, N.A. and Kelly, R.C. (1994) *Tetrahedron Lett.*, **35**, 7893–7896.

1171 Pulicani, J.-P., Bouchard, H., Bourzat, J.-D. and Commerçon, A. (1994) *Tetrahedron Lett.*, **35**, 9709–9712.

1172 Bouchard, H., Pulicani, J.-P., Vuilhorgne, M., Bourzat, J.-D. and Commerçon, A. (1994) *Tetrahedron Lett.*, **35**, 9713–9716.

1173 Pulicani, J.-P., Bezard, D., Bourzat, J.-D., Bouchard, H., Zucco, M., Deprez, D. and Commerçon, A. (1994) *Tetrahedron Lett.*, **35**, 9717–9720.

1174 Rogers, H.J. (1995) *J. Chem. Soc., Perkin Trans. 1*, 3073–3075.

1175 Sone, H., Kondo, T., Kiryu, M., Ishiwata, H., Ojika, M. and Yamada, K. (1995) *J. Org. Chem.*, **60**, 4774–4781.

1176 Kuwahara, S., Moriguchi, M., Miyagawa, K., Konno, M. and Kodama, O. (1995) *Tetrahedron*, **51**, 8809–8814.

1177 Chen, S.-H., Huang, S. and Roth, G.P. (1995) *Tetrahedron Lett.*, **36**, 8933–8936.

1178 Kelly, R.C., Wicnienski, N.A., Gebhard, I., Qualls, S.J., Han, F., Dobrowolski, P.J., Nidy, E.G. and Johnson, R.A. (1996) *J. Am. Chem. Soc.*, **118**, 919–920.

1179 Schlessinger, R.H. and Li, Y.-J. (1996) *J. Am. Chem. Soc.*, **118**, 3301–3302.

1180 Nicolaou, K.C., Postema, M.H.D., Yue, E.W. and Nadin, A. (1996) *J. Am. Chem. Soc.*, **118**, 10335–10336.

1181 Salamonczyk, G.M., Han, K., Guo, Z. and Shi, C.J. (1996) *J. Org. Chem.*, **61**, 6893–6900.

1182 Solladie, G. and Gressot-Kempf, L. (1996) *Tetrahedron: Asymmetry*, **7**, 2371–2379.

1183 Marder, R., Bricard, L., Dubois, J., Guenard, D. and Gueritte-Voegelein, F. (1996) *Tetrahedron Lett.*, **37**, 1777–1780.

1184 Andrus, M.B. and Argade, A.B. (1996) *Tetrahedron Lett.*, **37**, 5049–5052.

1185 Corey, E.J., Guzman-Perez, A. and Lazerwith, S.E. (1997) *J. Am. Chem. Soc.*, **119**, 11769–11776.

1186 Nicolaou, K.C., He, Y., Vourloumis, D., Vallberg, H., Roschangar, F., Sarabia, F., Ninkovic, S., Yang, Z. and Trujillo, J.I. (1997) *J. Am. Chem. Soc.*, **119**, 7960–7973.

1187 Murakata, M., Tamura, M. and Hoshino, O. (1997) *J. Org. Chem.*, **62**, 4428–4433.

1188 Nakayama, K., Terasawa, H., Mitsui, I., Ohsaki, S., Uoto, K., Imura, S. and Soga, T. (1998) *Bioorg. Med. Chem. Lett.*, **8**, 427–432.

1189 Matovic, R. and Saicic, R.N. (1998) *Chem. Commun.*, 1745–1746.

1190 Lebel, H. and Jacobsen, E.N. (1998) *J. Org. Chem.*, **63**, 9624–9625.

1191 Wiph, P. and Yokokawa, F. (1998) *Tetrahedron Lett.*, **39**, 2223–2226.

1192 Ojima, I., Bounaud, P.-Y. and Ahern, D.G. (1999) *Bioorg. Med. Chem. Lett.*, **9**, 1189–1194.

1193 Han, L. and Razdan, R.K. (1999) *Tetrahedron Lett.*, **40**, 1631–1634.

1194 Cristofoli, W.A. and Benn, M. (1991) *J. Chem. Soc., Perkin Trans. 1*, 1825–1831.

1195 Justus, K. and Steglich, W. (1991) *Tetrahedron Lett.*, **32**, 5781–5784.

1196 Smith, A.B. III, Sulikowski, G.A., Sulikowski, M.M. and Fujimoto, K. (1992) *J. Am. Chem. Soc.*, **114**, 2567–2576.

1197 Smith, A.B. III, Leahy, J.W., Noda, I., Remiszewski, S.W., Liverton, N.J. and Zibuck, R. (1992) *J. Am. Chem. Soc.*, **114**, 2995–3007.

1198 Pons, J.-M., Pommier, A., Lerpiniere, J. and Kocienski, P. (1993) *J. Chem. Soc., Perkin Trans. 1*, 1549–1551.

1199 Matsushita, Y., Furusawa, H., Matsui, T. and Nakayama, M. (1994) *Chem. Lett.*, 1083–1084.

1200 Hidalgo-Del Vecchio, G. and Oehlschlanger, A.C. (1994) *J. Org. Chem.*, **59**, 4853–4857.

1201 Rychnovsky, S.D. and Hwang, K. (1994) *J. Org. Chem.*, **59**, 5414–5418.

1202 Imaeda, T., Hamada, Y. and Shioiri, T. (1994) *Tetrahedron Lett.*, **35**, 591–594.

1203 Couladouros, E.A. and Soufli, I.C. (1994) *Tetrahedron Lett.*, **35**, 4409–4412.

1204 Noda, A., Aoyagi, S., Machinaga, N. and Kibayashi, C. (1994) *Tetrahedron Lett.*, **35**, 8237–8240.

1205 Shin, C., Nakamura, Y., Yamada, Y., Yonezawa, Y., Umemura, K. and Yoshimura, J. (1995) *Bull. Chem. Soc. Jpn.*, **68**, 3151–3160.

1206 Pommier, A., Pons, J.-M. and Kocienski, P.J. (1995) *J. Org. Chem.*, **60**, 7334–7339.

1207 Xu, Y.-M. and Zhou, W.-S. (1996) *Tetrahedron Lett.*, **37**, 1461–1462.

1208 Yamamoto, I. and Narasaka, K. (1997) *Bull. Chem. Soc. Jpn.*, **70**, 3327–3333.

1209 Yang, H.W. and Romo, D. (1997) *J. Org. Chem.*, **62**, 4–5.

1210 Gomez, A.M., De Uralde, B.L., Varverde, S. and Lopez, J.C. (1997) *Chem. Commun.*, 1647–1648.

1211 Mukai, C., Hirai, S. and Hanaoka, M. (1997) *J. Org. Chem.*, **62**, 6619–6626.

1212 Goti, A., Fedi, V., Nannelli, L., De Sarlo, F. and Brandi, A. (1997) *Synlett*, 577–579.

1213 Matsushita, Y., Sugamoto, K., Nakama, T., Matsui, T., Hayashi, Y. and Uenakai, K. (1997) *Tetrahedron Lett.*, **38**, 6055–6058.

1214 Couladouros, E.A., Soufli, I.C., Moutsos, V.I. and Chadha, R.K. (1998) *Chem. Eur. J.*, **4**, 33–43.

1215 Smith, A.B. III and Ott, G.R. (1998) *J. Am. Chem. Soc.*, **120**, 3935–3948.

1216 Tanaka, K. and Sawanishi, H. (1998) *Tetrahedron*, **54**, 10029–10042.

1217 Yadav, J.S., Barma, D.K. and Dutta, D. (1998) *Tetrahedron Lett.*, **39**, 143–146.

1218 Wen, J.J. and Crew, C.M. (1998) *Tetrahedron Lett.*, **39**, 779–782.

1219 Varadarajan, S., Mohapatra, D.K. and Datta, A. (1998) *Tetrahedron Lett.*, **39**, 1075–1078.

1220 Murahashi, S.-I., Ohtake, H. and Imada, Y. (1998) *Tetrahedron Lett.*, **39**, 2765–2766.

1221 Shimano, M., Shibata, T. and Kamei, N. (1998) *Tetrahedron Lett.*, **39**, 4363–4366.

1222 Kaisalo, L., Koskimies, J. and Hase, T. (1999) *Synth. Commun.*, **29**, 399–407.

1223 Shen, R., Lin, C.T. and Porco, J.A. Jr. (2002) *J. Am. Chem. Soc.*, **124**, 5650–5651.

1224 Paterson, I. and Tudge, M. (2003) *Tetrahedron*, **59**, 6833–6849.

1225 Narasaka, K., Yamaguchi, M. and Mukaiyama, T. (1977) *Chem. Lett.*, 959–962.

1226 Denis, J.-N. and Greene, A.E. (1988) *J. Am. Chem. Soc.*, **110**, 5917–5919.

1227 Cooper, J., Knight, D.W. and Gallagher, P.T. (1992) *J. Chem. Soc., Perkin Trans. 1*, 553–559.

1228 Miki, Y. and Hachiken, H. (1993) *Synlett*, 333–334.

1229 White, J.D. and Johnson, A.T. (1994) *J. Org. Chem.*, **59**, 3347–3358.

1230 Pu, Y., Lowe, C., Sailer, M. and Vederas, J.C. (1994) *J. Org. Chem.*, **59**, 3642–3655.

1231 Nicolaou, K.C., Koide, K. and Bunnage, M.E. (1995) *Chem. Eur. J.*, **1**, 454–466.

1232 King, A.G., Meinwald, J., Eisner, T. and Blankespoor, C.L. (1996) *Tetrahedron Lett.*, **37**, 2141–2144.

1233 Uoto, K., Ohsuki, S., Takenoshita, H., Ishiyama, T., Iimura, S., Hirota, Y., Mitsui, I., Terasawa, H. and Soga, T. (1997) *Chem. Pharm. Bull.*, **45**, 1793–1804.

1234 Miyabe, H., Torieda, M., Kiguchi, T. and Naito, T. (1997) *Synlett*, 580–582.

1235 Tanner, D., Tedenborg, L., Almario, A., Petterson, I., Csoeregh, I., Kelly, N.M., Andersson, P.G. and Hoegberg, T. (1997) *Tetrahedron*, **53**, 4857–4868.

1236 Shiina, I., Saito, K., Frechard-Ortuno, I. and Mukaiyama, T. (1998) *Chem. Lett.*, 3–4.

1237 Miyabe, H., Torieda, M., Inoue, K., Tajiri, K., Kiguchi, T. and Naito, T. (1998) *J. Org. Chem.*, **63**, 4397–4407.

1238 Sankaranarayanan, S., Sharma, A. and Chattopadhyay, S. (1996) *Tetrahedron: Asymmetry*, **7**, 2639–2643.

1239 Wood, H.B. and Ganem, B. (1990) *J. Am. Chem. Soc.*, **112**, 8907–8909.

1240 Kluge, M., Hartenstein, H., Hantschmann, A. and Sicker, D. (1995) *J. Heterocycl. Chem.*, **32**, 395–402.

1241 Shing, T.K.M. and Yang, J. (1995) *J. Org. Chem.*, **60**, 5785–5789.

1242 Gibson, C.L. and Handa, S. (1996) *Tetrahedron: Asymmetry*, **7**, 1281–1284.

1243 Riehs, G. and Urban, E. (1997) *Monatsh. Chem.*, **128**, 281–289.

1244 Paulsen, H. and Hoppe, D. (1992) *Tetrahedron*, **48**, 5667–5670.

1245 Takano, S., Setoh, M. and Ogasawara, K. (1992) *Tetrahedron: Asymmetry*, **3**, 533–534.

1246 Chavan, S.P., Zubaidha, P.K. and Ayyangar, N.R. (1992) *Tetrahedron Lett.*, **33**, 4605–4608.

1247 Tu, C. and Berchtold, G.A. (1993) *J. Org. Chem.*, **58**, 6915–6916.

1248 Mori, Y. and Furukawa, H. (1994) *Chem. Pharm. Bull.*, **42**, 2161–2163.

1249 Marshall, J.A. and Welmaker, G.S. (1994) *J. Org. Chem.*, **59**, 4122–4125.

1250 Soti, F., Kajtar-Peredy, M., Kardos-Balogh, Z., Incse, M., Keresztury, G., Czira, G. and Szantay, C. (1994) *Tetrahedron*, **50**, 8209–8226.

1251 Appendino, G., Varese, M., Gariboldi, P. and Gabetta, B. (1994) *Tetrahedron Lett.*, **35**, 2217–2220.

1252 Yamamoto, N., Nishikawa, T. and Isobe, M. (1995) *Synlett*, 505–506.

1253 Kucherenko, A., Flavin, M.T., Boulanger, W.A., Khilevich, A., Shone, R.L., Rizzo, J.D., Sheinkman, A.K. and Xu, Z.-Q. (1995) *Tetrahedron Lett.*, **36**, 5475–5478.

1254 Hamelin, O., Depres, J.-P. and Greene, A.E. (1996) *J. Am. Chem. Soc.*, **118**, 9992–9993.

1255 Rehder, K.S. and Kepler, J.A. (1996) *Synth. Commun.*, **26**, 4005–4021.

1256 Suri, S.C. (1996) *Tetrahedron Lett.*, **37**, 2335–2336.

1257 Spino, C., Mayes, N. and Desfosses, H. (1996) *Tetrahedron Lett.*, **37**, 6503–6506.

1258 Tsuboi, S. (1998) *J. Org. Chem.*, **63**, 1102–1108.

1259 Schreiber, S.L. and Meyers, H.V. (1988) *J. Am. Chem. Soc.*, **110**, 5198–5200.

1260 Schreiber, S.L., Desmaele, D. and Porco, J.A. Jr. (1988) *Tetrahedron Lett.*, **29**, 6689–6692.

1261 Tan, D.S., Foley, M.A., Srockwell, B.R., Shair, M.D. and Schreiber, S.L. (1999) *J. Am. Chem. Soc.*, **121**, 9073–9087.

1262 Trotter, N.S., Takahashi, S. and Nakata, T. (1999) *Org. Lett.*, **1**, 957–959.

1263 Mandai, T., Kuroda, A., Okumoto, H., Nakanishi, K., Mikuni, K., Hara, K.-J. and Hara, K.-Z. (2000) *Tetrahedron Lett.*, **41**, 243–246.

1264 Ha, J.D., Lee, D. and Cha, J.K. (1997) *J. Org. Chem.*, **62**, 4550–4551.

1265 Epsztajn, J., Jozwiak, A. and Szczesniak, A.K. (1998) *J. Chem. Soc., Perkin Trans. 1*, 2563–2567.

1266 Sinha, S.C. and Keinan, E. (1993) *J. Am. Chem. Soc.*, **115**, 4891–1892.

1267 Frey, B., Wells, A.P., Rogers, D.H. and Mander, L.N. (1998) *J. Am. Chem. Soc.*, **120**, 1914–1915.

1268 Krische, M.J. and Trost, B.M. (1998) *Tetrahedron*, **54**, 3693–3704.

1269 Shing, T.K.M., Zhou, Z.-H. and Mak, T.C.W. (1992) *J. Chem. Soc., Perkin Trans. 1*, 1907–1910.

1270 Oveman, L.E., Rabinowitz, M.H. and Renhowe, P.A. (1995) *J. Am. Chem. Soc.*, **117**, 2657–2658.

1271 Riehs, G. and Urban, E. (1995) *Tetrahedron Lett.*, **36**, 7233–7234.

1272 Bonini, C., Pucci, P., Racioppi, R. and Viggiani, L. (1992) *Tetrahedron: Asymmetry*, **3**, 29–32.

1273 Pawar, A.S., Chattopadhyay, S., Chattopadhyay, A. and Mamdapur, V.R. (1993) *J. Org. Chem.*, **58**, 7535–7536.

1274 Bull, S.D. and Carman, R.M. (1994) *Aust. J. Chem.*, **47**, 1661–1672.

1275 Sundram, H., Golebiowski, A. and Johnson, C.R. (1994) *Tetrahedron Lett.*, **35**, 6975–6976.

1276 Pawar, A.S., Sankaranarayanan, S. and Chattopadhyay, S. (1995) *Tetrahedron: Asymmetry*, **6**, 2219–2226.

1277 Sharma, A., Sankaranarayanan, S. and Chattopadhayay, S. (1996) *J. Org. Chem.*, **61**, 1814–1846.

1278 Sakagami, H., Samizu, K., Kamikubo, T. and Ogasawara, K. (1996) *Synlett*, 163–164.

1279 Yamane, T. and Ogasawara, K. (1996) *Synlett*, 925–926.

1280 Khmelnitsky, Y.L., Budde, C., Arnold, J.M., Usyatinsky, A., Clark, D.S. and Dordick, J.S. (1997) *J. Am. Chem. Soc.*, **119**, 11554–11555.

1281 Yamamura, I., Fujiwara, Y. and Yamato, T. (1997) *Tetrahedron Lett.*, **38**, 4121–4124.

1282 Shimoma, F., Kusaka, H., Wada, K., Azami, H., Yasunami, M., Suzuki, T., Hagiwara, H. and Ando, M. (1998) *J. Org. Chem.*, **63**, 920–929.

1283 Honzumi, M., Kamikubo, T. and Ogasawara, K. (1998) *Synlett*, 1001–1003.

1284 Yoshida, N., Kamikubo, T. and Ogasawara, K. (1998) *Tetrahedron Lett.*, **39**, 4677–4678.

1285 Yamada, O. and Ogasawara, K. (1998) *Tetrahedron Lett.*, **39**, 7747–7750.

1286 Nagarajan, M. (1999) *Tetrahedron Lett.*, **40**, 1207–1210.

1287 Lu, W. and Sih, C.J. (1999) *Tetrahedron Lett.*, **40**, 4965–4968.

1288 Ho, T.-L. (1992) *Can. J. Chem.*, **70**, 1375–1384.

1289 Holton, R.A., Zhang, A., Clarke, P.A., Nadizadeh, H. and Procter, J. (1998) *Tetrahedron Lett.*, **39**, 2883–2886.

1290 Damen, E.W.P., Braamer, L. and Scheeren, H.W. (1998) *Tetrahedron Lett.*, **39**, 6081–6082.

1291 Iida, H., Tanaka, M. and Kibayashi, C. (1984) *J. Org. Chem.*, **49**, 1909–1912.

1292 Rosen, T. and Heathcock, C.H. (1985) *J. Am. Chem. Soc.*, **107**, 3731–3733.

1293 Niwa, H., Nishiwaki, M., Tsukada, I., Ishigaki, T., Ito, S., Wakamatsu, K., Mori, T., Ikagawa, M. and Yamada, K. (1990) *J. Am. Chem. Soc.*, **112**, 9001–9003.

1294 Mancini, I., Guella, G. and Pietra, F. (1991) *Helv. Chim. Acta*, **74**, 941–950.

1295 Migliuolo, A., Piccialli, V. and Sica, D. (1991) *Tetrahedron*, **47**, 7937–7950.

1296 Cameron, D.W., Crosby, I.T. and Feutrill, G.I. (1992) *Aust. J. Chem.*, **45**, 2025–2035.

1297 Comins, D.L. and LaMunyon, D.H. (1992) *J. Org. Chem.*, **57**, 5807–5809.

1298 Matsumoto, T., Tanaka, Y., Terao, H., Takeda, Y. and Wada, M. (1993) *Bull. Chem. Soc. Jpn.*, **66**, 3053–3057.

1299 Holmes, A.B., Hughes, A.B. and Smith, A.L. (1993) *J. Chem. Soc., Perkin Trans. 1*, 633–643.

1300 Grandjean, D., Pale, P. and Chuche, J. (1993) *Tetrahedron*, **49**, 5225–5236.

1301 Nicolaou, K.C., Nantermet, P.G., Ueno, H. and Guy, R.K. (1994) *J. Chem. Soc., Chem. Commun.*, 295–296.

1302 Nicolaou, K.C., Claiborne, C.F., Nantermet, P.G., Couladouros, E.A. and Sorensen, E.J. (1994) *J. Am. Chem. Soc.*, **116**, 1591–1592.

1303 Leonard, J. and Hussain, N. (1994) *J. Chem. Soc., Perkin Trans. 1*, 61–70.

1304 Mori, K. and Ogita, H. (1994) *Liebigs Ann. Chem.*, 1065–1068.

1305 Senokuchi, K., Nakai, H., Kawamura, M., Katsube, N., Nonaka, S., Sawaragi, H. and Hamanaka, N. (1994) *Synlett*, 343–344.

1306 Swarts, H.J., Verstegen-Haaksma, A.A., Hansen, B.J.M. and De Croot, A. (1994) *Tetrahedron*, **50**, 10083–10094.

1307 Zhu, J., Klunder, J.H. and Zwanenburg, B. (1994) *Tetrahedron*, **50**, 10597–10610.

1308 Rao, A.V.R., Rao, B.V., Bhanu, M.N. and Kumar, V.S. (1994) *Tetrahedron Lett.*, **35**, 3201–3204.

1309 Kim, N.-S., Kang, C.H. and Cha, J.K. (1994) *Tetrahedron Lett.*, **35**, 3489–3492.

1310 Nakamura, S. and Shibasaki, M. (1994) *Tetrahedron Lett.*, **35**, 4145–4148.

1311 Yang, Z.-C. and Zhou, W.-S. (1995) *J. Chem. Soc., Chem. Commun.*, 743–744.

1312 Rao, A.V.R., Gurjar, M.K. and Vasudevan, J. (1995) *J. Chem. Soc., Chem. Commun.*, 1369–1370.

1313 Hoye, T.R. and Vyvyan, J.R. (1995) *J. Org. Chem.*, **60**, 4184–4195.

1314 Parker, K.A. and Resnick, L. (1995) *J. Org. Chem.*, **60**, 5726–5728.

1315 Gravier-Pelletier, C., Merrer, Y.L. and Depezay, J.-C. (1995) *Tetrahedron*, **51**, 1663–1674.

1316 Crawforth, J.M. and Rawlings, B. (1995) *Tetrahedron Lett.*, **36**, 6345–6346.

1317 Horie, T., Kitou, T., Kawamura, Y. and Yamashita, K. (1996) *Bull. Chem. Soc. Jpn.*, **69**, 1033–1041.

1318 Shiina, I., Saito, M., Nishimura, K., Saito, K. and Mukaiyama, T. (1996) *Chem. Lett.*, 223–224.

1319 Wang, T.-Z., Pinard, E. and Paquette, L.A. (1996) *J. Am. Chem. Soc.*, **118**, 1309–1318.

1320 Tse, B. (1996) *J. Am. Chem. Soc.*, **118**, 7094–7100.

1321 Toyooka, N., Yotsui, Y., Yoshida, Y. and Momose, T. (1996) *J. Org. Chem.*, **61**, 4882–4883.

1322 Fernandez, A.-M. and Duhmel, L. (1996) *J. Org. Chem.*, **61**, 8698–8700.

1323 Surivet, J.-P., Gore, J. and Vatele, J.-M. (1996) *Tetrahedron*, **52**, 14877–14890.

1324 Shen, C.-C., Chou, S.-C. and Chou, S.-O. (1996) *Tetrahedron: Asymmetry*, **7**, 3141–3146.

1325 Gao, Y., Wu, W.-L., Ye, B., Zhou, R. and Wu, Y.-L. (1996) *Tetrahedron Lett.*, **37**, 893–896.

1326 Hatakeyama, S., Fukuyama, H., Mukugi, Y. and Irie, H. (1996) *Tetrahedron Lett.*, **37**, 4047–4050.

1327 Tatsuta, K., Miura, S. and Gunji, H. (1997) *Bull. Chem. Soc. Jpn.*, **70**, 427–436.

1328 Yang, Z.-C., Jiang, X.-B., Wang, Z.-M. and Zhou, W.-S. (1997) *J. Chem. Soc., Perkin Trans. 1*, 317–321.

1329 Andrus, M.B., Li, W. and Keyes, R.F. (1997) *J. Org. Chem.*, **62**, 5542–5549.

1330 Mukai, C., Moharram, S.M., Azukizawa, S. and Hanaoka, M. (1997) *J. Org. Chem.*, **62**, 8095–8103.

1331 Paquette, L.A., Sun, L.-Q., Watson, T.J.N., Friedrich, D. and Freeman, B.T. (1997) *J. Org. Chem.*, **62**, 8155–8161.

1332 McDonald, F.E. and Schultz, C.C. (1997) *Tetrahedron*, **53**, 16435–16448.

1333 Lohray, B.B. and Venkateswarlu, S. (1997) *Tetrahedron: Asymmetry*, **8**, 633–638.

1334 Muratake, H., Matsumura, N. and Natsume, M. (1998) *Chem. Pharm. Bull.*, **46**, 559–571.

1335 Horita, K., Nagasawa, M., Sakurai, Y. and Yonemitsu, O. (1998) *Chem. Pharm. Bull.*, **46**, 1199–1216.

1336 Mukai, C., Sugimoto, Y., Miyazawa, K., Yamaguchi, S. and Hanaoka, M. (1998) *J. Org. Chem.*, **63**, 6281–6287.

1337 Mori, K., Audran, G. and Monti, H. (1998) *Synlett*, 259–260.

1338 Schlessinger, R.H. and Gillman, K.W. (1999) *Tetrahedron Lett.*, **40**, 1257–1260.

1339 Lan, J., Liu, Z., Cen, W., Xing, Y. and Li, Y. (1999) *Tetrahedron Lett.*, **40**, 1963–1966.

1340 Evans, D.A., Ng, H.P. and Rieger, D.L. (1993) *J. Am. Chem. Soc.*, **115**, 11446–11459.

1341 Rao, A.V.R., Yadav, J.S. and Valluri, M. (1994) *Tetrahedron Lett.*, **35**, 3613–3616.

1342 Hareau-Vittini, G., Kocienski, P.J. and Reid, G. (1995) *Synthesis*, 1007–1013.

1343 Hareau-Vittini, G. and Kocienski, P.J. (1995) *Synlett*, 893–894.

1344 Sharma, G.V.M., Rao, A.V.S.R. and Murthy, V.S. (1995) *Tetrahedron Lett.*, **36**, 4117–4120.

1345 Corey, E.J. and Kania, R.S. (1996) *J. Am. Chem. Soc.*, **118**, 1229–1230.

1346 Roush, W.R. and Works, A.B. (1996) *Tetrahedron Lett.*, **37**, 8065–8068.

1347 Burton, J.W., Clark, J.S., Derrer, S., Stork, T.C., Bendall, J.G. and Holmes, A.B. (1997) *J. Am. Chem. Soc.*, **119**, 7483–7498.

1348 Nicolaou, K.C., Ninkovic, S., Sarabia, F., Vourloumis, D., He, Y., Vallberg, H., Finlay, M.R.V. and Yang, Z. (1997) *J. Am. Chem. Soc.*, **119**, 7974–7991.

1349 Toshima, K., Jyojima, T., Yamaguchi, H., Noguchi, Y., Yoshida, T., Murase, H., Nakata, M. and Mastumura, Y. (1997) *J. Org. Chem.*, **62**, 3271–3284.

1350 Evans, D.A. and Kim, A.S. (1997) *Tetrahedron Lett.*, **38**, 53–56.

1351 Mohapatra, D.K. and Datta, A. (1998) *J. Org. Chem.*, **63**, 642–646.

1352 Nicolaou, K.C., Finlay, M.R.V., Ninkovic, S. and Sarabia, F. (1998) *Tetrahedron*, **54**, 7127–7166.

1353 Wipf, P. and Graham, T.H. (2004) *J. Am. Chem. Soc.*, **126**, 15346–15347.

1354 Gogoi, S., Barua, N.C. and Kalita, B. (2004) *Tetrahedron Lett.*, **45**, 5577–5579.

1355 Mlynarski, J., Ruiz-Caro, J. and Fürstner, A. (2004) *Chem. Eur. J.*, **10**, 2214–2222.

1356 Smith, A.B. III and Simov, V. (2006) *Org. Lett.*, **8**, 3315–3318.

1357 Du, Y., Chen, Q. and Linhardt, R.J. (2006) *J. Org. Chem.*, **71**, 8446–8451.

1358 Chou, C.-Y. and Hou, D.-R. (2006) *J. Org. Chem.*, **71**, 9887–9890.

1359 Va, P. and Roush, W.R. (2007) *Org. Lett.*, **9**, 307–310.

1360 Ogawa, T., Fang, C.-L., Suemura, H. and Sakai, K. (1991) *J. Chem. Soc., Chem. Commun.*, 1438–1439.

1361 Earle, M.J., Abdur-Rashid, A. and Priestley, N.D. (1996) *J. Org. Chem.*, **61**, 5697–5700.

1362 Konoike, T., Takahashi, K., Araki, Y. and Horibe, I. (1997) *J. Org. Chem.*, **62**, 960–966.

1363 Guillier, F., Nivoliers, F., Bourguignon, J., Dupas, G., Marsais, F., Godard, A. and Queguignon, G. (1992) *Tetrahedron Lett.*, **33**, 7355–7356.

1364 Tsunoda, T., Tatsuki, S., Kataoka, K. and Ito, S. (1994) *Chem. Lett.*, 543–546.

1365 Kende, A.S., Liu, K., Kaldor, I., Dorey, G. and Koch, K. (1995) *J. Am. Chem. Soc.*, **117**, 8258–8270.

1366 Kelly, T.R. and Xie, R.L. (1998) *J. Org. Chem.*, **63**, 8045–8048.

1367 Ojima, I., Habus, I., Zhao, M., Zucco, M., Park, Y.H., Sun, C.M. and Brigaud, T. (1992) *Tetrahedron*, **48**, 6985–7012.

1368 Chen, S.-H., Huang, S., Kant, J., Fairchild, C., Wei, J. and Farina, V. (1993) *J. Org. Chem.*, **58**, 5028–5029.

1369 Chen, S.-H., Wei, J.-M. and Farina, V. (1993) *Tetrahedron Lett.*, **34**, 3205–3206.

1370 Ojima, I., Sun, C.M., Zucco, M., Park, Y.H., Duclos, O. and Kuduk, S. (1993) *Tetrahedron Lett.*, **34**, 4149–4152.

1371 Chen, S.-H., Wei, J.-M., Vyas, D.M., Doyle, T.W. and Farina, V. (1993) *Tetrahedron Lett.*, **34**, 6845–6848.

1372 Georg, G.I., Boge, T.C., Cheruvallath, Z.S., Harriman, G.C.B., Hepperle, M., Park, H. and Himes, R.H. (1994) *Bioorg. Med. Chem. Lett.*, **4**, 335–338.

1373 Georg, G.I., Harriman, G.C., Park, H. and Himes, R.H. (1994) *Bioorg. Med. Chem. Lett.*, **4**, 487–490.

1374 Ojima, I., Fenoglio, I., Park, Y.H., Pera, P. and Bernacki, R.J. (1994) *Bioorg. Med. Chem. Lett.*, **4**, 1571–1576.

1375 Ojima, I., Fenoglio, I., Park, Y.H., Sun, C.M., Appendino, G., Bernacki, R.J. and Pera, P. (1994) *J. Org. Chem.*, **59**, 515–517.

1376 Lampe, J.W., Hughes, P.F., Biggers, C.K., Smith, S.H. and Hu, H. (1994) *J. Org. Chem.*, **59**, 5147–5148.

1377 Chen, S.-H., Kadow, J.F. and Farina, V. (1994) *J. Org. Chem.*, **59**, 6156–6158.

1378 Holton, R.A., Somoza, C. and Chai, K.-B. (1994) *Tetrahedron Lett.*, **35**, 1665–1668.

1379 Kant, J., O'Keeffe, W.S., Chen, S.-H., Farina, V., Fairchild, C., Johnston, K., Kadow, J.F., Long, B.H. and Vyas, D. (1994) *Tetrahedron Lett.*, **35**, 5543–5546.

1380 Mukai, C., Moharram, S.M., Kataoka, O. and Hanaoka, M. (1995) *J. Chem. Soc., Perkin Trans. 1*, 2849–2854.

1381 Nicolaou, K.C., Nantermet, P.G., Ueno, H., Guy, R.K., Couladouros, E.A. and Sorensen, E.J. (1995) *J. Am. Chem. Soc.*, **117**, 624–633.

1382 Nicolaou, K.C., Renaud, J., Nantermet, P.G., Couladouros, E.A., Guy, R.K. and Wrasidlo, W. (1995) *J. Am. Chem. Soc.*, **117**, 2409–2420.

1383 Young, W.B., Masters, J.J. and Danishefsky, S. (1995) *J. Am. Chem. Soc.*, **117**, 5228–5234.

1384 Mukai, C., Kataoka, O. and Hanaoka, M. (1995) *J. Org. Chem.*, **60**, 5910–5918.

1385 Georg, G.I., Harriman, G.C.B., Hepperle, M., Clowers, J.S. and Vander Velde, D.G. (1996) *J. Org. Chem.*, **61**, 2664–2676.

1386 Kant, J., Schwartz, W.S., Fairchild, C., Gao, Q., Huang, S., Long, B.H., Kadow, J.F., Langley, D.R., Farina, V. and Vyas, D. (1996) *Tetrahedron Lett.*, **37**, 6495–6498.

1387 Wender, P.A., Lee, D., Lal, T.K., Horwitz, S.B. and Rao, S. (1997) *Bioorg. Med. Chem. Lett.*, **7**, 1941–1944.

1388 Gennari, C., Carcano, M., Donghi, M., Mongelli, N., Vanotti, E. and Vulpetti, A. (1997) *J. Org. Chem.*, **62**, 4746–4755.

1389 Kim, S.-C., Moon, M.-S., Choi, K.-M., Jun, S.-J., Kim, H.-K., Jung, D.-I, Lee, K.-S. and Chai, K.-B. (1998) *Bull. Korean Chem. Soc.*, **19**, 1027–1028.

1390 Soto, J., Mascarenas, J.L. and Castedo, L. (1998) *Bioorg. Med. Chem. Lett.*, **8**, 273–276.

1391 Morihira, K., Nishimori, T., Kusama, H., Horiguchi, Y., Kuwajima, I. and Tsuruno, T. (1998) *Bioorg. Med. Chem. Lett.*, **8**, 2977–2982.

1392 Tsuboi, S., Takeda, S., Yamasaki, Y., Sakai, T., Utaka, M., Ishida, S., Yamada, E. and Hirano, J. (1992) *Chem. Lett.*, 1417–1420.

1393 Georg, G.I., Cheruvallath, Z.S., Himes, R.H., Mejillano, M.R. and Burke, C.T. (1992) *J. Med. Chem.*, **35**, 4230–4237.

1394 Corey, E.J. and Lee, J. (1993) *J. Am. Chem. Soc.*, **115**, 8873–8874.

1395 Wang, Z., Warder, S.E., Perrier, H., Grimm, E.L. and Bernstein, M.A. (1993) *J. Org. Chem.*, **58**, 2931–2932.

1396 Angle, S.R. and Louie, M.S. (1993) *Tetrahedron Lett.*, **34**, 4751–4754.

1397 Yoshifuji, S. and Kaname, M. (1995) *Chem. Pharm. Bull.*, **43**, 1617–1620.

1398 Szewcyk, J., Wilson, J.W., Lewin, A.H. and Carroll, F.I. (1995) *J. Heterocycl. Chem.*, **32**, 195–199.

1399 Tamagnan, G., Gao, Y., Bakthavachalam, V., White, W.L. and Neumeyer, J.L. (1995) *Tetrahedron Lett.*, **36**, 5861–5864.

1400 Ohmori, K., Suzuki, T., Nishiyama, S. and Yamamura, S. (1995) *Tetrahedron Lett.*, **36**, 6515–6518.

1401 Nakata, T., Nomura, S. and Matsukura, H. (1996) *Chem. Pharm. Bull.*, **44**, 627–629.

1402 Krohn, K., Roemer, E. and Top, M. (1996) *Liebigs Ann.*, 271–277.

1403 Lee, E., Lim, J.W., Yoon, C.H., Sung, Y.-S. and Kim, Y.K. (1997) *J. Am. Chem. Soc.*, **119**, 8391–8392.

1404 Singh, S., Basmadjian, G.P., Avor, K., Pouw, B. and Seale, T.W. (1997) *Synth. Commun.*, **27**, 4003–4012.

1405 Yoshizawa, J., Obitsu, K., Maki, S., Niwa, H., Hirano, T. and Ohashi, M. (1997) *Synlett*, 1387–1388.

1406 Hirai, G., Oguri, H. and Hirama, M. (1999) *Chem. Lett.*, 141–142.

1407 Monn, J.A. and Valli, M.J. (1994) *J. Org. Chem.*, **59**, 2773–2778.

1408 Green, R.H. (1997) *Tetrahedron Lett.*, **38**, 4697–4700.

1409 Drioli, S., Felluga, F., Forzato, C., Nitti, P., Pitacco, G. and Valentin, E. (1998) *J. Org. Chem.*, **63**, 2385–2388.

1410 Holton, R.A. (1991) Method for preparation of taxol. *Chemical Abstract, Eur. Pat. Appl.*, EP 400,971, **114**, 164568.

1411 Nicolaou, K.C. *et al.* (1994) *Nature*, **367**, 630–634.

1412 Masters, J.J., Link, J.T., Snyder, L.B., Young, W.B. and Danishefsky, S.J. (1995) *Angew. Chem. Int. Ed. Engl.*, **34**, 1723–1726.

1413 Hoemann, M.Z., Vander Velde, D., Aube, J. and Georg, G.I. (1995) *J. Org. Chem.*, **60**, 2918–2921.

1414 Shimizu, I. and Omura, T. (1993) *Chem. Lett.*, 1759–1760.

1415 Hikota, M., Tone, H., Horita, K. and Yonemitsu, O. (1990) *J. Org. Chem.*, **55**, 7–9.

1416 Miyaji, A., Echizen, T., Nagata, K., Yoshinaga, Y. and Okuhara, T. (2003) *J. Mol. Catal. A*, **201**, 145–153.

1417 Yamamatsu, S. (2001) *Catalysts & Catalysis*, **43**, 549–554.

1418 Ito, H., Kume, M. and Baba, T. (2005) JP 2005-126346A.

1419 Tamura, S. and Kaneki, Y. (2004) JP 2004-035873A.

1420 Kitagawa, N., Yonemoto, T., Kuribayashi, H. and Takayanagi, H. (2007) PCT WO/2007/114441.

1421 Saka, S. and Kusdiana, D. (2004) *Appl. Biochem. Biotechnol.*, **115**, 781–791.

1422 Kaieda, M., Samukawa, T., Matsumoto, T., Ban, K., Kondo, A., Shimada, Y., Noda, H., Nomoto, F., Ohtsuka, K., Izumoto, E. and Fukuda, H. (1999) *J. Biosci. Bioeng.*, **88**, 627–631.

1423 Oda, M., Kaieda, M., Hama, S., Yamaji, H., Kondo, A., Izumoto, E. and Fukuda, H. (2005) *Biochem. Eng. J.*, **23**, 45–51.

1424 Toda, M., Takagi, A., Okamura, M., Kondo, J., Hayashi, S., Domen, K. and Hara, M. (2005) *Nature*, **438**, 178.

1425 Mazur, R.H., Schlatter, J.M. and Goldkamp, A.H. (1969) *J. Am. Chem. Soc.*, **91**, 2684–2691.

1426 Ariyoshi, Y., Nagao, M., Sato, N., Shimizu, A. and Kirimura, J. (1974) U.S. Patent 3786039.

1427 Hisamitsu, K., Takemoto, T., Hijiya, T. and Takahashi, S. (1987) U.S. Patent 4680403.

1428 Takemoto, T. and Ariyoshi, Y. (1987) U.S. Patent 4684745.

1429 Sakakibara, H., Okekawa, O., Fujiwara, T., Otani, M. and Omura, S. (1981) *J. Antibiot.*, **34**, 1001–1010.

1430 Omoto, S., Iwamatsu, K., Inouye, S. and Niida, T. (1976) *J. Antibiot.*, **29**, 536–548.

1431 Fukuda, Y., Arai, Y., Iinyma, K. and Yamamoto, H. (1984) JP 59-222500.

1432 Su, Q., Beeler, A.B., Lobkovsky, E., Porco, J.A. Jr. and Panek, J.S. (2003) *Org. Lett.*, **5**, 2149–2152.

1433 Trost, B.M., Papillon, J.P.N. and Nussbaumer, T. (2005) *J. Am. Chem. Soc.*, **127**, 17921–17937.

1434 Trost, B.M. and Stiles, D.T. (2007) *Org. Lett.*, **9**, 2763–2766.

1435 Xia, J., Hui, Y. (1995) *Synth. Commun.*, **25**, 2235–2251.

1436 Tanaka, K., Yoshikawa, R., Ying, C., Kita, H., Okamoto, K. (2002) *Chem. Eng. Sci.*, **57**, 1577–1584.

Index

Esterification. Methods, Reactions, and Applications. 2nd Ed. J. Otera and J. Nishikido

ISBN: 978-3-527-32289-3

d

e

p

q

r